T0233774

Progress in metamorphic and magmatic petrology

Progress in metamorphic and magmatic petrology

A memorial volume in honor of
D.S. Korzhinskiy

Edited by
L.L. Perchuk

Institute of Experimental Mineralogy
USSR Academy of Sciences, Moscow

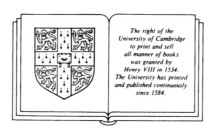

The right of the
University of Cambridge
to print and sell
all manner of books
was granted by
Henry VIII in 1534.
The University has printed
and published continuously
since 1584.

CAMBRIDGE UNIVERSITY PRESS

Cambridge

New York   Port Chester   Melbourne   Sydney

PUBLISHED BY THE PRESS SYNDICATE OF THE UNIVERSITY OF CAMBRIDGE
The Pitt Building, Trumpington Street, Cambridge, United Kingdom

CAMBRIDGE UNIVERSITY PRESS
The Edinburgh Building, Cambridge CB2 2RU, UK
40 West 20th Street, New York NY 10011–4211, USA
477 Williamstown Road, Port Melbourne, VIC 3207, Australia
Ruiz de Alarcón 13, 28014 Madrid, Spain
Dock House, The Waterfront, Cape Town 8001, South Africa

http://www.cambridge.org

First published 1991
First paperback edition 2003

A catalogue record for this book is available from the British Library

Library of Congress cataloguing in publication data

Progress in metamorphic and magmatic petrology: a memorial volume in
honor of D.S. Korzhinskiy/ edited by L.L. Perchuk.
    p. cm.
ISBN 0 521 39077 X hardback
1. Rocks, Metamorphic.   2. Rocks, Igneous.   3. Magmatism.
4. Metamorphism (Geology)   5. Korzhinskiĭ, Dmitriĭ Sergeevich, 1899–
I. Korzhinskiĭ, Dmitriĭ Sergeevich, 1899–   .   II. Perchuk, L.
L. (Leonid L'vovich)
QE475.A2P77   1990
552′.4—dc20   90–1807   CIP

ISBN 0 521 39077 X hardback
ISBN 0 521 54812 8 paperback

Contents

Contents

Contributors

L.Ya. Aranovich	Institute of Experimental Mineralogy, USSR Academy of Sciences, 142432 Chernogolovka, Moscow district, USSR
V.N. Balashov	Institute of Experimental Mineralogy, USSR Academy of Sciences, 142432 Chernogolovka, Moscow district, USSR
T. Baller	Institut für Mineralogie, Ruhr-Universität, Postfach 10 21 48 D-4630 Bochum 1, Federal Republic of Germany
S.L. Boettcher	Department of Earth & Space Sciences, UCLA, Los Angeles, California 90024–1567, USA
D.M. Burt	Department of Geology, Arizona State University, Tempe, AZ85281, USA
W.G. Ernst	Department of Earth and Space Sciences, 3806 Geology Building, Los Angeles, California, 90024, USA
V.I. Fonarev	Institute of Experimental Mineralogy, USSR Academy of Sciences, 142432 Chernogolovka, Moscow district, USSR
M.S. Ghiorso	University of Washington, Department of Geological Sciences, Seattle, Washington, 98195, USA
A.A. Graphchikov	Institute of Experimental Mineralogy, USSR Academy of Sciences, 142432 Chernogolovka, Moscow district, USSR
E.S. Grew	Department of Geological Sciences, University of Maine, 110 Boardman Hall, Orono, Maine, 04469, USA
R.A. Haugerud	US Geological Survey, 959 National Center, Reston, Virginia 22092, USA
I. Kushiro	University of Tokyo, Geological Institute, Faculty of Science Bldg. 5, Japan
M.I. Lebedeva	Institute of Chemical Physics, USSR Academy of Sciences, 143432 Chernogolovka, Moscow district, USSR
R.W. Luth	Department of Geology, University of Alberta, Edmonton, Alberta, Canada T6G 2E3
K.S. McBride	1519 North Bundy Drive, Los Angeles, California 90049, USA
J.M. McLelland	Department of Geology, Colgate University, Hamilton, N.Y. 13346, USA
A.A. Marakushev	Institute of Experimental Mineralogy, USSR Academy of Sciences, 142432 Chernogolovka, Moscow district, USSR
W.V. Maresch	Mineralogisches Institut der Universität, Corrensstr. 24, D-4400 Münster, W. Germany

N. Marquez — Materials Science Department, Aerospace Corporation, P.O. Box 92957, Los Angeles, CA 90009, USA

A. Montana — University of California, Los Angeles, Institute of Geophysics and Planetary Physics, Physics, Los Angeles, California 90024–1567, USA

P.B. Moore — Department of Geophysical Sciences, University of Chicago, Chicago, IL 60637, USA

B.O. Mysen — Geological Laboratory, 2801 Upton St., N.W., Washington D.C., 20008, USA

L.L. Perchuk — Institute of Experimental Mineralogy, USSR Academy of Sciences, 142432 Chernogolovka, Moscow district, USSR

K.K. Podlesskii — Institute of Experimental Mineralogy, USSR Academy of Sciences, 142432 Chernogolovka, Moscow district, USSR

J.F. Rice — Institut für Mineralogie der Ruhr-Universität, D-4630 Bochum, Postfach 2148, BRD

W. Schreyer — Institut für Mineralogie der Ruhr-Universität, D-4630 Bochum, Postfach 2148, BRD

G.H. Swihart — Department of Geology, Memphis State University, Memphis, TN 38152, USA

B.S. White — Division of Geological & Planetary Science, California Institute of Technology, Pasadena, California 91125, USA

P.J. Wyllie — California Institute of Technology, Division of Geological and Planetary Sciences 170–25, Pasadena, California, 91125, USA

M.G. Yates — Department of Geological Sciences, University of Maine, Orono, ME 04469, USA

G.P. Zaraisky — Institute of Experimental Mineralogy, USSR Academy of Sciences, 142432 Chernogolovka, Moscow district, USSR

E-an Zen — US Department of the Interior, Geological Survey, Reston, VA 22092, USA

V.A. Zharikov — Institute of Experimental Mineralogy, USSR Academy of Sciences, 142432 Chernogolovka, Moscow district, USSR

Preface

This volume is devoted to the memory of the Russian petrologist D.S. Korzhinskiy. The world community of geoscientists has highly valued his contributions to petrology, particularly the discovery and thermodynamic description of open systems with perfectly mobile components. Korzhinskiy's work reached this community's attention because his book *Physicochemical Basis of the Analysis of the Paragenesis of Minerals* (1959, Consultants Bureau, New York) was translated into English. However, in the Soviet Union D.S. Korzhinskiy is also highly regarded for his contributions to geology and study of ore deposits as well as to theoretical petrology. Stratigraphy and geology of the Precambrian Aldan shield in Eastern Siberia; origin of lapis lazuli and phlogopite deposits and iron formations; boron mineralization; genesis of skarns and general theory of metasomatic zoning; origin of granites, charnockites and anorthosites; theory of the acid–basic interaction of components in the dry silicate melts: all of these were subjects of his more than 200 papers and books. The papers in the present volume span Korzhinskiy's broad interests. These papers were selected with the aim of illustrating the progress made in physicochemical petrology since Korzhinskiy.

The volume is divided into three main parts. Each one is directly referred to the basic directions of Korzhinskiy's scientific activity: (I) general thermodynamics and mineral equilibria as applied to geology, (II) metasomatic and metamorphic petrology, (III) properties of melts and magmatic petrology. The papers were written by people who personally knew D.S. Korzhinskiy or are well acquainted with his work.

I would like to thank all the contributors to this volume for agreeing to pay their respects to the memory of D.S. Korzhinskiy. I also acknowledge the invaluable help of the following scientists for reviews of the papers in this volume: L. Aranovich, E. Artushkov, L. Ashwal, S. Banno, J. Brady, M. Brown, D. Burt, A. Chekhmir, V. Fonarev, M. Fonteilles, M. Frenkel, N. Epelbaum, E. Grew, S. Harley, H. Helgeson, H. Haselton, R. Joesten, A. Kadik, A. Kanilov, I. Kiseleva, I. Kushiro, P. Lichtner, A. Meunier, E. Persikov, K. Podlesskii, R. Powell, L. Plusnina, N. Pertsev, A. Reymer, I. Ryabchikov, F. Seifert, G. Schubert, J. Valley, B. Yardley, E-an Zen.

T. Holland, S. Harley and C. Roberts have done their best to accelerate publication of this volume.

I also thank E. Grew for his editorial assistance with the introductory sections of the volume. Technical help from G. Gonchar and E. Zabotina is much appreciated.

Leonid L. Perchuk
Editor

Dmitriy Sergeyevich Korzhinskiy

To the memory of Dmitriy Sergeyevich Korzhinskiy

LEONID L. PERCHUK

The distinguished petrologist D.S. Korzhinskiy has completed his life's work. I find it difficult to believe that this bearer of original ideas and striking scientific conceptions, this scientist of incredibly powerful analytical mind, this charming and kind-hearted person full of gentle humor, has ceased to exist.

Dmitriy Sergeyevich was born on 1(13) September, 1899, in St Petersburg, the only son of the chief botanist at the Petersburg Botanical Gardens, Academician Sergey Ivanovich Korzhinskiy. Sergey Ivanovich died at age 39 during an expedition in 1900. Dmitriy Sergeyevich thus did not know his father and was brought up by his mother. Although he entered the Gymnasium at the age of 12, his level of preparation was sufficient for him to be admitted directly to the fourth grade. A highly able and brilliant student, D.S. Korzhinskiy completed the Gymnasium in a short time. In 1918, after finishing at technical school, he worked as a foreman in the Urals. During the Civil War Dmitriy Sergeyevich served as a telephonist in the Red Army.

He was always fond of nature and he decided to devote his life to geology. While working at the Geological Survey of Russia and simultaneously studying at the Leningrad Mining Institute, he attracted the attention of the great scientists of that time, Ye.S. Federov, A.N. Zavaritskiy and V.N. Lodochnikov. Korzhinskiy always considered Prof. Lodochnikov his teacher. After graduating from the Leningrad Mining Institute in 1926 Dmitriy Sergeyevich worked for ten years at the Central Geological and Prospecting Institute (Leningrad), where he devoted himself fully to mapping several areas in Kazakhstan and eastern Siberia. During these years he first conceived his ideas concerning open systems with perfectly mobile components, bimetasomatic processes, the link between metamorphism and granitisation, and the deep-seated source of fluids for diaphthoresis. At this time Korzhinskiy also began to develop original methods for the physico-chemical analysis of mineral parageneses, which would later form the basis for his many studies in theoretical petrology. Dmitriy Sergeyevich was not yet a Kandidat of geological-mineralogical sciences (equivalent to a PhD in the US); he was awarded this degree in 1937 without a thesis defence.

Nonetheless, his publications were distinguished by such a high scientific level that they received wide recognition. In particular, in using original approaches to study conditions of the metamorphic formations in eastern Siberia, Dmitriy Sergeyevich discovered a number of fundamental relationships, for which he provided a sound theoretical basis. Impressed by Korzhinskiy's work, P. Eskola wrote that Korzhinskiy had developed his ideas to such an extent that petrologists at that time were unable to follow him. D.S. Korzhinskiy's geological investigations which he published in his papers 'Geological cross-section across the Stanovoy Ridge along the Amuro-Yakutian Road' (Trans. Central Geological and Prospecting Institute, No.41, 1935), 'Petrology of the Archean Complex of the Aldan Platform' (Trans. Central Geological and Prospecting Institute, No.86, 1936), 'Archaean Marbles of the Aldan Platform and the Problem of Depth Facies' (Trans. Central Geological and Prospecting Institute, No.71, 1936), formed the basis of a reliable stratigraphy for this extensive territory. These papers have not lost their significance even today.

The beginning of Korzhinskiy's profound theoretical investigations into the processes of mineral formation is tied to his move in 1937 to the Institute of Geological Sciences (predecessor of the present Institute of Geology of Ore Deposits (IGEM) and Geological Institute (GIN)) of the USSR Academy of Sciences in Moscow, where he was invited to work by A.E. Fersman. While working in this Institute, Korzhinskiy organized four theories that established his reputation:

(i) the theory of open systems with perfectly mobile components (including the theory for extremum states);
(ii) the theory for metasomatic zoning;
(iii) the theory for acid–base interaction;
(iv) the theory for magmatic replacement during granitization.

These theories were unified by an original approach which Korzhinskiy developed in his classic text published by the Academy of Sciences Press in 1957: *Physicochemical Basis of the Analysis of the Paragenesis of Minerals* (translation by the Consultants Bureau, New York, 1959), subsequently updated and amplified in his 1973 monograph *Theoretical Basis of the Analysis of the Parageneses of Minerals* (Nauka Press, Moscow, 1973). D.S. Korzhinskiy always attached particular importance to the methods of investigation. He maintained that new methods would in principle open up new possibilities for the understanding of natural laws. But the methods, in turn, should be based on rigorous scientific approaches and fundamental physical laws. He subjected to criticism any method or theory he felt was not sound scientifically. In disputes with his opponents (V.A. Nikolayev, I.V. Aleksandrov, W.S. Fyfe, D.F. Weill) Korzhinskiy was able to demonstrate unambiguously the validity of the concepts he had developed. Academician A.A. Polkanov once noted: 'Dmitriy Sergeyevich Korzhinskiy differs from the majority of geologists in that he practically makes no mistakes'. At the same time he carefully considered the advice and wishes of his colleagues to whom he often gave his manuscripts to review.

It should be pointed out that Korzhinskiy's concepts were distinguished by an unusual approach to solving diverse petrological problems. His creative thinking was characterized by deep penetration into the essence of each problem he studied, and a command of the facts; his memory was legendary. Korzhinskiy's papers are remarkable for well-honed thinking and completeness of ideas.

D.S. Korzhinskiy worked independently, that is, in the tradition of the solitary scholar, who had contributed so much to the sciences during the 19th and early 20th centuries. Korzhinskiy published more than 190 scientific papers, and practically none of them was co-authored. His productivity is all the more striking, because any one of his theories had sufficient significance to bring him recognition. Many of his works became classics in petrology while Dmitriy Sergeyevich was alive. The ideas expressed in his papers led to a re-evaluation of the nature of many crystalline rocks and ores, and provided a rigorous theoretical framework for treating metamorphic, metasomatic and plutonic processes applicable to any crystalline terrain. Examples are new principles in the classification of skarn deposits and a new understanding of their formation and a theoretical framework for estimating temperatures and pressures of metamorphism from rock-forming minerals. In other words, the influence of D.S. Korzhinskiy's works in shaping contemporary physico-chemical petrology was enormous and his authority was unquestioned. Even with this authority Korzhinskiy himself did not attach such a great significance to his work. In comparing his own achievements with those of physicists, technologists, chemists and biologists, he would maintain that his papers had only a transient significance and that their influence on geology was only temporary. Korzhinskiy's perception of his own place in the history of science resulted from his extraordinary modesty.

Dmitriy Sergeyevich was the recognized leader of Soviet petrological science, a position he never sought for himself. He was also widely perceived as the head of the so-called 'Korzhinskiy school'. However, he did not formally establish such a school. Young scientists simply flocked to him, and those sufficiently able and hard-working to 'stick it out' would ultimately contribute their bit to contemporary physico-chemical petrology. Many of these scientists justifiably regard D.S. Korzhinskiy as their teacher. He was indeed really a teacher, he was never a mentor. He lived by one simple rule: one must continually work and create every day, every minute, every second. This rule also applied willy-nilly to Korzhinskiy's students, for whom Korzhinskiy was always the model of an outstanding scientist.

I can confidently assert that Dmitriy Sergeyevich Korzhinskiy's scientific work is closely associated with the birth of the USSR Academy of Sciences Institute of Experimental Mineralogy. The advanced theoretical level of works by D.S. Korzhinskiy and his students enabled Academician D.I. Shcherbakov, who was Secretary for the Department of Geology and Geophysics in the Academy of Sciences, to propose to the Academy Presidium the founding of an institute of experimental mineralogy and petrology. The proposal was made in 1962, at which time Dmitriy Sergeyevich knew nothing about it. However, the Academy of Sciences President, Academician M.V. Keldysh, supported the proposal and asked to meet Korzhinskiy. The proposal was

To the memory of Dmitriy Sergeyevich Korzhinskiy

approved in 1963. From 1969 until 1979 Dmitriy Sergeyevich organized and directed the newly formed Institute of Experimental Mineralogy. The Institute subsequently became the world's largest research institution specializing in petrological science. Its staff now totals more than 300. Dmitriy Sergeyevich carried out his administrative tasks without interrupting the intense pace of his scientific research and publication. At the same time, he remained in charge of the Laboratory of Metamorphism and Metasomatism in the Institute for the Geology of Ore Deposits, Petrology, Mineralogy, and Geochemistry (IGEM) of the USSR Academy of Sciences.

During those years, while he was director of the Institute of Experimental Mineralogy, D.S. Korzhinskiy succeeded in inculcating in the scientists a spirit of good will, of high scientific exactitude, and of principled evaluation of scientific results, as well as with an intolerance towards a casual attitude in science. These principles even now determine the scientific and human relations within this relatively large group of like-minded persons.

Dmitriy Sergeyevich was permanent Chairman of the scientific council in the Institute of Experimental Mineralogy. Anyone who was preparing to submit a paper for publication realized that he or she had to meet Korzhinskiy's demanding standard of quality. Korzhinskiy could be abrupt and had a sharp tongue. At the same time, with his gentle humor, Korzhinskiy was able to reassure a young scientist and open his eyes to simple truths that are at times forgotten during days of hard work.

Contributions of Dmitriy Sergeyevich Korzhinskiy were valued highly by the Soviet government and by the scientific community in the USSR, and he was elected Academician, awarded the State and Lenin prizes, and given the title of Hero of Socialist Labor.

D.S. Korzhinskiy's recognition by the international scientific community resulted in his being elected an Honorary Member of numerous national academies, and geological and mineralogical societies, and in his being given several awards. In 1980 he was awarded the Roebling Medal, the Mineralogical Society of America's highest award and among the most prestigious awards in mineralogy and petrology. In presenting Korzhinskiy with this award Professor James B. Thompson, Jr, aptly summarized Korzhinskiy's contribution to the development of petrology.

A characteristic of Korzhinskiy's writing is the skill with which he applies sophisticated geometric and algebraic techniques to multicomponent polyphase systems, greatly extending the earlier methods developed by Federov, Niggli, Eskola and others. One of his major contributions has been to our understanding of the constraints that must be considered in the treatment of open systems, namely that rock systems respond to variations in the ambient values of certain chemical potentials as well as to variations in temperature and pressure. In dealing with such systems he also showed that energy functions other than the classic set of four presented by Gibbs are often convenient. To the conventionally minded of a decade and more ago this seemed an outrageous and unnecessary tampering with long-established procedures, certainly upsetting, and possibly downright wrong. Experimental petrologists, however, soon got the message. Experiments with systems buffered by controlled activities are now commonplace, whether or not a specific investigator may realize that the function that is minimized in the charge inside the capsule, when it equilibrates, is *not* its Gibbs free energy.

Korzhinskiy's honors in his native land have been many but it is fitting that we also honor him in recognition of the value of his contributions to the world at large. Mr President: I am pleased to present the Roebling Medalist for 1980: Dmitriy Sergeyevich Korzhinskiy.

No doubt Korzhinskiy's contribution to science will be greatly augmented and extended by his followers. In so doing, they will render homage and tribute to the memory of this MAN and SCIENTIST, DMITRIY SERGEYEVICH KORZHINSKIY.

Symbols

Thermodynamic functions

T, K:	temperature, K
t, °C:	temperature, °C
R:	ideal gas constant (1987 cal/mole/K)
P:	pressure, bar
C_p:	heat capacity, cal/mole/K
U:	internal energy, cal/mole (U^z–Korzhinskiy potential)
F:	Helmholtz energy, cal/mole (F^z–Korzhinskiy potential)
G:	Gibbs free energy, cal/mole (G^z–Korzhinskiy potential)
H:	enthalpy, cal/mole (H^z–Korzhinskiy potential)
S:	entropy, cal/mole/K
V:	volume, cal/bar
$\Delta G(i)$,	
$\Delta S(i)$,	
$\Delta H(i)$,	
$\Delta V(i)$:	change of corresponding parameters for reaction i
ΔH_{sol}:	heat of solution
ΔH_{ord}:	enthalpy of ordering
ΔG_{ord}:	Gibbs free energy of ordering
K_i:	partition coefficient of component i between phases α and β
K_D:	distribution coefficient

Minerals

Ab:	albite	Kln:	kaolinite
Acm:	acmite	Kls:	kalsilite
Act:	actinolite	Knr:	knorringite
Adr:	andradite	Ky:	kyanite
Alm:	almandine	Lct:	leucite
Am:	amphibole	Lmt:	laumontite
An:	anorthite	Lws:	lawsonite
And:	andalusite	Mag:	magnetite
Ann:	annite	Mgs:	magnesite
Ap:	apatite	Ms:	muscovite
Arg:	aragonite	Ne:	nepheline
Aug:	augite	Ol:	olivine
Bt:	biotite	Ocd:	'orthocorundum' ($AlAlO_3$)
Cal:	calcite	Opx:	orthopyroxene
Chl:	chlorite	Or:	orthoclase
Chr:	chromite	Pch:	pycrochromite
Cld:	chloritoid	Phl:	phlogopite
Cpx:	clinopyroxene	Pgt:	pigeonite
Crd:	cordierite	Po:	pyrrhotite
Crn:	corundum	Prl:	pyrophyllite
Crs:	crystabolite	Prp:	pyrope
Cum:	cummingtonite	Py:	pyrite
Czo:	clinozoisite	Qtz:	quartz
Di:	diopside	Rt:	rutile
Dol:	dolomite	Scp:	scapolite
En:	enstatite	Sid:	siderophyllite
Ep:	epidote	Sil:	sillimanite
Est:	eastonite	Spl:	spinel
Fo:	forsterite	St:	staurolite
Fsp:	alkali feldspar	Tlc:	talc
Ged:	gedrite	Tr:	tremolite
Gr:	graphite	Trd:	tridymite
Grt:	garnet	Tur:	tourmaline
Hem:	hematite	Usp:	ulvospinel
Hm:	hydromica	Uvr:	uvarovite
Hyp:	hypersthene	Wus:	wústite
Ilm:	ilmenite	Zrn:	zircon

Pl_{25}: plagioclase and molar % of anorthite in it

Px_{10}: solid solution of aegerine–diopside–hedenbergite and mole percent of aegerine in it

PART I

General thermodynamics and mineral equilibria including geothermobarometry

1

Mineral thermodynamicals and equilibria for geothermobarometry: an introduction

LEONID L. PERCHUK

To the general symbols listed in the preface of this volume are to be added the following specific symbols related to mineral solid solutions and their equilibria.

Partial thermodynamic functions for component i in phase α

$N_i^\alpha = N(i,\alpha)$: molar percent
$m_i^\alpha = m(i,\alpha)$: mass, g
$X_i^\alpha = X(i,\alpha)$: molar fraction
$\mu_i^\alpha = \mu(i,\alpha)$: chemical potential
$f_i^\alpha = f(i,\alpha)$: fugacity, bar
$P_\alpha^i = P(i,\alpha)$: pressure, bar
$a_i^\alpha = a(i,\alpha)$: activity
$\gamma_i^\alpha = \gamma(i,\alpha)$: activity coefficient
a_i^{id}: activity in ideal solution.

Mixing function

$G_i^m(\alpha) = G^m(i,\alpha)$: Gibbs free energy, cal/mole
$S_i^m(\alpha) = S^m(i,\alpha)$: entropy, cal/mole/K
$V_i^m(\alpha) = V^m(i,\alpha)$: volume, cal/bar/mole
$H_i^m(\alpha) = H^m(i,\alpha)$: enthalpy, cal/mole
S^{conf}: configuration entropy

Excess function

$G_i^e(\alpha) = G^e(i,\alpha)$: Gibbs free energy, cal/mole
$S_i^e(\alpha) = S^e(i,\alpha)$: entropy, cal/mole/K
$V_i^e(\alpha) = V^e(i,\alpha)$: volume, cal/bar/mole
$H_i^e(\alpha) = H^e(i,\alpha)$: enthalpy, cal/mole

$W_i^G(a)$: Margules parameter of i component for binary solution, cal/mole;
$W_i^G(a) = W_i^H(a) - W_i^S(a) \times T + W_i^V(a) \times P$

$W_{ij}(a)$: Margules interaction parameter for components i and j mixed in a ternary solid solution, cal/mole

$Q_{123}(a)$: Margules interaction parameter for three components in ternary system, cal/mole

---

Some general thermodynamic relations and formulae used

All the thermodynamic relations given in this chapter are generally known from Sryvalin & Esin (1959), Perchuk (1965, 1970a), Thompson (1967), Saxena (1973), Perchuk & Ryabchikov (1976), Wood & Fraser (1976), Powell (1978).

Ideal solution (id)

$$G_i^m(a) = RT\ln X_i^a, \tag{1}$$

$$S_i^m(a) = -R\ln X_i^a. \tag{2}$$

Non-ideal solution α

$$a_i^a = \gamma_i^a X_i^a, \tag{3}$$

$$G_i^e = RT\ln\gamma_i^a(a) = G_i^m(a) - RT\ln X_i^a. \tag{4}$$

Relation of integral functions $\Phi(a)$ to partial functions $\Phi(i,a)$:

$$\Phi^m(a) = \sum_i X_i^a \, \Phi_i^m(a), \tag{5}$$

$$\Phi^e(a) = \sum_i X^a \Phi_i^e(a), \tag{6}$$

$$\overline{K}_i = X_i^a / X_i^\beta, \tag{7}$$

$$K_D = \left\{\frac{X_i}{1-X_i}\right\}^a \times \left\{\frac{1-X_i}{X_i}\right\}^\beta, \tag{8}$$

$$RT\ln K_D = \Delta G_i^o + \Delta G_i^e + \Delta V_{ipy}^o + \Delta V_i^e P, \tag{9}$$

where any $\Delta\Phi$ is determined as the change of corresponding function due to a reaction.

If $\Delta G_\beta^e = G^e(1,\beta) - G^e(2,\beta)$ is known, then the following equations (Perchuk & Ryabchikov, 1968, 1976, Ryabchikov, 1975) can be used:

$$G^e(1,a) = -X_2^a(\Delta G_\beta + RT\ln K_D) + \int_0^{X_1^a}(\Delta G_\beta^e + RT\ln K_D)dX_1^a, \tag{10}$$

$$G^e(2,a) = -X_1^a(\Delta G_\beta + RT\ln K_D) - \int_0^{X_2^a}(\Delta G_\beta^e + RT\ln K_D)dX_2^a. \tag{11}$$

Excess molar mixing functions (Φ^e) for binary mixtures can also be described by the Margules formulation (Margules, 1895, Wagner, 1952):

$$\Phi^e = (1 - X_1)X_1 \sum B_n^\Phi (2X_1 - 1)^n, \tag{12}$$

$$\Phi_1^e = (1 - X_1)^2 \{B_1^\Phi + \sum B_n^\Phi [(2n+1)X_1 - (1-X_1)](2X_1 - 1)^{n-1}\} \tag{13}$$

$$\Phi_2^e = X_1^2 \{B_2^\Phi + \sum B_n^\Phi [X_1 - (2n+1)(1-X_1)](2X_1 - 1)^{n-1}\} \tag{14}$$

where $\Phi^e(\Phi_i^e)$ stands for $G^e(G_i^e)$, $H^e(H_i^e)$ etc. According to Gildebrandt a solution with $n = 0$ and $S^e = 0$ is called a regular or one-parameter solution. For symmetric solutions $B_0^G = 2RT_{crit}$ at $X_1 = 0.5$.

So-called asymmetric two-parameter models can be easily derived by combination of the Duhem equation ($X_1 + X_2 = 1$)

$$X_1 d\Phi_1 + X_2 d\Phi_2 = 0 \tag{15}$$

with Taylor exponential series $\Phi_i = f(X_j)$ at constant T and P.

According to the Margules model any molar excess thermodynamic functions may be represented by exponential series

$$\Phi_1^e = B_1^\Phi + B_{11}^\Phi X_2 + B_{21}^\Phi X_2^2 + B_{31}^\Phi X_2^3 \ldots , \tag{16}$$

$$\Phi_2^e = B_2^\Phi + B_{12}^\Phi X_1 + B_{22}^\Phi X_1^2 + B_{32}^\Phi X_1^3 \ldots , \tag{17}$$

where B_{ij}^Φ are expansion constants in the Taylor series. For example, according to (16) the following exponential series for component 1 can be written:

$$H_1^e = B_{01}^H + B_{11}^H X_2 + B_{21}^H X_2^2 + B_{31}^H X_2^3 \ldots ,$$

$$S_1^e = B_{01}^S + B_{11}^S X_2 + B_{21}^S X_2^2 + B_{31}^S X_2^3 \ldots , \tag{18}$$

$$V_1^e = B_{01}^V + B_{11}^V X_2 + B_{21}^V X_2^2 + B_{31}^V X_2^3 \ldots ,$$

where parameters B_{ij} on the one hand, and parameters (19) and (20), on the other hand, are in the following relationships:

$$B_{01}^H = W_1^H, \; B_{11}^H = 2(W_2^H - 2W_1^H) \tag{19}$$

$$B_{01}^S = W_1^S, \; B_{11}^S = 2(W_2^S - 2W_1^S) \tag{20}$$

$$B_{01}^V = W_1^V, \; B_{11}^V = 2(W_2^V - 2W_1^V) \tag{21}$$

$$B_{21}^H = 5W_1^H - 4W_2^H, \; B_{31}^H = 2(W_2^H - W_1^H) \tag{22}$$

$$B_{21}^S = 5W_1^S - 4W_2^S, \; B_{31}^S = 2(W_2^S - W_1^S) \tag{23}$$

$$B_{21}^V = 5W_1^V - 4W_2^V, \; B_{31}^V = 2(W_2^V - W_1^V) \tag{24}$$

With the help of equations (22)–(24) the following expressions for Gibbs excess free energies can be readily deduced:

$$G^e = W_1^G X_1 X_2^2 + W_2^G X_1^2 X_2, \tag{25}$$

$$G_1^e = X_2^2 [W_1^G + 2X_1(W_2^G - W_1^G)], \tag{26}$$

$$G_2^2 = X_1^2[W_2^G + 2X_2(W_1^G - W_2^G)]. \tag{27}$$

Thus, two-parameter equations like (18)–(24) or so-called asymmetric models of subregular solutions also belong to the Margules empiric formulation. At well-determined particles (molecules, ions, atoms etc.) a number of binary solid solutions can be described with the observed relations (10)–(27).

For ternary and multicomponent solid solutions a the ideal mixing can be described with the equations

$$S^m(a) = -R\sum_i X_i^a \ln X_i^a, \tag{28}$$

$$G^m(a) = RT\sum_i X_i^a \ln X_i^a. \tag{29}$$

For non-ideal ternary solid solutions, excess molar Gibbs free energies are expressed as the following (Wohl, 1953):

$$G^e = X_1(X_2^2 W_{12} + X_3^2 W_{13}) + X_2(X_1^2 W_{21} + X_3^2 W_{23}) + X_3(X_1^2 W_{31} + X_2^2 W_{32}) + X_1 X_2 X_3\{(W_{12} + W_{21} + W_{13} + W_{31} + W_{23} + W_{32})/2 + W_{123}\}, \tag{30}$$

$$G_1^e = X_2^2\{W_{12} + 2X_2(W_{22} - W_{12})\} + X_3^2\{W_{13} + 2X_1(W_{31} - W_{13})\} + X_2 X_3\{(W_{12} + W_{21} + W_{31} + W_{13} - W_{32} - W_{23})/2 + X_1(W_{21} - W_{12} + W_{31} - W_{13}) + (X_2 - X_3)(W_{23} - W_{32})\} + X_3 X_2^2 W_{123}, \tag{31}$$

$$G_2^e = X_1^2\{W_{21} + 2X_2(W_{12} - W_{21})\} + X_3^2\{W_{23} + 2X_2(W_{32} - W_{23})\} + X_1 X_3\{(W_{32} + W_{23} + W_{21} + W_{12} - W_{31} - W_{13})/2 + X_2(W_{32} - W_{23} + W_{12} - W_{21}) + (X_3 - X_1)(W_{31} - W_{13})\} + X_1 X_3^2 W_{123}, \tag{32}$$

$$G_3^e = X_2^2\{W_{32} + 2X_3(W_{23} + W_{32})\} + X_1^2\{W_{31} + 2X_3(W_{13} + W_{31})\} + X_1 X_1\{(W_{32} + W_{23} + W_{13} + W_{31} + W_{12} + W_{21})/2 + X_3(W_{23} - W_{32} + W_{13} - W_{31}) + (X_1 - X_2)(W_{12} - W_{21})\} + X_2 X_1^2 W_{123}, \tag{33}$$

where $X_1 + X_2 + X_3 = 1$, $W = W_{ij}^G$ and

$$W_{ij}^G = W_{ij}^H - W_{ij}^S T + W_{ij}^V P. \tag{34}$$

The following relations can be applied to a regular solid solution model ($S^e = 0$):

$$G_1^e = X_2^2 W_{12} + X_3^2 W_{13} + X_3 X_2(W_{12} + W_{13} - W_{23}), \tag{35}$$

$$G_2^e = X_1^2 W_{12} + X_3^2 W_{23} + X_1 X_3(W_{12} + W_{23} + W_{13}), \tag{36}$$

$$G_3^e = X_2^2 W_{32} + X_1^2 W_{13} + X_2 X_3(W_{13} + W_{23} + W_{12}). \tag{37}$$

Korzhinskiy's theory of extreme states

In the 30s, while studying petrology of the Aldan shield, D.S. Korzhinskiy found correlation between the succession of carbonation reactions of silicates and depth of metamorphism. Using this correlation and thermodynamic calculations of reactions of carbonation of some end-members of solid solutions Korzhinskiy (1935, 1937) graduated the granulite-facies rocks into several deep-seated units. Later Korzhinskiy (1940) added to this work calculation of some net-transfer equilibria. Actually these were the first steps in geothermobarometry, which has developed since that time in two directions: (i) estimation of $P-T$ parameters on the basis of distribution of components between coexisting phases, and (ii) studies on homogenization of gas–liquid inclusions in rock-forming minerals in respect to $P-T$ estimates.

From the point of view of the history of thermobarometry, Korzhinskiy's contribution to the theory of the rock-forming mineral equilibria is remarkable. D.S. Korzhinskiy (1958, 1960, 1963) believed that the Gibbs–Konovalov theory of extreme states of a system can be used to estimate the $P-T-f_i$ parameters. According to this theory thermodynamic potential (U) of system in equilibrium appears to be the algebraic sum of the products of extensive (E) and intensive (I) parameters

$$G^z = \sum_i^{k+2} EI, \tag{38}$$

where $i = 1, \ldots, k+2$. The full differential of the potential, in the case of a completely closed system, equals

$$dU = \sum_i I dE. \tag{39}$$

In the case of an open system consisting of i inert and k perfectly mobile components at constant S and V, the above should be written as follows:

$$dU^z = \sum_i \mu_i dm_i - \sum_k m_k d\mu_k. \tag{40}$$

Elementary change of Gibbs–Korzhinskiy free energy can be described by the equation

$$dG^z = - SdT + VdP + \sum_i \mu_i dm_i - \sum_k m_k d\mu_k, \tag{41}$$

which may be considered as an algebraic sum of thermal, mechanical and chemical energies, with each of them being the product of conjugate extensive (S, V, m_i) and intensive (P, T, μ_i) parameters.

For an equilibrium of solid solutions α, β, γ, $\ldots$, composed of k components, relationships between partial molar values can be described by a system of Duhem equations:

$$\left. \begin{aligned} - S^\alpha dT + V^\alpha dP - \sum_i m_i^\alpha d\mu_i &= 0, \\ - S^\beta dT + V^\beta dP - \sum_i m_i^\alpha d\mu_i &= 0, \\ - S^\phi dT + V^\phi dP - \sum_i m_i^\alpha d\mu_i &= 0, \end{aligned} \right\} \tag{42}$$

Table 1.1.

	$-\mathrm{d}T$	$\mathrm{d}P$	$-\mathrm{d}\mu_1$	$-\mathrm{d}\mu_2$	$\ldots$	$-\mathrm{d}\mu_k$
α	S^a	V^a	$m(1,a)$	$m(2,a)$	$\ldots$	$m(k,a)$
β	S^β	V^β	$m(1,\beta)$	$m(2,\beta)$	$\ldots$	$m(k,\beta)$
γ	S^γ	V^γ	$m(1,\gamma)$	$m(2,\gamma)$	$\ldots$	$m(k,\gamma)$
$\ldots$	$\ldots$	$\ldots$	$\ldots$	$\ldots$	$\ldots$	$\ldots$
ϕ	S^ϕ	V^ϕ	$m(1,\phi)$	$m(2,\phi)$	$\ldots$	$m(k,\phi)$

This expression may also be presented in the form of a table of intensive and extensive parameters (Table 1.1., Korzhinskiy, 1963).

The maximum number of columns or rows in a non-zero determinant that can be cut out of this matrix shows the rank r_A of the matrix with k components involved:

$$k+2 \geqslant r_A \leqslant \phi. \tag{43}$$

If i extensive parameters in ϕ phases turn out to be connected linearly, then a reaction between these phases takes place at the extremum (minimum or maximum conditions) values of intensive parameters. For example, let $k = \phi = 3$ in a matrix with $k+2 > \phi > r_A$. Then the determinant composed of extensive parameters of phases is equal to zero. For example, in the case where

$$\begin{vmatrix} m(1,a) & m(2,a) & m(3,a) \\ m(1,\beta) & m(2,\beta) & m(3,\beta) \\ m(1,\gamma) & m(2,\gamma) & m(3,\gamma) \end{vmatrix} = 0, \tag{44}$$

chemical potentials of perfectly mobile components 1, 2 and 3 show extremums at constant T and P. Fig. 1.1 illustrates the relations between phase compositions, temperature and pressure. T–X_i, P–X_i and μ_i–X_i diagrams can be calculated while all the extensive parameters excluding one are constant. Tie-lines in Fig. 1.1 indicate two versions of the following reaction:

$$A\beta_i = Aa_i + C_\gamma \tag{45}$$

where A and C are stoichiometric coefficients. In both cases the signs of ΔV and ΔH are the same. However, in the first version (the upper row of diagrams in Fig. 1.1) the relations between partial molar volume V_i^m and enthalpy H_i^m for the equilibrium under discussion determine $P = \max$ and $T = \min$ because of similar compositions (for example, Fe:Mg ratios of coexisting minerals may be equal). The diagrams of the lower row in Fig. 1.1 reflect the reverse relations.

Extreme relationships can be considered in terms of molar fractions of isomorphic components in coexisting minerals α and β. Fig. 1.2 shows the change in X_1^α/X_1^β while $K_D = 1$ for three possible reactions of type (45) at extreme P and T.

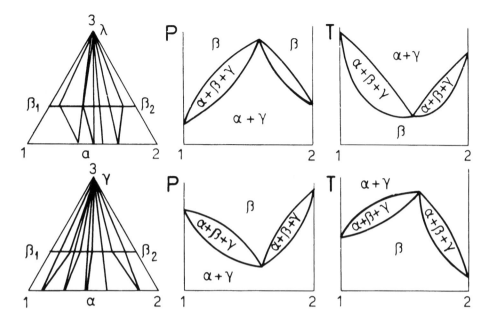

Fig. 1.1. Relationships between compositions of coexisting phases α, β and γ in the case of extreme states of external parameters. Slab of the tie-lines shows pre-extreme and post-extreme conditions for the system.

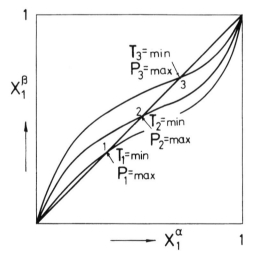

Fig. 1.2. Relationships of molar fractions of component *1* for two coexisting minerals α and β for three possible reactions of type (45) in the systems with extremums of T and P.

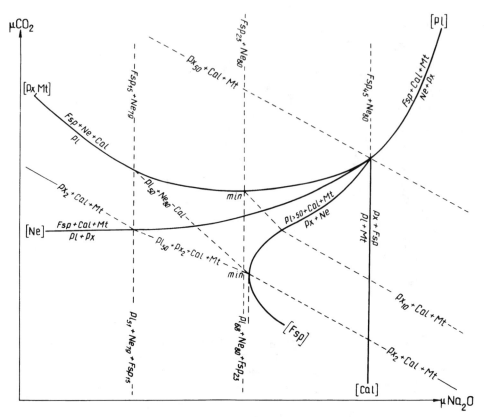

Fig. 1.3. The $\mu_{Na_2O} - \mu_{CO_2}$ diagram for the system $Cpx + Ne + Fsp + Mag =$
$Pl + Cal + Na_2O + K_2O + O_2$. Two of five equilibria show minima of μ_{H_2O}
and μ_{CO_2} for the univariant reaction $(_1 - 47)$ and $Pl_{>50} + Cal + Mag =$
$Px + Ne + CO_2 + O_2 + Na_2O$ at temperature 500 °C and pressure 1 kbar.
Values μ_{O_2} and μ_{K_2O} are constant at given T–P parameters. MT = Mag.

The principles of the extremum state theory presented here in brief were originated
by Gibbs, Konovalov, Vrevskiy, Jouguet, Saurel, Prigogine, Defay, Storonkin, and
Schultz. However, D.S. Korzhinskiy was the first to develop the theory of extremums
of chemical potentials of perfectly mobile components at constant T and P. Fig. 1.3.
shows phase relationships in the system Na_2O–K_2O–CaO–Fe_2O_3–Al_2O_3–SiO_2–CO_2
in coordinates of chemical potentials of sodium oxide and CO_2 at $T = 500$ °C, $P = 1$
kbar, and at constant values of potentials of oxygen and potassium oxide. Minima of
μ_{CO_2} and μ_{Na_2O} in a fluid are readily seen. The following reaction can be discussed in
more detail:

$$Na_xCa_{1-x}Al_{2-x}Si_{2+x}O_8 + \{[y(1-x) - 0.5x(1-z)]Na_2O + (1-x)CO_2$$
$$+ [(1-x)(1-y) + 0.5x(1-z)]K_2O\} = 2(1-x)Na_yK_{1-y}AlSiO_4$$
$$+ xNa_zK_{1-z}AlSiO_3O_8 + (1-x)CaCO_3 \quad (46)$$

or

$$Pl + (Na_2O + K_2O + CO_2) = Ne + Fsp + Cal.$$

10

Perfectly mobile components are shown in brackets. Applying Korzhinskiy potential G^z to the equilibrium (46), we get the following equation of shift equilibrium:

$$-\Delta S \mathrm{d}T + \Delta V \mathrm{d}P - \Delta m_{K_2O}\mathrm{d}\mu_{K_2O} - \Delta m_{Na_2O}\mathrm{d}\mu_{Na_2O} - \Delta m_{CO_2}\mathrm{d}\mu_{CO_2} = 0. \qquad (47)$$

At constant T, P and μ_{K_2O}

$$\left[\frac{\partial \mu_{CO_2}}{\partial \mu_{Na_2O}}\right]_{T,\,P,\,\mu_{K_2O}} = \frac{x(1-z) - 2y(1-x) \times 62}{2(1-x) \times 44}, \qquad (48)$$

where 62 and 44 are the molecular weights of Na_2O and CO_2, respectively. While

$$x(1-z) = 2y(1-x) \qquad (49)$$

the derivative (48) is equal to zero reflecting a minimum of μ_{CO_2}. This example was discussed in more detail by Perchuk (1966) and Perchuk & Ryabchikov (1968).

Marakushev (1964, 1965) has found a member of extreme relations of Fe–Mg minerals in metamorphic rocks. Many extremums of external parameters of magmatic and metamorphic systems were discussed by Perchuk (1969, 1970a) while calibrating geothermometers and geobarometers. An example of minimum of temperature at $\mu_{H_2O} = \max$ and $P = \max$ is shown in Fig. 1.4 deduced for an equilibrium of hornblende and clinopyroxene:

$$Am + Grt + Qtz = Cpx + Pl + Opx + H_2O. \qquad (50)$$

Extremums of external parameters in the system are achieved at $X_{Mg}^{Cpx} = X_{Mg}^{Am} = 0.45$ when its determinant $D(i, \phi) = 0$, where i are components of the system Na_2O–CaO–MgO–FeO–Al_2O_3–SiO_2–H_2O and ϕ is the number of phases in table 1.1. Fig. 1.4 shows an extremum at $X_{Mg} = 0.42$: while X_{Mg} is less than 0.42 the relation $X_{Mg}^{Cpx} > X_{Mg}^{Am}$ is valid (pre-extremum state); if X_{Mg} is more than 0.42 the relation $X_{Mg}^{Cpx} < X_{Mg}^{Am}$ appears to hold in the equilibrium (post-extremum state).

Extreme relations of minerals in equilibria are commonly explained by deviation of solid solutions from ideal mixtures. D.S. Korzhinskiy (1973) has shown that these relations may be characteristic for equilibria of ideal solid solutions in the case of considerable difference in dimensions of their end-member molecules. For example, extremum of intensive parameters is impossible in biotite–garnet exchange equilibrium because $(Fe,Mg)_{Grt} \cong (Fe,Mg)_{Bt}$. However, extremums are quite possible in amphibole–hypersthene or olivine–hypersthene exchange equilibria because of a considerable difference in dimensions of the end-members.

For non-ideal solutions, deviation from ideal mixtures is the greater the smaller is the dimension of end-members constituting the mixtures. For example, a solid solution of enstatite and tshermakite is non-ideal. Its deviation from ideal mixture is quite small because tshermakite $MgAl_2SiO_6$ is an intermediate compound in the mixture $MgSiO_3$–Al_2O_3 described with considerably larger values of mixing energies (Aranovich & Kosyakova, 1986).

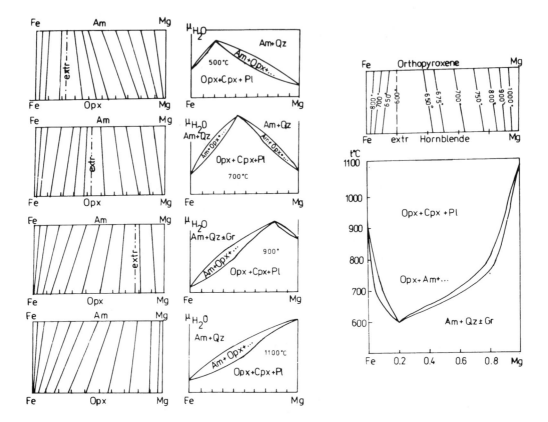

Fig. 1.4. Temperature minimum for equilibrium $Gr + Am + Qtz = Cpx + Opx + Pl + H_2O$. Varying with temperature, μ_{H_2O} achieves a maximum at molar fractions equal for both hornblende and clinopyroxene at minimum T; pressure varies with T and μ_{H_2O}, achieving a maximum at $X_{Mg}^{Am} = X_{Mg}^{Cpx}$.

Correct choice of end-members is important and rather complicated for magmatic systems. The unification of end-members is of primary importance especially while calibrating liquidus geothermometers and geobarometers.

Geothermobarometry of mineral equilibria and fluid inclusions

The theory of extreme states of thermodynamic systems is a constituent case in the general theory of phase equilibria. Friedman (1949) was the first who offered to use distribution of components between co-existing minerals aimed at mineralogical thermometry. Barth (1951) has calibrated the first geothermometer based on albite distribution between co-existing feldspars. Ramberg & De Vore (1951) calculated olivine–orthopyroxene equilibria on the basis of thermochemical data. Later Kretz (1963), Lindsley (1963), Ryabchikov (1965), and Perchuk (1965, 1969, 1970a,b,

1977a,b, 1988) contributed to mineralogical thermometry. Their calculations were based on experimental data on exchange equilibria as well as on empirical studies of some net-transfer reactions at high T–P parameters.

The first system of internally consistent mineralogical thermometers was created on the basis of *phase correspondence formulation*, i.e. the general theory of redistribution of cations between silicates as a function of temperature (Perchuk, 1968, 1971, 1977a). The theory predicts direction and relative value of the shift of cation exchange reaction. A dozen geothermometers based on the redistribution of Fe, Mg, Mn, Ca, K and Na were evaluated (Perchuk, 1970a) on the basis of magnetite–ilmenite (Lindsley, 1963), two-feldspar (Ryabchikov, 1965) and nepheline–feldspar (Perchuk, 1965) geothermometers following this theory. It should be mentioned that even in those years the Margules approach was used to calculate mixing properties of mineral solid solutions (Ryabchikov, 1965, Perchuk, 1965, Perchuk & Ryabchikov, 1968, 1976, Perchuk & Andrianova, 1968, Thompson & Waldbaum, 1968, 1969) and ordered Fe–Mg minerals with energetically non-equivalent sites (Grover & Orville, 1969, Perchuk, 1967, 1973, Perchuk & Surikov, 1970). Later these methods were modified for multicomponent multisite rock-forming minerals (see for example, Saxena, 1973, Wood & Fraser, 1976, Kurepin, 1976, Aranovich & Kosyakova, 1986).

Mineral barometry developed by Sobolev (1961), Marakushev (1965, 1968), Perchuk (1968, 1970a,b, 1973, 1977a,b), Hensen (1973), Hensen & Green (1971, 1972), Boyd (1973), Wood & Banno (1973), Ghent (1976) achieved the most impressive results by the middle of the 80s. Some of the most important and well developed barometers will be discussed in Chapters 5 and 7 of this book.

Beginning with papers by Kudo & Weil (1970) and Roeder & Emslie (1970), so-called liquidus thermometry started to develop (Leeman, 1978, Perchuk & Vaganov, 1977, Herzberg, 1979, Nilsen & Drake, 1979). Then attempts to calculate depth of crystallization of magmatic melts (Carmichael *et al.*, 1977, Nicholls *et al.*, 1971) and generation of basic and ultrabasic magmas (Perchuk *et al.*, 1982, Perchuk, 1987) were made. Thermobarometry of ultrabasic inclusions of peridotites and eclogites in kimberlite pipes has spread widely (see, for example, Davis & Boyd, 1966, Boyd, 1973, Wood & Banno, 1973). Along with mineralogical thermometry a geochemical direction based on temperature dependence of distribution coefficients of stable isotopes such as sulphur, oxygen and carbon (Koefs, 1980, Bottinga & Javoy, 1973, Suvorova & Perchuk, 1982) appeared. However, this direction has not been widely applied to petrology due to complex techniques of mass-spectrometric analyses and mineral inhomogeneity.

Strict thermodynamics of solid solutions and their equilibria has always been the main trend in mineralogical thermobarometry. The best geothermometers and geobarometers were made by combination of experimental data on equilibria and thermodynamics of solid solutions with analysis of mineral parageneses in natural rocks. Examples of such calibrations will be discussed in subsequent chapters of this part of the book.

At present, versions of one and the same mineralogical thermometer or barometer are sometimes counted in dozens. Moreover, in some cases, later versions repeat those ones published several years earlier within their accuracy, and authors of the new versions, for reasons obscure, do not refer to their predecessors. Such a situation has probably arisen from the lack of periodical reviews on mineralogical thermobarometry. The present part of the book is to compensate for a deficiency of such reviews and to show the present-day state of this branch of physico-chemical petrology.

Even in the previous century gas–fluid inclusions in transparent minerals were known. At that very time the ways of finding out those inclusions and a hypothesis of their formation were offered. But only in the year 1948 did N.P. Ermakov work out the classification of gas–fluid inclusions and proposed methods of their homogenization aimed at learning temperatures and pressures of magmatic and metamorphic process in the mineral-forming medium (Ermakov, 1948, 1950). Ermakov's pioneer works, which were highly supported by D.S. Korzhinskiy, have found a lot of followers, and his investigation of fluid and melt inclusions in different minerals led to a number of important discoveries in petrology and genetic mineralogy. But for a long time methods of investigation of fluid inclusions in metamorphic minerals remained a complicated problem. Only in the 60s and 70s did considerable progress in this field of petrology begin to show (for example, Dolgov, 1970, Dolgov *et al.*, 1965, Touret, 1976); recent progress is summarized in books by Tomilenko & Chupin (1983) and Berdnikov (1987).

Perchuk (Perchuk, 1989, Perchuk *et al.*, 1985, Perchuk & Fed'kin, 1986) has also demonstrated satisfactory agreement between *T–P* parameters of metamorphism based on mineral geothermobarometry with the results of homogenization of gas–fluid inclusions in minerals from the same rocks.

Chapters 2, 4, 5 and 7 present reviews of studies on thermodynamic properties of solid solutions of feldspars, nepheline, garnet, orthopyroxene, cordierite, epidote, biotite as well as their exchange and net-transfer equilibria. Two-feldspar and nepheline–feldspar equilibria and a set of geothermobarometic equilibria of Fe–Mg rock-forming minerals will be discussed in terms of consistency of thermodynamic data. Application of these instruments in combination with fluid inclusion data to analysis of evolution of some metamorphic terranes and magmatic rocks will be discussed.

Acknowledgements I thank K.K. Podlesskii who read the manuscript and offered useful suggestions which improved the paper.

References

Aranovich, L.Ya. & Kosyakova, N.A. (1986). Equilibrium cordierite = orthopyroxene + quartz: experimental data and thermodynamics of ternary Fe–Mg—Al orthopyroxene solid solution. *Geokhimia*, **8**, 1181–201.

Barth, T.F.W. (1951). The feldspar geologic thermometer. *Neues Jahrb. Miner. Abhandl.*, **82**, 143–50.

Berdnikov, N.V. (1987). *Thermobarogeochemistry of the Far East Precambrian metamorphic complexes.* Moscow: Nauka.

Bottinga, I. & Javoy, M. (1973). Comments on isotope geothermometry. *Earth Plant. Sci. Lett.,* **20**, 250.

Boyd, F.R. (1973). A pyroxene geotherm. *Geochim. Cosmochim. Acta,* **37**, 2533–46.

Carmichael, I.S., Nicholls, J., Spera, F.J., Wood, B.J. & Nelson, S.A. (1977). High temperature properties of silicate liquids: applications to the equilibration and ascent of basic magma. *Phil. Trans. Roy. Soc. London, A,* **286**, 373–431.

Davis, B.T.C. & Boyd, F.R. (1966). The join $Mg_2Si_2O_6$–$CaMgSi_2O_6$ at 30 kb pressure and its application to pyroxenes from kimberlites. *J. Geophys. Res.,* **71**, 3567–76.

Dolgov, Ju.A. (1970). Inclusions in minerals of metamorphic rocks as indicators of the thermodynamic conditions of metamorphism. In *Problems of petrology and genetic mineralogy,* pp. 272–81. Moscow: Nauka.

Dolgov, Ju.A., Bazarov, L.Sh. & Bakumenko, I.I. (1965). The method of determination of pressure in inclusions using homogenization and cryometry. In *Mineral thermometry and barometry,* vol. 2, pp. 9–17. Moscow: Nauka.

Ermakov, N.P. (1948). Origin and classification of liquid inclusions in minerals. *L'vov Geol. Soc., Mineral. Sbornik,* **3**, 53–73.

— (1950). Technique of thermometric analyses of minerals from hydrothermal deposits. *L'vov Geol. Soc. Mineral. Sbornik,* **4**, 45–70.

Friedman, I.I. (1949). A proposed method for the measurement of geologic temperatures. *J. Geology,* **57**, N 6.

Ghent, E.D. (1976). Plagioclase–garnet–Al_2SiO_5–quartz: a potential geobarometer–geothermometer, *Amer. Mineral.,* **61**, 710–14.

Grover, J.E. & Orville, P.M. (1969). The partitioning of cations between coexisting single and multi-site phases with application to the assemblages orthopyroxene–clinopyroxene and orthopyroxene–olivine. *Geochim. Cosmochim. Acta,* **33**, 205–26.

Hensen, B.J. (1973). Pyroxenes and garnets as geothermometers and barometers. *Carnegie Inst. Wash. Yearb.,* **72**, 527–34.

Hensen, B.J. & Green, D.H. (1971). Experimental study of the stability of cordierite and garnet in pelitic compositions at high pressures and temperatures. 1. Compositions with excess alumino-silicate. *Contrib. Mineral. Petrol.,* **33**, 309–30.

—— (1972). Experimental study of the stability of cordierite and garnet in pelitic compositions at high pressures and temperatures. 2. Compositions without excess alumino-silicate. *Contrib. Mineral. Petrol.,* **35**, 331–54.

Herzberg, C.T. (1979). The solubility of olivine in basaltic liquids: an ionic model. *Geochim. Cosmochim. Acta,* **43**, 1241–53.

Koefs, J. (1980). *Stable isotope geochemistry.* Springer-Verlag.

Korzhinskiy, D.S. (1935). Thermodynamics and geology of some metamorphic reactions involving a gas phase. *Zapiski Vserossiyskogo Mineralogicheskogo Obschestva,* **62**, N 1.

— (1937). Dependence of mineralforming on the depth. *Zapiski Vserossiyskogo Mineralogicheskogo Obschestva,* **66**, N 2.

— (1940). Factors of mineral equilibria and deep-seated mineral facies. *Trudy Instituta Geologicheskikh Nauk Akad. Nauk SSSR,* Vyp. **12**.

— (1958). The extremum states in the systems with perfectly mobile components. *Journal phys. khimii,* **32**, No.7.

— (1960). Additional comments to theory of extremum states. *Journal phys. khimii,* **34**, N 7.

— (1963). The theory of extremum states and its significance for mineral systems. In *Chemistry of the Earth's Crust,* ed. A.P. Vinogradov, 1, pp. 63–84. Moscow: USSR Acad. Sci. Press.

— (1973). *Theoretical principles for analyses of mineral parageneses.* Moscow: Nauka.

Kretz, R. (1963). Temperature dependence of the distribution coefficient. *Geol. Magazine,* **100,** 190–6.

Kudo, A. M. & Weill, D.F. (1970). An igneous plagioclase thermometer. *Contrib. Mineral. Petrol.,* **25,** 52–65.

Kurepin, V.A. (1976). To the technique of thermodynamic analyses of the mineral equilibria with solid solutions. *Geokhimia,* **2,** 289–98.

Leeman, W.P. (1978). Distribution of Mg^{2+} between olivine and silicate melt, and its implications regarding melt structure. *Geochim. Cosmochim. Acta,* **42,** 789–801.

Lindsley, D.H. (1963). Fe–Ti oxides in rocks as thermometers and oxygen barometers: equilibrium relations of coexisting pairs of Fe–Ti oxides. *Carnegie Inst. Wash. Yearb.,* **62,** 60–6.

Marakushev, A.A. (1964). Extremum states of some mineral equilibria in respect of geochemical regimes of metamorphism and metasomatism. In *Chemistry of the Earth's crust,* ed. A.P. Vinogradov, vol. 2, pp. 122–44. Moscow: Nauka.

— (1965). *The problem of mineral facies in metamorphic and metasomatic rocks.* Moscow: Nauka.

— (1968). *The thermodynamics of metamorphic hydration of minerals.* Moscow: Nauka.

Margules, M. (1895). Uber die Zusammensetzung der gesatigten Damphe von Mischungen. *Sitzungsber., Akad. Wiss. Mat. Naturwissenschaft,* **11a,** Bd. 104, 1243.

Nicholls, J., Carmichael, I.S.E. & Stromer, J. (1971). Silica activity and P_{total} in igneous rocks. *Contrib. Mineral. Petrol.,* **33,** 1–20.

Nilsen, R.L. & Drake, M.I. (1979). Pyroxene-melt equilibria. *Geochim. Cosmochim. Acta,* **43,** 1259–72.

Perchuk, L.L. (1965). The paragenesis of nepheline with alkali feldspar as the indicator of thermodynamic conditions of mineral equilibria. *Doklady Akademii Nauk SSSR,* **161,** 932–5.

— (1966). Certain mineral facies of alkalinity and the problem of deep-seated facies of nepheline syenites. *Izvestia Akad. Nauk SSR, Ser. Geol.,* **7.**

— (1967). A possiblity of using ordering in the crystalline solution for creating the mineralogical thermometers and barometers. *Doklady Akademii Nauk SSSR,* **174,** 934–6.

— (1968). Pyroxene–garnet equilibrium and the deep-seated facies of eclogites. *Inter. Geol. Rev.,* **10,** 280–318.

— (1969). The effect of temperature and pressure on the equilibria of natural Fe–Mg minerals. *Inter. Geol. Rev.,* **11,** 875–901.

— (1970a). *Equilibria of rock-forming minerals.* Moscow: Nauka.

— (1970b). Equilibrium of biotite with garnet in metamorphic rocks. *Geochim. Inter.,* **7,** 157–79.

— (1971). Crystallochemical problems in the phase correspondence theory. *Geokhimiya,* N 1, 23–38.

Perchuk, L.L. (1973). *Thermodynamical regime of deep-seated petrogenesis.* Moscow: Nauka.

— (1977a). Thermodynamical control of metamorphic processes. In *Energetics of geological processes,* ed. S.K. Saxena & S. Bhattacharji, pp. 285–352. Springer–Verlag.

— (1977b). Pyroxene barometer and pyroxene geotherms. *Doklady Akademii Nauk SSSR,* **233,** N 6, 1196–9.

— (1987). Studies of volcanic series related to the origin of some marginal sea floors. In *Magmatic processes: physicochemical principles,* ed. B.O. Mysen, pp. 319–8. Geochem. Soc., Special Publication.

— (1989). *P–T*–fluid regimes of metamorphism and related magmatism with specific reference to the Baikal granulites. In *Evolution of metamorphic belts,* ed. S. Daly, B.W. Yardley & B. Cliff. 275–291. Geol. Soc. London Special Publications.

Perchuk, L.L. & Andrianova, Z.S. (1968). Thermodynamics of equilibria of alkali feldspar $(K,Na)AlSi_3O_8$ with aqueous solution of $(K,Na)Cl$ at 500–800 °C and pressure 2000–1000 bar. In *Theoretical and experimental studies of mineral equilibria*, pp. 37–72. Moscow: Nauka.

Perchuk, L.L., Aranovich, L.Ya. & Kosyakova, N.A. (1982). Thermodynamic models of the origin and evolution of basalt magmas. *Vest. Moscow. Univ. Ser. Geol.*, **4**, 3–25.

Perchuk, L.L., Aranovich, L.Ya., Podlesskii, K.K. *et al.* (1985). Precambrian granulites of the Aldan Shield, eastern Siberia, USSR. *J. Metamorphic Geol.*, **3**, 265–310.

Perchuk, L.L. & Fed'kin, V.V. (1986). Evolution of *P–T* parameters during regional metamorphism. *Geologia i Geophysika*, N 7, 65–9.

Perchuk, L.L. & Ryabchikov, I.D. (1968). Mineral equilibria in the system nepheline – alkali feldspar – plagioclase and their petrological significance. *J. Petrology*, **9**, 123–67.

—— (1976). *Phase correspondence in the mineral systems*. Moscow: Nedra.

Perchuk, L.L. & Surikov, V.V. (1970). Thermodynamic problems of cation distributions between sublattices of multi-site minerals. *Izvestia Akademii Nauk SSSR. Geol.*, N 10, 3–18.

Perchuk, L.L. & Vaganov, V.I. (1977). *Temperature regime of formation of continental volcanic series*. Chernogolovka: USSR Acad. Sci., special issue.

Powell, R. (1978). *Equilibrium thermodynamics in petrology: an introduction*. Harper & Row.

Ramberg, H. & De Vore, G. (1951). The distribution of Fe^{++} and Mg^{++} in coexisting olivines and pyroxenes. *J. Geol.*, **59**, 193–210.

Roeder, P.L. & Emslie, R.E. (1970). Olivine–liquid equilibrium. *Contrib. Mineral. Petrol.*, **29**, 275–89.

Ryabchikov, I.D. (1965). New diagram for two-feldspar geological thermometer deduced with thermodynamic treatment of experimental data. *Doklady Akademii Nauk SSSR*, **165**, N 3.

— (1975). *Thermodynamics of fluids related to granite magmas*. Moscow: Nauka.

Saxena, S.K. (1973). *Thermodynamics of rock-forming crystalline solutions*. Springer-Verlag.

Sobolev V.S. (1961). On pressure at metamorphic process. In *Physico-chemical problems of formation of rocks and ores*, vol. I, ed. G.A. Sokolov, pp. 5–18. USSR Acad. Sci. Press.

Sryvalin, I.I. & Esin, O.A. (1959). On the sytem with reversible deviation from ideal solutions. In *Thermodynamics and structures of solutions*. Moscow: USSR Acad. Sci. Press.

Suvorova, B.A. & Perchuk, L.L. (1982). Effect of temperature on distribution coefficient of oxygen isotope O^{18} between garnet and biotite. *Doklady Akademii Nauk SSSR*, **269**, 460–3.

Thompson, J.B. (1967). Thermodynamic properties of simple solutions. In *Researches in geochemistry*, Pt 2, ed. P.H. Abelson, pp. 340–61. New York: Wiley.

Thompson, J.B. & Waldbaum, D.R. (1968). Mixing properties of sanidine crystalline solutions: 1. Calculations based on ion exchange data. *Amer. Mineral.*, **53**, 1965–99.

—— (1969). Mixing properties of sanidine crystalline solutions: 3. Calculations based on two phase data. *Amer. Mineral.*, **54**, 811–38.

Tomilenko, A.A. & Chupin, V.P. (1983). *Thermobarogeochemistry of metamorphic complexes*. Novosibirsk: Nauka.

Touret, J. (1976). The significance of fluid inclusions in metamorphic rocks. In *Thermodynamics in geology*, ed. D.G. Fraser, pp. 203–28. Dordrecht, Netherlands: Reidel.

Wagner, C. (1952). *Thermodynamics of alloys*. Cambridge, Mass.: Addison-Wesley.

Wohl, K. (1953). Thermodynamic evaluation of binary and ternary liquid systems. *Chem. Engin. Prog.*, **49**, 218–19.

Wood, B.J. & Banno, Sh. (1973). Garnet–orthopyroxene and orthopyroxene–clinopyroxene relationships in simple and complex systems. *Contrib. Mineral. Petrol.*, **42**, p. 109–124.

Wood, B.J. & Fraser D.G. (1976). *Elementary thermodynamics for geologists*. Oxford University Press.

2

Thermodynamic systems and factors of petrogenesis

A.A. MARAKUSHEV

The Gibbs method of thermodynamic potentials (1931) has been used and extended by Korzhinskii (1959, 1969, 1976) to endogenic mineral formation. Petrogenic systems may be distinguished by different types of thermodynamic potentials – the characteristic functions of state whose minimum values are the condition for minerals to attain equilibrium. The following thermodynamic potentials of isochemical equilibrium are recognized: the Gibbs $G(T, P)$ and Helmholtz $F(T, V)$ free energies, the enthalpy or thermal function $H(S, P)$ and the internal energy $U(S, V)$. The respective functions of allochemical equilibrium – the Korzhinskii thermodynamic potentials $G^z(T, P, \mu_m, \ldots, \mu_f)$, $F^z(T, V, \mu_m, \ldots, \mu_f)$, $H^z(S, P, \mu_m, \ldots \mu_f)$ and $U^z(S, V, \mu_m, \ldots, \mu_f)$ – characterize the reversible gain–loss of certain perfectly mobile components having constant chemical potentials.

Systems with perfectly mobile components thermodynamically can be called isopotential systems ($\mu_m, \ldots, \mu_f$ are constant). They can gain or lose perfectly mobile components ($m \ldots f$), in their equilibrium state ($dG^z = O$, $dF^z = O$), just the same as isothermal systems can gain or lose heat at constant temperature without disturbance of equilibrium ($dG = O$). Chemical potentials of perfectly mobile components ($\mu_m, \ldots, \mu_f$) are factors of mineral equilibria in corresponding systems in just the same way as temperature ('potential of heat') is a factor of mineral equilibrium in isothermal systems. Potentials G^z and F^z have been derived in order to describe a thermodynamic system of allochemical equilibria, in which the potentials of certain components are controlled by external conditions. This control is achieved by the infiltration through the systems of flows of volatile and perfectly mobile components (H_2O, CO_2, K_2O, Na_2O, HCl etc.). The potentials are very useful when applied to simple natural systems but they are not convenient for complex systems of osmotic type or systems in which chemical potentials of components vary from place to place and overall equilibrium may not be reached.

In their application it is necessary to distinguish isochemical and allochemical processes. Allochemical processes, which occur in solid state, belong to two types –

metamorphic (when lithostatic pressure is an important factor of mineral equilibrium) and metasomatic, developing at constant volume in isolation from lithostatic pressure (formation of pseudomorphous minerals for example). Thermodynamic potentials G (G^z) and F (F^z) are useful to describe the equilibrium states of these metamorphic and metasomatic systems correspondingly. The minimum values of thermodynamic potentials denote the equilibrium states of these systems, respectively metamorphic ($dG = O$, $dG^z = O$) and metasomatic ($dF = O$, $dF^z = O$). Magmatic systems may also be divided thermodynamically into two types – constant volume schlieren pegmatite systems and simple isobaric systems.

Metamorphism

Metamorphic systems that contain no volatile components (H_2O, CO_2) are isothermal-isobaric and conform to the Gibbs thermodynamic potential (G). The mineral facies of metamorphic rocks are separated in P–T diagrams by equilibria which correspond to the zero total differentials of this potential ($dG = O$) expressed as $dG = -S\,dT + V\,dP$ and $dG = -S\,dT + V\,dP + \mu_a\,dn_a + \ldots + \mu_k\,dn_k$, respectively, for closed ($dn_a = O, \ldots, dn_k = O$) and open systems (with the variable number of species $n_a \ldots, n_k$).

Metamorphic systems in which the minerals of constant or variable composition undergo polymorphic or other isochemcal transformations are classed as closed. The variance of mineral composition within the facies (the fields in the P–T diagrams), according to Goldschmidt's phase rule ($n = k - f$), determines the number of internal degrees of freedom (n) in mineral systems (k = no. of components, f = no. of phases). At $n > O$, the mineral compositions of metamorphic rocks are unconstrained thermodynamically and depend on the composition of sedimentary and other primary rocks. At $n = O$, the composition of minerals depends on the P and T conditions of metamorphism. The dependence is established using the method of equipotential lines (reactions of displaced equilibria) developed by Korzhinskii. For example, the paragenesis sillimanite + garnet + cordierite + quartz in the quaternary systems ($MgO + FeO + Al_2O_3 + SiO_2$) has no internal degrees of freedom, being divariant ($n = k + 2 - f$) according to Gibbs' phase rule, i.e. the $Fe \cdot 100 / Mg + Fe$ of the minerals cannot have an arbitrary value but strictly depends on the pressure and temperature (within the garnet–cordierite–sillimanite facies field in the P–T diagram). When the composition of minerals is set constant there remains only one degree of freedom, with the result that P and T can only vary according to the univariant relationship characterized by the equipotential lines of constant μ_{FeO}, or the lines depicting constant mineral compositions in the parageneses with no internal degrees of freedom (divariant). These lines are calculated using Korzhinskii's method of displaced equilibria often used in metamorphic studies. The method discloses the dependence of the mineral compositions in metamorphic rocks on the P–T conditions and furnishes a valuable addition

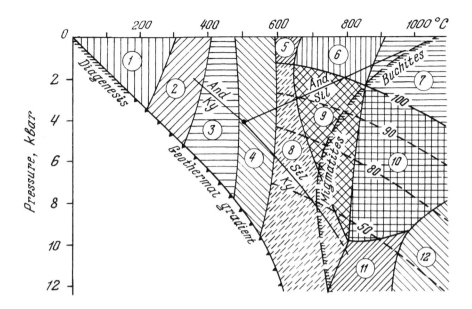

Fig. 2.1. Mineral facies of metapelites: 1 – clay shales; 2 – phyllite; 3 – chlorite–
muscovite schists (andalusite and kyanite); 4 – muscovite–biotite (two-
mica) schists and gneisses (andalusite, sillimanite and kyanite); 5–7 –
hornfels (5 – biotite, 6 – andalusite–cordierite, 7 – pyroxene–cordierite,
sanidinite); 8 – andalusite–biotite, sillimanite–biotite and kyanite–biotite
gneisses; 9 – andalusite–garnet–cordierite and sillimanite–garnet–cordierite
gneisses; 10 – garnet–hypersthene–cordierite gneisses; 11 – sillimanite–
hypersthene gneisses; 12 – quartz–sapphirine gneisses. Dashed lines refer
to the Fe·100/Mg + Fe ratio in the garnet from sillimanite–cordierite
gneisses.

to the isograd method. The latter had formerly been used only upon the appearance or
disappearance of the metamorphic minerals believed to be critical for the given isograd
(the chlorite, biotite, garnet etc. isograds), whereas the mineral composition variations
were largely disregarded. The variations in mineral compositions in divariant equilibria are calculated based on the shifted equilibrium reactions, such as

$$4n\,Fe_{1.8}Mg_{1.2}Al_2Si_3O_{12} + 8n\,Al_2SiO_5 + (4.2-6n)\,Fe_{0.5}Mg_{1.5}Al_4Si_5O_{18} + 10n\,SiO_2$$
$$= 4.2\,Fe_{0.5+n}Mg_{1.5-n}Al_4Si_5O_{18}.$$

i.e. garnet + sillimanite + cordierite + quartz = higher-iron cordierite. As n becomes
infinitesimal, the reaction follows the equipotential line of mineral constant composition in Fig. 2.1.

These reactions produce positive volume and thermal effects and move to the left
with increasing pressure and decreasing temperature (towards more magnesian
sillimanite–garnet–cordierite gneisses). In the diagram (Fig. 2.1) this is represented by
the set of dashed lines for constant (Fe·100/Mg + Fe) of the garnet from sillimanite–
cordierite rocks, as based on the experimental results of Aranovich & Podlesskii (1983)

21

for both 'dry' conditions and the pressure of water possibly present in the cordierite. The isocompositional lines have been drawn on the assumption of varying water content in the cordierite. In hornfels (Fig. 2.1, facies 5–7), cordierite occurs as sectionally twinned crystals that contain practically no water. During the transition from hornfels to deep-seated gneisses (Fig. 2.1, facies 8–10) the sectionally twinned cordierite ('hornfels' type) gives way to the polysynthetically twinned variety containing water and other volatiles that stabilize this mineral and make it difficult for the garnet + sillimanite paragenesis to replace it. For this reason, the garnet + cordierite paragenesis persists to very high pressures and, the Fe·100/Mg + Fe of garnet having reached 50 or so, is replaced by the hypersthene–sillimanite paragenesis, typical of the most deeply eroded Precambrian shields and crystalline massifs (facies 11–12). Consequently, the lines for the constant Fe·100/Mg + Fe in garnet and cordierite from garnet–cordierite–sillimanite gneisses delineate the rock depth facies relative to the lithostatic pressure and characterize the depth of erosion of metamorphic belts. As lithostatic pressure increases, metamorphic rocks develop that contain progressively greater proportions of minerals with dense structure which replace the low-density minerals.

The concept of 'pressure' in geology is often related to uniaxial stress leading to folding, thrusting and other dislocations. Stress was once considered an important factor of mineral equilibrium and was believed to promote the formation of minerals with dense structure known as 'stress-minerals'. It was presumed that many minerals are unstable under stress conditions (antistress minerals). Korzhinskii was the first to demonstrate (1940) that these concepts were untenable. In contrast to all-around (lithostatic) pressure which varies with depth, stress is not a factor of mineral equilibrium, even a factor of minor importance. It does not affect the development or loss of minerals or mineral parageneses. Consequently the mineral facies of metamorphic rocks are established with respect to lithostatic pressure, which in turn is determined by the depth (the mass of superposed rocks). Lithostatic pressure is therefore similar to hydrostatic pressure (determined by the mass of the water column) and is different in principle from the uniaxial pressure (stress) developed as the result of uniaxial compression. Stress does not belong to the mineral equilibrium factors, because metamorphic reactions do not involve any physico-chemical effects produced by stress. Nevertheless, stress is important for metamorphism in many other respects: as a catalyst of metamorphic reactions, a factor of the formation of schistose structures (schists, gneisses, amphibolites and others), metamorphic differentiation (the redistribution of minerals in metamorphic systems owing to irregular stress–strain) and cataclasis, mylonitization etc.

In addition to the mineral equilibrium factors by which rock mineral facies are established, it is necessary to consider the kinetic factors which promote the petrogenetic processes, determine rock textures, redistribute material within the systems etc. The directed pressure (stress) is a striking example of such a kinetic factor of metamorphism and is believed to influence the general geological structure of metamorphic sequences –

folds, overthrust sheets, and thrust faults. In some concepts of plate tectonics stress is regarded as a mineral formation factor: subduction-induced stress–strains are related to the development of metamorphic rocks within the glaucophane schist formation. This is highly improbable on thermodynamic grounds. Although glaucophane schist facies of metamorphism is not considered in the present paper, it has clearly originated within ophiolitic complexes in close association with hyperbasites through the agency of solutions of alkaline character (high μ_{Na_2O}). This metamorphism cannot be due to stress and strain developed from the interaction of lithosphere plates.

Granite-gneisses associated with the orogenic period in the development of folded belts display a particularly wide range of depths of formation. In this regard, they are grouped into several facies (Marakushev, 1965). Within the medium-depth facies, sillimanite–garnet–cordierite gneisses are formed as a result of metamorphism of iron-rich pelitic or argillaceous rocks and are distinctive for the high iron content of garnet (Alm_{70}–Alm_{90}). Their typical structures are granite-gneiss domes which develop in the Phanerozoic (more rarely, Precambrian) folded systems at the orogenic stage of their evolution. The granite-gneiss domes have zonal structure: metamorphism attains the maximum in the central parts and is only moderate at the sides.

With increasing depth of formation of metamorphic complexes, sillimanite–cordierite–garnet gneisses form as a result of metamorphism of sediments of increasing Mg:Fe ratio. In deep-seated metamorphic complexes exposed in Precambrian shields and crystalline massifs, the $Fe \cdot 100/Mg + Fe$ ratio in garnet of these rocks is in the range of 50–70. Examples of the more deeply eroded formations are found (Marakushev, 1964, 1965, Perchuk *et al.*, 1985) in the southern margin of the Aldan shield (Fig. 2.2), the Enderby Land in the Antarctic, the Limpopo folded belt in Africa, etc. Sillimanite–garnet–cordierite gneisses in these formations have magnesian compositions (the $Fe \cdot 100/Mg + Fe$ of garnet is 50–60) and are partly replaced by hypersthene–sillimanite (more rarely, hypersthene–kyanite) gneisses according to the reaction garnet ($Fe \cdot 100/Mg + Fe = 50$) + cordierite (17) = aluminous hypersthene (30) + sillimanite (kyanite)-+ quartz. Hypersthene–sillimanite gneisses are associated with quartz–sapphirine gneisses and are the most deep-seated rocks (some 35–40 km) exposed at the present erosion level of the earth's crust – facies 11 and 12 in Fig. 2.1. The diagram (Fig. 2.1) outlines the stability fields for the major types of metapelitic rocks and determines the mineral facies (at $P_{H_2O} \sim 1$ kbar), based on the extrapolation of the experimental data available for the hydration–dehydration of minerals, thermodynamic estimates of mineral equilibria as well as data on mineral thermometry–barometry (Perchuk, 1977, 1985). Results obtained with these methods are in good agreement and yield a generally consistent pattern of changes in the mineral assemblages of metamorphic rocks with depth, with the temperatures of formation varying within broad limits at all depth levels. However, with increasing depth of formation, the lower-temperature metamorphic facies are continually lost owing to increasing background temperature determined by the geothermal gradient. The temperature of metamorphism cannot be lower than the background temperature. According to the diagram (Fig. 2.1), the metamor-

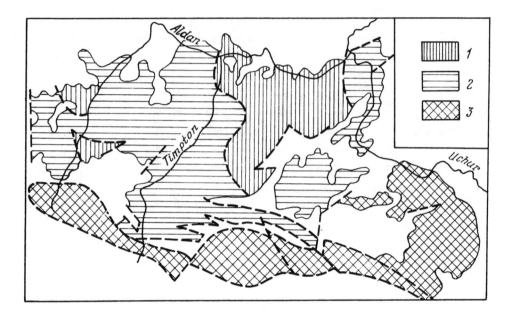

Fig. 2.2.   Occurrence of metamorphic complexes of different depth facies within the Aldan shield area, after Kitsul (Perchuk *et al.*, 1985). The complexes differ in their Fe·100/Mg + Fe ratios of garnets in garnet–sillimanite–cordierite gneisses. 1 – moderate depth (the Fe·100/Mg + Fe ratio is in excess of 70); 2 – great depths (60–70, the Aldan facies); 3 – very great depth (50–60, the Sutamskaya facies).

phic temperatures are always higher than the general background determined by the gradient. Consequently, metamorphism can only be of endogenic nature and results from the anomalous ascent of geoisotherms caused by upstreaming fluid flows within dislocation zones of the earth's crust, such as folds, deep faults, etc.

The fluid pressure, increasing with depth, stabilizes silicate melts (magmas) in the earth's crust and magmatic stability areas overlap the high-temperature facies of metamorphic rocks in deep-seated zones, giving rise to migmatites. Magmatic stability fields in Fig. 2.1 are defined by the position of the solidus of granite. In deep-seated zones, this process occurs at moderate temperatures of ~ 700 °C. In near-surface zones of very low fluid pressures, the partial melting of acidic rocks takes place at very high temperatures (~ 1000 °C) corresponding to the solidus of rhyolites. Such temperatures are attained in the pyroxene–cordierite hornfels of the sanidinite facies where buchites (hornfels-bearing acid glass) are formed. The development of the 'crustal' migmatites and granites within deep zones of metamorphism is attributed to the intensification of the fluid (water) effect on rocks on transition to the orogenic stage in the evolution of folded belts, as a result of an increase in the oxidation state of fluids ($H_2 \rightarrow H_2O$ etc.). As a consequence, p_{H_2O} becomes greater than the moderate values used in the calculation of mineral facies 1–12. Metamorphism which produces migmatites (ultrametamorphism) is allochemical in nature and involves rock debasification, with loss of strong bases (MgO, CaO and some FeO) and concomitant accumulation of silica and alkali metals (Marakushev, 1979). Concurrently, the debasification effect of fluids on rocks becomes more pronounced to start the process of decompaction of abyssal zones and the orogenic arising of folded belts.

Apart from migmatite production, allochemical metamorphism involves other transformations such as rock desilication, which predates the attainment of mineral equilibria corresponding to the corundum–plagioclase, jadeite or albite–nepheline rocks and nepheline gneisses, as well as alkaline metamorphism at high μ_{Na_2O} which involves the introduction of alkalis during the formation of glaucophane and riebeckite schists, aegirine iron quartzites, aegirine–jadeite schists, etc.

Clearly, these types of metamorphism can take place only in thermodynamically open systems. The appropriate functions to describe equilibrium in these metamorphic processes are Gibbs' potentials for allochemical metamorphic systems: $G(T, P, n_a, \ldots, n_k)$, $dG = -S\,dT + V\,dP + \mu_a\,dn_a + \ldots + \mu_k\,dn_k$. Components can be lost or gained in the systems only irreversibly with changes in the thermodynamic potential value ($\mu_a\,dn_a + \ldots + \mu_k\,dn_k$). The introduction of new components into these systems (above their solubility in minerals) results in an increase in the number of minerals.

Open systems of a different type are seen during the metamorphic rock alterations that involve volatile components (H_2O, CO_2). Their chemical potentials are independent variables along with temperature and pressure, T, P, μ_{H_2O}, μ_{CO_2}.

Metamorphism that is associated with a considerable gain or loss in perfectly mobile components (H_2O, CO_2), while variations in other component contents can be disregarded, is conventionally classified as isochemical, the same as metamorphism involving no volatile components. The term 'allochemical metamorphism' implies also the gain–loss of inert components ($SiO_2, \ldots, K_2O$) which usually occurs irreversibly with changes in the thermodynamic potential. Correspondingly there are two appropriate characteristic functions (G^z) and their differentials:

$G^z(T, P, \mu_{H_2O}, \mu_{CO_2})$ for 'isochemical' metamorphism;

$dG^z = -S\,dT + V\,dP - n_{H_2O}\,d\mu_{H_2O} - n_{CO_2}\,d\mu_{CO_2}$,

$G^z(T, P, \mu_{H_2O}, \mu_{CO_2}, n_{SiO_2}, \ldots, n_{K_2O})$ for allochemical metamorphism;

$dG^z = -S\,dT + V\,dP - n_{H_2O}\,d\mu_{H_2O} - n_{CO_2}\,d\mu_{CO_2} + \mu_{SiO_2}\,dn_{SiO_2} + \ldots + \mu_{K_2O}\,dn_{K_2O}$.

The independence of the chemical potentials of H_2O and CO_2, perfectly mobile components by the terminology of Korzhinskii (1959), is due to the arbitrary composition and the degree of oxidation of juvenile fluids, in particular such ratios as H_2O/H_2, CO/CH_4, CO_2/CO and $(H_2O + H_2)/(CO_2 + CO + CH_4)$. This predetermines the thermodynamics of hydration–carbonation metamorphic processes which evolve in the regime $P \geqslant P_F \geqslant P_{H_2O(CO_2)}$, where P is the lithostatic pressure depending on the depth metamorphism, P_F is the fluid pressure, and $P_{H_2O(CO_2)}$ is the H_2O (and CO_2) partial pressure (fugacities) in the fluid. This regime is different from hydration and carbonation under the experimental conditions $P = P_F = P_{H_2O(CO_2)}$, inasmuch as the experimental results must be recalculated for the metamorphic conditions (Marakus-

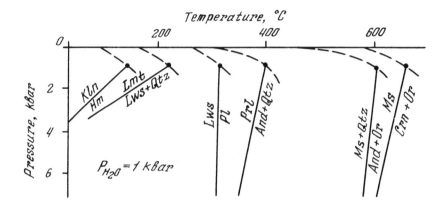

Fig. 2.3. Hydration–dehydration equilibria in a P–T diagram as recalculated to fit to metamorphic conditions at a moderate H_2O pressure (to 1 kbar) which is independent of the lithostatic pressure ($P = P_s$).

hev, 1968). For this purpose it is convenient to use the relative chemical potentials of H_2O and CO_2 which characterize the positive free energy, calculated at different temperatures from the chemical potential values at standard pressure:

$$\mu_{H_2O}^P = \int_1^{P_{H_2O}} (V_{H_2O})_T \, dP_{H_2O} \text{ and } \mu_{CO_2}^P = \int_1^{P_{CO_2}} (V_{CO_2})_T \, dP_{CO_2},$$

where $(V_{H_2O})_T$ and $(V_{CO_2})_T$ are the molar volumes of gases at given temperature.

These values are used to calculate the mineral equilibria of the hydration and carbonation. For the experimental modeling of these processes under the H_2O or CO_2 pressure ($P_{H_2O(CO_2)} = P_{total}$), the calculation formula is derived using the Gibbs thermodynamic potential (G) for reactions of the type $A + B = C + D + H_2O(CO_2)$, calculated per 1 mole of gas: $\Delta G_T^o + P \Delta V_s + \mu_{H_2O(CO_2)}^P = O$. This formula is used to calculate the relative chemical potentials of H_2O and CO_2 at each given temperature and the corresponding total pressure (dashed lines in Figs. 2.3 and 2.4). Mineral equilibria involving H_2O and CO_2 under metamorphic conditions are calculated with a different formula which contains Korzhinskii's thermodynamic potential (G^z) expressing independent changes in the lithostatic (P_s) and partial ($P_{H_2O(CO_2)}$) pressures:

$$\Delta G^z(T, P, \mu_{H_2O}, \mu_{CO_2}) = \Delta G_T^o + \Delta V_s(P - 1) + \mu_{H_2O(CO_2)}^{P = const},$$

$$\Delta G^z = \Delta G_T^o + P_s \Delta V_s + \mu_{H_2O(CO_2)}^{P = const} = O.$$

For the normal metamorphic dehydration reactions (calculated per 1 mole of H_2O) at $P_{H_2O(CO_2)} = 1000$ bar, the lithostatic pressure of the equilibria (solid lines in Figs. 2.3 and 2.4) at every given temperature is calculated by the formula

$$P_s = -\frac{\Delta G_T^o + \mu_{H_2O(CO_2)}^{1000}}{\Delta V_s},$$

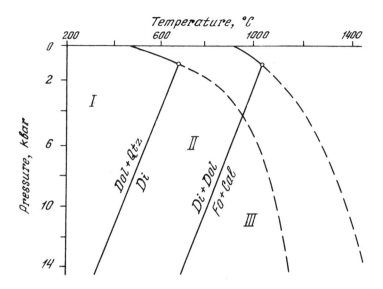

Fig. 2.4. Carbonation–decarbonation equilibria in a P–T diagram at $P = P_{CO_2}$ (dashed lines) and under metamorphic conditions at $P_{CO_2} = 1$ kbar (solid lines) separating the magnesian marble facies: I – quartz–dolomite, II – diopside–dolomite and III – forsterite–calcite.

where ΔG_T° is the free energy of the dehydration reaction (kcal per mole H_2O) at given temperature (T), $\mu_{H_2O(CO_2)}^{1000}$ is the relative chemical potential of the components at $P_{H_2O(CO_2)} = 1000$ bar at the same temperature (reference data), and ΔV_s is the change in volume of the solids, cal/bar.

Metamorphic rock facies are presented relative to the independent variables P, T and $P_{H_2O(CO_2)}$. For arbitrary values of these variables, metamorphic rocks conform to Korzhinskii's mineralogical phase rule $n = k - f$, where k does not include the perfectly mobile components (H_2O, CO_2). Accordingly, their variance increases with the result that the equipotential lines are transformed into planes characterized by the displaced equilibrium reactions of hydration or carbonation. For example, in the system containing five inert components (K_2O, MgO, FeO, Al_2O_3, SiO_2), the paragenesis of garnet with biotite, sillimanite, orthoclase and quartz has three degrees of freedom and appears as the corresponding facies in the three-dimensional diagram P–T–P_{H_2O} (μ_{H_2O}). At fixed $Fe \cdot 100/Fe + Mg$ of the minerals, the paragenesis becomes divariant and corresponds to the equipotential plane, characterized by the following reaction of displaced equilibrium:

$$(8.1–10n)\ Fe_{2.31}Mg_{0.69}Al_2Si_3O_{12} + 6n\ K_2Fe_{2.5}Mg_{2.5}Al_4Si_5O_{20}(OH)_4$$
$$+ 4n\ Al_2SiO_5 + 32n\ SiO_2 = 8.1\ Fe_{2.31-n}Mg_{0.69+n}Al_2Si_3O_{12}$$
$$+ 12n\ KAlSi_3O_8 + 12n\ H_2O,$$

i.e. garnet + biotite + sillimanite + quartz = more magnesian garnet + orthoclase + H_2O.

Sets of such planes demonstrate the effect of thermodynamic conditions on mineral composition. According to the above reaction, an increase in the chemical potential of water or decrease in the temperature results in higher iron contents of garnet–biotite–sillimanite gneisses whereas increases in the lithostatic pressure (depth) have the opposite effect.

The chemical potentials of fluid components (H_2O, CO_2 etc.) taking part in mineral reactions belong as noted above to the factors of mineral equilibrium whereas the actual fluid effect on rocks is among the kinetic factors of metamorphism and promotes metamorphic reactions even when fluid components are not involved. In general, fluids are indispensable for the action of metamorphism.

This is particularly well seen on contacts with igneous rocks that have long been subdivided into 'dry' (no traces of wall rock metamorphism) and those where hornfels developed under the action of fluid flows that had separated from magmatic melts although high-temperature hornfels contains practically no volatiles (H_2O etc.). Fluid effects in high-temperature contacts show up in the partial melting of rocks producing aqueous fluids which on cooling yield water-rich (pearlite) acidic glasses typical of buchites. The occurrence of relics of the original rocks or those from other mineral facies in metamorphic formations are also signs of the fluids' catalytic effect. These relics survived metamorphism only because the metamorphosing fluids had no access to them and the metamorphic reactions were therefore hindered kinetically. Similar conclusions have been drawn from observations on coronated pyroxene–plagioclase schists, amphibolites, gneisses, silicate marbles etc. The coronas are formed during metamorphic reactions which were evolving on the grain edges of the separate minerals reacting with the neighbouring minerals under the changing parameters of metamorphism. It is not uncommon to find grains with the newly developed rims side by side with grains of the same minerals with only partially developed rims (covering part of the grain) or no rims at all, since they were not brought in contact with the fluids. For example, in the Archean magnesian marbles from the shield on the Siberian platform some of the forsterite grains formed during the granulite metamorphism while others, having dolomite–diopside (and pargasite) rims, developed as a result of the superimposed regressive metamorphism, according to the reaction (Fig. 2.4):

$$2\,Mg_2SiO_4 + 4\,CaCO_3 + 2\,CO_2 = CaMgSi_2O_6 + 3\,CaMg(CO_3)_2.$$

Removal of carbon dioxide, accompanying the regressive alteration of marbles, is hampered in the absence of intergranular metamorphosing solution, or else may not take place at all. Examples of some such kinetically retarded metamorphic process are seen in the Archean quartz–calcite marbles of the Aldan massif which in places have undergone progressive transformations and developed wollastonite along the boundaries of quartz and calcite grains:

$$SiO_2 + CaCO_3 = CaSiO_3 + CO_2.$$

Kinetic retardation of the process results in the incompleteness of the reaction and the formation of the paragenesis of all three minerals (quartz, calcite, wollastonite). This amounts to CO_2 losing its perfectly mobile state which it normally has under metamorphic conditions, where the mineral equilibria are determined by the minimum value of the thermodynamic potential:

$$G^z(T, P, \mu_{H_2O}, \mu_{CO_2}, n_{SiO_2}, \ldots, n_{K_2O}),$$
$$dG^z = -S\,dT + V\,dP - n_{H_2O}\,d\mu_{H_2O} - n_{CO_2}\,d\mu_{CO_2} + \mu_{SiO_2}\,dn_{SiO_2} + \ldots + \mu_{K_2O}\,dn_{K_2O} = 0.$$

For these relations the thermodynamic potential is different:

$$G^z(T, P, \mu_{H_2O}, n_{CO_2}, n_{SiO_2}, \ldots, n_{K_2O}),$$
$$dG^z = -S\,dT + V\,dP - n_{H_2O}\,d\mu_{H_2O} + \mu_{CO_2}\,dn_{CO_2} + \mu_{SiO_2}\,dn_{SiO_2} + \ldots + \mu_{K_2O}\,dn_{K_2O} = 0.$$

Unlike carbon dioxide, water always seems to retain its perfect mobility during metamorphism. Earlier it was presumed that water behaves as an inert component (according to the 'water deficiency' concept) in assemblages of hydrous and anhydrous minerals such as hypersthene and biotite that commonly occur in hypersthene gneisses and charnockites. These minerals were thought to be related to one another by the following simplest reactions:

$$K(Mg, Fe)_3[Si_3AlO_{10}](OH)_2 + 3\,SiO_2 = 3\,(Mg, Fe)SiO_3 + KAlSi_3O_8 + H_2O.$$

According to the reaction, biotite and quartz, under the perfectly mobile behaviour of water (having the arbitrary chemical potential), form a stable assemblage with hypersthene only in the absence of potassium feldspar (orthoclase or microcline), whereas the paragenesis of all of these minerals (biotite, quartz, hypersthene, ortho-clase) is indicative of inert water (water deficiency). However, this reaction has been written for biotite of the phlogopite–annite series (extremely poor in alumina), not representative of metamorphic rocks, in which biotite is more aluminous (rich in eastonite–siderophyllite). Its decomposition reaction yielding high-temperature gneisses involves several other minerals, for example biotite + quartz = hypersthene + orthoclase + codierite + garnet (of the pyrope–almadine series) + H_2O. According to this reaction, during the formation of gneisses at excess quartz and orthoclase and mobile water, biotite forms stable parageneses with hypersthene, hypersthene and garnet, hypersthene and cordierite or, where no quartz or potassium feldspar are present, with hypersthene, garnet and cordierite. All these parageneses of biotite are widespread in gneisses of the high-temperature (granulite) facies of metamorphism. The chemical potential of water is an independent factor of equilibrium, along with the lithostatic pressure and temperature. An increase in the chemical potential gives rise to hydration processes of minerals in metamorphic systems, which are exothermic at the constant H_2O chemical potential and proceed with decreasing temperature. Therefore, the chemical potential of water and temperature are oppositely directed parameters of metamorphism, which should be taken into consideration in calculating mineral facies.

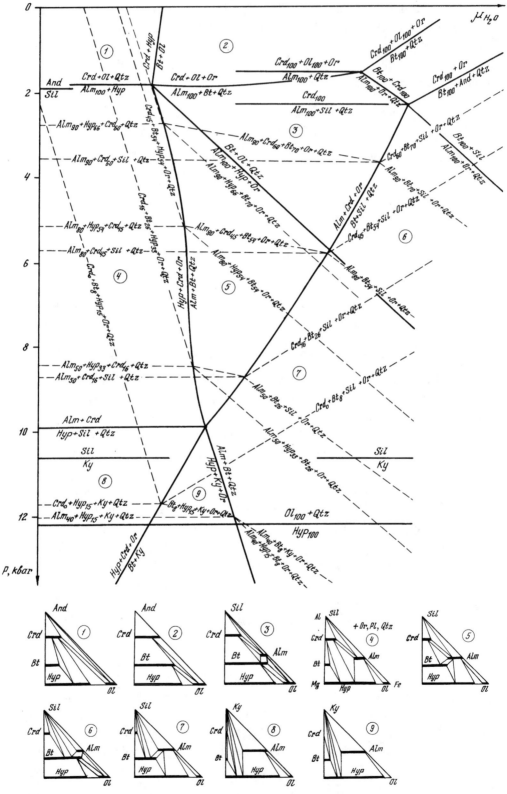

The chemical potential of water in its relative formulation $\mu_{H_2O}^P = RT \ln P_{H_2O}$, where P_{H_2O} is the partial thermodynamic pressure (fugacity) of water in the metamorphosing fluids, and the lithostatic pressure define the mineral facies of gneisses in Fig. 2.5, which are similar to the metapelite mineral facies considered above (Fig. 2.1).

Both diagrams, $P-\mu_{H_2O}^P$ ($T=$const; Fig. 2.5) and $P-T$ ($\mu_{H_2O}^P=$const; Fig. 2.1) constitute orthogonal sections of the general three-dimensional diagram $P-T-\mu_{H_2O}^P$ that represents the mineral facies of metamorphic rocks, where the boundaries between the facies are shown as surfaces of divariant equilibria, conforming to equations of the type:

$$\Delta G^z(P, T, \mu_{H_2O}) = \Delta V \, dP - \Delta S \, dT - \Delta n \, d\mu_{H_2O}^P = O.$$

Oxygen is another volatile component whose relative chemical potential $\mu_{O_2}^P = RT \ln P_{O_2}$ can be one of the independent intensive parameters of metamorphic systems, $G^z(P, T, \mu_{O_2}^P)$, in which mineral equilibria are attained by the relation $dG^z = \Delta V \, dP - \Delta S \, dT - \Delta n_{O_2} \, d\mu_{O_2}^P = O$.

The three-dimensional diagram for the redox mineral facies takes the form $P-T-\mu_{O_2}^P$. An orthogonal section of the diagram is presented in Fig. 2.6. Mineral facies in this diagram are outlined relative to native metals, their oxides, and sulphides. The composition of pyrrhotite in assemblages with pyrite changes systematically with temperature (increasing temperatures decrease the Fe:S ratio in pyrrhotite). Assemblages of iron sulphides and oxides are common in deposits of iron formations of the low and intermediate grades of regional metamorphism. Pyrite–pyrrhotite–magnetite and pyrite–magnetite–hematite ores are abundant in these deposits, and the enclosing schists may contain the pyrrhotite–pyrite assemblage.

The pyrrhotite stability limit is defined by its oxidation to pyrite and magnetite: $3 \, FeS + O_2 = 0.5 \, Fe_3O_4 + 1.5 \, FeS_2$. This reaction, which separates the pyrrhotite and pyrite–magnetite facies, is complicated by the compositional variability of pyrrhotite equilibrated with pyrite and magnetite; pyrrhotite's Fe:S ratio varies with the temperature and redox conditions. The pyrrhotite composition is shown in the diagram by sets of dashed lines for the constant-composition phases in the parageneses pyrrhotite + pyrite and pyrrhotite + magnetite. Although the composition of pyrrhotite in equilibrium with pyrite depends mainly on temperature, this relationship is not strong, given the compositional variations of both pyrrhotite and pyrite and structural variations in pyrrhotite.

Pyrrhotite–magnetite equilibria are also of interest. In this paragenesis, pyrrhotite is more iron-rich than the pyrrhotite in equilibrium with pyrite, but with increasing

Fig. 2.5. The $P-\mu_{H_2O}$ plot (lithostatic pressure – chemical potential of water) shows an isothermal section at $T = 800\ ^\circ$C of the diagram for metapelite mineral facies (Fig. 2.1). Solid univariant lines separate the mineral facies; dashed lines refer to the $\cdot 100\ Fe\cdot/Mg + Fe$ ratio of minerals in presumably divariant parageneses. Parageneses triangles correspond to the numbered locations in the diagram.

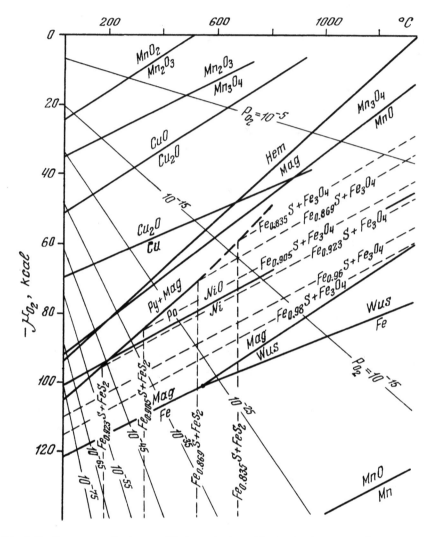

Fig. 2.6.   Summary of redox equilibria and mineral facies (pyrite–magnetite, pyrrhotite, etc.) on the diagram oxygen chemical potential – temperature.

oxygen potential, the pyrrhotite compositions in the two assemblages draw closer together to become identical at the point of the pyrrhotite decomposition into magnetite and pyrite. The Fe:S ratio in the pyrrhotite associated with magnetite is not very temperature-dependent, being defined chiefly by the oxygen chemical potential. At the stability limit of magnetite and the assemblage pyrrhotite + magnetite + native iron, the stable minerals are troilite, or pyrrhotite having a composition approaching stoichiometric FeS. Increasing oxygen potential causes a consistent decrease in the Fe:S ratio in the pyrrhotite from magnetite assemblages. On crystallization pyrrhotite develops monoclinic and hexagonal crystals. The monoclinic and hexagonal pyrrhotites form under very complex conditions, and it is not uncommon for the compositionally different monoclinic (magnetic) and hexagonal (paramagnetic) modifications to occur together. They are intergrown so intimately as to make separate phases practically indistinguishable, and the magnetic susceptiblity of the mineral changes steadily with the composition and temperature.

The perfect mobility of oxygen (the chemical potential is independent of temperature and pressure), on which the above diagram is based, is not characteristic of all metamorphism and only shows up to a degree in high-temperature facies. Oxygen is inert for the most part, which is most clearly seen in magnetite–hematite iron formations: during their formation the oxygen chemical potential is fully determined by the lithostatic pressure and temperature (FeO and Fe_2O_3 behave as independent inert components). However, in the aureoles of high-pressure contact metamorphism superimposed on magnetite–hematite iron formations, hematite is normally reduced to magnetite with the result that oxygen passes into the perfectly mobile state and the number of minerals in the rocks decreases.

Therefore chemical potentials of perfectly mobile components are very important factors of petrogenesis. We have considered diagrams where they are combined with lithostatic pressure ($P-\mu_{H_2O}$; Fig. 2.5) and temperature ($T-\mu_{O_2}$; Fig. 2.6) and $P-T$ diagrams where chemical potentials or partial pressure of perfectly mobile components are constant (Figs. 2.1, 2.3 and 2.4).

Diagrams of component chemical potentials (T, P constant) are another type of the diagram. The diagrams of the chemical potentials of water and carbon dioxide are of prime importance for metamorphism. Consider this diagram for the system $CaO-MgO-SiO_2$.

According to the phase rule, $n = k + 2 - f$. Provided that the arbitrary values of pressure and temperature are constant ($n = 2$), the invariant points in the diagram $\mu_{H_2O}-\mu_{CO_2}$ (Fig. 2.7) correspond to the assemblages of five minerals $n = 5 + 2 - 5 = 2$ (i.e. talc + magnetite + dolomite + quartz + tremolite, dolomite + calcite + talc + tremolite + quartz, etc.). They are located at the intersection of the univariant equilibrium of four minerals related by the reactions, such as

$$\underbrace{2\,Mg_3Si_4O_{10}(OH)_2}_{\text{talc}} + \underbrace{3\,CaCO_3}_{\text{calcite}} =$$

$$\underbrace{Ca_2Mg_5Si_8O_{22}(OH)_2}_{\text{tremolite}} + \underbrace{CaMg(CO_3)_2}_{\text{dolomite}} + CO_2 + H_2O,$$

$$\left\{\frac{d\mu_{H_2O}}{d\mu_{CO_2}}\right\}_{T,\,P_s} = -\frac{\Delta n_{CO_2}}{\Delta n_{H_2O}} = -1.$$

This derivative corresponds to the slope of the respective curve in the $\mu_{H_2O}-\mu_{CO_2}$ plot.

Univariant lines separate the mineral facies (Roman numbers) in which the stable parageneses containing three minerals are defined by the inert component ratios $CaO-MgO-SiO_2$, as is shown in the triangular (ternary) diagrams.

The $\mu_{H_2O}-\mu_{CO_2}$ diagrams have similarly been used to study the relations between the

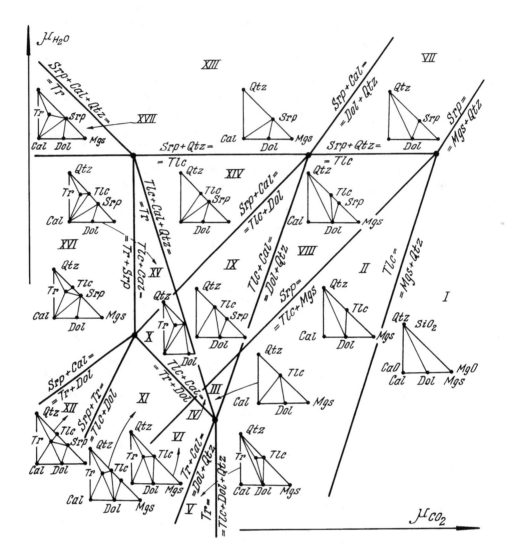

Fig. 2.7. Diagram of H_2O and CO_2 chemical potentials in the system CaO–MgO–SiO$_2$.

zeolite and prehnite–pumpelliite facies (Albee & Zen, 1969) as well as to resolve a number of other petrological problems. Diagrams of the chemical potentials of components are particularly important for the study of metasomatic processes.

Metasomatism

Metasomatism is different from metamorphism in that it produces more drastic changes in the chemistry of rocks, and involves a consecutive passage of the components into the perfectly mobile state and, as a possible result, a decrease in the number of minerals in metasomatic rocks to one along with the development of metasomatic zoning. The theory of metasomatic zoning has been developed by Korzhinskii (1969).

Unlike allochemical metamorphism that takes place in isobaric systems (thermodynamic potentials are G and G^z), metasomatism typically develops in systems at constant volume. Their thermodynamic potential is $F^z(T, V, \mu_m, \ldots, \mu_f, n_a, \ldots, n_k)$, $dF^z = -S\,dT - P\,dV - n_m\,d\mu_m - \ldots - n_f\,d\mu_f + \mu_a\,dn_a + \ldots + \mu_k\,dn_k$, where $m, \ldots, f$ are perfectly mobile components, $a, \ldots, k$ inert components. Typically metasomatism develops in hydrothermal ore deposits (at constant volume according to the Lindgren rule). Formation of pseudomorphs (pseudoleucite and others) is a characteristic feature of metasomatism. Metasomatic development of the constant volume systems in their ultimate state (all components are perfectly mobile) is defined by the thermodynamic potential $F^z = U - TS - n_m\mu_m - \ldots - n_f\mu_f = -PV$, where the internal energy $U = TS - PV + n_m\mu_m + \ldots + n_f\mu_f$. The differential of the function is $dF^z = -S\,dT - P\,dV - n_m\,d\mu_m - \ldots - n_f\,d\mu_f$. These conditions ensure the perfectly mobile state of the components and the development of monomineralic zones. The equilibrium state of metasomatic systems is achieved at the minimum value of the thermodynamic potential ($dF^z = 0$) and hence, at constant volume, under maximum pressure. Therefore, minerals (ultimately, one mineral) that are able, at given temperatures and chemical potentials of the components, to build up the highest crystallization pressure, will remain stable during metasomatic processes. This pressure increases as the solution becomes progressively supersaturated with the components constituting the mineral:

$$dP = \sum_{m}^{f} i\,d\mu_i,$$

where i is the component content in the unit volume. That is why the identity of phases which are capable of replacing all the other minerals varies with the geochemical conditions of metasomatism: garnet dominates in calcareous skarns, forsterite in magnesian skarns, jadeite in formations resulting from rock desilication in ultramafic massifs, quartz during acidic leaching etc.

Korzhinskii postulated a stepwise evolution of postmagmatic mineral-forming processes, an evolution which reflects regular changes in the acid–base character of hydrothermal solutions related to granites. He recognized the early alkaline, main-stage acidic, and late alkaline stages of the hydrothermal process. The early alkaline stage is observed in contacts of granites and carbonate rocks, since the originally acidic solutions formed within the granites become weakly alkaline upon interaction with these rocks. The typical products of this stage are skarns. The acidic stage fits well into the normal conditions of postmagmatic mineral formation which accompany the emplacement of granite plutons. At this stage the increase in the solution acidity with decreasing temperature reaches a maximum. The increase starts as early as the end of the magmatic process and can be traced by higher alumina and fluorine contents of biotites on transition from granites to aplites and pegmatites. The main rocks of this stage are greisens.

For a long time no satisfactory explanation had been given for the change in the

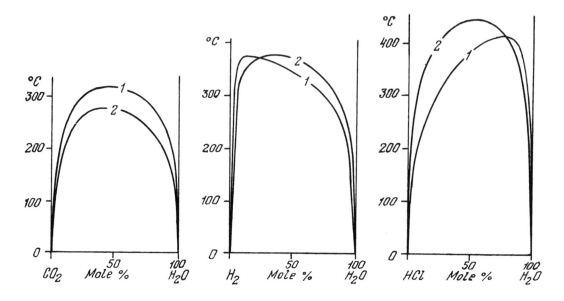

Fig. 2.8. Schematic representation of fluid unmixing with decreasing temperature at
0.5 kbar (1) and 1kbar (2), after Bach, 1977, and Franck, 1980.

regime of hydrothermal mineral formation on transition from the acidic to the late
alkaline stage. Only recently has a model of this transition begun taking shape, based
on experimental data on the splitting of hydrothermal solutions at temperatures below
critical (300–400 °C) into two phases – the aqueous (polar molecules) which contains
dissolved alkalies K(OH), Na(OH) etc., and the hydrogen phase (non-polar molecules)
which concentrates the acidic components (HCl, CO_2 etc.); Fig 2.8. Owing to the
remarkable migrating capacity of the hydrogen phase, the acidic components are
removed from hydrothermal solutions and their alkalinity and oxidation degree
increase. Accordingly, the hydrothermal solutions become more important as concen-
trators of ore material (in compounds such as Na_2WO_4, Na_2MoO_4 etc.). As the salt
content of hydrothermal solutions increases, the dense hydrous salt phases which are
particularly efficient in concentrating ore metals, begin to separate from the solutions.

Before we pass on to magmatic systems, let us consider pegmatites which, by
Zavaritsky's definition (1947), are formations intermediate between ore veins and
igneous rocks.

Formation of pegmatites

Many types of igneous rocks contain pegmatites enriched in the petrogenic components which accumulate in residual melts together with the fluid (water, fluorine,
chlorine), rare and trace elements. However, they are not direct products of the

crystallization of the residual melts – their formation is related to the liquid immiscibility in these melts and the separation of certain fluid (pegmatite) melts, as has long been noted by researchers (Smith, 1948) and supported by geological observations and experiments (Marakushev, 1979, Marakushev, Ivanov & Rymkevich, 1979, 1981, Marakushev & Gramenitsky, 1983).

Pegmatites are formed as a result of fluid immiscibility in fluid-rich residual granite, syenite, nepheline–syenite and other magmas. The parent magmas of pegmatites, although residual (related to crystal separation), are still above the liquidus temperature, since this temperature is depressed considerably due to the accumulation of volatiles (H_2O, HF, HCl, B_2O_3 etc.) in the magmas, which facilitate the evolution of liquid immiscibility. The separating pegmatite melts split off as ball-shaped bodies in the magmas (this form persists in schlieren pegmatites) and have a composition close to the pegmatite eutectic.

The term pegmatite was proposed by Haüy in 1822 as a synonym for graphic granite featuring the simultaneous crystallization of feldspar and quartz. The eutectic nature of pegmatite melts is apparent in the crystallization pattern of the schlieren bodies which begins in the margins as the formation of graphic (eutectoid) mineral intergrowths with sparse idiomorphic crystals of fluorine-rich biotite or feldspar, the minerals that are in excess relative to the eutectic. Crystallization is accompanied by increasing fluid pressure in pegmatite melts. Under the action of fluids, the quartz–feldspar melts split into the fluid-feldspar and fluid-quartz (silexite) varieties to define the internal structure of schlieren granite pegmatites which contain quartz cores surrounded by nearly pure feldspar zones followed by the quartz-feldspar zones of undifferentiated pegmatite rocks of the graphic structure (Fig. 2.9), their grain size diminishing outwardly to the margins of pegmatite bodies. Some of the pegmatite schlieren develop outer aplite quench rims as a result of the peripheral degassing of the pegmatite magma.

Many researchers relate the formation of pegmatite cores to postmagmatic leaching. However, data from schlieren pegmatite studies suggest that quartz cores had formed during the magmatic stage in the development of pegmatite bodies. The earliest high-temperature subsolidus leaching in these bodies is constrained by contacts between the feldspar zone and quartz core which recrystallizes along the contacts to yield a high-temperature (in excess of 580 °C) modification of quartz. Further evidence for the magmatic nature of the core is provided by the granular-sutured texture of the constituent quartz and liquid inclusions therein.

Pegmatite schlieren are often ideally round in shape, range in size from several cm to hundreds of metres and are more leucocratic and more fluid than their parent rocks. This is why they remain liquid (superliquid state) for some time after the enclosing country rocks have consolidated completely.

This suggests that schlieren pegmatites crystallize in nearly closed systems of constant volume that had been fixed by the size of the schlieren at the time of igneous country rock consolidation. This volume (V_1) is normally greater than that (V_2) occupied by the mass of crystallized pegmatite (graphic, feldspar and quartz zones, all

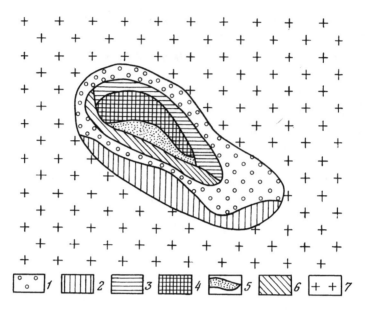

Fig. 2.9. Section across a schlieren pegmatite body in the central Volodarsko–
Volyinsky pegmatite region (after Panchencko, cited from Lazarenko *et
al.*, 1973). 1 – pegmatite of graphic texture; 2 – pegmatite of graphic–
pegmatoid texture; 3 – feldspar; 4 – grey quartz; 5 – cavity; 6 – leaching
zone; 7 – enclosing granites.

together) with the result that within the pegmatite schlieren there are fluid cavities
known as bonanzas or vugs, which gave the name 'cavity pegmatites' to pegmatites of
this type. The original size of these cavities is defined by the difference of the volumes:
$\Delta V = V_1 - V_2$ (shrinkage on crystallization). They are filled with fluids (dense supercritical gas mixtures) and salt solutions. These fluids vigorously attack the rocks mainly
along the boundary of the quartz and feldspar zones of the pegmatites (Fig. 2.9),
causing their leaching and collective mineral crystallization to produce crystals of
quartz, topaz, beryl, fluorite and other economically important minerals of the cavity
pegmatites.

Vein or dyke-shaped pegmatites follow a different path of development which
normally involves no formation of fluid–salt residual cavities but rather a large-scale
superimposed metasomatic development of muscovite and minerals containing F, Li,
B, Nb, Ta, U, Th, La, etc. These minerals are commonly concentrated in particular
layers. At the proper magmatic stage, vein pegmatites, unlike their schlieren varieties
and in common with all other intrusive rocks, crystallize under the isobaric conditions
of more open systems. The transition to these conditions is believed to be related to
pegmatite melt crystallization in closed schlieren magma chambers, which is accompanied by increasing fluid pressure. Under this pressure country rocks may develop
fissures which are intruded by pegmatite melts that are capable of travelling considerable distances from their source of generation in the schlieren. As a result, petrogenetic

systems transform into isobaric open systems which are characteristic not only of igneous rocks in dykes but of deep-seated magmatism in general.

Magmatism

Magmatism develops at depth as a result of the interaction of silicate melts with juvenile transmagmatic fluids or solutions (according to Korzhinskii, 1969) which cause the volatiles (H_2O, H_2, CO_2, etc.) and often several other components (K_2O, Na_2O, etc.) to pass into the perfectly mobile state. Thus the number of minerals in plutonic igneous rocks (granites, etc.) is usually less than in their metamorphic analogues (gneisses, etc.,). Under the influence of the fluid flows, magmatism at deep zones evolves mostly as the magmatic replacement (Korzhinskii, 1976) of the earth's crust or mantle – a special process of the selective incorporation by the magma of the substrate material, complicated by the acid–base interaction of the components. The physico-chemical process of infiltration of fluids from magma chambers into country rocks is accompanied by their alteration and magmatic replacement. The process is quite different from the assimilation of xenoliths which is diffusive by nature (Bowen, 1929).

As we have shown (Marakushev, 1987), when sialic rocks are replaced, magmatism acquires a 'crustal' character and is accompanied by the extensive development of granite magmas (migmatites, granite plutons etc.); when ultrabasic and carbonate rocks are replaced, magmatism follows an alkaline trend (alkaline-basalt, nephelenite, ijolite, carbonatite, kimberlite, etc.).

Magmatic rocks normally crystallize in several steps; the earliest are associated with the abyssal sources of magma generation and primary differentiation, the latest take place in magmatic bodies where the consolidation of magma is finally complete. In effusive and subvolcanic rocks, the early stage of magmatic crystallization is represented by phenocrysts known as intratelluric (deep-seated). They consist predominantly of minerals which form under high lithostatic pressure such as garnet, which occurs as phenocrysts in some types of liparites, dacites, andesites, basalts and porphyry peridotites. The phenocrysts of hydrous minerals in effusive rocks are also of the intratelluric nature (biotite, basaltic hornblende, occasionally cummingtonite) and crystallize under high fluid pressure.

Garnet and hydrous minerals belong to the abyssaphilic group (i.e. form at depth and are unstable under the conditions of near-surface magmatism) and during magmatic eruptions they are subjected to magmatic resorption (dissolution in magmas) or decomposition (replacement by other minerals, beginning from the grain edges to form characteristic kelyphytic, opacitic etc. rims). However, many intratelluric minerals (olivine, pyroxene, feldspar) remain stable under near-surface conditions as well, so that crystals formed at depth can still continue growing, often with accompanying changes in the mineral composition due to decreasing lithostatic pressure.

Early intratelluric minerals are more difficult to identify in intrusive (plutonic) rocks,

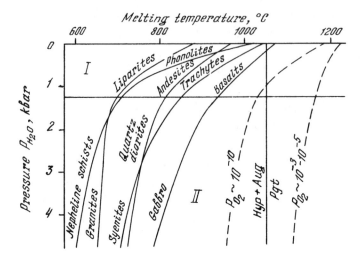

Fig. 2.10. Crystallization regimes of magmas of different types in the volcanic (I) and plutonic (II) facies. Complete consolidation of magmas is shown by solid solidus lines; the beginning of crystallization of basic magma is represented by dashed liquidus lines for reducing ($P_{O_2} = 10^{-10}$) and oxidizing ($P_{O_2} = 10^{-3} - 10^{-5}$) conditions. The pigeonite equilibrium separates the conditions under which pigeonite and hypersthene-type basalt complete crystallization (solidus); Kuno, 1968.

since the replacement reactions by newly formed minerals are more complete at depth than in the volcanic facies where unstable intratelluric minerals are preserved by quenching (Harris *et al.*, 1970).

Under normal conditions there is a clear distinction between the volcanic and plutonic facies of igneous rocks with respect to the fluid regime of magma consolidation. Volcanic systems, as distinct from plutonic ones, are confined to fracture systems which are responsible for the relatively easier escape of the fluid components from magmas and, accordingly, for the higher melt consolidation temperatures, characteristic of the low fluid pressure regime (Fig. 2.10). This conclusion is backed by observations on eruptions of magmas excessively rich in the fluid (gaseous) components ($H_2O, H_2, CO_2, CO, N_2, CH_4$, HF, HCl, etc.), their escape from the silicate melts during the eruptions and rapid consolidation of lavas being fixed by the formation of miarolytic or bubble textures.

This effect of fluids on magma crystallization temperatures brings about the major difference between volcanic and plutonic facies of igneous rocks, and is characteristic of crustal magmatic chambers. On transition to mantle conditions this effect is overlapped by the influence of total lithostatic pressure which can attain very high values (40–50 kbar) in the mantle chambers and can impose high crystallization temperatures (1000–1300 °C) on basic and ultrabasic magmas even though their volatile content might be very high (e.g. the eclogite and pyrope peridotite facies, especially diamondiferous types).

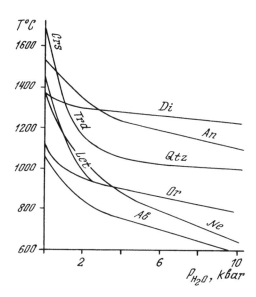

Fig. 2.11. Temperature–H₂O pressure dependence of the melting of rock-forming minerals.

Increasing temperatures of formation of magmatic rocks on transition from the plutonic to volcanic facies are related to decreasing fluid pressure in the magma on its way to the surface. The separation of fluids is accompanied by the supercooling of the magmas and the formation of zoned crystals and volcanic glass. The fluid components display differential migration from the melts as a result of their dissimilar affinity for silicate melts (differential separation from magmas). This is why, on transition from the plutonic to the volcanic facies, both fluid pressure and the fluid component composition undergo considerable changes.

A striking example is hydrogen which escapes from the melts much more readily than water with the result that at shallower depths the separating fluids are depleted in H_2 relative to H_2O (oxidation of fluids). This accounts for the basic distinction between rocks in terms of the oxidation state of the elements, which is higher in volcanic than in plutonic rocks.

The differences in the crystallization temperatures of volcanic and plutonic rocks are a measure of the fluid effect on the melting of the constituent minerals (Fig. 2.11). The effect of the water pressure on the melting temperatures of minerals is seen to increase in order: pyroxene < quartz < orthoclase < albite < anorthite < nepheline. Accordingly, the differences in the formation temperatures are greatest between the rocks rich in minerals from the right end of this sequence, both in the plutonic and volcanic facies. Consequently, the fluid effect is most pronounced for the nepheline rocks which, within the volcanic facies, have very high crystallization temperatures, commonly from 1200 °C (phonolites) to 1400 °C (leucite phonolites and nephelinites), whereas in the plutonic facies their crystallization temperature is much lower (600–780 °C). The difference is considerable for the feldspar rocks (rhyolites–granites, trachyts–syenites, etc.), where-

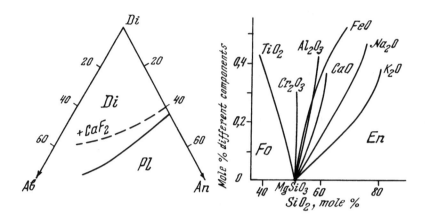

Fig. 2.12. Diagrams illustrating the effects of magma alkalinization and acidification. Left – effects produced by components of dissimilar basicity on the plagioclase and diopside crystallization fields (fluorine is added until fluorite appears (after Batanova, 1977); right – forsterite and enstatite (addition of different components), after Kushiro, 1973.

as the pyroxene-rich varieties (basalt–melanocratic gabbros, etc.) from different depths are much less distinct.

The fluids affect not only the rock crystallization temperatures but also the eutectic composition of magmas and consequently the direction of magmatic evolution. The composition on the eutectic is enriched in the minerals whose melting temperatures are most sensitive to the fluid of a particular composition.

The general mechanism by which various fluid components affect the eutectic composition of magmas in the course of intensive fluid activity has been formulated by Korzhinskii as the principle of acid–base interaction of components. It states that the acidic components of fluids enlarge the crystallization areas of minerals with relatively acidic properties while the alkaline components promote crystallization of essentially basic minerals. Consider the granitic systems as an example. The addition of fluorine and boron (acidic components) to the system results in a greater region of quartz crystallization; since quartz is more acidic than feldspars the eutectic compositions shift towards feldspar compositions.

Liquid immiscibility in magmas is related to the specific effect produced by fluid components – they depress crystallization temperatures of magmas (with decreasing temperatures, magmas no longer behave as ideal solutions) and displace their eutectic compositions towards excessively fluid 'monomineralic' compositions, which give rise to a contrasting component partitioning and the development of magmatic immiscibility regions. The widespread occurrence of these processes in nature is evidenced by eruptions of split (bimodal or 'mixed') lavas (chondrite, variolite, spherulite, ignimbrite). Note that the fluids exert a powerful action on magmas in this respect, especially if they contain components which considerably depress the magmatic crystallization temperatures and displace their eutectic components in opposite directions (Fig. 2.12),

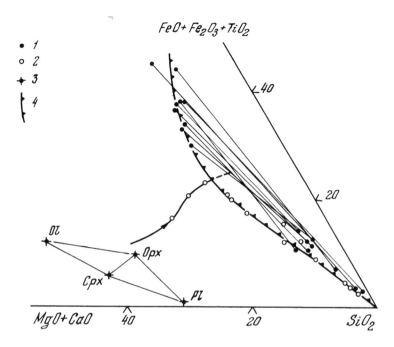

Fig. 2.13. Petrochemical diagram (mass %) of magmatic differentiation of basalt –
by crystallization, early in the process and by liquid immiscibility, in the
last stages: 1 – compositions of varioles (even numbers in Table 2.1) and
their matrix (odd numbers); 2 – basalts, ferrobasalts, icelandites,
ferrodacites and liparites (rhyolites) from the Thingmuli volcano,
Iceland; 3 – phenocrysts in basalts; 4 – contour of the magmatic
unmixing of ferrobasalt magma.

producing simultaneous alkalinity and acidity of the magma system. Ultimately, this is
caused by differential solubility of components in the fluids (A and B) into which the
original magma (A, B) is separated under the action of fluids of complex (C, D)
composition: $(A, B) + (C, D) = (A, C) + (B, D)$. Therefore, fluids which are composed of
different components in terms of their chemical affinities for petrogenic oxides promote
liquid immiscibility in magmas more effectively than do homogeneous fluids. Textures
formed by this immiscibility (variole-type and others) are well preserved under the non-
uniform distribution of fluid components, which gives rise to melt of different viscosity,
with the result that drops of more viscous solutions (chondrules, varioles) develop
within the less viscous fluid matrix.

Magmatism of spreading geotectonic structures proceeds under reducing conditions
and relatively low fluid pressure. Rock crystallization temperatures are relatively high
under these conditions, and basalts crystallize within the stability fields of pigeonite
which is found not only in basalt phenocrysts (it is typical of basalts of all types) but in
the groundmass as well. Kuno (1968) has recognized a special type of pigeonite basalt.
The scheme of rock differentiation, presented in Fig. 2.13 and in Table 2.1, is based on
fractional crystallization which gives rise to ferrobasalt melts, with subsequent liquid

Table 2.1. *Compositions of varioles (even numbers) and matrices (odd numbers) in iron- and titanium-rich variolites and separated residual glasses from basalts; mass %*

No. in Fig. 2.13	SiO_2	TiO_2	Al_2O_3	FeO + Fe_2O_3	MnO	MgO	CaO	Na_2O	K_2O	P_2O_5
1	47.59	4.04	7.34	21.65	0.46	2.89	10.02	2.14	0.62	3.25
2	70.29	1.01	10.73	9.46	0.09	0.67	3.06	2.94	1.48	0.27
3	45.67	7.66	7.39	20.90	0.90	1.95	10.87	1.63	0.12	2.91
4	70.79	2.75	10.59	7.28	0.15	0.82	4.11	2.19	0.80	0.52
5	43.40	4.08	7.30	24.11	0.36	2.23	9.39	1.83	0.77	6.53
6	68.75	2.01	11.73	6.54	0.15	0.92	4.28	2.86	1.06	1.70
7	40.13	3.49	6.13	26.29	0.51	2.51	11.42	1.57	0.20	7.75
8	70.66	1.18	10.92	7.92	0.19	0.66	3.92	2.08	1.56	0.91
9	43.44	4.37	6.74	28.40	0.59	2.17	9.76	2.11	0.25	2.17
10	66.90	1.49	11.03	10.00	0.23	0.98	4.03	3.18	0.97	1.19
11	36.36	—	19.54	27.41	0.22	3.23	6.30	4.10	2.84	—
12	64.75	—	13.54	8.34	—	1.59	4.63	5.39	1.76	—
13	44.86	3.22	7.96	30.76	—	0.30	10.09	2.32	0.49	—
14	75.61	0.29	13.11	2.02	—	0.19	0.51	4.34	3.93	—
15	44.06	3.97	3.36	37.61	0.61	0.81	8.77	0.41	0.10	0.30
16	73.76	0.62	12.81	5.00	—	—	2.50	4.06	1.25	—
17	39.80	4.24	2.46	37.65	0.33	0.56	11.28	—	0.33	3.35
18	78.49	0.21	12.74	2.78	—	—	1.50	3.21	0.96	0.11

Notes: Analyses as recalculated on an anhydrous basis
Sources: Rosenbush, 1934, Dixon & Rutherford, 1979, Philpotts & Doyle, 1983

immiscibility and the separation of ferro-andesites (icelandites), ferrodacites, rhyolites, normally of enhanced alkalinity. All intermediate and acidic oceanic rocks (Luchitsky, 1973) belong to this type and so do granites and granophyres (including tin-bearing) which occur in the roof zones of layered intrusions, with the ferrogabbroic differentiation trend (Bushveld, Skaergaard, etc.). This segregation of acidic melts reflects the liquid immiscibility typical only of residual magmas rich in iron and titanium. Matrices and varioles have compositions similar to many widespread rocks which are presumably (based on this analogy) formed by liquid immiscibility. The ratios of the variole-to-matrix compositions govern discontinuities, recognized in magmatic series of rocks (Marakushev, 1983). The distinctive feature of the commonest variolites are their more basic compositions and a much higher iron content in the matrix relative to the varioles. In this respect, variolites are similar to chondrites in which drop-like segregates (chondrules are analogues of varioles) of the olivine–pyroxene and pyroxene–plagioclase composition (both vitreous and crystallized) are enclosed in the iron-rich, essentially olivine matrix (Marakushev & Bezmen, 1983). In common with variolites, chondrites occur as lava and fragmental (pyroclastic) varieties. However, chondrites are basically different from variolites in their higher contents of native iron, nickel, etc., wider variations in the chondrule compositions and structure and more reduced conditions of formation, with the result that iron generally occurs in its native state in the chondrite matrices. The reducing conditions and rapid cooling facilitate the unmixing of iron-bearing melts to give rise to chondritic, variolitic and other unmixed magmas and, accordingly, to the apparent (discrete) layering of the intrusive complexes.

The variolites, summarized in the petrochemical diagram (Fig. 2.13), display the extreme unmixing of the parent magma, which is responsible for the enrichment of these rocks in iron, titanium, and phosphorus, their oxide contents reaching 30–40%. In fact, this is the ore-forming process which gives rise to titano-magnetite, ilmenite and magnetite (ilmenite)–apatite ores. The unmixing of magmas resulting from liquid immiscibility is a most effective mechanism of concentration of ore material, owing to which its content often increases by several orders of magnitude, as has been found in the experimental studies on niobium and tantalum (Marakushev, Ivanov & Rymke-vich, 1979), phosphorus (Skripnichenko, 1979), chromium, iron, titanium, rare earths (Watson, 1976) and many other elements. This effect explains the relationship normally observed between ore deposits and strongly differentiated magmatic complexes. The fluid ore-bearing magmas are usually separated at the last stages of magmatic evolution, and their crystallization, which proceeds at lower temperatures than that of the parent rocks, is invariably more or less concurrent with postmagmatic metasomatic processes.

Acknowledgements   The author is indebted to Professors D.M. Burt, M. Fonteilles and J.B. Brady for their constructive reviews and many helpful criticisms, which have been taken into consideration in the preparation of the paper for publication.

References

Albee, A.L. & Zen, E-an (1969). Dependence of the zeolitic facies of the chemical potentials of CO_2 and H_2O in *Ocherki fisiko-khimicheskoi petrologii*, vol. I, p. 249–60. Moscow: Nauka.

Aranovich, L.Ya. & Podlesskii, K.K. (1983). The cordierite–garnet–sillimanite–quartz equili-brium; experiments and applications. In *Kinetics and equilibrium in mineral reactions*, p. 173–98. New York: Springer–Verlag.

Bach, R.W., Friedrichs, H.A. & Rau, H. (1977). $p–V–T$ relations for $HCl–H_2O$ mixtures up to 500 °C and 1500 bars. In *High temperatures – high pressures, vol. 9*, p. 305–12.

Batanova, A.M. (1977). The effect of fluorine on crystallization of melts in the system albite–anorthite–diopside. In *Petrologic-mineralogical characteristics of rocks and technical stones*, pp. 152–65. Moscow: Nauka.

Bowen, N.L. (1928). *The evolution of the igneous rocks*. Princeton University Press.

Dixon, S. & Rutherford, M.J. (1979). Plagiogranites as late-stage immiscible liquids in ophiolite and mid-ocean suites: an experimental study. *Earth and Planet. Sci. Lett.*, **45**, No. 1, 45–57.

Franck, E.U. (1980). Sub- and supercritical water at high pressure – selected results. In *Water and steam: Proceedings of the 9th International Conference on the Properties of Steam*, pp. 465–76. Pergamon Press.

Gibbs, S.W. (1931). *The collected works. Thermodynamics, vol. I*. New York–London–Toronto.

Harris, P.G., Kennedy, W.Q. & Scarfe C.M. (1970). Volcanism versus plutonism – the effect of chemical composition. In *Mechanism of igneous intrusion* ed. G. Newell & N. Rast, *Geol. J. Spec. Issue*, No. 2, pp. 187–200.

Korzhinskii, D.S. (1940). Factors of mineral equilibria and mineralogical depth facies. *Trudy

Instituta Geol. Nauk AN SSSR, Seriya Petrograficheskaya, **12**, No. 5, 148.

— (1959). Acid–base interaction of components in silicate melts and direction of cotectic lines. *Doklady Akad. Nauk SSSR*, **128**, No. 2, 383–6.

— (1969). *Theory of metasomatic zoning*. Moscow: Nauka.

— (1976). Transmagmatic fluids and magmatic replacement. In *Petrography, part I*, pp. 269–87. Izdatelstvo Moskovskogo Univ.

Kuno, H. (1968). Differentiation of basalt magmas. In *Basalts* vol. 2. New York–London.

Kushiro, J. (1973). Regularities in the shift of liquidus boundaries in silicate systems and their significance in magma genesis. In *Annual Report*, Geophys. Lab., Carnegie Inst., Washington, DC, pp. 497–502.

Lazarenko, E.N., Pavlishin, V.I., Latyish, V.T. & Sorokin, Yu.G. (1973). *Mineralogy and genesis of chamber pegmatites in Volyin*. Lvov; Vysshaya Shkola.

Luchitsky, I.V. (1973). On acid magmatic oceanic rocks. *Geotektonika*, 1973, No. 5, 22–35.

Marakushev, A.A. (1964). Some mineral facies of metamorphic rocks poor in calcium. Report to the Twenty-second session. Intern. Geol. Congress, Petrology. New Delhi, India, 1964.

— (1965). *Problems of mineral facies in metamorphic and metasomatic rocks*. Moscow: Nauka.

— (1968). *Thermodynamics of metamorphic hydration of minerals*. Moscow: Nauka.

— (1979). *Petrogenesis and ore formation*. Moscow: Nauka.

— (1983). On the genesis of volcanic series of rocks. *Vestnik Moskovskogo Univ.*, seriya 4, Geologiya, No. 5, 3–21.

— (1987). Magmatic replacement and its petrogenetic role. In *Ocherki fisiko-khemicheskoi petrologii*, vol. 15, pp. 24–38. Moscow: Nauka.

Marakushev, A.A. & Bezman, N.I. (1983). *The evolution of meteoritic matter, planets and magmatic series*. Moscow: Nauka.

Marakushev, A.A. & Gramenitsky, Ye.N. (1983). Problem of the origin of pegmatites. *Intern. Geol. Rev.*, **25**, No. 10, 1179–86.

Marakushev, A.A., Ivanov, I.P., Rymkevich, V.S. (1979). The role of liquid immiscibility in the genesis of magmatic rocks. *Vestnik Moskovskogo Univ., seriya 4, Geologiya*, No. 1, 3–21.

—— (1981) Experimental modelling of rhythmic magmatic layering. *Doklady AN SSSR*, **258**, No. 1, 183–6.

Perchuk, L.L. (1977). Thermodynamic control of metamorphic processes. In *Energetics of geological processes*, pp. 285–352. New York: Springer–Verlag.

Perchuk, L.L., Aranovich, L.Ya., Podlesskii, K.K. *et al.* (1985). Precambrian granulites of the Aldan shield, eastern Siberia, USSR. *Journ. Metamorphic Geol.*, **3**, 265–310.

Philpotts, A.R. & Doyle, C.D. (1983). Effect of magma oxidation state on the extent of silicate liquid immiscibility in a tholeiitic basalt. *Amer. Journ. Sci.*, **283**, No. 9, 967–86.

Rosenbush, G. (1934). *Descriptive petrography*. Gosgeoltechizdat.

Skripnichenko, V.A. (1979). Phosphorus as a factor of the liquid immiscibility in silicate melts. *Doklady AN SSSR*, **245**, No. 4, 930–3.

Smith, F.G. (1948). Transport and deposition of nonsulphide vein materials. III. Phase relations at pegmatitic stage. *Econ. Geol.*, **43**, No. 7, 535–46.

Watson, E.B. (1976). Two-liquid partition coefficients: experimental data and geochemical implications. *Contrib. Mineral. and Petrol.*, **56**, No. 1, 119–34.

Zavaritsky, A.N. (1947). On pegmatites as formations intermediate between the igneous rocks and ore veins. *Zapiski Vsesojuznogo Mineral obshchestva*, **76**, No. 1, 3–18.

3

A new hydrous, high-pressure phase with a pumpellyite structure in the system MgO–Al$_2$O$_3$–SiO$_2$–H$_2$O

W. SCHREYER, W.V. MARESCH AND T. BALLER

Introduction: accidental synthesis

Pumpellyite is a complex hydrous Ca-silicate with additional (Al, Fe^{3+}) and (Mg, Fe^{2+}) (Coombs *et al.*, 1976). It occurs mainly in very-low-grade metamorphic rocks, preferably of basic compositions, for example in former amygdules of metabasalts. In the classical grid of metamorphic facies (Winkler, 1974), the 'pumpellyite–prehnite facies' is located in the temperature range of 300° ± 50 °C and at pressures below about 3 kbar. This is in general agreement with experimental data, such as those obtained by Schiffman & Liou (1980), who determined the stability field of the Fe-free pumpellyite end member Ca$_4$MgAl$_5$Si$_6$O$_{21}$(OH)$_7$, which they named MgAl-pumpellyite. They found the upper thermal stability of this phase to lie near 350 °C for a water pressure of 3 kbar. However, they also recognized that this temperature limit increases with water pressure, so that the stability field of MgAl-pumpellyite extends to at least 390 °C at 8 kbar.

The experimental petrology group at Bochum has recently conducted high-pressure studies in excess of 30 kbar, with the goal of determining the phase relations of unusually (Ca, Fe, Na)-poor, but Mg-rich metapelites, which had been found, through field and laboratory studies, to be particularly diagnostic for very deep-seated metamorphism of crustal rocks in subduction zones (Chopin & Schreyer, 1983, Chopin, 1984, Schreyer, 1985). Because of its fundamental importance, the system MgO–Al$_2$O$_3$–SiO$_2$–H$_2$O (MASH) was a primary objective. Among other topics, an attempt was made to synthesize a (Ti, Zr, P)-free end-member of the new high-pressure mineral ellenbergerite (Chopin *et al.*, 1986), which was extrapolated to lie in the MASH-system rather close to the composition of the garnet pyrope, Mg$_3$Al$_2$Si$_3$O$_{10}$, plus water. Despite initially positive indications, these attempts failed, even at water pressure as high as 50 kbar. On the other hand, the X-ray powder diffractograms of the run products obtained showed a multitude of reflections that clearly did not correspond to those of any of the known crystalline compounds of the MASH-system, including the new high-pressure compounds Mg-carpholite and Mg-chloritoid reported by Chopin & Schreyer (1983).

After a considerable and frustrating period of trial and error, the idea was found helpful that the Mg ion generally present in sixfold coordination in silicates might, at the high pressures applied here, proxy for larger cations with higher coordination with respect to oxygen, such as Ca. A comparison of the unknown diffraction pattern with those of CaAl-silicates such as zoisite, prehnite etc. showed similarities only with natural pumpellyite. However, the peaks of the unknown phase were found to lie at consistently higher 2θ values, the differences being as much as 2 degrees at 2θ near 60° CuK$_a$. On this basis, more than 40 observed reflections of the unknown phase below 70°2θ could be indexed successfully in analogy to natural pumpellyite, but with a much smaller monoclinic cell. A first report on the identification of the new synthetic hydrous Mg–Al silicate as a calcium-free pumpellyite phase was published by Schreyer *et al.* (1986), but its exact chemical composition was not yet known at that time.

In the present paper, the results of new synthesis and analytical work on the new phase are presented, which lead to a clarification of its composition, and also a fairly good knowledge of its stability relations. The latter are of interest with regard to the potential occurrence of this hydrous phase at the depths of the Earth's upper mantle.

Composition and planned synthesis

Following Coombs *et al.* (1976) the general formula of pumpellyite can be given as

$$W_4^{[7]}X_2^{[6]}Y_4^{[6]}Z_6^{[4]}O_{20+t}(OH)_{8-t},$$

where W = Ca, Mn; X = (Mg, Fe^{2+}, Mn)$_{2-t}$ (Fe^{3+}, Al)$_t$; Y = Fe^{3+}, Al; Z = Si. Note that the number of hydroxyl groups or hydrogen atoms can vary, and is linked to the occupancy of the X-position by either divalent or trivalent cations, or by both according to the equation $M^{2+} + H^+ = M^{3+}$. Thus t can, at most, vary between zero and 2. For the Ca-bearing MgAl-pumpellyite, Ca$_4$(MgAl)Al$_4$Si$_6$O$_{21}$(OH)$_7$, studied by Schiffman & Liou (1980), t equals one.

If one assumes the same type of compositional variation for the Ca-free pumpellyite of the MASH-system as well, the following hypothetical formulae are derived from t equalling 0, 1, and 2, respectively:

Pu 1: Mg$_4$Mg$_2$Al$_4$Si$_6$O$_{20}$(OH)$_8$
Pu 2: Mg$_4$(MgAl)Al$_4$Si$_6$O$_{21}$(OH)$_7$
Pu 3: Mg$_4$Al$_2$Al$_4$Si$_6$O$_{22}$(OH)$_6$

These compositions are plotted in Fig. 3.1 in a projection of the MASH-system. The interconnecting line between Pu 1 and Pu 3 would correspond to the substitution

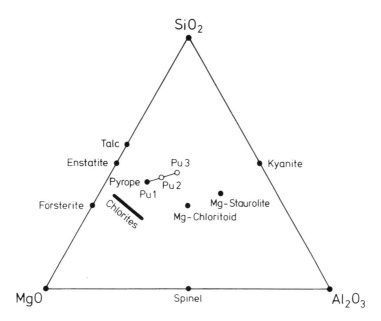

Fig. 3.1. Ternary projection of the system $MgO–Al_2O_3–SiO_2$ (MASH) from the H_2O apex showing potential compositions of the new high-pressure phase $MgMgAl$–pumpellyite (Pu_1, Pu_2, Pu_3) as discussed in the text. Pu_1 coincides in the projection with pyrope. Solid dots and bar represent the compositions of some other crystalline phases stable at high water pressures.

$Mg + H = Al$ and may indicate the maximum possible range of solid solution of the pumpellyite phase. Note that Pu 1 plots together with pyrope, because its hypothetical formula corresponds to the composition 2 pyrope, $Mg_3Al_2Si_3O_{12}$, plus 4 H_2O.

For the synthesis studies, 3 gels were prepared with Mg:Al:Si ratios corresponding to those of Pu 1–3. They were run, together with excess H_2O, in sealed gold capsules in a standard piston–cylinder apparatus at 50 kbar, 700 °C for 2 to 4 days. The ratio of solids to water in the capsules was kept at approximately 5:1. No evidence for significant leaching or incongruent dissolution of the starting gel during the runs was detected. It became clear that 100% yield of the pumpellyite phase could only be achieved from gel Pu 2. In gel Pu 1 considerable amounts of chlorite and possibly some enstatite appeared as additional phases, while gel Pu 3 yielded additional kyanite and possible traces of coesite. At 50 kbar, 800 °C, the composition Pu 1 + excess water crystallized to single-phase pyrope; nor did any of the other gels yield the pumpellyite phase at this temperature. It must also be mentioned that in some runs with gel Pu 2 the apparently metastable alternative assemblage enstatite + Mg-staurolite formed instead of the pumpellyite phase. This is discussed below.

As a result of this experimentation, it can be concluded that the pumpellyite phase of the MASH-system has the Mg:Al:Si ratio 5:5:6 in accordance with Pu 2, and that there are no indications of any appreciable solid solution within this system. Fig. 3.1 shows that the assemblages obtained from gels Pu 1 and Pu 3 (see above) are also consistent with a pumpellyite composition of Pu 2.

As a further constraint on the composition of the new MASH-phase, a single-phase run product of the Pu 2 gel was analysed for water at this Institute using Karl-Fischer-titration; the method was described by Johannes & Schreyer (1981). It was found that the pumpellyite phase contains 7.1 ± 0.2 weight % H_2O. The calculated H_2O content of the hypothetical formula Pu 2 given above is 7.16 weight %. This is corroborating evidence for the conclusion that the new phase does indeed have the formula $Mg_5Al_5Si_6O_{21}(OH)_7$ and is thus the pure Mg analogue of $Ca_4MgAl_5Si_6O_{21}(OH)_7$ synthesized by Schiffman & Liou (1980). Because the latter phase is known as MgAl-pumpellyite in the literature, the new Mg end-member had to be named *MgMgAl-pumpellyite* (Schreyer et al., 1986, 1987). It is worth mentioning in this connection that Schiffman & Liou (1980) achieved 100 % yield of MgAl-pumpellyite in the system $CaO–MgO–Al_2O_3–SiO_2–H_2O$ only with a starting material of the ideal composition given above. Thus, in both cases, the synthetic pumpellyite end-members contain 7 hydrogens per formula unit, for which there must be crystal-chemical reasons.

In order to extend the work of Schiffman & Liou (1980) to higher water pressures, a gel was prepared in the present study with the cation ratio Ca:Mg:Al:Si = 4:1:5:6 of MgAl-pumpellyite. This gel was preheated at 900 °C and could then be crystallized to 100 % pumpellyite at 40 kbar, 500 °C within 4 days. There is thus no doubt that the pumpellyite structure is stable to very high water pressures even for Ca-bearing compositions.

Physical properties of MgMgAl-pumpellyite

Under the petrographic microscope, the single-phase run products of MgMgAl-pumpellyite appear as extremely fine-grained felts of intimately intergrown, seemingly anhedral crystals with low interference colours. Only in exceptional cases, along the walls of the charges, very thin, euhedral, monoclinic blades with lengths up to 50 micrometers and occasional twinning were found (Fig. 3.2). However, scanning electron microscopy of the fine-grained bulk of the products shows that even those minute crystals well below one micrometer in size are actually of euhedral, monoclinic morphology as well (Fig. 3.3).

The optical refractive indices n_{x-z} and the axial angle of MgMgAl-pumpellyite, measured by O. Medenbach on the spindle stage, are listed in Table 3.1, together with the data on the Ca-analogue. Note that the refractive indices of the pure MgAl-silicate phase are considerably higher than those of the Ca-bearing end-member.

Powder X-ray diffraction data on $Mg_5Al_5Si_6O_{21}(OH)_7$ are compiled in Table 3.2. They were indexed on the basis of a monoclinic cell with the dimensions listed in Table 3.1. Note here that these cell constants are consistently smaller than those of the Ca-bearing synthetic phase of Schiffman & Liou (1980), resulting in a cell volume that is

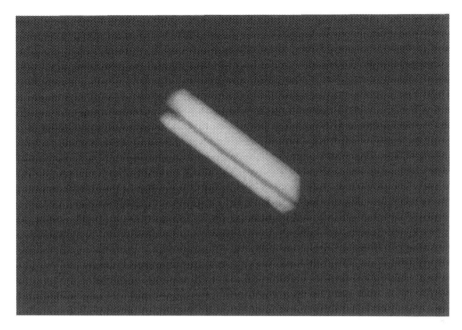

Fig. 3.2. Euhedral, twinned, monoclinic crystal of synthetic MgMgAl-pumpellyite obtained at 50 kbar, 700 °C. The photograph was taken under the polarizing microscope with crossed nicols. Length of crystal 50 micrometers.

Fig. 3.3. Scanning electron micrograph of fine-grained run product of gel Pu_2 (see Fig. 3.1) after treatment of 50 kbar, 700 °C, for 2 days in the presence of excess water. Despite small and variable grain size, only euhedral crystals of one kind can be discerned. Length of bar at bottom equals 4 micrometers.

Table 3.1. *Physical properties of MgMgAl-pumpellyite synthesized at 50 kbar, 700 °C, 4 days, compared to those of Ca-bearing MgAl-pumpellyite*

	$Mg_5Al_5Si_6O_{21}(OH)_7$ This work[a]	$Ca_4MgAl_5Si_6O_{22}(OH)_7$ Schiffman & Liou (1980)
n_x	1.674(2)	
n_y	1.6825(10)	} Average 1.624
n_z	1.699(2)	
$2V(°)$	72(1)	—
$a(Å)$	8.544(1)	8.825(8)
$b(Å)$	5.717(1)	5.875(5)
$c(Å)$	18.491(2)	19.10(1)
$\beta(°)$	97.75(1)	97.39(7)
$V(Å^3)$	894.9(1)	982.1
$D_x(g/cm^3)$	3.27	3.19

Note:

[a] Lattice constants refined according to the method described by Maresch & Czank (1985) on the basis of the data given in Table 3.2.

only 91% that of the Ca-phase. Thus, despite the fact that the lighter Mg atom substitutes for Ca, the density of the pure MgAl-phase is still higher by 2.5% than that of CaMgAl-pumpellyite. The 9% decrease in cell volume caused by the Mg-for-Ca substitution is comparable to the 12% difference observed in garnet between grossular and pyrope.

Allmann (1984) as well as Mellini *et al.* (1984) has drawn attention to the fact that the newly determined structure of the natural MnAl-silicate mineral *sursassite* is closely related to that of pumpellyite. In a recent experimental study, Reinecke *et al.* (1987) were able to synthesize an end-member phase with the composition $Mn_4^{2+}MgAl_5$ $Si_6O_{21}(OH)_7$, which they named MgAl-sursassite, and which is the chemical Mn-analogue of Schiffman & Liou's (1980) MgAl-pumpellyite. We wish to emphasize here that the powder X-ray diffraction data of our new synthetic phase MgMgAl-pumpellyite show distinct differences from those of natural sursassite as well as the synthetic sursassite phase (Reinecke, pers. commun., 1987). However, it is also clear that more detailed crystallographic and TEM studies are required for a better characterization of the structure and 'Realbau' of our new high-pressure MgAl-silicate phase, which could conceivably contain stacking faults with alternating units of both the pumpellyite and sursassite structures, as found by Mellini *et al.* (1984) in natural sursassite crystals.

Preliminary electron diffraction studies on our synthetic material by M. Pasero, Pisa (pers. comm., Dec. 1988) seem to confirm the occurrence of only minor amounts of sursassite-type structural units. It is probable that such units occur as very rare lamellae of some 500 Å thickness rather than as fine-scale intercalations within the pumpellyite unit, because one diffraction pattern of a *sursassite* unit alone could be obtained which showed no indications of stacking disorder.

Table 3.2. *X-ray powder diffraction data for MgMgAl-pumpellyite (CuK_{a_1}-radiation)*

h k l	2θ(obs)	2θ(calc)	d(obs)	d(calc)	$I^{\text{rel}}_{\text{obs}}$
ᵃ0 0 2	9.628	9.647	9.179	9.161	5
1 0 0	10.395	10.441	8.503	8.466	1
ᵃ1 1 $\bar{1}$	18.957	18.976	4.678	4.673	16
ᵃ0 0 4	19.347	19.362	4.584	4.581	54
ᵃ1 1 1	19.678	19.689	4.508	4.505	18
ᵃ2 0 0	20.978	20.970	4.231	4.233	4
ᵃ2 0 $\bar{2}$	21.904	21.894	4.055	4.056	4
ᵃ1 1 $\bar{3}$	22.835	22.853	3.891	3.888	4
ᵃ2 0 2	24.291	24.303	3.661	3.659	79
2 1 $\bar{1}$	26.183	26.097	3.401	3.412	9
ᵇ2 0 $\bar{4}$	—	26.657	—	3.341	—
ᵃ2 1 $\bar{3}$	28.625	28.596	3.116	3.119	4
ᵃ0 1 5	28.955	28.918	3.081	3.085	5
ᵃ0 0 6	29.215	29.221	3.054	3.054	12
ᵃ2 0 4	30.586	30.603	2.921	2.919	30
ᵃ0 2 0	31.257	31.268	2.859	2.858	37
2 1 3	31.377	31.431	2.849	2.844	39
3 0 $\bar{2}$ } 1 1 5 }	31.937	{31.874 31.957}	2.780	{2.805 2.798}	85
0 2 2	32.708	32.795	2.736	2.729	21
ᵃ2 0 $\bar{6}$	33.768	33.772	2.652	2.652	55
ᵃ3 1 $\bar{1}$	35.169	35.182	2.550	2.549	57
ᵃ3 1 1	36.380	36.395	2.468	2.467	3
ᵃ0 2 4	37.041	37.042	2.425	2.425	45
2 2 0 } 1 1 $\bar{7}$ }	37.951	{37.952 37.969}	2.369	{2.369 2.368}	100
2 2 $\bar{2}$ } 2 0 6 }	38.552	{38.498 38.577}	2.333	{2.337 2.332}	9
1 0 $\bar{8}$ } 1 2 4 }	39.305	{39.313 39.341}	2.290	{2.290 2.288}	4
ᵃ2 2 2	39.992	39.992	2.253	2.253	36
3 1 3	~40.15	40.155	~2.24	2.244	shoulder
6 0 $\bar{3}$ } 1 1 7 }	40.513	{40.462 40.582}	2.225	{2.228 2.221}	5
ᵃ3 1 $\bar{5}$	40.783	40.780	2.211	2.211	5
ᵃ2 2 $\bar{4}$	41.533	41.543	2.173	2.173	11
ᵃ2 0 $\bar{8}$	42.214	42.224	2.139	2.139	68
ᵃ4 0 $\bar{2}$	42.484	42.486	2.126	2.126	47
4 0 0	~42.70	42.686	~2.12	2.117	shoulder
ᵃ0 2 6	43.324	43.324	2.087	2.087	25
ᵃ2 2 4	44.305	44.319	2.043	2.042	21
ᵃ4 0 $\bar{4}$	44.645	44.642	2.028	2.028	19
ᵃ2 2 $\bar{6}$	46.666	46.685	1.9448	1.9441	7
3 1 $\bar{7}$	~46.80	46.783	~1.94	1.9402	shoulder
1 1 $\bar{9}$ } 3 2 2 }	47.176	{47.207 47.217}	1.9250	{1.9238 1.9234}	4
ᵃ4 0 $\bar{6}$	48.897	48.905	1.8612	1.8609	7
4 1 $\bar{5}$ } 1 3 1 }	49.388	{49.335 49.380}	1.8438	{1.8457 1.8441}	3

Table 3.2 (*cont.*)

h k l	2θ(obs)	2θ(calc)	d(obs)	d(calc)	I^{rel}_{obs}
1 1 9⎫	50.128	⎧50.041⎫	1.8183	⎧1.8213⎫	2
4 1 3⎭		⎩50.043⎭		⎩1.8212⎭	
[a]2 2 6	50.468	50.467	1.8069	1.8069	7
0 2 8⎫	51.058	⎧51.060⎫	1.7874	⎧1.7873⎫	25
1 2 $\bar{8}$⎭		⎩51.064⎭		⎩1.7872⎭	
3 2 4	~51.40	51.360	~1.78	1.7776	shoulder
2 0 $\overline{10}$	51.619	51.574	1.7693	1.7707	3
1 0 10⎫	52.449	⎧52.459⎫	1.7432	⎧1.7429⎫	2
2 3 $\bar{1}$⎭		⎩52.588⎭		⎩1.7389⎭	
[a]4 2 0	53.880	53.855	1.7002	1.7010	4
[a]4 2 $\bar{4}$	55.520	55.510	1.6538	1.6541	12
4 0 6⎫	56.061	⎧56.059⎫	1.6392	⎧1.6392⎫	16
1 3 5⎭		⎩56.086⎭		⎩6.6385⎭	
[a]5 1 $\bar{3}$	56.711	56.711	1.6219	1.6219	8
[a]1 1 $\overline{11}$	57.171	57.175	1.6099	1.6098	8
[a]2 0 10	57.394	57.382	1.6042	1.6045	11
0 1 11	57.550	57.591	1.6002	1.5992	9
[a]3 3 $\bar{1}$	58.231	58.209	1.5831	1.5837	13
[a]4 2 4	59.982	59.981	1.5410	1.5410	62
[a]0 0 12	60.612	60.596	1.5265	1.5269	32
[a]2 2 $\overline{10}$	61.553	61.559	1.50554	1.5053	11
2 0 $\overline{12}$⎫		⎧61.692⎫		⎧1.5024⎫	
3 3 3 ⎬	61.703	⎨61.719⎬	1.5021	⎨1.5018⎬	11
3 1 9 ⎭		⎩61.793⎭		⎩1.5001⎭	
[a]4 0 8	63.713	63.714	1.4595	1.4595	9
[a]4 2 $\bar{8}$	64.564	64.559	1.4424	1.4423	35

Notes:
[a] used for refinement of lattice constants.
[b] coincides with F-phlogopite standard.

Stability and compatibility relations

The question of the thermodynamic stability of the new MASH-phase MgMgAl-pumpellyite is, of course, only part of the overall problem of the phase relations in that system at high water pressures. Earlier knowledge on this topic relevant to the pressure range 10–35 kbar has been summarized by Schreyer (1968), Schreyer & Seifert (1969), as well as by Chopin & Schreyer (1983). A more recent summary extending the range up to 50 kbar was given by Schreyer (1988), who also dealt with the new phase described here.

Fig. 3.4 shows, in a ternary projection, those MASH-phases that are stable at high pressures, and which are thus of relevance as potential breakdown products of MgMgAl-pumpellyite outside its stability range. As a major result of the paper by Schreyer (1988), it should be noted here that assemblages of Mg-staurolite with either talc or enstatite have to be considered metastable at these high pressures, although they form very readily in non-seeded experiments, especially on compositions close to or

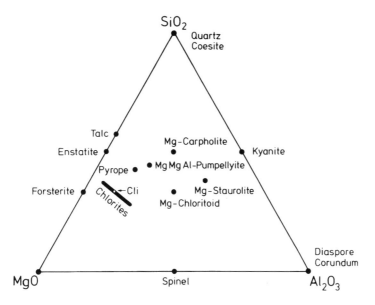

Fig. 3.4.   Ternary MASH projection as in Fig. 3.1 now showing the proper composition of MgMgAl-pumpellyite in relation to those of other crystalline phases (Chopin & Schreyer, 1983), which may possibly appear among its breakdown products. The composition 'Cli' indicated by the arrow is that of the chlorite clinochlore which was used in the stoichiometric calculations.

identical with that of MgMgAl-pumpellyite. Under equilibrium conditions, these assemblages are replaced by stable pairs such as chlorite–kyanite and pyrope–kyanite. On this basis, hypothetical reaction relations between MgMgAl-pumpellyite and its possible breakdown assemblages were worked out. Because of the high water content of MgMgAl-pumpellyite, it turned out that most of its potential breakdown reactions release H$_2$O, so that these can only be considered to mark the upper thermal stability limits of the new MASH-phase. The only terminal reaction in which H$_2$O is consumed, on the basis of stoichiometry, was found to be

$$14\,Mg_5Al_2Si_3O_{10}(OH)_8 + 3\,Mg_3Si_4O_{10}(OH)_2 + 51\,MgAl_2Si_2O_6(OH)_4$$
Clinochlore   Talc   Mg-Carpholite

$$= 26\,Mg_5Al_5Si_6O_{21}(OH)_7 + 70\,H_2O$$
MgMgAl-pumpellyite

(1)

In this equation, clinochlore is taken to be the representative member of the chlorite series, although this is by no means clear. Indeed, the extent of chlorite solid solution is virtually unknown, especially at the high pressures involved, and it is also to be expected that the composition of the chlorite in the above reaction will vary as a function of pressure and temperature.

    In order to initiate experiments on the pressure–temperature stability of MgMgAl-

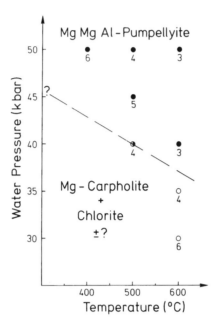

Fig. 3.5.   Water pressure/temperature plot of results of seeded runs bearing on the high-pressure/low-temperature stability of MgMgAl–pumpellyite and its breakdown into an assemblage containing Mg–carpholite, $MgAl_2Si_2O_6(OH)_4$. Solid dots indicate growth, open circles breakdown of MgMgAl-pumpellyite. Mixed symbol: no measurable reaction. Numbers give run durations in days. Pressure uncertainties are estimated to be less than $\pm 5\%$, temperature uncertainties ± 15 °C. For further discussion see text and reaction (1).

pumpellyite, reaction (1) was investigated by the seeding technique, using standard piston–cylinder apparatus with NaCl pressure cells.

Starting materials were:

1. Synthetic MgMgAl-pumpellyite prepared at 50 kbar, 700 °C, as described above.
2. Synthetic Mg-carpholite crystallized from a gel of the requisite composition at 25 kbar, 540 °C for 3 days, using small amounts of precrystallized Mg-carpholite provided by C. Chopin as seeds.
3. The crystallization product of a gel of the remaining composition of the left-hand side of equation (1), that is 14 clinochlore + 3 talc, obtained after running at 30 kbar, 650 °C for 6 days, as in all other cases in the presence of excess H_2O. To our surprise this product did not contain any talc, but consisted almost entirely of chlorite, with a possible trace of Mg-staurolite. The nature of this crystallization product seems to indicate that the chlorite obtained can hardly be of clinochlore composition.

The results of runs using an appropriate stoichiometric mixture of the MgMgAl-pumpellyite, Mg-carpholite, and chlorite solid solution synthesized as described above, together with additional H_2O, are shown graphically in Fig. 3.5. In all runs at 50 kbar, MgMgAl-pumpellyite became by far the dominant phase, while at 40–45 kbar the reaction to form this product was found to be much slower. At 30 kbar, 600 °C the pumpellyite phase disappeared completely, while chlorite and Mg-carpholite increased strongly. It is interesting to note that in all the runs shown in Fig. 3.5, small amounts of diaspore, AlOOH, appeared as an additional phase. This may indicate that the coexisting chlorite is actually less aluminous than clinochlore (compare Fig. 3.4).

The experimental results of Fig. 3.5, together with the stability data and chemographic constructions on other compositions in the MASH-system summarized by Schreyer (1988), lead to the conclusion that the new phase MgMgAl-pumpellyite does indeed have a lower temperature and pressure stability limit that lies, with a strongly negative slope, between about 400 °–600 °C and 37–45 kbar.

An upper thermal stability of MgMgAl-pumpellyite was already indicated by the synthesis results at 50 kbar, showing that the phase grew directly from gels at 700 °C, while only alternative phase assemblages including Mg-staurolite formed at 800 °C. Based on the equilibrium considerations mentioned earlier and the chemographic constructions of Schreyer (1988), the stable breakdown assemblage of MgMgAl-pumpellyite at 50 kbar and elevated temperature must be pyrope + kyanite + coesite. This assemblage was synthesized at 50 kbar, 1000 °C within 2 hours and used for bracketing runs, together with seeds of the pumpellyite phase. The results show clear breakdown of MgMgAl-pumpellyite at 50 kbar, 800 °C after 1 day, while there was virtually no reaction at this pressure and 750 °C after 1 day and 775 °C after 3 days. Since the growth of the pumpellyite phase at the expense of such refractory crystals as pyrope, kyanite, and coesite is likely to be kinetically less favoured than its breakdown, we tend to locate the curve at about 775 °C.

Although no further experimental studies on any other possible breakdown reaction of MgMgAl-pumpellyite have been conducted thus far, it is possible, on the basis of the overall MASH phase relations at high water pressures as presented by Schreyer (1988), to constrain the PT-stability field of the new phase even more. The results are shown in the PT-plot of Fig. 3.6, which also gives the relevant compatibility triangles for each breakdown assemblage. It can be seen from this figure that, with increasing temperature, the breakdown reaction (1) studied experimentally (see Fig. 3.5) leading to Mg-carpholite + chlorite ($\pm$ talc) should be followed by a reaction producing Mg-carpholite + talc + Mg-chloritoid, but not involving any gain or loss of water. It is for this reason that for the latter reaction, that is

$$14 \text{ Mg-chloritoid} + 5 \text{ talc} + 1 \text{ Mg-carpholite} = 6 \text{ MgMgAl-pumpellyite}, \qquad (2)$$

a horizontal slope was assumed in Fig. 3.6. The reactions ensuing at still higher temperatures are as follows:

$$8 \text{ MgMgAl-pumpellyite} = 19 \text{ Mg-chloritoid} + 7 \text{ talc} + 1 \text{ kyanite} + 2H_2O; \qquad (3)$$

$$12 \text{ MgMgAl-pumpellyite} = 1 \text{ talc} + 11 \text{ kyanite} + 19 \text{ pyrope} + 41H_2O; \qquad (4)$$

$$6 \text{ MgMgAl-pumpellyite} = 10 \text{ pyrope} + 5 \text{ kyanite} + 1 \text{ coesite} + 21 H_2O. \qquad (5)$$

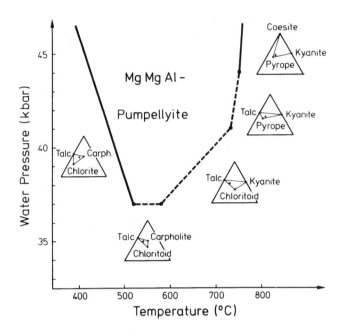

Fig. 3.6. Water pressure/temperature diagram exhibiting the approximate stability field of MgMgAl–pumpellyite in the range 35–45 kbar bounded by five univariant reaction curves as determined experimentally (Fig. 3.5) and/or constrained by theoretical work. Triangles outside the stability range show the respective breakdown assemblage in the ternary projection of Fig. 3.4, crosses indicating the bulk composition of MgMgAl–pumpellyite. The sequence of the reaction curves with increasing temperature is that of equations (1) to (5) as discussed in the text.

These release an increasing amount of H_2O and, for this as well as for chemographic reasons, have increasingly steeper dP/dT-slopes.

The compatibility relations of the new high-pressure MASH-phase along its stability limits can obviously be directly taken from the phase triangles shown in Fig. 3.6. More important, however, are the compatibilities valid for PT-conditions further inside the stability field of MgMgAl-pumpellyite. According to the preliminary experimental data and theoretical deductions of Schreyer (1988), the appearance of the pumpellyite phase at pressures above some 40 kbar leads to drastic changes in the phase relations at these pressures. Most importantly, the high-pressure phases Mg-carpholite and Mg-chloritoid (Chopin & Schreyer, 1983) themselves become unstable, so that MgMgAl-pumpellyite attains the significance of a central phase in the system that may coexist with many other phases of widely different compositions. Fig. 3.7 serves as an example for the inferred topology at 50 kbar and 600 °C. Here, MgMgAl-pumpellyite coexists with coesite, kyanite, diaspore, chlorite and probably talc. Uncertainty exists on the Al-poor side, where phase relations for the system $MgO–SiO_2–H_2O$ were extrapolated from the 20 kbar results calculated by Berman et al. (1986). At higher temperatures, the development cannot be foreseen at this stage of the experimental work. Either pyrope

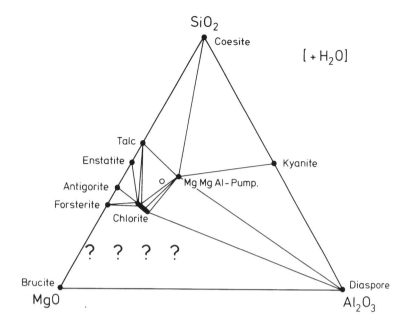

Fig. 3.7. Inferred compatibility relations in the MASH-system at 50 kbar water
pressure and 600 °C with MgMgAl-pumpellyite representing the central
ternary phase coexisting stably with many others. The open circle shows
the composition of pyrope, a phase which is not stable under these
conditions.

forms at the expense of chlorite + talc + MgMgAl-pumpellyite, or the chlorite + talc
pair reacts to form the pumpellyite phase in coexistence with enstatite. In the latter case,
pyrope would crystallize from the assemblage enstatite + chlorite + MgMgAl-
pumpellyite. Future experimental work should be directed towards the open problem
of a compatibility of MgMgAl-pumpellyite with the less siliceous magnesian phases
antigorite and forsterite, perhaps also with the MSH high-pressure phases synthesized
by Yamamoto & Akimoto (1977), which will be discussed in the final section of this
paper.

Crystal-chemical and petrologic discussion

Provided that the new high-pressure phase Mg$_5$Al$_5$Si$_6$O$_{21}$(OH)$_7$ is indeed truly
isostructural with pumpellyite, a conclusion that has yet to be confirmed by structural
analysis, it represents, for the MASH-system, the second case (for the pressure range
considered), in which the coordination of Mg against oxygen increases as a result of
increasing pressure. While MgMgAl-pumpellyite is expected to contain four out of five
Mg per formula unit in sevenfold coordination like Ca in pumpellyite, all the Mg-
bearing, lower-pressure breakdown products of this phase given in Fig. 3.6 exclusively
show sixfold coordination of Mg. The other well-known case is, of course, pyrope (with

Mg in eightfold coordination), which forms at some 15 kbar (Schreyer & Seifert, 1969), again at the expense of phases with $Mg^{[6]}$. It is thus in accordance with this empirical pressure/coordination rule that, at pressures in the order of 50 kbar, most of these hydrous, lower-pressure MgAl-silicates (see Fig. 3.4) have become unstable (Schreyer, 1988), notable exceptions being chlorite (see Fig. 3.7) and the unusual phase Mg-staurolite (with Mg located in a large tetrahedron). It would seem that at still higher pressures, MgMgAl-pumpellyite and pyrope should represent the only ternary phases in the MAS triangle, unless other new high-pressure phases make their appearance. In this connection, one might think of a high-pressure MgAl-silicate with an epidote or zoisite structure and a composition $Mg_2Al_3Si_3O_{12}(OH)$, which would also be based on the complete Mg-for-Ca substitution documented in the present work.

While the existence of the MgMgAl-pumpellyite phase is clearly of considerable interest in crystal chemistry, it remains to be discussed as to whether or not it may also be of any petrological significance. Because its chemical composition is so close to that of pyrope (Fig. 3.4), which is a common mineral in the Earth's upper mantle, the immediate answer is a positive one, and one might consider the new phase as a potential receptacle of water for the upper mantle. However, the problem obviously hinges on the thermal stability of MgMgAl-pumpellyite relative to the possible temperature distributions, i.e. geotherms, in the uppermost 150–200 km of the mantle.

The *PT*-diagram of Fig. 3.8 reproduces and extrapolates the stability range of MgMgAl-pumpellyite (Fig. 3.6 and Schreyer, 1988); the pressure axis is also calibrated for depths, assuming a mean rock density of 3.3 g/cm^3. This value happens to be the density of MgMgAl-pumpellyite itself (Table 3.1), but it may also be close to the mean density of the lithosphere (crust + uppermost mantle). It can thus be seen from Fig. 3.8 that MgMgAl-pumpellyite requires a minimum depth for formation in the Earth of about 115 km. Fig. 3.8 also shows a continental geotherm calculated for a heat flow of 40 mWm^{-2} by Pollack & Chapman (1977), and it contains idealized linear geotherms with 5° to 20 °C temperature rise per km. One must conclude from these comparisons that MgMgAl-pumpellyite cannot form along the continental geotherm indicated, because the mantle temperatures are too high by at least 150°C. However, the new MASH-phase could appear, over a depth interval of some 35 km, along a linear gradient of 5 °C/km.

It should be noted that in earlier discussions of the high-pressure mineralogy of the MASH-system in relation to subduction zone metamorphism (Chopin & Schreyer, 1983, Schreyer, 1985, 1988), the linear geotherms were computed for densities of crustal rocks of $2.7–2.8 \text{ g/cm}^3$. In this case, the stability field of MgMgAl-pumpellyite falls into the range with geotherms < 5 °C/km, which metamorphic petrologists do not consider to be verifiable within the Earth. Indeed, the highest reported *PT*-condition for a metamorphic crustal rock, the pyrope–coesite rock of Chopin (1984), clearly lies, with some 30 kbar, 750 °C, outside the MgMgAl-pumpellyite stability field.

For further comparison, Fig. 3.8 also shows the stability fields of some high-pressure phases that were synthesized by other workers in the system $MgO–SiO_2–H_2O$ and that

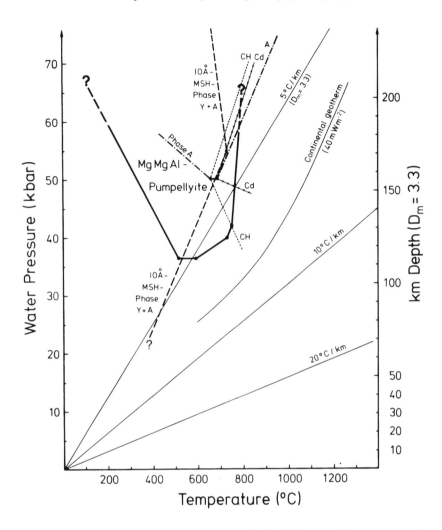

Fig. 3.8.   Summarizing pressure/temperature plot including possible temperature/
depth distributions within the Earth's upper mantle, calculated for a mean
rock density of 3.3 g/cm^3. Light solid lines denote linear geotherms
ranging from 5° to 20 °C/km, and a continental geotherm as given by
Pollack & Chapman (1977). In addition to the extrapolated stability field
of MgMgAl-pumpellyite (see Fig. 3.6), those of some synthetic high-
pressure phases of the system MgO–SiO$_2$–H$_2$O reported by various
authors are indicated as well: 1. the 10 Å phyllosilicate phase (Yamamoto
& Akimoto, 1977); 2. phase A (Akimoto & Akaogi, 1984); 3. hydroxyl-
clinohumite (abbreviated as CH); and 4. hydroxyl-chondrodite (Cd), both
after Akimoto & Akaogi (1984).

were ignored in the phase triangles of Figs. 3.1, 3.4, and 3.7, because they were not
obtained in our experimental study on Al-bearing compositions. These phases are:
1. A 10 Å phyllosilicate phase first synthesized by Sclar *et al.* (1965). According to
   Yamamoto & Akimoto (1977), this phase has a composition near Mg$_3$Si$_4$O$_8$(OH)$_6$,

that is, the composition of talc with additional hydroxyl groups. Based on the synthesis diagrams of these authors, the phase has an upper thermal stability attaining some 700 °C at about 55 kbar. As shown in Fig. 3.8, this limit lies well below the upper thermal stability of MgMgAl-pumpellyite. Thus the 10 Å-phase would appear to be a less likely candidate mineral in a hydrated upper mantle than MgMgAl-pumpellyite.

2. Phase A was first synthesized by Ringwood & Major (1967). According to Yamamoto & Akimoto (1977), it has the very silica-poor composition $Mg_7Si_2O_{14}H_6$. In a more recent abstract, however, Benimoff & Sclar (1984) gave the composition of phase A as $Mg_2SiO_3(OH)_2$, that is forsterite + H_2O. The stability field of phase A as reproduced in Fig. 3.8 from the review by Akimoto & Akaogi (1984) is also located below the linear 5 °C/km geotherm of Fig. 3.8. This would not make its occurrence in a hydrated mantle very likely either. Nevertheless, Benimoff & Sclar (1985) consider 'phase A' important 'with respect to subduction zone processes and the storage and release of water at high pressures', because of the temperature decrease due to endothermic dehydration reactions.

3. There are two hydroxylated phases of the normally F-bearing humite group, which were first synthesized by Yamamoto & Akimoto (1977); these have the compositions $Mg_9Si_4O_{18}H_2$ (= OH-clinohumite) and $Mg_5Si_2O_{10}H_2$ (= OH–chondrodite). At 50 kbar, their stability ranges begin at minimum temperatures of about 650°–700 °C, where they overlap with the field of MgMgAl-pumpellyite (Fig. 3.8). Although no upper thermal stability limits are given in the review of Akimoto & Akaogi (1984), these two phases were shown in the synthesis diagrams of Yamamoto & Akimoto (1977) to persist up to temperatures as high as 1000°–1100 °C. However, according to the compatibility-diagrams presented by Akimoto & Akaogi (1984), these phases cannot coexist with enstatite, due to the existence of the stable assemblage of forsterite with H_2O-fluid. Since the normal material of the upper mantle is known to be more, and not less, siliceous than the forsterite composition, one should expect these hydroxyl humite group phases to be without significance for the mineralogy of the upper mantle. Nevertheless, they are considered by Yamamoto & Akimoto (1977) to represent possible sites for H_2O in the upper mantle. As pointed out by these authors, this would require that the mantle be depleted in silica by rising hydrous fluids, as was first proposed by Nakamura & Kushiro (1974). Since so far only Ti-bearing clinohumites were found in mantle-derived kimberlites and its garnet xenocrysts (McGetchin et al., 1970), and since no samples of mantle rocks could ever be shown to contain less SiO_2 than olivine, it seems rather doubtful that the hydroxyl humites should play the important role implied by earlier authors.

Summarizing the discussion of phase relations in the Al-poor portion of the MASH-system, which is chemically significant as a model for the upper mantle, it would seem that the new phase MgMgAl-pumpellyite actually has the highest probability, if any, of appearing as a mineral in hydrated, low-temperature slabs of the mantle, which are believed to exist, at least as transitional stages, in subduction zones. To substantiate these speculations, obviously more experimental work is necessary on the compatibility relations of MgMgAl-pumpellyite with the MSH-phases. Moreover, the influence of additional components present in mantle material on the stability field of the pumpellyite phase must be studied. One is tempted to predict that partial replacement of Mg in MgMgAl-pumpellyite by Fe^{2+}, and of only the seven-coordinated Mg by some Ca, would extend the stability field towards lower pressures. Thus this field would approach or even overlap the geotherm shown in Fig. 3.8, and a Mg-rich, Ca- and Fe-bearing pumpellyite phase might indeed form as an important mineral in subducted slabs of hydrated mantle, such as those occurring especially within the oceanic lithosphere.

Acknowledgements   We thank G. Andersen for maintaining the high-pressure apparatus, O. Medenbach for optical measurements, G. Werding for water analysis, C. Chopin for providing seeds of synthetic Mg-carpholite, and R. Allmann for helpful advice on X-ray properties. B. Dutrow prepared the SEM picture of Fig. 3.3. M. Pasero did preliminary electron diffraction work on the new phase. F. Seifert and L.P. Plyusnina reviewed the manuscript.

References

Akimoto, S. & Akaogi, M. (1984). Possible hydrous magnesian silicates in the mantle transition zone. In *Material science of the Earth's interior*, ed. I. Sunagawa, pp. 477–80, Tokyo: Terra Scientif. Publ. Comp.

Allmann, R. (1984). Die Struktur des Sursassits und ihre Beziehungen zur Pumpellyit- und Ardennitstruktur. *Fortschr. Mineral.*, **62**, Beih. 1, 3–4.

Benimoff, A.I. & Sclar, C.B. (1984). Further characterization of 'Phase A' in the system MgO–SiO_2–H_2O. *Abstract EOS*, **65**, 16, 308.

—— (1985). The *PT* stability and geophysical significance of phase A, a hydrous pressure-dependent phase in the system MgO–SiO_2–H_2O. *Abstract EOS*, **66**, 18, 422.

Berman, R.G., Engi, M., Greenwood, H.J. & Brown, T.H. (1986). Derivation of internally-consistent thermodynamic data by the technique of mathematical programming: a review with applications to the system MgO–SiO_2–H_2O. *J. Petrol.*, **27**, 1331–64.

Chopin, C. (1984). Coesite and pure pyrope in high-grade blueschists of the Western Alps: a first record and some consequences. *Contrib. Mineral. Petrol.*, **86**, 107–18.

Chopin, C., Klaska, R., Medenbach, O. & Dron, D. (1986). Ellenbergerite, a new high-pressure Mg-Al-(Ti, Zr)-silicate with a novel structure based on face-sharing octahedra. *Contrib. Mineral. Petrol.*, **92**, 316–21.

Chopin, C. & Schreyer, W. (1983). Magnesio-carpholite and magnesiochloritoid; two index minerals of pelitic blueschists and their preliminary phase relations in the model system MgO–Al_2O_3–SiO_2–H_2O. *Amer. J. Sci.*, **283-A**, 72–96.

Coombs, D.C., Nakamura, Y. & Vuagnat, M. (1976). Pumpellyite–actinolite facies schists of the Taveyanne Formation near Loèche, Valais, Switzerland. *J. Petrol.*, **17**, 440–71.

Johannes, W. & Schreyer, W. (1981). Experimental introduction of CO_2 and H_2O into Mg-cordierite. *Amer. J. Sci.*, **281**, 299–317.

Maresch, W.V. & Czank, M. (1985). Optical and X-ray properties of $Li_2Mg_2[Si_4O_{11}]$, a new type of chain-silicate. *N. Jb. Mineral. Mh.*, 1985, 289–97.

McGetchin, T.R., Silver, L.T. & Chodos, A.A. (1970). Titano-clinohumite: a possible mineralogical site for water in the upper mantle. *J. Geophys. Res.*, **75**, 255–9.

Mellini, M., Merlino, S. & Pasero, M. (1984). X-ray and HRTEM study of sursassite: crystal structure, stacking disorder, and sursassite–pumpellyite intergrowth. *Phys. Chem. Minerals*, **10**, 99–105.

Nakamura, Y. & Kushiro, I. (1974). Composition of the gas phase in Mg_2SiO_4–SiO_2–H_2O at 15 kilobars. In *Carnegie Institution, Washington, Year Book 73*, pp. 255–8.

Pollack, H.N. & Chapman, D.S. 1977). On the regional variation of heat flow, geotherms and lithospheric thickness. *Tectonophys.*, **38**, 279–96.

Reinecke, T., Koch-Müller, M. & Langer, K. (1987). Synthesis of sursassite in the system MnO–MgO–Al_2O_3–SiO_2–H_2O. *Terra Cognita*, **7**, 391.

Ringwood, A.E. & Major, A. (1967). High-pressure reconnaissance investigations in the systems Mg_2SiO_4–MgO–H_2O. *Earth Planet. Sci. Lett.*, **2**, 130–3.

Schiffman, P. & Liou, J.G. (1980). Synthesis and stability relations of Mg-Al pumpellyite, $Ca_4Al_5MgSi_6O_{21}(OH)_7$. *J. Petrol.*, **21**, 441–74.

Schreyer, W. (1968). A reconnaissance study of the system MgO–Al_2O_3–SiO_2–H_2O at pressures between 10 and 25 kbar. In *Carnegie Institution, Washington, Yearbook 66*, pp. 380–92.

— (1985). Metamorphism of crustal rocks at mantle depths: high-pressure minerals and mineral assemblages in metapelites. *Fortschr. Mineral.*, **63**, 227–61.

— (1988). Experimental studies on metamorphism of crustal rocks under mantle pressures, *Mineral. Mag.* **52**, 1–26.

Schreyer, W., Maresch, W.V. & Baller, T. (1987). MgMgAl-pumpellyite: a new hydrous, high-pressure, synthetic silicate resulting from Mg-for-Ca substitution, *Terra Cognita*, **7**, 385.

Schreyer, W., Maresch, W.V., Medenbach, O. & Baller, T. (1986). Calcium-free pumpellyite, a new synthetic hydrous Mg–Al–silicate formed at high pressures. *Nature*, **321**, 510–11.

Schreyer, W. & Seifert, F. (1969). High-pressure phases in the system MgO–Al_2O_3–SiO_2–H_2O. *Amer. J. Sci.*, **267-A**, 407–43.

Sclar, C.B., Carrison, L.C. & Schwartz, C.M. (1965). High-pressure synthesis and stability of a near hydronium-bearing layer silicate in the system MgO–SiO_2–H_2O. *Abstract Am. Geophys. Union Trans.*, **46**, 184.

Winkler, H.G.F. (1974). Petrogenesis of metamorphic rocks 3rd ed, Berlin: Springer Verlag.

Yamamoto, K. & Akimoto, S. (1977). The system MgO–SiO_2–H_2O at high pressures and temperatures – stability field for hydroxyl-chondrodite, hydroxyl–clinohumite and 10 Å-phase. *Amer. J. Sci.*, **277**, 288–312.

4

Two-pyroxene thermometry: a critical evaluation

V.I. FONAREV and A.A. GRAPHCHIKOV

Introduction

The widespread natural occurrence of two-pyroxene assemblages and the temperature effect produced by the component redistribution between the constituent minerals have long been used to obtain temperature information, and quite a number of two-pyroxene geothermometers have been derived over the past decade or so. At first there was considerable optimism about applying these thermometers to determine the physico-chemical conditions of natural mineral formation. However, as new experimental data (especially on Fe–Mg minerals) were accumulated and more thorough investigations on particular geological objects became possible, it was found that various thermometers differed considerably and did not fit to the actual geological situation. Rightly, two-pyroxene thermometry was criticised and there were even some quite pessimistic conclusions (Bohlen & Essene, 1979, Lindsley, 1983, Saxena, 1983 and oth.) concerning its potentials. Indeed, a paradoxical situation has developed where the very abundance of temperature values yielded by the many types of two-pyroxene thermometers has resulted in their devaluation. Moreover, this leads to errors in the petrogenetic interpretation, especially when different complexes are compared on the basis of results obtained with different thermometers (depending on the authors' preferences). Clearly, a situation has arisen where it has become necessary to evaluate critically all the thermometers available, choosing only those which yield more reliable and consistent results, relegating the rest to history. Only then will it become possible to achieve correct petrogenetic comparisons of the data obtained by different authors on different geological objects. Experimental evidence accumulated in recent years on component partitioning between ortho- and clinopyroxenes at relatively low temperatures (Fonarev & Graphchikov, 1982a, Lindsley, 1983, and oth.) has formed the basis for such evaluation. These and earlier data for the high-temperature region (Mori, 1978, Turnock & Lindsley, 1981) and end-member compositions (Lindsley & Munoz,

1969, Boyd & Shairer, 1964, Davis & Boyd, 1966, Warner & Luth, 1974, Nehru & Wyllie, 1974, Mori & Green, 1975, 1976, Lindsley & Dixon, 1976, and oth.) allow for fairly accurate interpolations over a wide range of temperatures and compositions and also reasonable extrapolations to temperatures for which no experimental data are available. This approach is far more reliable than broad extrapolations from high-temperature data for limited compositional ranges that are commonly used in deriving two-pyroxene thermometers. Such extrapolations, though they appear reliable thermodynamically, may generate significant errors, since various approximations and simplifying assumptions – often for the same experiments – are required to derive most of the thermodynamic parameters used in the formulation of the thermometers. Therefore, these are not entirely independent parameters but they are largely similar to the usual approximation-based ones that have typical limitations to yield accurate extrapolations.

New two-pyroxene (bi- and monomineralic) geothermometers

Based on experimental results first obtained at $T^* = 750$ °C and 800 °C (Fonarev & Graphchikov, 1982a) a two-pyroxene thermometer was derived and used to determine the physico-chemical conditions of metamorphism in several pre-Cambrian iron formations (Fonarev *et al.*, 1983a, b, 1986, Fonarev, 1987). The thermometer was obtained by empirical interpolation of the authors' experimental results for $T = 750$, 800 °C, earlier high-pressure high-temperature data (Mori, 1978) and the data for end-member compositions (Mori & Green, 1975, Lindsley & Dixon, 1976, Lindsley & Munoz, 1969). At the time, the calibrations could not be refined over the 800–1200 °C range because the $T = 900°$ and 1000 °C results had not yet been published (Turnock & Lindsley, 1981) and the 810–815°, 910° and 990 °C data only became available in 1983 (Lindsley, 1983).

All these data were processed statistically while deriving a new analytical equation for the two-pyroxene geothermometer.

To this end, the following end-member reactions have been used:

(a) $FeSiO_3 = FeSiO_3$
    Cpx   Opx

and

(b) $FeSiO_3 + CaSiO_3 = FeSiO_3 = CaSiO_3$
    Cpx   Opx   Opx   Cpx

---

* Abbreviations: T, temperature; P, pressure; Opx, orthopyroxene (Pbca); Cpx, clinopyroxene (C2/c); Di, diopside; Hd, hedenbergite; En, enstatite; Fs, ferrosilite.

For reaction (a), the dependence of $K_{D(a),P} = X_{Opx}^{Fe}/X_{Cpx}^{Fe}$ on temperature and Cpx composition is defined using the equation

$$K_{D(a),P} = Y_1/X = [(a_0 + b_0 t + c_0 t^2) + (a_1 + b_1 t + c_1 t^2)X + (a_2 + b_2 t + c_2 t^2)X^2 \\ + (a_3 + b_3 t + c_3 t^2)X^3]/n, \tag{1}$$

where $X = X_{Cpx}^{Fe} = Fe/(Fe + Mg + Ca)$ or $Fe/(Fe + Mg + Ca + \sum$ of impurities), $Y_1 = X_{Opx}^{Fe} = Fe/(Fe + Mg + Ca)$ or $Fe/(Fe + Ca + Mg + \sum$ of impurities), a_i, b_i, c_i are regression coefficients, n is empirical correction for pressure, $t = 10^3/T(K)$.

For reaction (b), the Ca-content of clinopyroxene as a function of its iron content and temperature was determined:

$$Y_2 = [(a_0 + b_0 t + c_0 t^2) + (a_1 + b_1 t + c_1 t^2)X + (a_2 + b_2 t + c_2 t^2)X^2 + (a_3 + b_3 t + c_3 t^2)X^3 \\ + (a_4 + b_4 t + c_4 t^2)X^4]/n, \tag{2}$$

where $Y_2 = X_{Cpx}^{Ca} = Ca/(Fe + Mg + Ca)$.

Equation (1) outlines the common thermodynamic relationship

$$K_{D(a),P} = \left(K_{e(a),P=1} \cdot \frac{\gamma_{Cpx}^{Fe}}{\gamma_{Opx}^{Fe}} \right) / \exp \frac{\Delta V \cdot P}{RT}, \tag{3}$$

where $K_{e(a),P=1}$ is the equilibrium constant for reaction (a) at $P = 1$, γ_{Cpx}^{Fe} and γ_{Opx}^{Fe} are the activity coefficients of Cpx and Opx iron end-members, ΔV is volume change of reaction (a).

Equation (3) was used earlier (Graphchikov & Fonarev, 1987) to estimate thermodynamic functions of clinopyroxene. However, its use in thermometry seriously complicates the analytical representation of the geothermometer, provided no significant simplifications are adopted. This applies to models for nonideal minerals, temperature dependences of the reaction constant (or enthalpy and entropy) and some others. In view of this, the present authors favour equations (1) and (2). The regression coefficients a_i, b_i and c_i for the equations concerned have been obtained by least squares as in Table 4.1. Some statistical characteristics of these regressions are also presented in this table (see table footnote). In order to obtain quite a reasonable fit with experimental data, the coefficients a_i, b_i and c_i for equation (1) were obtained for temperatures in the ranges 600–850 °C and 900–1200 °C. The n-parameter value was obtained as follows:

$n = 1 - (0.35 - X) \cdot P \cdot 10^{-5}$ for the lower-temperature hand side of equation (1) and $n = 1 - (0.5 - X) \cdot P \cdot 10^{-5}$ for the higher-temperature hand side of equation (1). For equation (3), $n = 1 - (0.5 - X) \cdot P \cdot 1.2 \cdot 10^{-5}$, where P is pressure, in bars. Rearranging equations (1) and (2) yields the following expression:

$\alpha + \beta t + \gamma t^2 = 0$, where for equation (1)

$$\alpha = a_0 X + a_1 X^2 + a_2 X^3 + a_3 X^4 - Y_1 n, \tag{4}$$

$$\beta = b_0 X + b_1 X^2 + b_2 X^3 + b_3 X^4, \tag{5}$$

Table 4.1. *Regression coefficient values for equations (1) and (2)*

Coefficients	Equation (1) at 600–850 °C	Equation (1) at 900–1200 °C	Equation (2) at 900–1200 °C
a_0	5.6066	− 15.9135	0.0435
a_1	13.6767	112.8452	− 2.0422
a_2	− 23.6366	− 232.7006	14.2661
a_3	− 60.7866	156.3120	− 38.4323
a_4	—	—	34.4206
b_0	− 9.9203	44.2986	0.6859
b_1	− 25.1949	− 313.6564	1.2269
b_2	30.5070	660.4657	− 37.7330
b_3	148.9344	− 446.5655	119.7173
b_4	—	—	− 109.3990
c_0	6.9830	− 27.1202	− 0.2832
c_1	10.0505	212.7500	0.9887
c_2	− 9.1097	− 462.5763	27.5921
c_3	− 85.1719	316.4015	− 97.3991
c_4	—	—	88.8668

Note:
In the temperature range 850–900 °C, equation (1) yields average values obtained for geothermometer equations at low and high temperatures. Equation (1): at $T = 600$–900 °C, $N = 270$, $\Delta Y_1 = 0.006$, $\sigma = \pm 0.029$; at $T = 900$–1200 °C, $N = 167$, $\Delta Y_1 = -0.003$, $\sigma = \pm 0.014$. Equation (2): $N = 183$, $\Delta Y_2 = -0.0003$, $\sigma = \pm 0.012$. N is array of experimental data, $\Delta Y_{1,2}$ are average deviations of X_{Opx}^{Fe} and X_{Cpx}^{Ca} experimental values from calculated ones; σ is mean-square deviation in estimating $\Delta Y_{1,2}$. The correlation coefficients for the a_i, b_i and c_i regression parameters are close to 0.9.

$$\gamma = c_0 X + c_1 X^2 + c_2 X^3 + c_3 X^4, \tag{6}$$

for equation (2)

$$\alpha = a_0 + a_1 X + a_2 X^2 + a_3 X^3 + a_4 X^4 - Y_2 n, \tag{7}$$

$$\beta = b_0 + b_1 X + b_2 X^2 + b_3 X^3 + b_4 X^4, \tag{8}$$

$$\gamma = c_0 + c_1 X + c_2 X^2 + c_3 X^3 + c_4 X^4, \tag{9}$$

While two solutions of this expression exist, only one may be actually used as established from experimental data:

$$T(\mathrm{K}) = \frac{2 \cdot \gamma \cdot 10^3}{-\beta + \sqrt{(\beta^2 - 4 \cdot \gamma \cdot \alpha)}} \tag{10}$$

This relation is an analytical representation of the geothermometers (bimineral and monomineral). Regression (1) and (2) appear to be of limited extrapolation abilities. This is manifested by high correlation coefficients for the a_i, b_i and c_i parameters (in the order of 0.9). The latter experience no marked independent changes, which are bound to occur on extrapolation. It must be noted that the parameters defined simultaneously from the data on mineral equilibria using relatively rigorous thermodynamic expressions also suffer from this disadvantage. However, for the proposed version of the two-pyroxene geothermometer, the errors of extrapolation are only possible at lower temperatures (< 750 °C), since the range 750–1200 °C is characterized by experimental data. In order to minimize these possible errors, at temperatures less than 900 °C, the coefficients a_i, b_i and c_i of equation (1) were calculated in several successive stages. The entropy change (ΔS_a) of reaction (a) was evaluated at the first stages, along with the coefficients concerned. At further stages, regression coefficients were corrected with due regard for linear temperature dependence of the entropy change in the temperature range 600–750 °C. The value obtained, $\Delta S_a = -1.379 \pm 0.376$, is in good agreement with that calculated from independent data (Helgeson *et al.*, 1978, Fonarev, 1987). The geothermometer (bimineralic) should first undergo a test in its low-temperature part and the values are considered valid if they correspond to $T < 850$ °C. Otherwise, the high-temperature portion is calculated and the values of $T \geq 900$ °C are used. Over the 850–900 °C range, the most accurate results are obtained using the averages of the low- and high-temperature portions. Inasmuch as the Ca content of Cpx is largely insensitive to T below 850–900 °C, regression equation (2) was solved only for the 900–1200 °C range. Consequently, this thermometer (monomineralic) is appropriate for temperatures no less than 850–900 °C. Fig. 4.1 shows the iron distribution isotherms between Opx and Cpx at $P = 5$ kbar; Fig. 4.2 presents the change in the composition of Cpx in assemblage with Opx as a function of temperature ($P = 5$ kbar). These diagrams are based on equations (1)–(9) and Table 4.1.

Comparison with experimental data

Various types of two-pyroxene thermometers were compared to evaluate their accuracy and precision. The principal and independent criterion is probably provided by the fit of the calculated temperature to the experimental data available. Petrogenic estimates are also a useful means of testing the thermometer accuracy. However it is often impossible to avoid being subjective owing to the inaccuracy and inconsistency of independent temperature measurements of the conditions of natural mineral formation.

Table 4.2 presents a number of two-pyroxene thermometers that can find a potential quantitative use in metamorphic and relatively shallow magmatic rocks. Not included in the table and the discussion to follow are numerous thermometers derived for low-iron deep-seated mantle rocks (Finnerty & Boyd, 1984, Nickel *et al.*, 1985, Perchuk, 1977, Ryabchikov, personal communication, and oth.). Owing to their specificity these thermometers warrant a special discussion. Neither are discussed most of the earlier graphical thermometers (Ross & Huebner, 1975, and oth.) as well as those that the

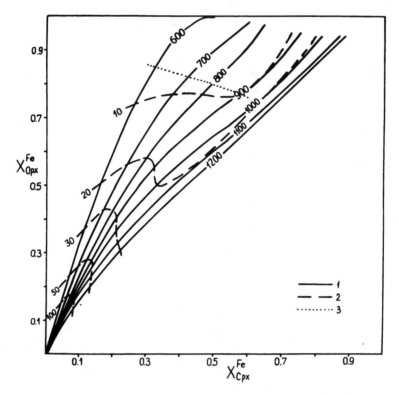

Fig. 4.1. Isotherms (°C) for iron distribution between the coexisting ortho- and clinopyroxenes at $P = 5$ kbar. 1 – distribution isotherms; 2 – isolines for probable errors (°C) of the two-pyroxene thermometer (bimineralic, equations (4)–(6), (10)) with the uncertainty in the clinopyroxene iron content of (± 1 mol. %); 3 – stability limit for the assemblage orthopyroxene + clinopyroxene + olivine + quartz.

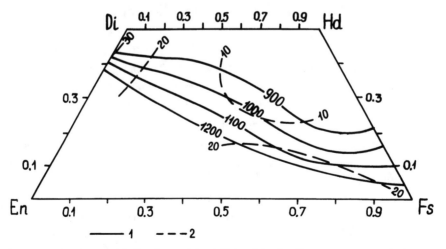

Fig. 4.2. Isotherms (°C) for the Fe, Mg and Ca ratios in clinopyroxenes coexisting with orthopyroxenes at $P = 5$ kbar. 1 – composition isotherms; 2 – isolines for probable errors (°C) of the thermometer (monomineralic, equations (7)–(10)) with the uncertainty in the clinopyroxene Ca content of (± 1 mol %).

authors themselves had revised (Lindsley, 1983, and oth.) or no longer considered to be correct (Saxena, 1976, 1983). The only exception has been made for the thermometers of Henry & Medaris (1976) who found their calibrations to be incorrect and decided against their application (Bohlen & Essene, 1979). In the present work, it was useful to compare them with similar thermometers which still continue in wide use. The thermometers in Table 4.2 fall into two main groups. The first is based largely on experimental results obtained for end-matter compositions (to varying accuracy) and for natural samples at high temperatures (1, 2, 3, 4).* Only limited use of experimental results on component partitioning in ferromagnesian compositions has been made for this group. The thermometers in the second group (5–13) were derived with a fuller coverage of experimental data for both the end-member compositions and CFMS systems. The fullest possible use of such data was made in the formulation of thermometers (12, 13) and (9, 10, 11). It will be noted that for the last three the results from the relatively low-temperature studies (Fonarev & Graphchikov, 1982a) were either not used at all (9) or used only in part (10, 11). Much of the experimental evidence for the CFMS system was not used in the formulation of thermometers (5, 7, 8). Dividing the thermometers into these groups is believed to reflect the basic difference between them in that the first-group thermometers were derived using a considerable extrapolation of the experimental results with respect to temperature and compositions of the coexisting minerals whereas the second group was obtained mainly by interpolation, which is normally more reliable. Although there were attempts to use the thermodynamic approach (of varying degree of complexity) for the handling of experimental data, hardly anyone has managed to do without empirical relationships. Obviously, this results from the thermodynamic apparatus being not perfect enough to give a correct representation and reliable extrapolation of the mineral compositions in systems featuring multicomponent isomorphism. Finally, most of the analytical expressions for the thermometers in Table 4.2 are written explicitly in terms of temperature, whereas the others require solution of a set of non-linear equations, which undoubtedly introduces additional complexities in their application.

In Fig. 4.3, the results obtained from several pyroxene thermometers are compared with experimental data in the coordinates $T_{th}^{\circ C}$ (the thermometer value) $- X_{Opx}^{Fe}$ (the iron content of Opx in each run). The reference values used were the results from the experimental studies in the system CFMS at 750° and 800 °C (Fonarev & Graphchikov, 1982a), 810–990 °C (Lindsley, 1983) and 1200 °C (Mori, 1978). These studies cover a fairly large range of temperatures and compositions of coexisting minerals and conform to modern experimental standards. However, not all run products can be regarded as equilibrium compositions. This reflects the actual situation and the complexities involved into minerals – solid solutions studies, particularly the difficulty of establishing the equilibrium compositions in those run products that contain both stable and compositionally metastable minerals. The interpretation of the results is

* Numbers of thermometers as in Table 4.2.

Table 4.2. *Characterization of some two-pyroxene geothermometers*

No. reference in the text 1	Author(s) 2	Experimental basis[a] 3	Method of experimental data processing[b] 4	Type of geothermometer[c] 5	Authors' constraints 6
1	Wood & Banno (1973)	CMS(2); CFS(1); NS; NK; CFMS(1)	1 + 2	Aa	Accurate to ±70°
2	Henry & Medaris (1976)	CMS(4); CFS(1); CFMS(1); NS; NK	1 + 2	Aa	
3	Henry & Medaris (1976)	CMS(7); CFS(1); CFMS(1); NS; NK	1 + 2	Aa	
4	Wells (1977)	CMS(1–7); CFS(1); CFMS(1); CMAS(1); NS; CFMATS(1)	1 + 2	Aa	Accurate to ±70° in the 785–1500 °C range
5	Fonarev & Graphchikov (1982b)	CMS(5,7); CFS(1); CFMS(3,7)	2	Aa	P to 30 kbar; the T range 600–1200 °C
6	Kretz (1982)	CMS(7); CFMS(4); NK	1 + 2	Aa	Accurate to ±60°
7	Slavinskii (1983)	CMS(4,5,7); CFS(1); CFMS(2,3,5,6); CMAS(2)	1	Aa	
8	Slavinskii (1983)	CMS(4,5,7); CFS(1); CFMS(2,3,5,6); CMAS(2)	1 + 2	Ab	
9	Bertrand & Mercier (1985)	CMS(3,4,5,7,8); CFMS(3,5,8); CMAS(2,3); NS	1 + 2	Ab	for mantle lherzolites
10,11	Davidson & Lindsley (1985)	CMS(1–7); CFMS(3,4,5,7,8,9)	1 + 2	G	
12	Fonarev & Graphchikov (the present work, equation (1))	CMS(7); CFS(1); CFMS(3,5,7,8)	1 + 2	Aa	P to 20 kbar; the T range 600–1200 °C
13	Fonarev & Graphchikov (the present work, equation (2))	CMS(7); CFS(1); CFMS(3,5,7,8)	1 + 2	Aa	P to 20 kbar; the T range 900–1200 °C

Notes:

[a] *System CMS:* (1) Boyd & Schairer (1964), $T = 800–1400$ °C, $P = 0.001–1$ kbar; (2) Davis & Boyd (1966), $T = 950–1800$ °C, $P = 30$ kbar; (3) Warner & Luth (1974), $T = 900–1300$ °C, $P = 2, 5, 10$ kbar; (4) Nehru & Wyllie (1974), $T = 1100–1500$ °C, $P = 30$ kbar; (5) Mori & Green (1975), $T = 900–1500$ °C, $P = 5–40$ kbar; (6) Mori & Green (1976), $T = 1500–1700$ °C, $P = 20$ and 30 kbar; (7) Lindsley & Dixon (1976), $T = 850–1400$ °C, $P = 15$ and 20 kbar; (8) Brey & Huth (1983), $T = 1100–1500$ °C, $P = 40–60$ kbar. *System SFS:* (1) Lindsley & Munoz (1969), $T = 700–1000$ °C, $P = 20$ kbar. *System CFMS:* (1) Smith (1972), $T = 850–950$ °C, $P = 15$ kbar; (2) Lindsley *et al.* (1974), $T = 810$ °C, $P = 15$ kbar; (3) Mori (1978), $T = 1200$ °C, $P = 30$ kbar; (4) Podpora & Lindsley (1979), $T = 970, 1040$ °C, $P = 1$ kbar and $T = 825$ °C, $P = 2$ kbar; (5) Turnock & Lindsley (1981), $T = 900, 1000$ °C, $P = 1$ kbar; (6) Lindsley (1981), $T = 980$ °C, $P = 15$ kbar; (7) Fonarev & Graphchikov (1982a), $T = 750$ and 800 °C, $P = 2.9$ kbar; (8) Lindsley (1983), $T = 810–815$ °C, 910 and 990 °C, $P = 15$ kbar; (9) Lindsley & Andersen (1983), $T = 800$ °C, $P = 1$ and 2 kbar, $T = 1100$ and 1200 °C, $P = 0$. *System CMAS:* (1) Akella (1976), $T = 1000–1500$ °C, $P = 26–44$ kbar; (2) Perkins & Newton (1980), $T = 900–1100$ °C, $P = 20$ and 30 kbar; (3) Yamada & Takahashi (1983), $T = 1200–1500$ °C, $P = 50–100$ kbar. *System CFMATS:* (1) Akella & Boyd (1972): $T = 1050$ °C, $P = 25$ kbar; $T = 1100$ °C, $P = 40$ kbar. *System NS:* experiments on natural samples (mainly at $T > 1000$ °C). See references in the original papers. *System NK:* analysis of natural complexes using the independent temperature estimates.

[b] 1 – thermodynamic treatment of the results was performed by the author; 2 – no thermodynamic treatment, the relations are empirical.

[c] A – analytical; a – solved in terms of temperature; b – a system of equations to be solved; G – graphical.

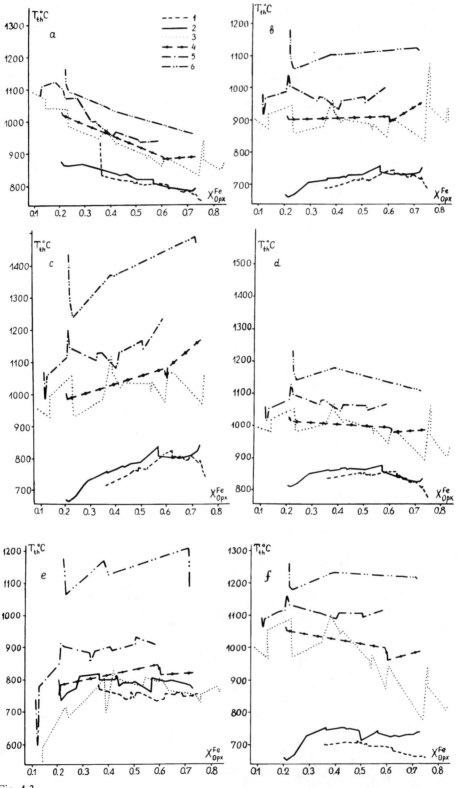

Fig. 4.3.

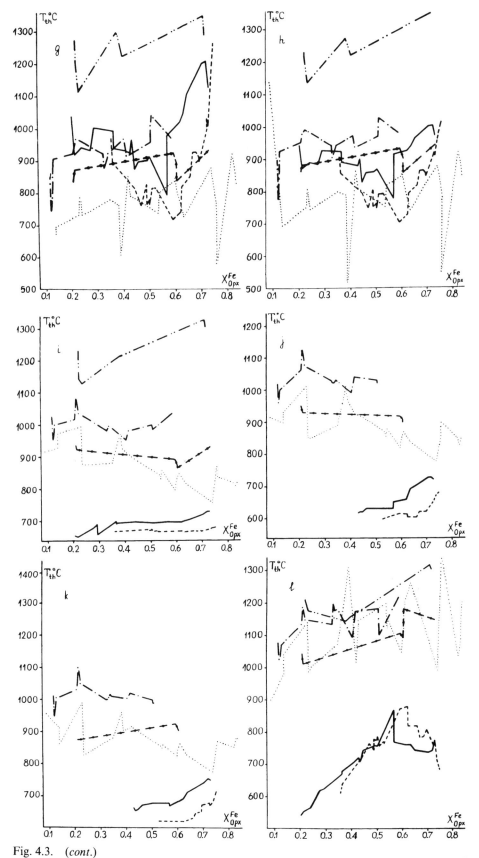

Fig. 4.3. *(cont.)*

75

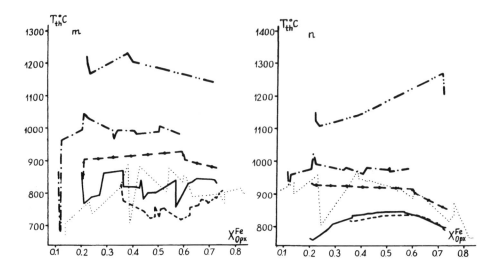

Fig.4.3.
(*cont.*)   Comparison of the temperatures produced by two-pyroxene thermometers with experimental results. 1, 2 – 750 and 800 °C, respectively, $P = 2.9$ kbar (Fonarev & Graphchikov, 1982a); 3, 4, 5 – 810–815, 910 and 990 °C, respectively, $P = 15$ kbar (Lindsley, 1983); 6 – 1200 °C, $P = 30$ kbar (Mori, 1978). (a)–(n) – dependences for different thermometers (numbers as in Table 4.2): (a) – 1; (b) – 2; (c)– 3; (d) – 4; (e) – 5; (f) – 6; (g) – 7; (h) – 8; (i) – 9; (j) – 10; (k) – 11; (l) – Ryabchikov (personal communication); (m) – 12; (n) – 13.

further complicated by their probable ('imaginary') heterogeneity – the grains are too small to allow accurate microprobe sampling (Fonarev & Konilov, 1986). As a result, the compositions of the adjacent heterogeneous sections within the same or several crystals become averaged (to a varying degree depending on their relative abundances in the microprobe excitation zone. All this shows up even at a brief inspection of Fig. 4.3, as temperature fluctuations that can be considerable even for Opx having close iron contents.

Different researchers take different approaches to the interpretation of results from such experiments as regards the equilibrium–non-equilibrium of phases. However, allowing for systematic errors inherent in the experimental methods, Fig. 4.3 on the whole gives a true statistical representation (within the experimental uncertainty) of the mineral equilibrium relations and can be used to evaluate the accuracy of various thermometers. The most important findings are outlined below.

Geothermometer 1 (Wood & Banno, 1973), covers the experimental data for $T = 800, 910$ and 990 °C (accurate to within ± 40 °C and only for intermediate and iron-rich compositions of coexisting pyroxenes ($X_{Opx}^{Fe} > 0.4$). The overall temperature range 750–1200 °C is too narrow and at $X_{Opx}^{Fe} > 0.2$ the thermometer is inaccurate below 800 °C (at least 50–100 °C too high) and above 1200 °C (100–200 °C too low). But even within this range, the difference between the 810–815°, 910 and 990 °C data is not distinct enough. In general, the visible slope of the isolines in Fig. 4.3 towards the X_{Opx}^{Fe}

axis of the orthopyroxene implies a considerable variance between the analytical expression of the thermometer and experimentally determined ratios of component partition between the minerals. Recommended application: the 800–1000 °C range (accurate to ± 40 °C), $X_{Opx}^{Fe} > 0.4$.

Geothermometer 2 (Henry & Medaris, 1976). It is formulated on the basis of the Nehru & Wyllie (1974) solvus data and has a fairly accurate fit only to the experimental results at 910° and 990 °C and also at 750 °C over a limited compositional range of $X_{Opx}^{Fe} \approx 0.5$–0.7. The thermometer tends to underestimate temperatures for the 800 and 1200 °C experiments by 80–100 °C and overestimates the 810–815 °C results by the same value. Recommended application: the 900–1000 °C range.

Geothermometer 3 (Henry & Medaris, 1976). It is based on the Lindsley & Dixon data (1976) and is appropriate only for $T = 800$ °C at $X_{Opx}^{Fe} > 0.4$. All other isotherms give greatly overestimated values. Recommended application: not recommended for use.

Geothermometer 4 (Wells, 1977). It corresponds to the experimental data over the 750–990 °C range, and generally gives temperatures 60–100 °C too high. At 1200 °C, the results are underestimated by approximately the same value although the uncertainty is reasonable for this temperature. Recommended application; the temperatures close on 1200 °C.

Geothermometer 5 (Fonarev & Graphchikov, 1982b). It provides a highly accurate fit to the isothermal experiments at 750° and 800 °C and 810–815 °C at $X_{Opx}^{Fe} > 0.35$. For 1200 °C, the values, although somewhat underestimated, remain within the uncertainty limits for this temperature range (to 60 °C). The thermometer is not appropriate for the 910 and 990 °C data. Recommended application: below 800–820 °C.

Geothermometer 6 (Kretz, 1982) (clinopyroxene). It is only appropriate for the 1200 °C data. All other isotherms are greatly overestimated except the 750 and 800 °C ones for which the thermometer produces lower temperatures. Kretz's other thermometer based on the Mg and Fe distributions between Opx and Cpx yields temperatures that grossly exceed all the permissible values (by 300–700 °C on the average). Recommended application: not recommended for use.

Geothermometers 7 (iron–magnesian) and 8 (iron–magnesian–calcium) (Slavinskii, 1983). They yield widely varying results when applied to experimental isotherms. Nevertheless, except for 750° and 800 °C, the agreement between the calculated and experimental results is satisfactory for all temperatures. It will be noted that the 750 and 800 °C experimental data unlike other temperatures were not used in the formulation of these thermometers. It thus follows from the comparison that the extrapolations from these thermometers are of only limited accuracy and the temperatures for the region $T < 800$ °C are greatly overestimated. Recommended application: temperatures in excess of 900 °C where the statistical representation for the material is large.

Geothermometer 9 (Bertrand & Mercier, 1985). It gives a good fit to experimental data at 900–1200 °C. However, extrapolations to lower temperatures produce considerably underestimated values (experimental results for 750 and 800 °C). Nor had the authors been able to obtain a satisfactory fit to the 810–815 °C experiments for the region of $X_{Opx}^{Fe} < 0.5$ although these results were used in their analytical derivations. Recommended application; temperatures in excess of 900 °C.

Geothermometers 10 and 11 (Davidson & Lindsley, 1985) (monomineralic and bimineralic). They are practically identical to the previous one: they shows a good agreement with the experimental results for 900–1200 °C but underestimate the values in the lower temperature region (750–800 °C). For the 800–815 °C data, the agreement is good only for $X_{Opx}^{Fe} > 0.5$. In the more magnesian region, the values calculated for the 810–815 °C data coincide with those for 910 °C. Recommended application: above 900 °C.

Geothermometer 12 (the present work, equations (4)–(6), (10)). The only thermometer that has a good fit to all experimental results available for the range 750–1200 °C. Although some of the experimental data (990, 750 °C; highly magnesian and iron-rich compositions; 910 °C and oth.) fall off the general pattern, this does not disturb the overall picture and is certainly due to experimental errors. Even the 810–815 °C isotherm shows a reasonably good fit (except for the magnesian compositions at $X_{Opx}^{Fe} < 0.2$) unlike most of the other thermometers. This stems from a close correspondence between the 800 °C (Fonarev & Graphchikov, 1982a) and 810–815 °C (Lindsley, 1983) experimental results on the iron content in the coexisting pyroxenes, used to derive thermometer 12. In this regard, this thermometer is different from several others which are based on the Ca content of the clinopyroxenes (for example, 9, 11, 13). Recommended application: the interpolation range of 750–1200 °C with extrapolations to lower (by no more than 100–150 °C) and slightly higher temperatures.

Geothermometer 13 (the present work, equations (7)–(9), (10)). It provides highly accurate fit to experimental results within 900–1200 °C. The 750 and 800 °C isotherms place closely together and thus imply that the thermometer becomes less accurate in this region. As with other thermometers, the values for the 810–815 °C isotherms are greatly overestimated especially for intermediate and magnesian compositions. Recommended application: temperatures in excess of 850–900 °C.

As was noted above, we are not considering the numerous two-pyroxene thermometers largely based on experimental data for the CMS system and natural material, intended for application to deep-seated high-temperature rocks of mantle origin. Clearly, these thermometers can not be used in lower-temperature (less than 1000–1200 °C) terranes with fairly high Fe contents of the minerals. As an illustration, in Fig. 4.3 the Ryabchikov (personal communication) calibration is compared with experimental data. The figure is self-evident and does not call for a special comment. Consequently, the following conclusions can be drawn from the comparison of different geothermometers with experimental calibrations.

(1) It appears that thermometers 1 (Wood & Banno, 1973), 2 and 3 (Henry & Medaris, 1976), 4 (Wells, 1977), 6 (Kretz, 1982) and also thermometers derived specifically for mantle rocks can not yield accurate quantitative estimates in shallow magmatic and metamorphic terranes with fairly high Fe contents of the minerals. Some

of these thermometers do not agree with experimental results for the system CFMS while others are only appropriate within a narrow composition and temperature range (1, 2). However, the warning from Wood & Banno (1973) is all too often disregarded: their thermometer is only valid within the composition and temperature range of the experiments from which the thermometer was derived. Outside that range there will be errors.

(2) Several thermometers produce highly accurate values for temperatures above 900 °C: 13 (the present work, monomineralic), 12 (the present work, the high-temperature portion of the bimineralic thermometer), 9 (Bertrand & Mercier, 1985), 10 and 11 (Davidson & Lindsley, 1985). The values obtained with these thermometers agree closely with experimental data and within the error limits any of them or their averages can be used. It will be noted, however, that the equations for thermometers 10 and 11 do not permit direct temperature estimates and in Davidson & Lindsley (1985) they are given in graphical form which complicates the temperature calculations. Thermometers 7 and 8 (Slavinskii, 1983) may also be used for these temperatures. However, their performance is not reliable due, as will be shown later, to their high sensitivity to errors (even though permissible) in analysed compositions of equilibrium minerals. This can significantly affect the results of these thermometers, especially where the statistical representation of the material is not large.

(3) Below 850(900) °C, consistent temperatures are obtained only with thermometers 5 (Fonarev & Graphchikov, 1982b) and 12 (the present work, the low-temperature part of the bimineralic thermometer). These thermometers agree well with experimental data of Fonarev & Graphchikov (1982a) and Lindsley (1983) (for the iron-contents of coexisting ortho- and clinopyroxene).

(4) At 810–815 °C (Lindsley, 1983), the calcium content was much lower than at 800 °C (Fonarev & Graphchikov, 1982b, Lindsley & Andersen, 1983) and practically identical to that at 910 °C (Lindsley, 1983). This results from systematic errors in the 810–815 °C experiments (analytical, determination of equilibrium compositions etc.). None of the thermometers that depend on the Ca content could produce a good fit to these data. This conclusion holds for the thermometers based on the 810-815 °C experiments including the calibrations of Davidson & Lindsley (1985), and Lindsley himself has noted so in his paper (1983).

Sensitivity of geothermometers

By sensitivity of a geothermometer we understand deviations in the temperature pattern it produces with variations in the compositions of coexisting minerals. The possible reasons for such variations could be the usual uncertainties of experimental data and the approximations used to calibrate the thermometers, as well as analytical errors (even within the error limit) in mineral compositions. Current analytical methods for mineral compositions do not normally yield results better than $\pm 1 - 2$

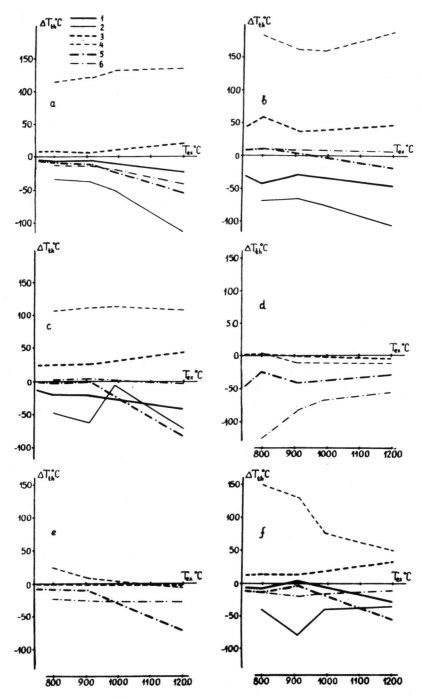

Fig. 4.4. 'Sensitivity' of geothermometers (see explanations in the text). 1–6 – deviations in the thermometer values with a ±1 mass % change in the element concentration (solid lines for experiments with $X_{Opx}^{Fe} \approx 0.72$; fine lines – for experiments with $X_{Opx}^{Fe} \approx 0.23$): 1,2 – FeO in Opx; 3,4 – FeO in Cpx; 5,6 – CaO in Cpx. (a)–(f) – dependences for different thermometers (numbers as in Table 4.2): (a) 5; (b) 7; (c) 8; (d) 9; (e) 13; (f) 12.

mass % for the major rock-forming components. Hence, the accuracy of two-pyroxene thermometers is, like that of others, limited both for temperature and mineral composition. As an example, Fig. 4.4 presents the calculated sensitivities for some of the most reliable (see above) thermometers as applied to certain magnesian ($X_{Opx}^{Fe} \approx 0.23$) and iron ($X_{Opx}^{Fe} \approx 0.71$) Opx and Cpx assemblages from experimental studies aimed at establishing coexisting compositions. It was interesting to assess how each thermometer (ΔT_{th}) would respond to variations in the starting mineral compositions. The calculations were separate for the following cases: (1) ± 1 mass % change in the FeO concentration of clinopyroxenes; (2) ± 1 mass % change in the FeO concentration of orthopyroxenes; (3) ± 1 mass % change in the CaO concentration of clinopyroxenes. Clearly, each of the thermometers responded differently to changing chemical composition, in accordance with its analytical expression. For example, the finite formulas for thermometers 9 and 13 do not contain the term for the Fe content of Opx, therefore variations in Fe of the mineral would not affect the temperature pattern these thermometers yield. In contrast, the thermometers 5, 7, 8 and 12 should record the variations (more or less accurately) of all the components because they enter directly or indirectly (when recalculated) into the analytical expression. Fig. 4.4 presents the calculated values. The accuracy of thermometers 9 and 13 depends practically only on the Ca content in clinopyroxenes. Thermometer 13 yields fairly accurate results for magnesian compositions (to 20–30 °C) whereas in the high iron portion $\geqslant 1100$ °C, errors may be 50° or greater. Thermometer 9 is accurate to ± 30–40° for iron compositions whereas in the magnesian portion scatter in the results becomes considerable (50–70 °C at $T > 1000$ °C and > 80–100° at lower temperatures). It is seen from Fig. 4.4 that thermometers 5, 7, 8 and 12 are not sufficiently precise in the magnesian region of compositions where errors can be in excess of 100 °C (with only 1 mass % uncertainty in the mineral composition). These thermometers are particularly sensitive to the Fe content of clinopyroxenes but variations in the orthopyroxene composition also affect the accuracy of these thermometers (~ 50 °C and more). In the region of high-iron compositions and relatively low temperatures (< 1000 °C) the best results (with only ± 20 °C error) are obtained from thermometers 5 and 12. Higher uncertainties for this region were recorded in the values produced by thermometer 8 (to 30 °C) and especially 7 (to 40–50 °C). With increasing temperature, the accuracy and precision of these thermometers deteriorate and errors may become as high as 50 °C (8) at 1200 °C. Note that most of the errors of thermometers 5, 8 and 12 at these temperatures stem from analytical errors in the Ca content of the clinopyroxenes. Fig. 4.1 presents the isolines of possible temperature uncertainties of thermometer 12 for a ± 1 mass % departure of the Opx iron content from the equilibrium. It is seen that for compositions with $X_{Opx}^{Fe} < 0.1$, this thermometer can be accurate to 50–100 °C at best. The same holds to a greater or lesser degree for all other modifications of the thermometer. Therefore, numerous attempts to apply two-pyroxene thermometry to highly-magnesian compositions seem hardly worthwhile even though certain refinements were made to account for the effect of additional components (Na, Al, etc.). With more iron-rich compositions and at relatively low temperatures (< 1000 °C), the thermometer becomes less sensitive to analytical errors in equilibrium compositions with the result that the precision and accuracy of the thermometer increase (Fig. 4.1).

The effect of additional components

Another important aspect of the present analysis is assessment of the effects of additional components, especially Al, Cr, Mn, Na, on the accuracy and precision of thermometers. There have been attempts (thermometers 1, 2, 3, 4 and oth.) to allow for the impurities using the component redistribution on the non-equivalent cation positions in pyroxenes. However, there is no correct solution to this problem at present because the necessary thermodynamic and experimental data are lacking. Studies involving natural samples are unfortunately of little value as they mainly concentrated on high magnesian compositions and high temperatures (> 1050 °C) and were not systematic. Therefore an empirical approach has been used to evaluate the effect of additional components on the accuracy of thermometers. Fig. 4.5 shows the dependence of the thermometer values on the sum of additional compounds $\sum M = \sum(Al_2O_3,$ MnO, TiO_2, Na_2O, K_2O etc.) in Cpx from the Adirondack granulites (analytical data by Bohlen & Essene, 1979). The data were processed statistically to obtain the mean measured temperatures (T_{mean}), standard deviations (σ_y), correlation coefficients for the thermometer values and the sum of impurities (r_{xy}) as well as linear regression coefficients (a and b). Using feldspar and Fe-Ti oxide thermometry, Bohlen & Essene (1979) have determined the temperature zoning of metamorphism and located the 650–750 °C isotherms. These authors strongly criticized thermometers (1, 2, 3, 4) which yield too high and inconsistent values. Fig. 4.5 shows the data separately for the samples which, according to Bohlen & Essene (1979), conform to temperatures 600–700 °C, 700–750 °C and > 750 °C. Two-pyroxene thermometers (5, 7, 8, 9, 12) also record the temperature differences between the first two groups of analyses.* These differences are not too large though and fall within the standard error limits (Fig. 4.5). Thermometers 5, 8, 12 yield metamorphic temperatures of about 700 °C. Thermometers 7 and 9, respectively, yield markedly lower (~ 650 °C) and slightly higher (~ 730 °C) values.† However, the fact that two-pyroxene thermometry fails to record regular and consistent temperature differences across the area provides no ground for criticism: it relates on the one hand to uncertainties in the values obtained with the feldspar and Fe–Ti thermometers used to determine the metamorphic zoning in the region, and on the other to analytical errors in the mineral equilibrium compositions. For example, specimens SR-29 and SR-31 (Bohlen & Essene, 1979) from closely spaced outcrops,

---

* The number of analyses is not representative enough to allow a statistical treatment for the latter (Bohlen & Essene, 1979).

† To compare with the same samples (Bohlen & Essene, 1979), average temperatures from other thermometers for the same area are: 1 – 785 °C; 2 – 703 °C; 3 – 760 °C; 4 – 817 °C.

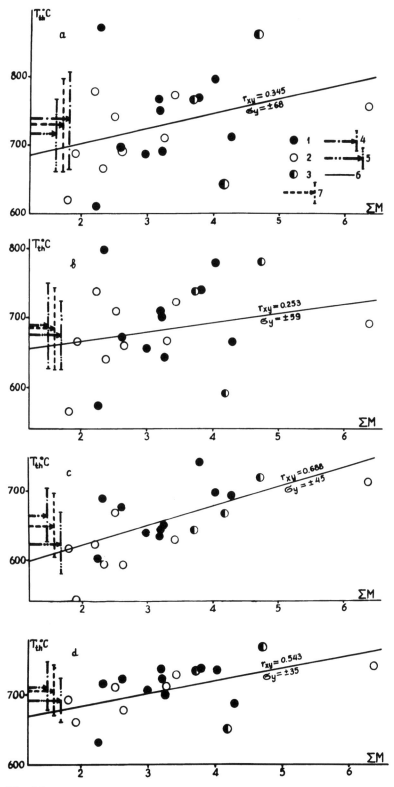

Fig. 4.5.

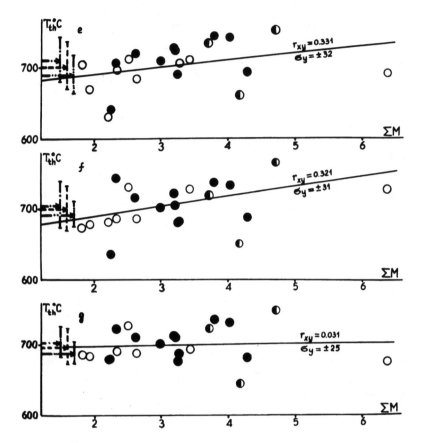

Fig. 4.5. Correlation between the thermometer values and the sum of additional
(*cont.*) components in Cpx ($\sum$M, mass %) for Adirondack granulites. 1–3–
thermometer values for samples grouped with respect to the temperatures
of formation (Bohlen & Essene, 1979): above 750 °C (1), 700–750 °C (2)
and 600–700 °C (3). 4,5 – average temperatures with standard deviations
obtained for the thermometers in the first and second groups, respectively.
6,7 – linear regression (6) and average temperatures with standard
deviations (7) for all samples. Thermometer values (Table 4.2): (a) 7; (b) 8;
(c) 9; (d) 5 (with $X_{Opx(Cpx)}^{Fe(1)}$); (e) 5 (with $X_{Opx(Cpx)}^{Fe(2)}$); (f) 12 (with $X_{Opx(Cpx)}^{Fe(1)}$); (g) 12
(with $X_{Opx(Cpx)}^{Fe(2)}$).

while having only moderate differences in their iron contents of Cpx (3.1 mole %) are
quite different in their iron contents of Opx (12.5 mole %) and calcium contents of Cpx
(3.1 mole %) which naturally leads to dissonant temperatures for these specimens (Fig.
4.5). A similar picture has been observed in some other cases (samples BM-13 and TL-
4, ET-1 and ET-10 and oth.). Moreover, in certain cases the CaO content of
clinopyroxene is too high (to 50.4 mass % of the wollastonite component).

Since two-pyroxene thermometry does not recognize a well-defined metamorphic
zoning across the region under study it seems reasonable to examine the effect the sum
of additional components in minerals (clinopyroxenes) has on the calculated temperatures for all samples analysed in Bohlen & Essene (1979).

Calculations show (Fig 4.5) that all thermometers including those for which there were attempts to allow for these effects show only slight positive correlation with the content of other components. The correlation coefficient is greatest for thermometer 9 and lowest for 7, 8 and 12. The weak correlation in the first two cases is believed to stem from a large scatter in the temperatures for which standard errors are greatest (± 68 and $\pm 59\,°C$, relatively). It will be noted that for practically all thermometers the effect of 3–4 mass % contamination (in Cpx) is well within the standard error of temperature measurements. In other words, the resulting errors are comparable to (thermometers 5, 9) or much smaller than (thermometers 7, 8, 12) those due to analytical errors in the major components (Fe, Ca and Mg) or non-equilibrium in mineral pairs. The errors generally depend on the 'sensitivity' of the thermometers and, according to Fig. 4.5, are greatest for thermometers 7 and 8 and smallest for thermometers 5 and 12. The latter two were used to calculate temperatures with the iron content of the minerals taken not only as $X_{Opx(Cpx)}^{Fe(1)} = Fe/(Fe+Mg+Ca)$ (Fig. 4.5) but also as $X_{Opx(Cpx)}^{Fe(2)} = Fe/(Fe+Mg+Ca+\sum M)$, where $M = Ti$, Al^{VI}, Mn, Na, K, i.e. allowing for the additional components. It is seen in Fig. 4.5 that this representation makes the correlation between the calculated temperatures and the sum of the impurities in Cpx disappear completely in thermometer 12; the correlation coefficient for thermometer 5 becomes much lower. The standard errors of measurement tend to become smaller whereas the mean temperatures remain practically unaltered (698 °C instead of 701 °C in the former and 703 °C instead of 705 °C in the latter).

Qualitatively close relations have been obtained for Bushveld pyroxenes from the Bethal area (analytical data of Buchanan, 1979). Nearly all thermometers except 13 reveal a positive correlation with the contents of additional components in Cpx (Fig. 4.6). The strongest correlation is found in thermometers 1 and 9 whereas thermometer 13 shows no correlation at all. Note also that standard deviations (σ) differ markedly with the thermometers, due to uncertainties in the mineral equilibrium compositions. The uncertainty is greatest for thermometers 7, 8, 9 and smallest for thermometer 13. When the iron content is expressed as $X_{Cpx(Opx)}^{Fe(2)}$, the values of r_{xy} and σ_y for thermometer 12 become a little smaller (Fig. 4.5, 4.6). In contrast, expressing the Ca content of Cpx as $X_{Cpx}^{Ca(1)} = Ca/(Fe+Mg)Ca+\sum M)$ instead of $X_{Cpx}^{Ca(1)} = Ca/(Fe+Mg+Ca)$ for thermometer 13 results in greater r_{xy} and σ_y (Fig. 4.6) which makes this correction inappropriate. As for the Adirondack metamorphic rocks, the average temperatures for thermometers 8, 9, 12 and 13 for the Bethal area on the whole agree among themselves despite the specific effects of additional components (890–910 °C). Somewhat higher temperatures are obtained from thermometer 7 (930 °C) and particularly 1 and 4 (960° and 1000 °C, respectively). However it is clear that the close agreement of the results from thermometers 8, 9, 12 and 13 has only become possible owing to satisfactory statistical representation of the stony material. Separate or poorly-represented determinations using the more 'sensitive' thermometers (in this case, involving the highest σ_y) can be inaccurate and imprecise. In this regard, thermometer 13 (with $X_{Cpx}^{Ca(1)}$) seems to yield the most reliable values for the Bethal area. It is

V.I. Fonarev and A.A. Graphchikov

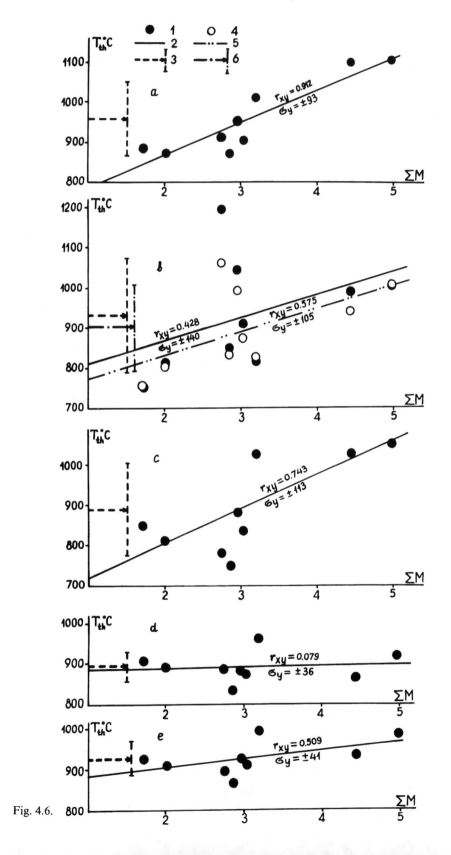

Fig. 4.6.

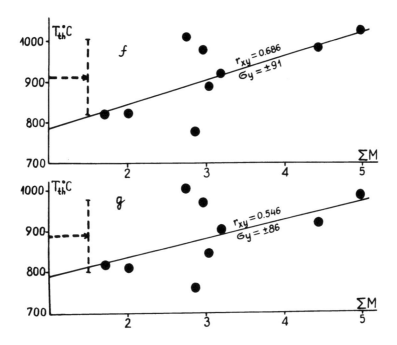

Fig. 4.6.   Correlation between the thermometer values and the sum of additional
(cont.)   components in Cpx ($\sum M$, mass %) for samples from the Bethal area,
Bushveld. 1–3: temperature values (1), linear regression relations (2) and
average temperatures with standard deviations (3) for the thermometers
(Table 4.2): 1(a), 7(b), 9(c), 13(d) with $X_{Cpx}^{Fe(1)}$ and $X_{Cpx}^{Ca(1)}$, 13(e) with $X_{Cpx}^{Fe(2)}$
and $X_{Cpx}^{Ca(2)}$, 12(f) with $X_{Opx(Cpx)}^{Fe(1)}$, 12(g) with $X_{Opx(Cpx)}^{Fe(2)}$. 4–6: correspond to 1–3
for thermometer 8(b).

noteworthy that here as in the preceding case, the effect of impurities on most
thermometers is within the standard deviations of the means (at least, at 'additional'
contents not higher than 3–4 mass %).

In some rare cases, several thermometers were characterized by a negative correlation between the calculated values and impurity contents. This has been found in thermometers 1, 4, 7, 8 as a result of the statistical treatment of the Immega & Klein (1976)
data and is explained by a higher impurity content in orthopyroxene relative to
clinopyroxene. For the same reason, the iron content of minerals expressed as $X_{Cpx(Opx)}^{Fe(2)}$
instead of $X_{Cpx(Opx)}^{Fe(1)}$ leads to considerably higher values of r_{xy} and σ_y in the temperatures
yielded by thermometer 12 and therefore this approach becomes irrational.

Conclusion

The material presented in this paper shows that two-pyroxene thermometry is a fairly
efficient method for determining the conditions of mineral formation in metamorphic
and certain magmatic complexes. For this method to be successful it is necessary, based
on a comprehensive evaluation, to discard some of the conventional two-pyroxene

thermometers that fail to fit with the latest experimental data and to replace them with accurately calibrated thermometers yielding consistent results over their respective 'performance' ranges. Only then will it become possible to avoid both erroneous conclusions concerning the thermal regimes of natural processes in different areas and the more general conclusions on these regimes being heterogeneous in different regions, as inferred by different authors using various thermometers.

A thorough analysis has shown that thermometers 1 (Wood & Banno, 1973), 2 and 3 (Henry & Medaris, 1976), 4 (Wells, 1977), 6 (Kretz, 1982) and others which had been constructed before much of the more recent experimental data on the Fe–Mg pyroxenes became available, are not particularly useful as quantitative thermometers over the 700–1200 °C range. The fact that some of them (1, 4) show a good fit to experimental results over limited temperature and mineral composition regions does not guarantee against errors outside these limits. Temperatures in excess of 900 °C can be measured using thermometers 12 and 13 (the present work), 9 (Bertrand & Mercier, 1985), 7 and 8 (Slavinskii, 1983), 10 and 11 (Davison & Lindsley, 1985). However, application of thermometers 7 and 8 is complicated by their high sensitivity to uncertainties in the analyses of coexisting minerals (even within the error limits). For this reason, these thermometers should not be used for single measurements or where the statistical representation of samples is limited. Application of thermometers 10 and 11 also presents some difficulty because their analytical expressions do not permit direct temperature measurements while the graphical representation has its limitations. For temperatures below 850 °C (900 °C) only thermometers 5 (Fonarev & Graph-chikov, 1982b) and 12 (the present work) are appropriate. Extrapolations to temperatures below 700–750 °C require, like any extrapolation, experimental verification.

Geothermometers should be applied taking into account their 'performance' in different regions of composition and temperature. On the whole, nearly all thermometers under study become less accurate with increasing temperature and for magnesian compositions so that at $X_{Cpx}^{Fe} < 0.1$ the accuracy will at best be 50–100 °C.

It has been found that the calculated temperatures depend little on the sum of additional components in minerals. The errors that are likely to arise, for the most part, are commensurable and often much lower than errors which commonly occur in the determinations of equilibrium mineral pairs or in their analysis for the major components (Fe, Ca, Mg). Therefore, the effect of additional components on the reliability of two-pyroxene thermometry, though being perfectly obvious in theory, is almost never felt in practice especially where the content of impurity does not exceed 3–5 mole %. For thermometers 5 and 12, the mineral iron content can be taken as $X_{Cpx(Opx)}^{Fe(1)}$ or $X_{Cpx(Opx)}^{Fe(2)}$. For the latter, the correlation between the temperatures and the impurity content decreases somewhat and so do the mean-square errors if the amount of these impurities in clinopyroxene is higher than in orthopyroxene. Otherwise, the use of the second ratio is not reasonable.

Application of the recommended thermometers to natural objects removes much of the criticism against two-pyroxene thermometry because the values they yield conform satisfactorily to the actual geological situation. For example, Griffin *et al.*'s (1980) temperatures of formation of granulites in the South Ameralik (West Greenland) are 630–700 °C against 850 °C given by thermometer 4. Thermometer 12 fixes 665–690 °C. According to Jayawardena & Carswell (1976), the mineral formation temperature of charnockites in Sri Lanka is 700 ± 50 °C in contrast to 858 ± 11 °C recorded by thermometer 1. Thermometer 12 yields 690 ± 20 °C for the same samples. For the Arendala granulites, Norway (Lamb *et al.*, 1986), two-pyroxene thermometry (thermometers 1, 4, 10 and 11) yields temperatures within the 800–940 °C range – too high in their view. Rollinson (1981) arrived at a similar conclusion for granulites in North-West Scotland for which thermometers 1 and 4 show 835 ± 45 °C and 850 ± 60 °C, respectively. Thermometer 12 produces much lower values of 705 ± 10 °C for Norwegian granulites and 735 ± 50 °C for granulites from Scotland.

In conclusion, we would like to note that it is by no means right to assert that two-pyroxene thermometry defies further refinement. But this refinement can only be achieved with additional, more accurate experimental data especially for temperatures below 700–750 °C. So it is important to improve methods for determining equilibrium mineral relations and mineral compositions in both experimental products and natural materials. Ultimately, each individual estimate will be more reliable and two-pyroxene thermometry will gain a wider application in determining the conditions of mineral formation in zoned and thermally heterogeneous complexes. So far, however, increasing statistical representation of the samples to be analysed remains the major method for achieving more reliable results. We do not share Saxena's (Saxena *et al.*, 1986) scepticism regarding components ever attaining an equilibrium distribution between coexisting pyroxenes. This scepticism stems from the observed differences between temperature patterns produced by the bimineralic and monomineralic thermometer types (Saxena *et al.*, 1986). In our view, these discrepancies are due to an inadequate thermodynamic processing of experimental data and as a consequence insufficient accuracy of the thermometers rather than to a drastic non-equilibrium partitioning of the components between ortho- and clinopyroxenes.

Acknowledgements   We are grateful to Prof. Perchuk, Prof. Banno and Dr Holland for critically reviewing the manuscript. We also wish to thank G.B. Lakoza for translating the chapter into English.

References

Akella, J. (1976). Garnet pyroxene equilibria in the system $CaSiO_3$–$MgSiO_3$–Al_2O_3 and in a natural mineral mixture. *Amer. Mineral.*, **61**, 589–98.

Akella, J. & Boyd, F.R. (1972). Partitioning of Ti and Al between pyroxenes, garnets and oxides. In *Carnegie Institution, Washington, Year Book 71*, pp. 378–84.

Bertrand, P. & Mercier, J.-C.C. (1985). The mutual solubility of coexisting ortho- and clinopyroxene; toward an absolute geothermometer for the natural system. *Earth Planet. Sci. Lett.*, **76**, 109–22.

Bohlen, S.R. & Essene, E.J. (1979). A critical evaluation of two-pyroxene thermometry in Adirondack granulites. *Lithos*, **12**, 335–45.

Boyd, F.R. & Schairer, J.E. (1964). The system $MgSiO_3$–$CaMgSi_2O_6$. *J. Petrol.*, **5**, 275–309.

Brey, G. & Huth, J. (1983). The enstatite–diopside solvus to 60 kbar. *Dev. Petrol.*, **9**, 257–64.

Buchanan, D.L. (1979). A combined transmission electron microscope and electron microprobe study of Bushveld pyroxenes from the Bethal area. *J. Petrol.*, **20**, 327–54.

Davidson, P.M. & Lindsley, D.H. (1985). Thermodynamic analysis of quadrilateral pyroxenes. Part II: model calibration from experiments and applications to geothermometry. *Contrib. Mineral. Petrol.*, **91**, 390–404.

Davis, B.T.C. & Boyd, F.R. (1966). The join $Mg_2Si_2O_6$–$CaMgSi_2O_6$ at 30 kbar pressure and its application to pyroxenes from kimberlites. *J. Geophys. Res.*, **71**, 3567–78.

Finnerty, A.A. & Boyd, F.R. (1984). Evaluation of thermobarometer for garnet peridotites. *Geochim. Cosmochim. Acta.*, **48**, 15–27.

Fonarev, V.I. (1987). Mineral equilibria in the Pre-Cambrian iron formations (experiment, thermodynamics and petrology). Moscow: Nauka.

Fonarev, V.I., Bogatyrev, V.F., Van, K.V., Korolkov, G.Ya. & Ionis, G.I. (1983a). Iron quartzites from Sredneye Pobuzhje (physico-chemical conditions of metamorphism). *Geokhimia*, **10**, 1413–24.

Fonarev, V.I. & Graphchikov, A.A. (1982a). Experimental study of Fe, Mg and Ca distribution between coexisting ortho- and clino-pyroxenes at $P = 294$MPa, $T = 750$ and 800 °C. *Contrib. Mineral. Petrol.*, **79**, 311–18.

—— (1982b). A two-pyroxene geothermometer. *Mineral. J.*, **5**, 3–12.

Fonarev, V.I., Graphchikov, A.A., Bogach, D.I. & Van, K.V. (1986). Physico-chemical conditions of metamorphism in iron-rocks in the Belotserkovskaya–Odesskaya zone. *Geologiya rudnykh mestorozhdenii*, **3**, 3–18.

Fonarev, V.I. & Konilov, A.N. (1986). Experimental study of Fe–Mg distribution between biotite and orthopyroxene at $P = 490$ MPa. *Contrib. Mineral. Petrol.*, **93**, 227–35.

Fonarev, V.I., Polunovskii, R.M., Korolkov, G.Ja., Van, K.V. & Krivonos, V.P. (1983b). Iron quartzites from the Mariupolskoje ore field (physico-chemical conditions of metamorphism). *Geokhimia*, **8**, 1184–202.

Graphchikov, A.A. & Fonarev, V.I. (1987). Thermodynamic functions of Fe–Mg–Ca clinopyroxenes. *Geokhimia*, **4**, 563–73.

Griffin, W.L., McGregor, V.R., Nutman, A., Taylor, P.N. & Bridgwater, D. (1980). Early archaean granulite-facies metamorphism south of Ameralik, West Greenland. *Earth Planet. Sci. Lett.*, **50**, 1, 59–74.

Helgeson, H.C., Delany, J.M., Nesbitt, H.W. & Bird, D.K. (1978). Summary and critique of the thermodynamic properties of rock-forming minerals. *Amer. J. Sci.*, **278-A**, 1–229.

Henry, D.J. & Medaris, L.G. (1976). Applications of pyroxene and olivine–spinel geothermometers to the alpine peridotites in south-western Oregon. *Geol. Soc. Amer. Abstracts with Programs*, **8**, 913–14.

Immega, J.P. & Klein, C., Jr. (1976). Mineralogy and petrology of some metamorphic Precambrian iron-formations in southwestern Montana. *Amer. Mineral.*, **61**, 1117–44.

Jayawardena, D.E.De & Carswell, D.S. (1976). The geochemistry of 'charnockites' and their constituent ferromagnesian minerals from the Precambrian of south-east Sri Lanka (Ceylon). *Mineral. Mag.*, **40**, 541–54.

Kretz, R. (1982). Transfer and exchange equilibria in a portion of the pyroxene quadrilateral as deduced from natural and experimental data. *Geochim. Cosmochim. Acta*, **46**, 411–21.

Lamb, R.C., Smalley, P.C., & Field, D. (1986). *P–T* conditions for the Arendal granulites, southern Norway: implications for the roles of *P*, *T* and CO_2 in deep crustal LILE-depletion. *J. Metamorphic Geol.*, **4**, 143–60.

Lindsley, D.H. (1981). The formation of pigeonite on the join hedenbergite–ferrosilite at 11.5 and 15 kbar: experiments and a solution model. *Amer. Mineral.*, **66**, 117–44.

— (1983). Pyroxene thermometry. *Amer. Mineral.*, **68**, 477–93.

Lindsley, D.H. & Andersen, D.J. (1983). A two-pyroxene thermometer. In *Proceedings of the 13th Lunar and Planetary Sciences Conference J. Geol. Res.*, Suppl. 88, A887–A906.

Lindsley, D.H. & Dixon, S.S. (1976). Diopside–enstatite equilibria at 850–1400 °C, 5–35 kbar. *Amer. J. Sci.*, **276**, 1285–301.

Lindsley, D.H., King, H.E., Jr. & Turnock, A.C. (1974). Compositions of synthetic augite and hypersthene coexisting at 810 °C: application to pyroxenes from lunar highlands rocks. *Geophys. Res. Lett.*, **1**, 134–6.

Lindsley, D.H. & Munoz, J.L. (1969). Subsolidus relations along the hedenbergite–ferrosilite. *Amer. J. Sci.*, **276-A**, 295–324.

Mori, T. (1978). Experimental study of pyroxene equilibria in the system CaO–MgO–FeO–SiO_2. *J. Petrol.*, **19**, 45–65.

Mori, T. & Green, D.H. (1975). Pyroxenes in the system $Mg_2Si_2O_6$–$CaMgSi_2O_6$ at high pressure. *Earth Planet. Sci. Lett.*, **26**, 277–86.

—— (1976). Subsolidus equilibria between pyroxene in the CaO–MgO–SiO_2 system at high pressures and temperatures. *Amer. Mineral.*, **61**, 616–25.

Nehru, C.E. & Wyllie, P.J. (1974). Electron microprobe measurement of pyroxenes coexisting with H_2O-undersaturated liquid in the join $CaMgSi_2O_6$–$Mg_2Si_2O_6$–H_2O at 30 kbar, with application to geothermometry. *Contrib. Mineral. Petrol.*, **48**, 249–63.

Nickel, K.G., Brey, G.P. & Kogarko, L. (1985). Orthopyroxene–clinopyroxene equilibria in the system CaO–MgO–Al_2O_3–SiO_2 (CMAS): new experimental results and implications for two-pyroxene thermometry. *Contrib. Mineral. Petrol.*, **91**, 44–53.

Perchuk, L.L. (1977). Modification of a two-pyroxene thermometer for deep-seated peridotites. *Dokl. Akad. Nauk SSSR*, **223**, 3, 456–9.

Perkins, D., III, & Newton, R.C. (1980). The composition of coexisting pyroxenes and garnet in the system CaO–MgO–Al_2O_3–SiO_2 at 900–1100 °C and high pressures. *Contrib. Mineral. Petrol.*, **75**, 291–300.

Podpora, C. & Lindsley, D.H. (1979). Fe-rich pigeonites: minimum temperatures of stability in the Ca–Mg–Fe quadrilateral (abstract). *EOS* **60**, 420–1.

Rollinson, H.R. (1981). Garnet–pyroxene thermometry and barometry in the Scourie granulites, NW Scotland. *Lithos*, **14**, 226–38.

Ross, M. & Huebner, J.S. (1975). A pyroxene geothermometer based on composition–temperature relationships, of naturally occurring orthopyroxene, pigeonite, and augite. In *Extended Abstract, International Conference on Geothermometry and Geobarometry*. Penn. State Univ.

Saxena, S.K. (1976). Two-pyroxene geothermometer: a model with an approximate solution. *Amer. Mineral.*, **61**, 643–52.

— (1983). Problems of two-pyroxene geothermometry. *Earth Planet. Sci. Lett.*, **65**, 382–8.

Saxena, S.K., Sykes, J. & Eriksson, G. (1986). Phase equilibria in the pyroxene quadrilateral. *J. Petrol.*, **27**, 843–52.

Slavinskii, V.V. (1983). Two-pyroxene thermometry. *Mineral. J.*, **6**, 29–37.

Smith, D. (1972). Stability of iron-rich pyroxene in the system $CaSiO_3$–$FeSiO_3$–$MgSiO_3$. *Amer. Mineral.*, **57**, 1413–28.

Turnock, A.C. & Lindsley, D.H. (1981). Experimental determination of pyroxene solvi for $\leqslant 1$

kbar at 900 and 1000 °C. *Canad. Mineral.*, **19**, 255–67.

Warner, R.D. & Luth, W.C. (1974). The diopside–orthoenstatite two-phase region in the system $CaMgSi_2O_6$–$Mg_2Si_2O_6$. *Amer. Mineral.*, **59**, 1/2, 98–109.

Wells, P.R.A. (1977). Pyroxene thermometry in simple and complex systems. *Contrib. Mineral. Petrol.*, **62**, 2, 129–39.

Wood, B.J. & Banno, S. (1973). Garnet–orthopyroxene and orthopyroxene–clinopyroxene relationships in simple and complex systems. *Contrib. Mineral. Petrol.*, **42**, 2, 109–24.

Yamada, H. & Takahashi, E. (1983). Subsolidus phase relations between coexisting garnet and two-pyroxenes at 50 to 100 kbar in the system CaO–MgO–Al_2O_3–SiO_2. *Develop. Petrol.*, **9**, 247–56.

5

Derivation of a thermodynamically consistent set of geothermometers and geobarometers for metamorphic and magmatic rocks

LEONID L. PERCHUK

Introduction

The first set of internally consistent geothermometers and barometers (Perchuk, 1969a, 1970b, 1977) requires modifications in the light of more recent experimental data on phase relationships and thermodynamic properties of solid solution that have appeared during the last ten years. These modifications refer mainly to the relatively high-temperature field ($T > 750$ °C) in which the mutual solubility of components in many complex solid solutions increases significantly affecting some exchange equilibria which just slightly depend on temperature at its lower values. This rule can be experimentally exemplified by equilibria with orthopyroxene involved (Harley, 1984; Aranovich & Kosyakova, 1986, 1987).

A number of geothermometers based on exchange equilibria depend insignificantly on additional components (mainly on high Ca content in garnet) involved in the coexisting solid solutions. This conclusion is supported by experimental studies of biotite–garnet, cordierite–garnet, amphibole–garnet and some other equilibria (Aranovich & Podlesskii, 1983, Lavrent'eva & Perchuk, 1981, 1989, 1990). On the other hand, there are some equilibria which exhibit remarkable deviations from the Nernst law (for example, epidote–garnet, spinel–garnet, spinel–cordierite etc.). There are now enough experimentally calibrated exchange equilibria for a useful revision of the first version of the internally consistent system of geothermometers to be attempted and to be extended to include some geobarometers.

The consistency is tested here on the basis of the standard thermodynamic equation

$$\Delta H_T^0 - T\Delta S_T^0 + P\Delta V^0 + RT\ln K_r + \sum(W^H - W^S T)X_i^{\Phi} = O, \qquad (1)$$

used for each equilibrium r (reaction) in Table 5.1 at given pressure and temperature within the field of its stability.

Table 5.1 shows the thermodynamic characteristics for 24 exchange equilibria which

Table 5.1. *Internally consistent standard thermodynamic data for some equilibria evaluated with equation (1) per one exchanging atom*

No.	Exchange equilibrium	ΔH_T^{0a}	ΔS_T^{0a}	$\Delta V^0_{.298}{}^a$	$L_c,\%$
1	$Chl_{Mg} + Grt_{Fe} = Grt_{Mg} + Chl_{Fe}$	7895	5.510	0.018	1.36
2	$Bt_{Mg} + Grt_{Fe} = Grt_{Mg} + Bt_{Fe}$	7843	5.699	0.025	0.00
3	$Crd_{Mg} + Grt_{Fe} = Grt_{Mg} + Crd_{Fe}$	6002	2.557	0.035	0.20
4	$Cpx_{Mg} + Grt_{Fe} = Grt_{Mg} + Cpx_{Fe}$	5543	3.447	0.010	2.50
5	$St_{Mg} + Grt_{Fe} = Grt_{Mg} + St_{Fe}$	5391	5.561	0.015	3.64
6	$Hb_{Mg} + Grt_{Fe} = Grt_{Mg} + Hb_{Fe}$	6617	4.630	—	3.38
7	$Opx_{Mg} + Grt_{Fe} = Grt_{Mg} + Opx_{Fe}$	4066	2.143	0.023	—
8	$Cld_{Mg} + Grt_{Fe} = Grt_{Mg} + Cld_{Fe}$	4273	3.523	—	3.10
9[b]	$Grt_{Fe} + Spl_{Mg} = Grt_{Mg} + Spl_{Fe}$	3909	3.216	0.009	1.20
10	$Chl_{Mg} + Cld_{Fe} = Cld_{Mg} + Chl_{Fe}$	3622	1.987	—	3.48
11	$Bt_{Mg} + Cld_{Fe} = Cld_{Mg} + Bt_{Fe}$	3570	2.176	—	1.40
12	$Chl_{Mg} + Crd_{Fe} = Crd_{Mg} + Chl_{Fe}$	1893	2.953	−0.017	1.56
13[c]	$Czo_{Al} + Grt_{Fe} = Grt_{Al} + Czo_{Fe}$	2713	0.426	0.193	—
14	$Chl_{Mg} + St_{Fe} = St_{Mg} + Chl_{Fe}$	2504	−0.051	—	2.10
15	$Bt_{Mg} + St_{Fe} = St_{Mg} + Bt_{Fe}$	2452	0.138	0.007	—
16[b]	$Crd_{Mg} + Spl_{Fe} = Crd_{Fe} + Spl_{Mg}$	2225	−0.659	0.026	1.0
17	$Crd_{Fe} + Bt_{Mg} = Bt_{Fe} + Crd_{Mg}$	1841	3.142	−0.010	—
18	$Crd_{Mg} + Opx_{Fe} = Opx_{Mg} + Crd_{Fe}$	1936	0.410	0.012	—
19	$Chl_{Mg} + Hb_{Fe} = Hb_{Mg} + Chl_{Fe}$	1278	0.879	—	—
20	$Bt_{Mg} + Hb_{Fe} = Hb_{Mg} + Bt_{Fe}$	1226	1.069	—	—
21	$Hb_{Mg} + St_{Fe} = St_{Mg} + Hb_{Fe}$	1226	−0.931	—	—
22	$St_{Mg} + Cld_{Fe} = Cld_{Mg} + St_{Fe}$	1118	2.037	—	—
23	$Crd_{Mg} + St_{Fe} = St_{Mg} + Crd_{Fe}$	611	−3.004	—	—
24	$Chl_{Mg} + Bt_{Fe} = Bt_{Mg} + Chl_{Fe}$	52	−0.189	—	—

Notes:

[a] ΔH_T^0 and ΔS_T^0 are calculated for reactions 2–4, 7, 9, 10 and 18 in the temperature range 600–1200 °C in which ΔC_p^0 is assumed to be constant. The values of ΔH_T^0 and ΔS_T^0 for other reactions are valid over the interval 350–700 °C. ΔV_{298}^0 is given at pressure 1 bar and temperature 298 K.

[b] Mixing properties of the spinel solid solution $(Fe, Mg, Zn) (Al, Cr)_2O_4$ are calculated by Perchuk & Gerya (1989).

[c] $Fe = Fe^{3+}$. Mixing properties for *clinozoisite* $Ca_2Al_3Si_3O_{12}(OH)$ – *pistacite* $Ca_2Fe_3Si_3O_{12}(OH)$ solid solution are calculated by Perchuk & Aranovich (1979).

Sources: Original sources for the exchange reactions are the following: 1–3, 5–10, 15, 16, 18 and 22 – Perchuk (1967a, b, 1968, 1969a, b, 1970a, b); 2 and 3 – Perchuk & Lavrent'eva (1983); 13 – Perchuk & Aranovich (1979). Other data were calculated on the basis of the Gess law (linear combination) without using the ΔV_{298}^0 P term.

have been computed with reference to standard ΔH_T^0 and ΔS_T^0 for reactions 3 and 2 studied within the temperature range 600–1000 °C at 6 kbar (Perchuk & Lavrent'eva, 1983). For many reactions from Tables 5.1 and 5.2 each term in equation (1) has been calculated using the experimental and empirical data on the compositions of coexisting minerals from the metamorphic and magmatic rocks of different mineral facies. All reactions from these tables were thoroughly tested as a subject of equilibria. Unfortunately the length of this paper is limited and I am unable to discuss details of textural and thermodynamical approaches used. However, examples of the methods (including dataset) were demonstrated in my paper of 1977 and in the recent papers by Perchuk *et al.* (1989) and Perchuk & Gerya (1989).

Consistent exchange equilibria as geothermometers calibrated on the basis of the ideal distribution law

Statistics symbols

n: number of parameters

t: arithmetic mean value of temperature

ΔT: arithmetic mean deviation of temperature from that chosen as standard

r: linear correlation coefficient

σ: mean square deviation from the standard chosen:

$$\sigma_n = \sqrt{\left(\frac{\sum_{i=1}^{n}(t_i - \bar{t})^2}{n}\right)}$$

$$\sigma_{n-1} = \sqrt{\left(\frac{\sum t_i^2 - (\sum \bar{t})^2/n}{n-1}\right)}$$

These values reflect *an accuracy* of each geothermometer or geobarometer.

The *level of consistency* for each given exchange equilibrium with respect to the biotite–garnet (No. 2 in Tables 5.1 and 5.2) or garnet–cordierite (No. 3 in Tables 5.1 and 5.2) geothermometers (taken as standards) can be estimated using the following criterion (Perchuk, 1989):

$$L_c(\%) = 100\sigma_{n-1}/\bar{t}, \tag{2}$$

where σ_{n-1} is the mean square deviation for *n* estimates of temperature estimated using the given geothermometer with respect to the mean square value $\bar{t}$ estimated by the standard geothermometer. Thus, L_c shows statistical accuracy of the geothermometer used in comparison to the mean arithmetic values of biotite–garnet and/or cordierite–garnet geothermometers for rocks which contain assemblages of these minerals. L_c is only an indicator of coexistence and only partly shows an accuracy which depends on many factors. For example, from experimental data (Perchuk & Lavrent'eva, 1983) an accuracy of the cordierite–garnet geothermometer is $\sigma_n = \pm 11°$ and for the biotite–garnet geothermometer is about $\sigma_n = \pm 13.5°$. However their consistency (L_c) is close to zero: compositions of Grt, Bt and Crd were determined from the same run products using similar approaches.

Values of L_c for several equilibria are shown in Table 5.1. Thermodynamic characteristics of the other reactions have been obtained by a linear combination. Only 11 reactions from Table 5.1 with $\Delta H_T^0 \geqslant 3500$ cal can be considered as efficient

Table 5.2. *Parameters* A, B *and* C *in equation (26) for mineral thermometers based on the ideal distribution of* Mg *and* Fe *between coexisting minerals*

No.	Exchange equilibrium	A	B	C
1	$Chl_{Mg} + Grt_{Fe} = Grt_{Mg} + Chl_{Fe}$	3973	2.773	0.0086
2	$Bt_{Mg} + Grt_{Fe} = Grt_{Mg} + Bt_{Fe}$	3947	2.868	0.0126
3	$Crd_{Mg} + Grt_{Fe} = Grt_{Mg} + Crd_{Fe}$	3020	1.287	0.0176
5	$St_{Mg} + Grt_{Fe} = Grt_{Mg} + St_{Fe}$	2713	2.799	0.0075
6	$Hb_{Mg} + Grt_{Fe} = Grt_{Mg} + Hb_{Fe}$	3330	2.333	—
8	$Cld_{Mg} + Grt_{Fe} = Grt_{Mg} + Cld_{Fe}$	2150	1.773	—
9	$Chl_{Mg} + Cld_{Fe} = Cld_{Mg} + Chl$	1822	1.000	—
11	$Bt_{Mg} + Cld_{Fe} = Cld_{Mg} + Bt_{Fe}$	1797	1.095	—

mineralogical thermometers. The geothermometer equations for these exchange equilibria are shown in Tables 5.2. Other reactions in Table 5.1 were used to control a consistency of thermodynamic values derived from data on distribution of Fe and Mg between coexisting minerals in metamorphic and magmatic rocks.

Thermodynamic data from Table 5.1 have been simultaneously tested for consistency on the basis of equation (1) for 24 reactions by minimization of L_c for all equilibria. The mean level of consistency is 2.645 or 0.952%. In the course of this computation the linear regression coefficient of $\ln K_D$ with $1/T$ for some equilibria can decrease and, as a result, the L_c for some pairs of geothermometers appear to decrease. For example, an empirical correlation of $\ln K_D$ with $1/T$ for chlorite–garnet equilibrium (No. 1 in Table 5.1) calibrated against the biotite–garnet geothermometer on the basis of 27 three-mineral assemblages yields the following polybaric equation (discussion follows):

$$RT\ln K_D(1) = 7895 - 5.51T \tag{3}$$

with $r = 0.979$ and $L_c = 1\%$. However this correlation does not fulfill the criterion of minimization of L_c for other equilibria in which chlorite and garnet coexist with staurolite, chloritoid etc. Minimization of L_c for all such coupled equilibria leads to a negligible increase of the level of consistency for the chlorite–garnet geothermometer up to the value $L_c = 1.36\%$.

Exchange equilibria are written in Table 5.1 for the condition $K_D > 1(\Delta G_T^0 > 0)$, so that the equation

$$\left(\frac{d\ln K_D}{d(1/T)}\right)_P = \frac{\Delta H_T^0}{R} \tag{4}$$

reflects a relative accuracy of each given geothermometer.

According to equation (4) the majority of equilibria in Table 5.1 have the tendencies (i) to be shifted with temperature to the right and (ii) for ΔH_T^0 of the reactions involving garnet to be larger than that for any equilibrium between a pair of hydrous minerals. Both of these tendencies follow from the *phase-correspondence formulation* (Perchuk, 1969a, 1971a, 1977). Evaluating the thermodynamic values for reactions in Table 5.1 Perchuk (1969a) used mean statistical compositions of coexisting silicate solid solutions as the characteristic of a given mineral facies. In doing this, the effect of additional isomorphic components (Ca, Al, Cr, Zn, F etc.) on $K_D(Fe-Mg)$ is automatically taken into account because all components are redistributed between minerals with temperature according to the general formulation mentioned above. The effect of Mn and Fe^{3+} substituting mainly for Fe^{2+} in the silicates is partly accounted for by the following formulation of the Mg molar fraction:

$$X_{Mg} = Mg/Mg + Fe^{\Sigma} + Mn. \tag{5}$$

Thermodynamic values for reaction 2 in Table 5.1, which were calculated on the basis of experimental data, appear to be very similar to those from empirical calibrations (Perchuk, 1967a, 1970a, 1977): $\Delta H = 7642$ cal and $\Delta S = 6.181$ e.u. This example shows that the rule of 'automatic' account of redistribution for all isomorphic elements between coexisting minerals with temperature in many cases yields quite reliable results. Special corrections for non-binary mixtures in the minerals involved are not therefore often required for the purposes of geothermometry.

First let us consider the most important geothermometric equilibria involving garnet listed in Table 5.2. From equation (1) a general geothermometric equation for the equilibria can be written as follows:

$$T = \frac{A + C \times P + X_{Ca}^{Grt} W^H / R}{\ln K(i)_D + B + X_{Ca}^{Grt} W^S / R} \tag{6}$$

where i is any reaction in Table 5.2. The formula (6) shows that the Ca content in garnet may affect some equilibria at a given temperature due to Ca–Mg and Ca–Fe interactions in the garnet solid solution Grs–Alm–Prp. However, twenty years' experience in the application of the geothermometers discussed to metamorphic rocks shows the reliability of the form chosen in equation (6) with the revised parameters A, B, C at the arithmetic mean value $L_C = 2.645 \pm 0.953\%$ given above.

The biotite–garnet geothermometer (equation 2 in Table 5.2) was first calibrated by Perchuk (1967a, 1970a). Later Perchuk (1977) transferred the thermometer to the form (26). Ferry & Spear (1978) were the first who experimentally studied this equilibrium. Many other calibrations were thoroughly discussed in the paper of Perchuk & Lavrent'eva (1983). Experimental calibration of the thermometer by Lavrent'eva & Perchuk (1981) within the temperature range 575–950 °C at a pressure of 6 kbar yielded the following result:

$$T = (3947.1 + 0.0126P)/\{\ln K_D^{(2)} + 2.868\} \tag{7}$$

where

$$K_D^{(2)} = \left(\frac{X_{Mg}}{1 - X_{Mg}}\right)^{Bt} \times \left(\frac{1 - X_{Mg}}{X_{Mg}}\right)^{Grt}. \qquad (8)$$

Precision of the calibration is restricted within 7–27 °C (depending on the run temperature) with the mean arithmetic value 13.5%.

A number of other calibrations have appeared during the last 12–15 years. Many of these have been reviewed in the papers of Perchuk & Lavrent'eva (1983) and Chipera & Perkins (1988). Equation (27) can be applied to the metamorphic and magmatic rocks including xenoliths from alkali basalts (Perchuk, 1987b) if the molar fraction of Mg is defined by equation (25) for Fe(Mn) $\rightleftharpoons$ Mg isomorphism. In the case of Mn $\rightleftharpoons$ Mg a special correction coefficient is required (Perchuk & Lavrent'eva, 1983). Perchuk & Aranovich (1984) have found an increase of $\ln K_D^{(2)}$ with increased contents of fluorine in biotite, due to the high chemical affinity of the element to magnesium. The possibility of using a special correction to avoid an 'artificial' decrease of temperature for equilibrium (2) was investigated for granulite-facies rocks from the Sutam complex of the Aldan shield metamorphic formation (Perchuk et al., 1985).

Aranovich et al. (1988) have suggested a correction for Ca–Mg and Fe–Mg isomorphism in garnet and also a correction for aluminum content in biotite:

$$T = \frac{3873.1 + 2871 X_{Ca}^{Grt} + 0.0124P - 957 N_{Al}^{Bt}}{\ln K_D^{(2)} + 2.609 + 1.449 X_{Ca}^{Grt} + 0.287 N_{Al}^{Bt} + 0.503 X_F^{Bt}} \qquad (9)$$

where $X_{Ca}^{Grt} = Ca/3$, $X_{Ca}^{Grt} = Mg/3$ and $X_{Fe}^{Grt} = Fe/3$, N_{Al}^{Bt} is the number of Al in the octahedral coordination of biotite formula calculated per 11 oxygens, $X_F^{Bt} = F/(F + OH)$ and P is in bar.

For the majority of metamorphic and magmatic rocks the biotite–garnet geothermometer can be used without any compositional corrections, i.e. in the form (7) (Perchuk, 1987a, Perchuk et al., 1984, Chipera & Perkins, 1988, Aggarwal & Nesbitt, 1987).

The cordierite–garnet geothermometer (equation 3 in Table 5.2) was first calibrated by Perchuk (1969a) in a graphical form. Later Perchuk (1977) deduced an analytical form of this geothermometer. Perchuk & Lavrent'eva (1983) discussed other calibrations and presented experimental data on the cordierite–garnet equilibrium within a temperature range of 600–1000 °C at a pressure of 6 kbar.

The geothermometric equation

$$T = (3020 + 0.0176P)/\{\ln K_D^{(3)} + 1.287\} \qquad (10)$$

describes these experimental data with a correlation coefficient $r = 0.998$ and, as mentioned above, an accuracy of 11 °C. Equation (10) agrees with previous formulations: discussion has been given in the paper of Perchuk & Lavrent'eva (1983). Application of equation (10) to metapelites of different metamorphic facies shows a well-repeated consistency with the biotite–garnet geothermometer within experimental accuracy.

The amphibole–garnet geothermometer (equation 6 in Table 5.2) was first calibrated by Perchuk (1976b) in the form of a phase correspondence (Roseboom-type) diagram which has been widely applied in the estimation of equilibrium temperatures in granulite-facies metabasites, ecologites, garnet amphibolites and blue schists (for example, Mottana, 1970, Udovkina, 1971, Hubregtse, 1973). In a recent paper Perchuk (1987b) has demonstrated an effective applicability of equation (10) to the deep-seated xenoliths in alkali basalts from Salt Lake Crater, Hawaii, Oahu, where $X_{Mg}^{Grt}/X_{Mg}^{Hb} = 1$. Above 500 °C the exchange equilibrium (6) in Table 5.1 is practically independent of mineral compositions and the temperature of amphibole–garnet equilibria can be estimated with the equation

$$T = 3330/\{\ln K_D^{(6)} + 2.33\} \tag{11}$$

where

$$K_D^{(6)} = \left(\frac{X_{Mg}}{1 - X_{Mg}}\right)^{Hb} \times \left(\frac{1 - X_{Mg}}{X_{Mg}}\right)^{Grt}.$$

Below 500 °C linear relationships between $\ln K_D(6)$ and $1/T$ are also valid but the correlation coefficient decreases. We have recalibrated the amphibole–garnet geother-mometer on the basis of 48 triplet (Bt + Grt + Hb) assemblages from handbooks (Perchuk, 1971, 1976) relative to the biotite–garnet geothermometer (equation 7). Statistical treatment of the data yields practically the same results as (equation 11) with $r = 0.907$, within the temperature range 400–1050 °C. Statistics of the compositional characteristics of coexisting minerals are as follows: $X_{Mn}^{Grt} = Ca/Ca + Mg + Fe = 0.212 + 0.05$, $X_{Mn}^{Grt} = 0.08$. Fe/Mg ratio in amphiboles varies within wide limits about the following mean formula of hornblende:

$$(Ca_{1.54}Na_{0.55}K_{0.15})_{2.24}(Mg,Fe,Fe_{0.49}^{3+})_{4.5}Ti_{0.16}Al_{0.34}[Al_{1.82}Si_{6.18}]_8O_{22}(O,OH,F)_2.$$

The efficiency of the amphibole–garnet geothermometer in the framework of equation (11) has been supported by Graham & Powell (1984) whose version of this geothermometer is described by the following equation:

$$T = (2880 + 3280X_{Ca}^{Grt})/\{\ln K_D^{(6)} + 2.426\}. \tag{12}$$

At $X_{Ca}^{Grt} = 0.212$ (arithmetic mean value for equation 11) this equation, within uncertainties, transforms to one very close to equation (11):

$$T = 3575/(\ln K_D^{(6)} + 2.426). \tag{13}$$

Our experience in application of the amphibole–garnet geothermometer in the form of equation (11) during the last 20 years shows the negligible effect of additional components in the coexisting minerals on the calculated temperatures. This conclusion is supported by the recent experimental results of Lavrent'eva & Perchuk (1989) on the exchange equilibrium

$$\text{grunerite} + \text{pyrope} = \text{almandine} + \text{cupferite} \tag{14}$$

studied at a pressure of 6 kbar. Distribution coefficients appear to be very similar to those calculated after equation (11):

Temperature	700 °C	800 °C
Experimental	2.86	2.12
After equation (11)	2.98	2.17

These results reflect the alternative effect of Ca content in hornblende and garnet on the interphase distribution of Fe and Mg. Integration of these results with data on cation sites occupancies in the natural amphiboles leads to the following equation after thermodynamic treatment using a regular solution model:

$$T, \text{K} = \frac{3366 - 1444[\text{Na}]^{\text{Hb}} - 323[\text{Al}]^{\text{Hb}} + 188[\text{Ca}]^{\text{Hb}} + [\text{Ca}]^{\text{Gr}}(728[\text{Ca}]^{\text{Gr}} - 148)}{\ln K_{\text{D}} + 2.417 - 1.393[\text{Na}]^{\text{Hb}} - 0.266[\text{Al}]^{\text{Hb}} + 0.051[\text{Ca}]^{\text{Hb}}}$$

where $\text{Gr} = \text{Grt}$, $[\text{Na}]^{\text{Hb}}$ content of Na + K in M4 $[\text{Al}]^{\text{Hb}} = \sum \text{Al} - [\text{Na}]^{\text{Hb}}$, $[\text{Ca}]^{\text{Hb}}$ and $[\text{Ca}]^{\text{Gr}}$ are coefficients in formulae Am (calculated per 23 oxygens) and Gr (calculated per 12 oxygens), respectively. A temperature accuracy of $\sigma_n = \pm 24°$ is estimated for this version of the thermometer. The effects of Na, Al and Ca on k_{D} largely cancel, so that equation (11) is a reasonable approximation.

The clinopyroxene–garnet geothermometer (equation 4 in Table 5.2). An empirical version was calibrated by Perchuk (1968) and later this equilibrium was experimentally investigated by Raheim & Green (1974) and Ellis & Green (1979). Powell (1985) revised their experimental results using the robust regression and jack-knife diagnostics approaches. He has demonstrated the high statistical accuracy of the experimental data used and suggested the application of Ellis and Green's geothermometer in the following form:

$$T = \frac{2790 + 10P(\text{kbar}) + 3140X_{\text{Ca}}^{\text{Grt}}}{\ln K_{\text{D}}^{(4)} + 1.735}. \tag{15}$$

On the basis of this equation and 21 triplets of coexisting amphiboles, garnets and clinopyroxene Powell (1985) modified the amphibole–garnet geothermometer to the form

$$T, \text{K} = (2580 + 3340X_{\text{Ca}}^{\text{Grt}})/\{\ln K_{\text{D}}^{(4)} + 2.20\} \tag{16}$$

Substituting $X_{\text{Ca}}^{\text{Grt}} = 0.212$, i.e. the mean arithmetic value of $X_{\text{Ca}}^{\text{Grt}}$ for amphibole–garnet rocks used above, into equation (16) we can write

$$T, \text{K} = 3288/\{\ln K_{\text{D}}^{(4)} + 2.20\} \tag{17}$$

which is almost identical to equation (11). In other words independent experimental calibrations of the cordierite–garnet geothermometer (equation 30), the biotite–garnet geothermometer (equation 7) and the clinopyroxene–garnet geothermometer (equation 16) *all* appear to be internally consistent with the amphibole–garnet geothermometer in the form of equation (11). The level of consistency L_C of equation (11) with equation (15) and (7) is estimated as 2.5% (see Table 5.1). This statistic follows from treatment of the mineral chemistry of 40 triplet assemblages (Hb + Grt + Bt and Hb + Grt + Cpx) taken from the handbooks of Perchuk (1971b, 1976).

Recently Krogh (1988) has revised existing experimental data on the clinopyroxene–garnet geothermometer and suggested using the equation

$$T, \text{K} = [1879 + 6731 X_{Ca}^{Grt} - 6173 (X_{Ca}^{Grt})^2 + 10 \ P(kb)]/\ln K_D + 1.393,$$

which is particularly useful for blueschists and eclogites.

The results discussed above mean that, despite the wide variation of mineral compositions in the natural rocks, a high level of consistency can, in fact, be achieved between four experimentally calibrated geothermometers (2–4 and 6 in Table 5.2).

The chlorite–garnet geothermometer (equation 1 in Table 5.2) first calibrated by Perchuk (1970b, p. 205) has been modified on the basis of equation (7) and new analytical data on coexisting biotite, garnet and chlorite published during the past 20 years (Perchuk, 1971b, 1976; Korikovskiy, 1979; Ghent *et al.*, 1987; Lang & Rice, 1985; Klaper & Bucher-Nurminen, 1987; Guiraud *et al.*, 1987). As shown in Table 5.1 the chlorite–garnet geothermometer in the form

$$T = 3973/\{\ln K_D^{(1)} + 2.773\} \tag{18}$$

where

$$K_D^{(1)} = \left(\frac{X_{Mg}}{1 - X_{Mg}}\right)^{Chl} \times \left(\frac{1 - X_{Mg}}{X_{Mg}}\right)^{Grt}$$

has the largest $\Delta H_T^0 = 7895$ cal at $r = 0.979$ at $n = 27$. According to the derivative of equation (4) this value reflects a significant effect of temperature on the redistribution of Mg and Fe between chlorite and garnet. High precision of the thermometer follows from the *phase correspondence formulation* as well.

Equation (18) is different from the analytical version of the thermometer developed by Ghent *et al.* (1987) on the basis of coexisting minerals from one metamorphic complex. However, temperature estimates made with equation (18) show similar results to those from a graphical version of the thermometer mentioned above. This version has been used for a long time for geothermometry of blueschists and greenschists and some epidote–amphibolite facies rocks. Equation (3) shows high statistical parameters, characteristic of its consistency with other exchange equilibria in Table 5.1: $\Delta T = 9.3^\circ$, $\sigma_n = 6.81^\circ$, $\sigma_{n-1} = 6.94^\circ$ at $n = 27$, $L_C = 1.36\%$ at $\bar{T} = 501.4\ ^\circ\text{C}$.

The staurolite–garnet geothermometer (equation 5 in Table 5.2) was first calibrated by Perchuk (1969b) and since that time revised periodically (see, for example, Fed'kin, 1975, Perchuk, 1977) with the continued accumulation of mineral analyses of coexisting phases. Thermodynamic values in Table 5.1 were obtained from linear correlation of $\ln K_D(5)$ and $1/T$ for 60 three-mineral assemblages Grt–Bt–St taken from papers of Perchuk (1971b, 1976), Korikovskiy (1979), Fed'kin (1975), Lang & Rice (1985) and Perchuk et al. (1984). This correlation allows derivation of an equation – geothermometer for staurolite–garnet paragenesis:

$$T = (2675 + 0.0075P)/(\ln K_D^{(5)} + 2.799). \tag{19}$$

In the previous papers mentioned above the volume change for reaction (5) was not considered because data on the Mg-staurolite end-member was lacking. Now we can use data on the molar volumes of the staurolite solid solution published by Griffen (1981). As shown in Table 5.1, $\Delta V(5) = 0.015$ cal/bar is too small to indicate a dramatic pressure effect on the reaction discussed.

The maximum deviation of temperature estimates with equation (19) in comparison with the biotite–garnet geothermometer (equation 7) is about 50 °C (10%) at $t = 520$ °C. However, the statistical characteristics of an accuracy for equation (13) are the following: $\Delta T = 26.43^0$, $\sigma_n = 18.95^0$, $\sigma_{n-1} = 19.11^0$ at $n = 60$, $L_C = 3.64\%$.

The garnet–chloritoid geothermometer (equation 8 in Table 5.2) was first calibrated by Perchuk (1970b, p. 205) on the basis of the chlorite–garnet geothermometer. Fed'kin (1975) discussed this geothermometer using very restricted data on the compositions of coexisting minerals from this paragenesis and concluded that the thermometer should be recalibrated because it yielded a very high temperature (570 °C) for an assemblage of St + Grt + Cld + Qtz.

Thermodynamic values in Table 5.1 for equilibrium (8) were deduced by applying equation (8) to the distribution of Fe and Mg among chlorite, garnet and chloritoid for 16 paragenesis. The correlation coefficient $\sigma = 0.962$ allows us to consider the equation

$$T = 2150.5/(\ln K_D^{(8)} + 1.773) \tag{20}$$

as a mineral geothermometer, because other statistics yield reasonable results: $\Delta T = 17.7^0$, $\sigma n = 14.25^0$, $\sigma n_{-1} = 14.75^0$ at $n = 16$ and $L_C = 3.1\%$ at $\bar{T} = 458.6$ °C.

Application of equation (20) to high aluminum and iron-rich greenschist facies metamorphic rocks results in a relatively narrow stability field for the paragenesis Grt + Chl + Cld + Qtz in terms of temperature. These assemblages may include the following minerals: muscovite, paragonite, ilmenite, rutile and apatite. Zoisite may be an additional phase.

Chlorite–chloritoid geothermometer. Equilibrium 10 in Table 5.1 was considered (Perchuk, 1970b) as a potential geothermometer but was never calibrated because of the poor analytical data set on coexisting chlorite and chloritoid. Even now the data on the mineral chemistry is not sufficient to calculate this thermometer by the method used in this paper. Its calibration and consistency have been obtained simultaneously with the garnet–chlorite geothermometer using microprobe data (Perchuk, 1970b, 1976, Ghent et al., 1987) on coexisting minerals from $Grt + Cld + Chl$ triple assemblages, equilibria of which have been proved by probe analyses of centers and rims of several grains. According to these data the enthalpy of the exchange reaction is found with a relatively high correlation coefficient ($r = 0.938$), which leads to the following equation

$$T = 1822/(\ln K_D^{(9)} + 1.000). \tag{21}$$

Statistics of this equation with respect to equation (38) reveal the following parameters: $\Delta T = 19^0$, $\sigma_n = 15.74^0$, $\sigma_{n-1} = 16.74^0$ at $n = 19$ and $L_C = 3.48\%$ at $\bar{T} = 460\,°C$.

The biotite–chloritoid geothermometer   (equation 11 in Table 5.2) can be described by the following linear equation:

$$T = 1796.7/(\ln K_D^{(10)} + 1.095). \tag{22}$$

In this polybaric version equation (22) reveals the same statistics as the chlorite–chloritoid geothermometer. However applicability of equation (22) to geothermometry of metamorphic rocks is quite limited because chloritoid and biotite coexist only rarely at the breakdown isograd of chloritoid in the prograde course of metamorphism (Korikovskiy, 1979, Guiraud et al., 1987, Klaper & Bucher–Nurminen, 1987). For example, a paragenesis of chloritoid and biotite in the Precambrian quartzite and schists from northern New Mexico was described by Grambling (1983). Garnet, chloritoid, muscovite and quartz were found in a sample of staurolite schist (No 76–445). Application of the above geothermometers (see also Table 5.2) reveals satisfactory consistency:

biotite–garnet (equation 7)   555 °C
biotite–chloritoid (equation 22)   567 °C
staurolite–garnet (equation 19)   576 °C
chloritoid–garnet (equation 20)   576 °C

The geothermometers listed in Table 5.2 can be applied to calculations of temperatures of mineral equilibria on the basis of the Nernst law (ideal distribution). Equilibria 7 and 13 are also potential geothermometers if excess functions for orthopyroxene and epidote, respectively, are known. We will discuss both of these equilibria in the next section of this paper. Other equilibria from Table 5.1 cannot be considered as precise geothermometers. However, they can be used for the empirical estimation of L_C for the reactions listed in Table 5.2 because all of them have been tested for internal consistency with the 24 reactions of Table 5.1. For example, equilibrium 14 ($St + Chl$) in Table 5.1 cannot be used as a 'sensitive' geothermometer due to a low $\Delta H_T^0 \cong 2.5$ kcal. With the equation

$$\ln K_D(12) = 0.0257 + 1260/T \tag{23}$$

we can check ΔT for 15 three-mineral assemblages taken from Perchuk (1971b, 1976), Grambling (1983), Klaper & Bucher–Nurminen (1987) and Korikovskiy (1979). According to the mineral compositions, the following statistics for these parageneses have been found: $\ln K_D(12) = 1.631$ at $\sigma_n = 0.1136^0$, $\sigma_{n-1} = 0.1176^0$ and $\bar{t} = 522.7\,°C$ with parameters $\sigma_n = 18.22^0$ and $\sigma_{n-1} = 18.86^0$. Substituting $T = \bar{t} + 273$ in equation (23) we can calculate $\ln K_D(12) = 1.609$. The difference between these theoretical and empirical (1.631) values corresponds to $\Delta T = 11^0$, i.e. $L_C = 2.1\%$ which is considerably less than the mean square deviation (18.22^0). This result reflects a good level of consistency of geothermometers involving chlorite and staurolite.

Consistent exchange equilibria as geothermometers involving nonideal mineral solid solutions

The epidote–garnet geothermometer   (reaction 13 in Table 5.1) was experimentally calibrated by Aranovich (1976), using (Fe, Al)Cl$_3$ aqueous solutions and synthetic starting materials. Perchuk & Aranovich (1979) revised these results and proposed a model for epidote solid solution with the aim of deducing an analytical version of the geothermobarometer for metamorphic and metasomatic rocks. On the basis of a thermodynamic treatment of experimental data, revised mixing properties of epidote have been calculated with equations (10) and (11) of Chapter 1. Assuming an ideal mixing in FeCl$_3$–AlCl$_3$ aqueous solutions and taking into account mixing properties of epidote solid solutions and data for reaction (13) in Table 5.1 the following equations can be written:

$$T = \frac{2713 + c - b - 0.193 P(\text{bar})}{1.987 \ln K_D - 0.426 - d + a} \tag{24}$$

where

$$K_D = \left(\frac{X}{1-X}\right)^{\text{Grt}} \times \left(\frac{1-X}{X}\right)^{\text{Ep}} \tag{25}$$

and $X_{Fe} = Fe/Fe + Al$;

$a = 37.3x^3 - 20.047x^2 - 22.8998x^4$;

$b = 16966.5x^2 - 30188.4x^3 + 18266.7x^4 + 0.193606Px^2$;

$c = 7961.4 + u \times P - 33933x + 62248.6x^2 - 54543.9x^3 + 18266.7x^4$;

$d = 40.0942x - 75.9957x^2 + 67.832x^3 - 22.9x^4 - 9.031$;

$u = 0.193606 - 0.387213x + 0.193606x^2$;

and

$$P = \frac{1.987 T \ln K_D - 2713 + 0.426(T - 298) + q + T(d - a) - z}{0.193606 + 0.387213 X_{Fe}^{Ep}} \tag{46}$$

where P is in kbar, $q - c + Ed - 0.193606 \times P$, $z = b - 0.193606 X^2 P$. Equation (46) shows quite satisfactory results in application to epidote–garnet assemblages for scarns and blue schists.

On the basis of the one Al site avoidance model for epidote, Bird & Helgeson (1981) proposed another version of the thermometer discussed. However, in terms of applicability to the natural rocks that version shows similar estimates of equilibrium temperatures to those obtained after equation (24).

Orthopyroxene–garnet geothermometer (equation 7 in Table 5.1). A first attempt to evaluate this thermometer in 1969 was unsuccessful (Perchuk, 1969c, 1970b) because $\Delta H_T^0(7) = 2980$ cal/mole had appeared to be too small for its satisfactory application even to low Ca garnet–orthopyroxene rocks. In addition, at that time we could not find an isothermal correlation between $\ln K_D(7)$ and Ca content in garnet on the basis of 11 datapoints within the temperature range 620–700 °C: sources of initial data were too poor. However, the exchange equilibrium was suggested as a potential geothermometer for granulite facies rocks and deep-seated xenoliths.

Saxena (1971, 1973) and Saxena & Ghose (1971) studied Mg–Fe order–disorder and the thermodynamics of the orthopyroxene solid solution and found positive deviations from an ideal mixture. Using a two-site model for orthopyroxene solid solution and its equilibrium with Fe–Mg garnet at 750 °C and 9 kbar pressure Perchuk (1971, 1973) deduced a negative deviation of the solution from an ideal mixture. Later Perchuk & Vaganov (1977) corrected this model to calculate the sign-variable deviation. Wood (1974) used the same approach and suggested an analytical version of the orthopyroxene-garnet geothermometer.

Kawasaki & Matsui (1982) were the first to publish the results of experimental study of the equilibrium (7) at pressure 50 kbar. They proposed an equation for the orthopyroxene–garnet geothermometer which takes account of Fe–Mg–Ca interaction parameters. An extensive experimental dataset in the $P–T$ ranges 5–30 kbar and 800–1200 °C was obtained by Harley (1984), who considered only Ca–Mg(Fe) interactions in the garnet by the term $W_{CaMg}^H - W_{CaFe}^H = 1400 \pm 500$ cal/mole site and proposed the following geothermometric equation:

$$T = \frac{3740 + 1400 X_{Ca}^{Grt} - 0.02286 P(\text{bar})}{1.987 \ln K_D + 1.96}, \tag{27}$$

where $K_D > 1$ and $X_{Ca}^{Grt} = \text{Ca}/(\text{Ca} + \text{Mg} + \text{Fe})$ and other values in equation (27) relate to standard thermodynamic functions $\Delta H_T^0 = 3740 + 610$, $\Delta S_T^0 = 1.96 \pm 0.04$ and $\Delta V^0 = 0.02286$ for the exchange reaction 7 in Table 5.1. Sen & Bhattacharya (1985) deduced a garnet–orthopyroxene geothermometer with the following parameters: $\Delta H_T^0 = 2713$, $\Delta S_T^0 = 0.787$, $W_{CaMg}^G - W_{CaFe}^G = 3300 - 1.5T$. Application of this version of the orthopyroxene–garnet geothermometer to the Madras charnockites yielded reasonable results, but unrealistically high temperatures are obtained for xenoliths and high-grade granulite facies rocks. Aranovich & Kosyakova (1986, 1987) have evaluated the orthopyroxene–garnet geothermometer on the basis of detailed experimental investigation of the orthopyroxene–cordierite equilibria in the system $FeO–MgO–Al_2O_3–SiO_2$, and the calorimetric data of Haselton & Newton (1980) on the garnet solid solution. Using a regular solid solution model for Fe–Mg–Al orthopyroxene and interaction parameters for Fe–Ca, Mg–Ca and Al–Cr in garnet (of $W^G_{FeMg} = W^G_{MgFe} = 0$) the authors proposed a system of equations for estimating the P–T parameters. Despite the bulky forms the system shows satisfactory results in respect of equations (2) and (3) and quite reasonable temperatures for metamorphic rocks and deep-seated xenoliths (Aranovich & Kosyakova, 1987).

Lavrent'eva & Perchuk (1990) have recently reinterpreted new and existing experimental data on equilibrium (7) in Table 5.1 and deduced the following equation:

$$T, \mathrm{K} = \frac{4066 - 347(X^{Opx}_{Mg} - X^{Opx}_{Fe}) - 17484X^{Opx}_{Al} + 5769X^{Grt}_{Ca} + 23.42P}{1.987\ln K_D + 2.143 + 0.0929(X^{Opx}_{Mg} - X^{Opx}_{Mg}) - 12.8994X^{Opx}_{Al} + 3.846X^{Grt}_{Ca}}, \quad (28)$$

where $X^{Grt}_{Ca} = Ca/(Ca + Mg + Fe + Mn)$, $K_D = (X_{Mg}/X_{Fe})^{Opx}(X_{Fe}/X_{Mg})^{Grt}$, and where $X^{Gr}_{Mg} = Mg/(Ca + Mg + Fe + Mn)$, $X^{Opx}_{Mg} = Mg/(Mg + Fe + 0.5Al)$, $X^{Opx}_{Fe} = Fe/(Fe + Mg + 0.5Al)$. P is pressure in kbar. Equation (48) shows an accuracy of $\sigma_n = \pm 42°$.

A set of internally consistent **geothermometers involving spinel** (see Table 5.1) was calibrated (Perchuk & Gerya, 1989, Perchuk et al., 1989) on the basis of equations (7) and (10). Temperatures for spinel–garnet and cordierite–spinel geothermometers can be calculated using the following formula:

$$T = \frac{\Delta H^0_r + \Delta V^0_r P + H^e_{Mg} - H^e_{Fe} + H^e_{Cr} + H^e_{Zn} + H^e_{Ca(Grt)}}{1.987\ln K^{(1,2)}_D + \Delta S^0 + S^e_{Mg} - S^e_{Fe}}, \quad (29)$$

where P is in bars, excess functions (Mg, Fe, Zn and Cr) are for spinel and

$$K^{(9)}_D = \left(\frac{X_{Mg}}{1 - X_{Mg}}\right)^{Grt} \times \left(\frac{1 - X_{Mg}}{X_{Mg}}\right)^{Spl} \quad \text{or} \quad K^{(15)}_D = \left(\frac{X_{Mg}}{1 - X_{Mg}}\right)^{Spl} \times \left(\frac{1 - X_{Mg}}{X_{Mg}}\right)^{Crd}.$$

Standard values are listed in Table 5.1 (see reactions 9 and 16). Other terms in equation (1) can be calculated with the following equations:

$$H^e_{Mg} = [(2W^H_{Mg} - W^H_{Fe}) - 2(W^H_{Fe} - W^H_{Mg})X^{Spl}_{Fe}](X^{Spl}_{Fe})^2,$$
$$H^e_{Fe} = [(2W^H_{Fe} - W^H_{Mg}) + 2(W^H_{Fe} - W^H_{Mg})X^{Spl}_{Mg}](X^{Spl}_{Mg})^2,$$

$$S^e_{Mg} = [(2W^S_{Mg} - W^S_{Fe}) - 2(W^S_{Fe} - W^S_{Mg})X^{Spl}_{Fe}](X^{Spl}_{Fe})^2,$$
$$S^e_{Fe} = [(2W^S_{Fe} - W^S_{Mg}) + 2(W^S_{Fe} - W^S_{Mg})X^{Spl}_{Mg}](X^{Spl}_{Mg})^2,$$

$$V^e_{Mg} = [(2W^V_{Mg} - W^V_{Fe}) - 2(W^V_{Fe} - W^V_{Mg})X^{Spl}_{Fe}](X^{Spl}_{Fe})^2,$$
$$V^e_{Fe} = [(2W^V_{Fe} - W^V_{Mg}) + 2(W^V_{Fe} - W^V_{Mg})X^{Spl}_{Mg}](X^{Spl}_{Mg})^2,$$

Table 5.3. *Net-transfer geobarometric reactions and related standard thermodynamic data*

No. (r)	Reaction	ΔH^0_{1000} cal	ΔS^0_{1000} cal/K	ΔV^0 cal/bar
30	$\frac{1}{3}Grt_{Mg} + \frac{2}{3}Sil + \frac{5}{3}Qtz = \frac{1}{2}Crd_{Mg}$	6^a	4.598^a	0.6383^a
31[b]	$Grt_{Mg} + 2Sil = Crd_{Mg} + Spl_{Mg}$	-1243	9.730	1.4387
32[b]	$Grt_{Mg} + Sil = Qtz + Spl_{Mg}$	-1255	0.534	0.1622
33[b]	$Crd_{Mg} = Spl_{Mg} + Qtz$	-1261	-4.064	-0.4760

Notes:
[a] after Aranovich & Polesskii (1983)
[b] after Perchuk *et al.*, (1989)
Source: Perchuk & Gerya (1989).

where $W^H_{Mg} = -469$, $W^S_{Mg} = -1.300$, $W^H_{Fe} = -1219$, $W^S_{Fe} = -1.579$, $W^V_{Mg} = 0.02$, $W^V_{Fe} = 0$, $H^e_{Ca(Grt)} = -2134 X^{Grt}_{Ca}$, $H^e_{Cr} = 3636 X^{Spl}_{Cr}$, $H^e_{Zn} = -865 X^{Spl}_{Cr}$, $X^{Spl}_{Cr} = Cr/2$, $X^{Grt}_{Ca} = Ca/(Ca + Fe + Mn + Mg)$, $X^{Spl}_{Mg} = Mg/(Mg + Fe)$.

Geobarometers A number of geobarometers calibrated on the basis of net-transfer reactions were recently observed by E. Essene (1989). One of these, (see reaction (30) in Table 5.3) which has been studied experimentally and used for calibration of some geobarometers involving spinel (Perchuk & Gerya, 1989, Perchuk *et al.*, 1989).

Geobarometric equations for reactions (31)–(33) in Table 5.3 can be derived from the following fundamental formula:

$$P = \frac{\Delta H^0_r + T\Delta S^0_r + RT\ln K_r + G^e_{Mg} + H^e_{Cr} + H^e_{Zn} + 3mH^e_{Ca(Grt)} + nA^0_G}{\Delta V^0_r + V^e_{Mg} + nA^0_V} \tag{34}$$

where values of all terms are given in the previous part of the paper and in Table 5.3 and

$$K_{(31)} = (X^{Crd}_{Mg})^2 X^{Spl}_{Mg}/(X^{Grt}_{Mg})^3,$$
$$K_{(32)} = X^{Spl}_{Mg}/X^{Grt}_{Mg},$$
$$K_{(33)} = X^{Spl}_{Mg}/X^{Crd}_{Mg},$$

m and n in equation (37) correspond to the number of Grt and Crd moles, respectively, in reaction r from Table 5.3 and A are the following:

System	A^0_G	A^0_V
Crd (dry)	0	0
Crd–CO$_2$	$18 + 0.8286\,T$	-0.37228
Crd–H$_2$O	$-2414 + 1.9552\,T$	-0.37228

Equation (35) can be used for geobarometry of spinel-bearing metamorphic rocks with garnet, cordierite, quartz and sillimanite. An accuracy of the estimates is 1 kbar.

Conclusion

A significant number of exchange equilibria of garnet with biotite, amphibole, cordierite, chlorite, chloritoid, staurolite as well as chloritoid–chlorite, chloritoid–staurolite and chloritoid–biotite can be used as mineralogical geothermometers in terms of ideal distribution of Fe and Mg between minerals involved (the Nernst law). All the 'instruments' can be applied to the metamorphic paragenesis without any compositional corrections with an accuracy of 3–8%. The level of consistency of thermodynamic values (ΔH_T^0, ΔS_T^0 and ΔV^0) for all the exchange equilibria is about 2.5%. Equilibria involving spinel, epidote, ortho- and clinopyroxene need some correction for deviation of these minerals from the ideal solid solution model and/or additional components. However their accuracy varies in practice between the same limits at similar levels of consistency (1–3.5%). A set of four internally consistent geobarometers listed in Table 5.3 can be successfully applied to pressure estimation for metapelites within an accuracy of ± 1 kbar.

Acknowledgements I thank M. Brown, S. Harley and B. Yardley for fruitful discussions and helpful suggestions which I used in the course of revision of the paper.

References

Aggarwal, P.K. & Nesbitt, B.E. (1987). Pressure and temperature conditions of metamorphism in the vicinity of three massive sulphide deposits, Flin Flon – Snow Lake belt, Manitoba. *Can. J. Earth Sci.*, **24**, 2305–15.

Aranovich, L.Ya. (1976). Phase correspondence in the system epidote-garnet in connection with experimental data. In *Contributions to physico-chemical petrology*, **6**, pp. 14–33. Moscow: Nauka.

Aranovich, L.Ya. & Kosyakova, N.A. (1986). Equilibrium cordierite = orthopyroxene + quartz: experimental data and thermodynamics of ternary Fe–Mg–Al orthopyroxene solid solution. *Geokhimiya*, **8**, 1181–201.

—— (1987). The garnet–orthopyroxene geothermobarometer: thermodynamics and examples of application. *Geokhimiya*, **10**, 1363–77.

Aranovich, L.Ya. & Podlesskii, K.K. (1983). The cordierite–garnet–sillimanite–quartz equilibrium: experiments and applications. *Advances in physical geochemistry*, **3**, pp. 173–98. New York: Springer–Verlag.

Aranovich, L.Ya., Lavrent'eva, I.V. & Kosyakova, N.A. (1988). Biotite–garnet and biotite–orthopyroxene geothermometers: calibrations accounting for the Al variations in biotite. *Geokhimiya*, **5**, 668–76.

Bird, D.K. & Helgeson, H.C. (1981). Chemical interaction of aqueous solutions in epidote-feldspar mineral assemblages in geologic systems. II. Equilibrium constraints in metamorphic/geothermal processes. *Amer. J. Sci.*, **281**, 576–615.

Chipera, S.J. & Perkins, D. (1988). Evaluation of biotite–garnet geothermometers: application to the English River Subprovince, Ontario. *Contrib. Mineral. Petrol.*, **98**, 40–8.

Ellis, D.J. & Green, D.H. (1979). An experimental study of the effect of Ca upon garnet–clinopyroxene Fe–Mg exchange equilibria. *Contrib. Mineral. Petrol.*, **71**, 13–22.

Essene, E.J. (1989). The current status of thermobarometry in metamorphic rocks. In *Evolution of metamorphic belts*, ed. J.S. Daly, P.A. Cliff and B.W. Yardley. Geol. Soc. London. Special Publication. No XX, 1–44.

Fed'kin, V.V. (1975). *Staurolite*. Moscow: Nauka.

Ferry, J.M. & Spear, F.S. (1978). Experimental calibration of the partitioning of Fe and Mg between biotite and garnet. *Contrib. Mineral. Petrol.*, **66**, 113–17.

Ghent, E.D., Stout, M.Z., Black, P.M. & Brothers, R.N. (1987). Chloritoid-bearing rocks associated with blueschists and eclogites, northern New Caledonia. *J. Metamorphic Geol.*, **5**, 239–54.

Graham, C.M. & Powell, R. (1984). A garnet–hornblende geothermometer: calibration, testing and application to the Pelona schist. Southern California. *J. Metamorph. Geol.*, **2**, 13–31.

Grambling, J.A. (1983). Reversals in Fe–Mg partitioning between chloritoid and staurolite. *Amer. Mineral.*, **68**, 373–88.

Griffen, D.T. (1981). Synthetic Fe/Zn staurolites and the ionic radius of $^{IV}Zn^{27}$. *Amer. Mineral.*, **66**, 932–7.

Guiraud, M., Burg, J.-P. & Powell, R. (1987). Evidence for Variscan suture zone in the Vendée, France; a petrological study of blueschist facies rocks from Bois de Cene. *J. Metamorph. Geol.*, **5**, 225–37.

Harley, S.L. (1984). An experimental study of the partitioning of Fe and Mg between garnet and orthopyroxene. *Contrib. Mineral. Petrol.*, **86**, 359–73.

Haselton, H.T. & Newton, R.C. (1980). Thermodynamics of pyrope-grossularite garnets and their stabilities at high temperatures and high pressures. *J. Geophys. Res.*, **85**, 6973–82.

Hubregtse, J.J.M.W. (1973). *Distribution of elements in some basic granulite facies rocks*. Amsterdam, London: North-Holland.

Kawasaki, T. & Matsui, Y. (1983). Thermodynamic analyses of equilibria involving olivine, orthopyroxene and garnet. *Geochim. Cosmochim. Acta*, **47**, 1661–80.

Klaper, E.M. & Bucher-Nurminen, K. (1987). Alpine metamorphism of pelitic schists in the Nufenen Pass area, Lepontine Alps. *J. Metamorph. Geol.*, **5**, 175–94.

Korikovskiy, S.P. (1979). *Metamorphic facies of metapelites*. Moscow: Nauka.

Krogh, E.J. (1988). The garnet–clinopyroxene Fe–Mg geothermometer – a reinterpretation of existing experimental data. *Contrib. Mineral. Petrol.*, **99**, 44–8.

Lang, H.M. & Rice, J.M. (1985). Regression modelling of metamorphic reactions in metapelites, Snow Peak, Northern Idaho. *J. Petrol.*, **26**, Pt4, 857–87.

Lavrent'eva, I.V. & Perchuk, L.L. (1981). Phase correspondence in the system biotite–garnet: experimental data. *Doklady Akademii Nauk SSSR*, **260**, 731–4.

—— (1989). Experimental study of amphibole–garnet equilibrium (Ca-free system). *Doklady Akademii Nauk SSSR*, **306**, 173–175.

—— (1990). Orthopyroxene–garnet geothermometer: experiments and theoretical treatment of the data base. *Doklady Akademii Nauk SSSR*, **310**, 179–182.

Mottana, A. (1970). Distribution of elements among coexisting phases in amphibole-bearing eclogites. *Neues Jahrb. Mineral. Abhandl.*, **112**, N 2.

Perchuk, L.L. (1967a). Biotite–garnet geothermometer. *Doklady Akademi Nauk SSSR*, **177**, 411–14.

— (1967b). Analysis of thermodynamic conditions of mineral equilibria in the amphibole–garnet rocks. *Prossed. USSR Acad. Sci.*, **3**, 57–83.

— (1968). Pyroxene–garnet equilibrium and the deep-seated facies of eclogites. *Inter. Geol. Rev.*, **10**, 280–318.

— (1969a). The effect of temperature and pressure on the equilibria of natural Fe–Mg minerals. *Inter. Geol. Rev.*, **11**, 875–901.

— (1969b). The staurolite–garnet thermometer. *Doklady Akademii Nauk SSSR*, **186**, 1405–7.

— (1969c). The parageneses of rhombic pyroxene with garnet in metamorphic rocks. In *Contributons to physico-chemical petrology*, **1**, 261–86. Moscow: Nauka.

— (1970a). Equilibrium of biotite with garnet in metamorphic rocks. *Geochim. Inter.*, **7**, 157–79.

— (1970b). *Equilibria of rock-forming minerals*. Moscow: Nauka.

— (1971a). Crystallochemical problems in the phase correspondence theory. *Geokhimiya*, **1**, 23–38.

— (1971b). *Coexisting minerals*. Leningrad: Nedra.

— (1973). *Thermodynamical regime of deep-seated petrogenesis*. Moscow: Nauka.

— (1976). *Paragenesis and compositions of coexisting minerals*. Moscow: Nauka.

— (1977). Thermodynamical control of metamorphic processes. In *Energetics of geological processes*, ed. S.K. Saxena & S. Bhattacharji, pp. 285–352. Springer–Verlag.

— (1987a). The course of metamorphism. *Inter. Geol. Rev.*, **28**, 1377–400.

— (1987b). Studies of volcanic series related to the origin of some marginal sea floors. In *Magmatic processes: physicochemical principles*, ed. B.O. Mysen, pp. 319–58. Washington, DC: US Geochem, Soc., Special Publication.

— (1989). *P–T*–fluid regimes of metamorphism and related magmatism with specific reference to the Baikal granulites. In *Evolution of metamorphic belts*, ed. J.S. Daly, B.W. Yardley & B. Cliff. Geol. Soc. London Special Publication, No. XX, 274–91.

Perchuk, L.L. & Aranovich, L.Ya. (1979). Thermodynamics of minerals of variable compositions: andradite–grossularite and pistacite–clinozoisite solid solutions. *Phys. Chem. Minerals*, **5**, 1–14.

—— (1984). Improvement of biotite–garnet geothermometer: correction for fluorine content in biotite. *Doklady Akademii Nauk SSSR*, **277**, 471–5.

Perchuk, L.L. & Lavrent'eva, I.V. (1983). Experimental investigation of exchange equilibria in the system cordierite–garnet–biotite. In *Advances in Phys. Geochemistry*, **3**, pp. 199–239. New York: Springer–Verlag.

Perchuk, L.L. & Gerya, T.V. (1989). A set of internally consistent spinel-bearing geothermometers and geobarometers. In *Granulite metamorphism* (program and abstracts). University of New South Wales. School of Mines. 42–44.

Perchuk, L.L. & Vaganov, V.I. (1977). *Temperature regime of formation of continental volcanic series*. Chernogolovka: USSR Acad. Sci. Press.

Perchuk, L.L., Gerya, T.V. & Nozhkin, A.D. (1989). Petrology and geotherm shift recorded in granulites of the Kanskaya formation, Yenisey range, eastern Siberia *J. Metamorph Geol*., to appear.

Perchuk, L.L., Lavrent'eva, I.V., Kotel'nikov, A.R. & Petric, I. (1984). Comparable characteristics of thermodynamic regimes of metamorphism in rock of the Caucasian Ridge and Eastern Carpathians. *Geol. Zb.-Geol. Carpatica* (*Bratislava*), **35**, 105–55.

Perchuk, L.L., Aranovich, L.Ya., Podlesskii, K.K. *et al.* (1985). Precambrian granulites of the Aldan shield, eastern Siberia, USSR. *J. Metamorph. Geol.*, **3**, 265–310.

Powell, R. (1985). Regression diagnostics and robust regression in geothermometer/geobarometer calibration: the garnet–clinopyroxene geothermometer revisited. *J. Metamorph. Geol.*, **3**, 231–43.

Raheim, A. & Green, D.H. (1974). Experimental determination of the temperature and pressure dependence of the Fe–Mg partition coefficient for coexisting garnet and clinopyroxene. *Contrib. Mineral. Petrol.*, **48**., 179–203.

Saxena, S.K. (1971). Mg^{2+}–Fe^{2+} order-disorder in orthopyroxene and the Mg^{2+}–Fe^{2+} distribution between coexisting minerals. *Lithos*, **4**, N 3.

— (1973). *Thermodynamics of rock-forming crystalline solutions*. New York: Springer–Verlag.

Saxena, S.K. & Ghose, S. (1971). Mg^{2+}–Fe^{2+} order-disorder and the thermodynamics of the orthopyroxene solid solution. *Amer. Mineral.*, **56**, 532–59.

Sen, S.K. & Bhattacharya, A. (1985). An orthopyroxene–garnet thermometer and its application to the Madras charnockites. *Contrib. Mineral. Petrol.*, **88**, 64–71.

Udovkina, N.G. (1971). *Eclogites of the Polar Urals.* Moscow: Nauka.

Wood, B.J. (1974). Solubility of Al_2O_3 in orthopyroxene coexisting with garnet. *Contrib. Mineral. Petrol.*, **46**, 1–15.

6

Vector representation of lithium and other mica compositions

DONALD M. BURT

Introduction

In mathematics, a vector of one component has a direction and magnitude, and defines a line (actually, a family of lines unless we define a starting point), two vectors define a plane, three a space, and so on. In chemical petrology, on the other hand, a one component system is a point, two components are needed for defining a line, three for a plane, and four for a space. Mathematicians and chemical petrologists appear to use the word 'component' somewhat differently. In this paper I shall attempt to show how these differences can be resolved, using the lithium micas as an example.

Isomorphic substitutions in minerals have both a direction (or sense) and a magnitude, and thus can be thought of as vector quantities. The analytical and graphical representation of such substitutions is facilitated by the use of exchange operators, or components that contain negative quantities of certain elements and that perform the operation of exchange (if added to a mineral formula: Burt, 1974, 1976, 1979; these are sometimes also called 'exchange components'). Simple examples of exchange operators (isomorphic substitutions) that are important in micas include NaK_{-1}, $MgFe2^+_{-1}$, $MnFe_{-1}$, $Fe^{3+}Al_{-1}$, and $Fe(OH)_{-1}$.

If we simultaneously considered all of such possible exchanges in the mica group, a graphical representation would become impossible. Inasmuch as we are mainly interested in complicated coupled substitutions, especially those involving lithium, we can simplify our task considerably by 'condensing' down vectors that involve simple substitution, such as those listed above. Once we understand the numerous coupled substitutions in this 'condensed' system, it should be relatively simple to separate out Na from K, Mg and Mn^{2+} from Fe^{2+}, Fe^{3+} from Al, or F from (OH) in natural micas. A similar approach has been demonstrated for the amphiboles by Smith (1959) and Thompson (1981); the general method is described by Thompson (1982).

Application

Using the vector approach, all of the more than 10 end-members of the lithium micas in 'condensed' mica composition space can be represented in terms of only three vectors (or exchange operators or coupled substitutions), starting from a single composition or point (this point is the 'additive component' in the terminology of Thompson, 1982). Preliminary results were given by Burt & Burton (1984) and in a section of Cerny & Burt (1984). The choice of vectors and of initial composition is somewhat arbitrary (although many of what appear to be independent coupled substitutions turn out to be coplanar vectors). A convenient choice is to use annite (Fe-phlogopite) as the starting composition (additive component) and $[\ ]Al_2Fe_{-3}$, $FeSiAl_{-2}$, and $LiAlFe_{-2}$ as the linearly independent exchange vectors, where [ ] designates an octahedral vacancy. The first operator converts trioctahedral into dioctahedral micas, the second represents the Tschermak substitution (e.g., Thompson, 1979, Miyashiro & Shido, 1985), and the third progressively changes annite into 'protolithionite', zinnwaldite, and trilithionite (Ginzburg & Berkhin, 1953, cited in Foster, 1960, p. 128). In a sense we are changing from the old components Li_2O, FeO, Al_2O_3, and SiO_2 into the new components annite, $[\ ]Al_2Fe_{-3}$, $FeSiAl_{-2}$, and $LiAlFe_{-2}$ (cf. Korzhinskii, 1959, Thompson, 1982), but we are using the exchange operators as vectors, rather than as points on a barycentric plane or tetrahedron.

The justification for using Fe^{2+} in place of Mg in these operators is twofold – first, the micas in Li-rich pegmatites, granites, and greisens are normally quite Mg-depleted (and Mn-poor as well) and second, they seem to grow under rather reducing conditions, so that Fe^{3+} contents are small.

Inasmuch as exchange operators or vectors intrinsically involve only elements, rather than ions (they normally are chosen to be electrically neutral, however), lines of equal Al on the diagrams discussed below do not depend on whether the Al is octahedral or tetrahedral, and lines of equal Fe do not necessarily depend on its valence. This feature should facilitate the plotting of mica compositions for which only partial analytical data are available; if complete data are available it should facilitate checking for internal consistency.

Planar subsystems

Li-free plane

The basal or Li-free portion of the condensed mica composition space (with names for $K-Fe^2-Al-Si$ micas) is shown on Fig. 6.1, with formulas expressed per 24 (O plus OH). A somewhat similar but less complete diagram, without the vector notation, has been presented by Green (1981; cf. Stern, 1979). The accessible (to micas) compositions

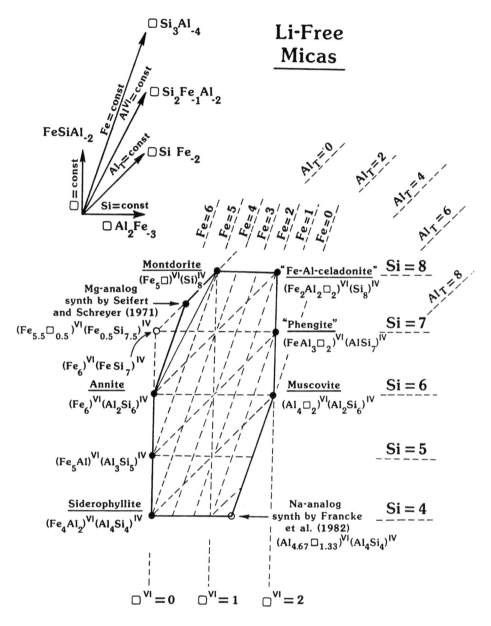

Fig. 6.1. Schematic depiction of the Fe–Al–Si mica composition plane (Li-free micas), obtained by 'condensation' down the compositional exchange vectors KNa_{-1}, $Fe^{2+}Mg_{-1}$, $FeMn_{-1}$, $AlFe^{3+}_{-1}$, and $F(OH)_{-1}$. The basis vectors, drawn to scale in the upper left, are $FeSiAl_{-2}$ ([ ] = constant) and $[\,]Al_2Fe_{-3}$ (Si = constant). Other vectors, of constant Fe, Al, and Al(IV) are linear combinations of the two basis vectors. All of the mica compositions shown can be derived from a single composition such as annite by the operation of the exchange vectors. Lines of equal Fe, Al, Si, and [ ] (octahedral vacancies) can be drawn on the composition plane because they are parallel to their respective vectors. See text for further discussion.

115

define an irregular polygon, all points of which can be derived from a single point such as the annite composition by application of the exchange vectors drawn to scale in the upper left. This is a part of the plane $K_2Fe_{21}O_{20}(OH)_4$–$K_2Al_{14}O_{20}(OH)_4$–$K_2Si_{10.5}O_{20}(OH)_4$, for which the end-member compositions are derived by applying those same exchange vectors to their logical extremes. The composition of the extreme Al-extreme mica was earlier derived by Rutherford (1973), Miyashira & Shido (1985), and presumably others.

The vertical axis is the Tschermak exchange vector $FeSiAl_{-2}$ that relates siderophyllite (lower left corner) to annite above it and muscovite to phengite and Fe–Al celadonite (leucophyllite) above it. The horizontal axis is the exchange vector $[\ ]Al_2Fe_{-3}$ that relates annite to muscovite and montdorite to Fe–Al celadonite. We could call this the 'dioctahedral converter' vector. These two vectors respectively coincide with vertical lines of constant numbers of octahedral vacancies ('[ ]') and with horizontal lines of constant Si. Lines of constant Fe have a slope of $+3$ and coincide with the vector Si_3Al_{-4}, whereas lines of constant total Al have a slope of $+1$ and coincide with the vector $SiFe_{-2}$, as shown on the vector diagram.

Normally, one might assume that a line between annite and montdorite would mark the outer limit of accessible compositions on the polygon, but Seifert & Schreyer (1971) synthesized an Mg mica with nearly 0.5 moles of tetrahedral Mg per 24 (O plus OH); the plotted composition is its Fe analog. (Recall that this, a condensed diagram, so that we are not differentiating among Fe^{2+}, Mn^{2+}, and Mg, Al and Fe^{3+}, K and Na, or F and OH.) Note that the fact that some of the Fe (or Mg) is tetrahedral does not detract from our ability to plot the resulting composition in the same plane. Similarly, Francke et al. (1982) reportedly synthesized an Na mica of the composition shown at the lower right, derived from the siderophyllite composition by the substitution $[\ ]Al_2Fe_{-3}$.
$[\ ]Al_2Fe_{-3}$.

In general, one might define the Li-free composition plane as that resulting from the simultaneous restrictions that, on the basis of 24 (O plus OH), Fe lie between 0 and 6, Al between 0 and 8.67, Si between 4 and 8, and vacancies between 0 and 2. These restrictions turn out to be too broad when applied to the K–Fe–Al–Si micas themselves. For example, Fe–Al celadonite (leucophyllite) is unstable (Velde, 1972), as are, due to geometric restrictions (Hazen & Wones, 1978), end-member annite and siderophyllite or 'eastonite' (Hewitt & Wones, 1975). In nature such micas are stabilized by numerous additional substitutions, including those used in the operation of condensation.

Fe-free Plane

The 'condensed' composition plane for the Fe-free lithium micas (Li–Al–Si micas) is shown on Fig. 6.2. Inasmuch as this plane does not include annite, muscovite has been used as the starting composition. The exchange vectors chosen are again those of constant octahedral vacancy level ($Al_3Li_{-1}Si_2$, shown increasing downwards, with vacancies decreasing to the right) and of constant Si ($Li_3[\ ]_{-2}Al_{-1}$, increasing to the

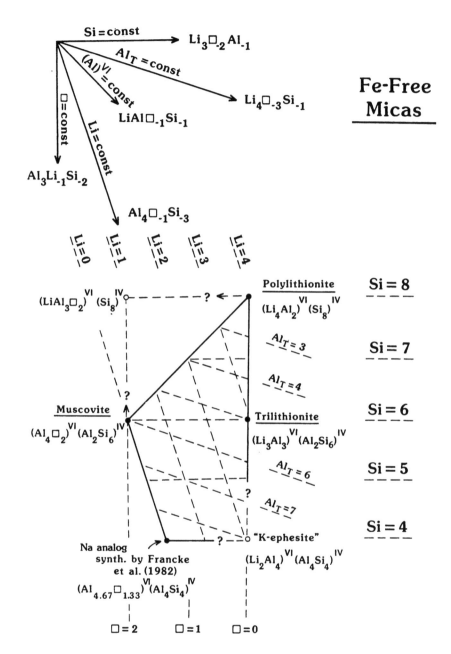

Fig. 6.2. Schematic depiction similar to Fig. 6.1 of the 'condensed' Li–Al–Si mica composition plane (Fe-free micas). The basis exchange vectors, drawn to scale in the upper left, are $Al_3Li_{-1}Si_{-2}$ ([ ] = constant) and $Li_3[ ]_{-2}Al_{-1}$ (Si = constant). The other mica compositions can be derived from any single composition, such as that of muscovite, by the operation of these vectors. The silica-deficient micas along the base, ephesite and the synthetic product produced by Francke *et al.* (1982), are sodic. The silica-rich tetrasilicic composition to the upper left is only a hypothetical component of natural micas. See text for further discussion.

117

right, with Si increasing upwards). Linear combinations of these produce vectors of constant total Al and of constant Li, as shown to scale at the upper left of the diagram. Not shown is the vector $Li_4Si_2[\ ]_{-2}Al_{-4}$, or 2 times $Li_2Si[\ ]_{-1}Al_{-2}$, that relates muscovite to polylithionite. It is equal to the horizontal vector minus the vertical vector. Applying these vectors to their logical extremes show that this is a part of the plane $K_2Li_{42}O_{20}(OH)_4-K_2Al_{14}O_{20}(OH)_4-K_2Si_{10.5}O_{20}(OH)_4$.

The result is again an irregular polygon, rather than the triangle muscovite –trilithionite–polylithionite shown by Foster (1960), although most analyses fall in this triangle (Foster, 1960, Hawthorne & Cerny, 1982). Additional points are again the mica synthesized by Francke *et al.* (1982), 'K-ephesite' (remember, this is a condensed diagram that does not differentiate Na from K), and possibly a mica of composition $K_2LiAl_3[\ ]_2Si_8O_{20}(OH, F)_4$ derived from muscovite or polylithionite by the reverse vertical or horizontal exchange vectors, respectively. Such an end-member is probably unstable (it is compositionally equivalent to a mixture of eucryptite, feldspar, quartz, and H_2O), but it might be a component of low temperature white mica formed under conditions of low temperature and high silica activity, such as in a hot spring. Bargar *et al.* (1973) estimated that a lepidolite from such an environment in Yellowstone National Park was very close to polylithionite in composition.

Vacancy-free plane (Trioctahedral micas)

Of equal interest is the composition plane of the trioctahedral lithium micas (the plane of zero vacancies), shown in Fig. 6.3. The vertical exchange vector is $FeSiAl_{-2}$, parallel to lines of constant Li (Li increases to the right), and the horizontal exchange vector is $LiAlFe_{-2}$, progressively relating annite to 'protolithionite', zinnwaldite, and trilithionite, as mentioned above. This vector is parallel to lines of constant Si (Si increases upwards). As shown to scale at the upper left, lines of constant Fe have a slope of $+2$ and are parallel to the vector $LiSi_2Al_{-3}$; lines of constant total Al have a slope of $+\frac{1}{2}$ and are parallel to the vector Li_2SiFe_{-3}. Also shown is the vector of constant octahedral Al, $LiSiFe_{-1}Al_{-1}$, of slope $+1$. This is a part of the plane $K_2Al_{14}O_{20}$ $(OH)_4-K_2Fe_7Si_7O_{20}(OH)_4-K_2Li_{4.67}Si_{9.33}O_{20}(OH)_4$.

The accessible part is again an irregular polygon (a pentagon), with vertices at siderophyllite, annite, 'Fe–taeniolite', polylithionite, and 'K-ephesite'. This polygon simultaneously satisfies the restrictions that Li lie between 0 and 4, Fe between 0 and 6, Al between 0 and 8, and Si between 4 and 8. The fact that this is a condensed representation excuses the fact that 'K-ephesite' is presumably unstable and that an Fe analog of taeniolite has not yet been described. (It might be expected in a peralkaline rock, in which most Fe, however, is normally present as Fe^{3+}.) According to Rieder *et al.* (1970) most greisen micas fall between siderophyllite and polylithionite (approximately along the vector $LiSiFe_{-1}Al_{-1}$, of constant Al(VI), rather than along $LiAlFe_{-2}$. The composition of 'protolithionite' on Fig. 6.3 does not lie on this join, and Rieder *et al.* (1970) report this composition as absent in their rocks; perhaps for this reason 'protolithionite' is no longer included as a valid mineral species in most compilations.

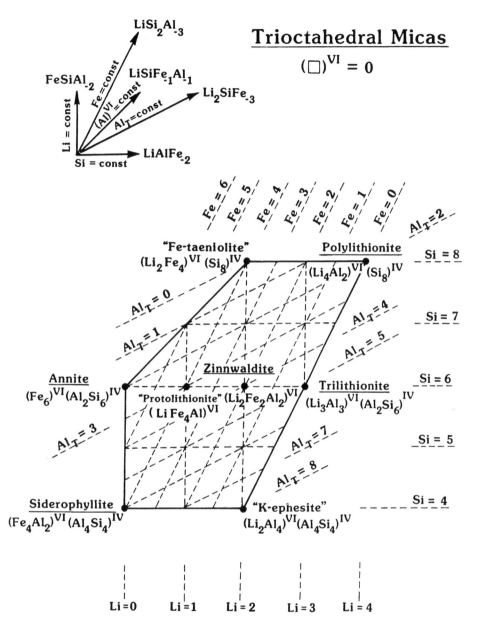

Fig. 6.3. Schematic depiction similar to Fig. 6.1 of the 'condensed' trioctahedral Li–Fe–Al–Si mica composition plane (micas with zero octahedral vacancies). The basis exchange vectors, drawn to scale at the upper left, are $FeSiAl_{-2}$ (Li = constant) and $LiAlFe_{-2}$ (Si = constant). The other mica compositions can be derived from any single composition, such as that of annite, by the operation of these vectors. Ephesite, to the lower right, is a sodium mica, and taeniolite, to the upper left, is a magnesium mica. See text for further discussion.

119

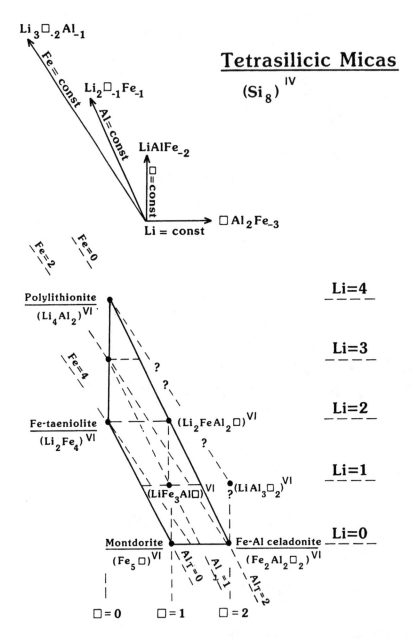

Fig. 6.4. Schematic depiction similar to Fig. 6.1 of the 'condensed' tetrasilicic Li–Fe–Al–Si mica composition plane (micas with 8 Si atoms per 24 O + OH). The basis exchange vectors, drawn to scale at the upper left, are $LiAlFe_{-2}$ ([ ] = constant) and [ ]Al_2Fe_{-3} (Li = constant). The other mica compositions can be derived from any single composition, such as that of montdorite, by the operation of these vectors. Taeniolite, to the left, is a magnesium mica and celadonite, to the lower right, is a magnesium–ferric iron mica. The composition immediately above celadonite is only a hypothetical component of natural micas. See text for further discusson.

The name protolithionite is, however, still widely used in eastern Europe and the USSR for micas intermediate between siderophyllite and zinnwaldite, which itself is an intermediate composition (e.g., Gottesmann & Tischendorf, 1980, Pavlishin *et al.*, 1981, Rub *et al.*, 1986).

Tetrasilicic plane (Si = 8)

A final composition plane of interest is that of the 'tetrasilicic micas', in which there is only Si in the tetrahedral position. Such micas tend to occur only in peralkaline rocks, in which there is very little unattached Al to fill this site. Fig. 6.4 shows the results of letting the two exchange vectors $LiAlFe_{-2}$ (vertical) and $[\,]Al_2Fe_{-3}$ (horizontal) operate on montdorite as a starting composition. This is a part of the plane $K_2Li_{10}Si_8O_{20}(OH)_4 - K_2Fe_5Si_8O_{20}(OH)_4 - K_2Al_{3.33}Si_8O_{20}(OH)_4$. The accessible part is an irregular polygon (a quadrilateral) for which the vertices are montdorite, Fe–Al celadonite, polylithionite, and Fe-taeniolite. As mentioned in the discussion related to Fig. 6.2, there may be some extension towards a hypothetical mica of octahedral occupancy $(LiAl_3[\,]_2)$. This polygon satisfies the restrictions that Li be between 0 and 4, Fe between 0 and 5, Al between 0 and 2 (or possibly 3), vacancies between 0 and 2, and Si remain constant at 8.

The 3-D polyhedron

Other composition planes could be drawn, but those depicted in Figs. 6.1, through 6.4 are probably sufficient to assist in understanding the condensed mica composition space, depicted in Fig. 6.5. The three exchange vectors are as depicted to the upper left. The result is an irregular polyhedron – part of the tetrahedron $K_2Li_{42}O_{20}(OH)_4 - K_2$ $Fe_{21}O_{20}(OH)_4 - K_2Al_{14}O_{20}(OH)_4 - K_2Si_{10.5}O_{20}(OH)_4$. The Li-free plane of Fig. 6.1 is the base, the Fe-free plane of Fig. 6.2 is the oblique front face, the vacancy-free (trioctahedral) plane of Fig. 6.3 is the back face on the left, and the tetrasilicic plane of Fig. 6.4 is the back face on the right. Values of Li increase upwards, octahedral vacancies increase from the rear to the front, and Si increases from the left front to the right rear, as shown.

Other coupled substitutions

The model presented assumes that all of the substitutions are occurring in the tetrahedral and octahedral layers of the mica structure – that is, that the interlayer or K sites are completely filled, as are the (OH) sites. Neither of these assumptions is completely justified in natural micas, and in fact numerous other minor coupled substitutions occur (cf. Bailey, 1980, 1984, Guidotti, 1984, Hewitt & Abrecht, 1986, Grew *et al.*, 1986); strictly speaking, each of these substitutions would add a dimension

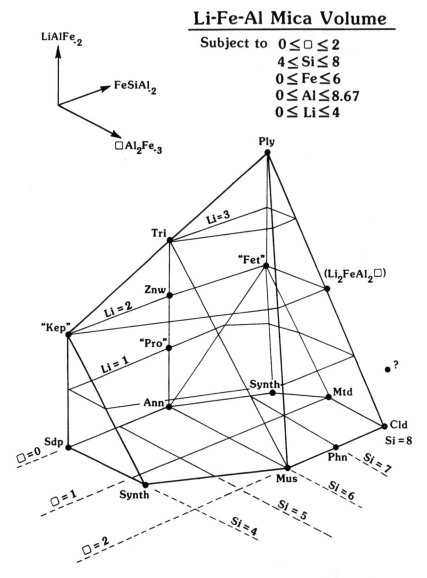

Li-Fe-Al Mica Volume

Subject to $0 \leq \square \leq 2$
$4 \leq Si \leq 8$
$0 \leq Fe \leq 6$
$0 \leq Al \leq 8.67$
$0 \leq Li \leq 4$

Fig. 6.5. Schematic depiction of the 'condensed' Li–Fe–Al–Si mica composition polyhedron, derived from any single mica composition, such as that of annite, by the operation of the three exchange vectors $LiAlFe_{-2}$, $FeSiAl_{-2}$, and $[\,]Al_2Fe_{-3}$. These vectors, drawn to scale in the upper left, determine planes of constant Li (horizontal), Si (vertical, trending to the right) and octahedral vacancies (vertical, trending to the left). Planes of constant Al and Fe (not shown, except in the sections in Fig. 6.1 through 6.4) are likewise determined by linear combinations of these vectors. In this figure Fig. 6.1 is the basal plane, Fig. 6.2 the large front face, Fig. 6.3 the large back face, and Fig. 6.4 the right face. See text for further discussion. Abbreviations: Sdp = siderophyllite, Synth = synthetic mica, Kep = ephesite, Ann = annite, Pro = 'protolithionite', Znw = zinnwaldite, Tri = trilithionite, Mus = muscovite, Phn = phengite, Cld = celadonite, Mtd = montdorite, Fet = taeniolite, and Ply = polylithionite.

to our composition space. Nevertheless, the major features can be seen in only three dimensions.

An example of these other coupled substitutions is $[i]Si(KAl)_{-1}$, where [i] is an interlayer vacancy; this leads towards K-deficient micas such as illite, then smectites and eventually to pyrophyllite and talc (cf. Yoder & Eugster, 1955, Rosenberg, 1987). Under acid conditions, micas can also become K-deficient through the minor substitution of H_3O^+ (hydronium) for interlayer K^+ (cf. Levillain, 1980, Bailey, 1980, p. 59); this also leads towards illite and smectites. Robert et $al.$ (1983) note that K-deficiency can also occur through the substitution of interlayer Li, although in different sites from K due to the difference in ionic radius.

Another important coupled substitution, best known in plagioclase feldspars, is $CaAl(KSi)_{-1}$; this leads to the formation of the brittle micas margarite and clintonite from muscovite and siderophyllite (eastonite) respectively. Ba commonly substitutes for Ca in this operator. Divalent interlayer cations such as Ca^{2+} could alternatively be introduced via other substitutions such as $CaAl_2([i]Si_2)_{-1}$ or $Ca[i]K_{-2}$; only one of these Ca^{2+}-vectors can be linearly independent.

Many micas are (OH)-deficient; this can occur through $SiO(AlOH)_{-1}$ or $AlO(Mg OH)_{-1}$ couples in Fe-free micas (cf. Forbes, 1972). In Fe-bearing micas the major couple is $Fe^{3+}O^{2-}(Fe^{2+}OH^-)_{-1}$ (e.g., Eugster & Wones, 1962), equivalent to minus H^0 (atomic hydrogen); that is, Fe-micas can gain or lose hydrogen without changing their crystal structure. Some micas contain S (e.g., Norman & Palin, 1982), presumably in the OH site by a coupled substitution such as $Fe^{3+}S^{2-}(Fe^{2+}OH^-)_{-1}$, written by analogy with the oxy-biotite couple above; this could be represented simply as the vector SO_{-1} combined with the oxy-biotite vector itself.

A common octahedral substituent in mica is Ti^{4+}, about which another paper could be written. The choice of vectors to represent Ti-substitution is again somewhat arbitrary; I would probably choose $TiFeAl_{-2}$ (analogous to $LiAlFe_{-2}$, in that it only involves octahedral sites; this is called the 'spinel substitution' by Dymek, 1983, although it also relates ilmenite to corundum). This can be combined with the Tschermak substitution $FeSiAl_{-2}$ to yield $TiSi_{-1}$, which is actually $TiAl(SiAl)_{-1}$, and the so-called Ti-Tschermak substitution, $TiAl_2(FeSi_2)_{-1}$. These and other derivable Ti-substitutions, such as $[\ ]TiFe_{-2}$ and $TiO_2(Fe(OH)_2)_{-1}$, are discussed analytically, but not represented graphically as vectors, by Dymek (1983), Labotka (1983), and Hewitt & Abrecht (1986).

Summary and conclusions

The condensed composition space theoretically accessible to Li–Fe–Al micas is shown to be an irregular three-dimensional polyhedron (Fig. 6.5), rather than the triangle Al^{3+}–Fe^{2+}–Li^+ used by Foster (1960) and later researchers. This result is demonstrated starting with the annite composition, and using the exchange operators

[ ]Al$_2$Fe$_{-3}$, FeSiAl$_{-2}$, and LiAlFe$_{-2}$ graphically as vectors that generate all of the other mica compositions. Individual element (and octahedral vacancy) values per 24 (O plus OH) are planes through the resulting polyhedron and are intersecting straight lines on planar sections through it. Another way of stating this result is to say that micas lie within the tetrahedron Li–Fe–Al–Si, or K$_2$Li$_{42}$O$_{20}$(OH)$_4$–K$_2$Fe$_{21}$O$_{20}$(OH)$_4$–K$_2$Al$_{14}$O$_{20}$(OH)$_4$–K$_2$Si$_{10.5}$O$_{20}$(OH)$_4$.

Any other mica composition could have been used as the starting point (additive component), and any other linearly independent set of exchange vectors used to generate the other micas; the result should have been the same. In order to emphasize this point, I give you Fig. 6.6, a barycentric representation in which the three exchange operators that I used as basis vectors are the corners on a triangle. The other derived exchange operators (combinations of the basis operators) are points on the same plane; any non-collinear set of three operators could be used as vectors instead, or you could derive your own operators. The ones shown lie along special lines. The base of the triangle is thus a line of constant (zero) Li; the operators along it were used as vectors in drawing Fig. 6.1. The oblique rising line on the right side of the figure is a line of constant (zero) Fe; its operators were used in drawing Fig. 6.2. The right side of the triangle is a line of constant (zero) octahedral vacancies; the operators along it were used as vectors in drawing Fig. 6.3. The left side of the triangle is a line of constant Si; the operators along it were used in drawing Fig. 6.4. Two other special lines on the figure, of constant total Al and of constant octahedral Al, could have been used to draw planar sections through the mica polyhedron, but were not. Incidentally, if Fig. 6.6 seems like a step backwards, it is. I derived it in fall of 1981 to show the relations among possible substitutions in the Li-micas; a few days later I hit upon the (in my opinion) far superior method of using exchange operators as vectors. This was while I was in China; when I came back to the USA I discovered that my former professor J.B. Thompson, Jr, had made the same breakthrough for the amphiboles (Thompson, 1981, preceded by Smith, 1959).

Exchange vectors need not (and generally do not) correspond to unique ionic substitutions in minerals. Those derived above were mainly chosen as lines of constant [ ], Li, Fe, and Al, in order to facilitate the interpretation of mineral compositions; there is nothing else unique about them. When the accessible composition space is a plane, as in Fig. 6.1 for example, it is unreasonable to expect that natural or synthetic solid solutions will correspond to either of the two basis vectors, or to any other single vector, unless compositions lie along a side-line and are forbidden (via a compositional degeneracy) to vary otherwise. In Fig. 6.1, Fe-free muscovite is constrained to vary along the line Fe $=0$ (the vector [ ]Si$_3$Al$_{-4}$), although it could also vary along some other vector outside the plane, such as [i]Si(KAl)$_{-1}$ or SiO(AlOH)$_{-1}$. Micas containing Fe are less constrained. Thus coexisting muscovite and biotite might be expected to demonstrate the substitution [ ]Al$_2$Fe$_{-3}$, but in nature (cf. Miyashiro & Shido, 1985) the biotite is always somewhat Si-depleted (towards siderophyllite) and the muscovite always somewhat Si-enriched (phengitic). Similarly, the compositions of phengite itself do not lie exactly along the Tschermak vector (Evans & Patrick, 1987).

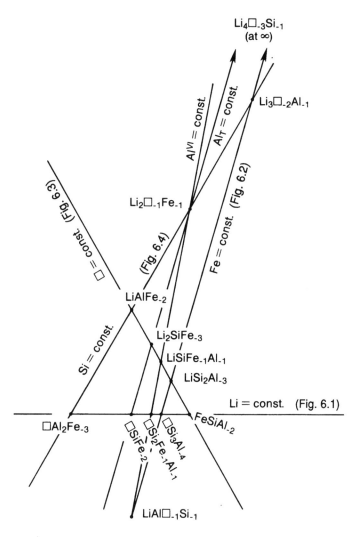

Fig. 6.6. Barycentric triangular representation of the plane [ ]Al$_2$Fe$_{-3}$–FeSiAl$_{-2}$–LiAlFe$_{-2}$, to demonstrate that all of the other exchange operators are coplanar with (can be represented as combinations of) those three taken as a basis. The labelled special lines, except the two of constant Al(VI) and Al-T(otal), contain the exchange operators used as vectors to draw Figs. 6.1 through 6.4, as marked. Thus the base of the triangle is a line of constant (zero) Li and contains the exchange operators used as vectors to draw Fig. 6.1. See text for further discussion.

The same phenomenon occurs even more notably in the Li–Al micas (Fig. 6.2) and in the Li–Fe–Al trioctahedral micas (Fig. 6.3), both discussed in detail by Foster (1960). Observed analyses, in general, do not correspond to any simple solid solution series. Why should they? They are pulled around the plane (or volume) by the compositions of coexisting minerals or fluids. Numerous papers (including many cited above) seem to express a misunderstanding of the exchange operator or coupled substitution concept, in that the authors expect that the operators (vectors) should correspond to unique and real ionic substitutions. This expectation is obviously naive. No one expects a ternary garnet solid solution to behave in a binary fashion; why should micas behave differently?

I also note that exchange vectors should not be used in a vacuum. By themselves, they tell you nothing about mineral structures or stabilities relative to other phases, and are just as good at generating metastable compositions as at elucidating the possible range of solid solutions in nature or experiment. Even stable compositions may be stable only under a very limited range of conditions. The Na–Li brittle mica ephesite, for example, although stable on its own composition, is unstable with quartz at any P and T (Chatterjee & Warhus, 1984). Stable quartz-bearing assemblages in pegmatites instead include feldspars, muscovite, and lithium aluminosilicates such as spodumene or petalite.

One useful aspect, emphasized elsewhere (Burt, 1974, 1979), is that exchange operators should be intrinsically acidic or basic, according to the Lewis (1938) definition. It should therefore be possible to correlate the compositions of various micas with the acidity of their environment. As shown by Marakushev & Perchuk (1970), however, this task is not a simple one, due to the fact that, in addition to accounting for the relative acidity of the complicated coupled substitutions, you must also consider the identity of the coexisting phases other than mica, and how they are affecting mica compositions. Nevertheless, a useful generalization, stated by Nockolds (1947; cf. Speer, 1984), is that the Al-content in igneous biotite is related to how peraluminous vs peralkaline the parent rock is. Similarly, Marakushev & Tararin (1967) obtained a better resolution by plotting Si/Al vs (Mg + Fe)/Al in igneous biotites; siderophyllite from greisens plots closest to the origin, whereas tetrasilicic micas such as montdorite, always from peralkaline rocks, plot furthest from the origin. There is a continuous gradation of mica compositions and rock types in between.

Finally, I note that Jolliff et al., (1987) successfully used the vector approach (in Cerny & Burt, 1984) to plot the compositions of lithium micas in pegmatite. An important advantage of the vector approach is that analyses can be plotted on standard rectangular graph paper (rather than triangles) or using standard computer plotting routines for generating $X-Y$ or 'scatter diagrams'.

Acknowledgements I am grateful to the late D.S. Korzhinskii for inspiring me to learn Russian so that I could better understand his work, and to J.B. Thompson, Jr, of Harvard University for first suggesting that I use the exchange operator F_2O_{-1} in doing my Ph.D. thesis. I also thank Sue Selkirk and Deborah Barron for the drafting, S.W. Bailey and P. Cerny for helpful comments on the early (1984) version of this paper, and A.N. Konilov for his comments on a more recent version. Most of the results on lithium micas were derived during the fall of 1981, while I was studying Chinese greisen deposits, during an exchange visit supported by the Committee for Scholarly Communication with the People's Republic of China, National Academy of Sciences (USA). The graphical work was completed in 1984 (Cerny & Burt, 1984) and the text was expanded and brought up to date in 1987 while I was Visiting Scientist at the Lunar and Planetary Institute, Houston, which is operated by the Universities Space Research Association under Contract No. NASW–4066 with the National Aeronautics and Space Administration. This paper is Lunar and Planetary Institute Contribution No. 664.

References

Bailey, S.W. (1980). Structures of layer silicates. In *Mineralogical Society Monograph no. 5*, 1–123.

Bailey, S.W. (1984). Crystal chemistry of the true micas. *Rev. Mineral., 13*, 13–60.

Bargar, K.E., Beeson, M.H., Fournier, R.O. & Muffler, L.J.P. (1973). Present-day deposition of lepidolite from thermal waters in Yellowstone National Park. *Am. Mineral., 58*, 901–4.

Burt, D.M. (1974). Concepts of acidity and basicity in petrology – the exchange operator approach. *Geol. Soc. Am., Abstr. w. Progr., 6*, 674–6.

——(1976). Hydrolysis equilibria in the system $K_2O-Al_2O_3-SiO_2-H_2O-Cl_2O_{-1}$: comments on topology. *Econ. Geol., 71*, 665–71.

—— (1979). Exchange operators, acids, and bases. In *Problemy fiziko-khimischeskoi petrologii*, vol. 2, ed. V.A. Zharikov, V.I. Fonarev, & S.P. Korikovskii, pp. 3–15. Moscow: Nauka. Nauka.

Burt, D.M. & Burton, J.H. (1984). Vector representation of lithium and other mica compositions using exchange operators. *Geol. Soc. Am., Abstr. w. Progr., 16*, 460.

Cerny, P. & Burt, D.M. (1984). Paragenesis, crystallochemical characteristics, and geochemical evolution of micas in granitic pegmatites. *Rev. Mineral., 13*, 257–97.

Chatterjee, N.D. & Warhus, U. (1984). Ephesite, $Na(LiAl_2)[Al_2Si_2O_{10}](OH)_2$: II. Thermodynamic analysis of its stability and compatibility relations, and its geologic occurrence. *Contrib. Mineral. Petrol., 85*, 80–4.

Dymek, R.F. (1983). Titanium, aluminum and interlayer cation substitutions in biotite from high-grade gneisses, West Greenland, *Am. Mineral., 68*, 880–99.

Eugster, H.P. & Wones, D.R. (1962). Stability relations of the ferruginous biotite, annite. *J. Petrol., 3*, 82–125.

Evans, B.W. & Patrick, B.E. (1987). Phengite-3T in high pressure metamorphosed orthogneisses, Seward Peninsula, Alaska. *Canad. Mineral., 25*, 141–58.

Forbes, W.C. (1972). An interpretation of the hydroxyl contents of biotites and muscovites. *Mineral. Mag., 38*, 712–20.

Foster, M.D. (1960). Interpretation of the composition of lithium-micas. U.S. Geol. Surv. Prof. Paper 354-E, 115–47.

Francke, W., Jelinksi, B. & Zarei, M. (1982). Hydrothermal synthesis of an ephesite-like sodium mica. *N. Jahrb. Mineral., Monatsh.*, 1982, 337–40.

Ginzburg, A.I. & Berkhin, S.I. (1953). On the composition and chemical constitution of the lithium micas. *Trudy Mineral. Muzeya Akad. Nauk S.S.S.R., 5*, 90–131.

Gottesmann, B. & Tischendorf, F. (1980). Uber Protolithionit. *Z. geol. Wiss. (Berlin), 8*, 1365–73.

Green, T.H. (1981). Synthetic high-pressure micas compositionally intermediate between the dioctahedral and trioctahedral mica series. *Contrib. Mineral. Petrol., 78*, 452–8.

Grew, E.S., Hinthorne, J.R. & Marquez, N. (1986). Li, Be, B, and Sr in margarite and paragonite from Antarctica. *Am. Mineral., 71*, 1129–34.

Guidotti, C.V. (1984). Micas in metamorphic rocks. *Rev. Mineral., 13*, 357–467.

Hawthorne, F.C. & Cerny, P. (1982). The mica group. In *Mineralogical Association of Canada, Short Course Handbook, No. 8*, 63–98.

Hazen, R.M. & Wones, D.R. (1978). Predicted and observed compositional limits of trioctahedral micas. *Am. Mineral.*, **63**, 885–92.

Hewitt, D.A. & Wones, D.R. (1975). Physical properties of some synthetic Fe–Mg–Al trioctahedral biotites. *Am. Mineral.*, **60**, 854–62.

Hewitt, D.A. & Abrecht, J. (1986). Limitations on the interpretation of biotite substitutions from chemical analyses of natural samples. *Am. Mineral.*, **71**, 1126–8.

Jolliff, B.L., Papike, J.J. & Shearer, C.J. (1987). Fractionation trends in mica and tourmaline as indicators of pegmatite internal evolution: Bob Ingersoll Pegmatite, Black Hills, South Dakota. *Geochim. Cosmochim. Acta*, **51**, 519–34.

Korzhinskii, D.S. (1959). *Physicochemical basis of the analysis of the paragenesis of minerals.* (translated from 1957 Russian edn). New York: Consultants Bureau.

Labotka, T.C. (1983). Analysis of the compositional variations in biotite in pelitic hornfeldes from northeastern Minnesota. *A. Mineral.*, **89**, 900–14.

Levillain, C. (1980). Etude statistique des variations de la teneur en OH et F dans les micas. *Tschermaks Mineral. Petrogr. Mitt.*, **27**, 209–23.

Lewis, G.N. (1938). Acids and bases. *J. Franklin Institute*, **226**, 293–313.

Marakushev, A.A. & Tararin, I.A. (1967). Mineralogical criteria of alkalinity of granitoids. *Internat. Geol. Rev.*, **9**, 78–91 (translated from *Izv. Akad. Nauk SSSR, Ser. Geol.*, 1965, no. 3, 20–37).

Marakushev, A.A. & Perchuk, L.L. (1970). On the influence of the acidity and temperature of postmagmatic solutions on the compositions of micas and chlorite. In *Problems of hydrothermal ore deposition*, ed. Z. Pouba & N. Stemprok, Internat. Union Geol. Sci., A, No. 2, pp. 274–8. Stuttgart: Schwiezerbart.

Miyashiro, A. & Shido, F. (1985). Tschermak substitution in low- and middle-grade pelitic schists. *J. Petrol.*, **26**, 49–487.

Nockolds, S.R. (1947). The relation between chemical composition and paragenesis in the biotite micas of igneous rocks. *Am. J. Sci.*, **245**, 401–20.

Norman, D.I. & Palin, J.M. (1982). Volatiles in phyllosilicate minerals. *Nature*, **296**, 551–3.

Pavlishin, V.I., Semenova, T.F. & Rozhedestvenskaya, I.V. (1981). Protolithionite – 3T: structure, typomorphism, and practical importance. *Mineral. Zh.*, **3**, 37–60.

Rieder, M., Huka, M., Kucerova, D., Minarik, L., Obermajer, J. & Povondra, P. (1970). Chemical composition and physical properties of lithium–iron micas from the Krusne Hory Mts. (Erzgebirge). Part A: Chemical composition. *Contrib. Mineral. Petrol.*, **27**, 131–58.

Robert, J.L., Volfinger, M., Barrandon, J.N. & Basutcu, M. (1983). Lithium in the interlayer space of trioctahedral micas. *Chem. Geol.*, **40**, 337–51.

Rosenberg, P.E. (1987). Synthetic muscovite solid solutions in the system $K_2O–Al_2O_3–SiO_2–H_2O$. *Am. Mineral.*, **72**, 716–32.

Rub, M.G., Rub, A.K. & Akimov, V.M. (1986). Rare-metal granites of central Sikhote-Alin. *Izv. Akad. Nauk SSR, Ser. Geol.*, no. 7, 33–46.

Rutherford, M.J. (1973). Phase relations of aluminous iron biotites in the system $KAlSi_3O_8–KAlSiO_4–Al_2O_3–Fe–O–H$. *J. Petrol.*, **14**, 159–80.

Seifert, F. & Schreyer, W. (1971). Synthesis and stability of micas in the system $K_2O–MgO–SiO_2–H_2O$ and their relations to phlogopite. *Contrib. Mineral. Petrol.*, **30**, 196–215.

Smith, J.V. (1959). Graphical representation of amphibole compositions. *Am. Mineral.*, **44**, 437–40.

Speer, J.A. (1984). Micas in igneous rocks. *Rev. Mineral.*, **13**, 299–356.

Stern, W.B. (1979). Zur Strukturformelberechnung von Glimmermineralien. *Schweiz. Mineral. Petrogr. Mitt.*, **59**, 75–82.

Thompson, J.B., Jr. (1979). Tschermak component and reactions in pelitic schists. In *Problemy fiziko-khimicheskoi petrologii*, vol. 1, ed. V.A. Zharikov, V.I. Fonarev & S.P. Korikovskii, pp. 146–59. Moscow: Nauka.

Thompson, J.B., Jr. (1981). An introduction to the mineralogy and petrology of the biopyriboles. *Rev. Mineral.*, **9A**, 141–88.

Thompson, J.B., Jr. (1982). Composition space: an algebraic and geometric approach. *Rev. Mineral.*, **10**, 1–31.

Velde, B. (1972). Celadonite mica: solid solution and stability. *Contrib. Mineral. Petrol.*, **37**, 235–47.

Yoder, H.S. & Eugster, H.P. (1955). Synthetic and natural muscovites. *Geochim. Cosmochim. Acta.*, **8**, 225–80.

7

Thermodynamics of some framework silicates and their equilibria: application to geothermobarometry

LEONID L. PERCHUK, KONSTANTIN K. PODLESSKII and
LEONID Ya. ARANOVICH

Introduction

A review of thermodynamic properties of plagioclase, nepheline and alkali feldspar solid solutions is the subject of this paper with the aim of revising two-feldspar and nepheline–feldspar geothermometers. Sources of experimental data on equilibria of the solid solutions in the system $NaAlSiO_4$–$KAlSiO_4$–$CaAl_2Si_2O_8$–SiO_2 and data on the calorimetry of $NaAlSi_3O_8$–$KAlSi_3O_8$ and $NaAlSi_3O_8$–$CaAl_2Si_2O_8$ joins are presented in Table 7.1. The table shows that the volume of data on the equilibria exceeds that of the calorimetric data by an order of magnitude. Thermodynamic properties of the nepheline–kalsilite solid solution have not been studied calorimetrically at all. Data on equilibria of ternary solid solutions in the systems $CaAl_2Si_2O_8$–$NaAlSi_3O_8$–KAl Si_3O_8 and $KAlSiO_4$–$NaAlSiO_4$–SiO_2 are very limited. Temperature estimates from different versions of two-feldspar thermometer (Ryabchikov, 1965, Perchuk & Ryabchikov, 1968, Stormer, 1975, Perchuk & Aleksandrov, 1976, Whitney & Stormer, 1977, Brown & Parsons, 1981, 1985, Haselton *et al.*, 1983) based on the binary and ternary solution models show considerable divergence.

Attempts to develop a thermodynamic model of the ternary solid solution in the system $CaAl_2Si_2O_8$–$NaAlSi_3O_8$–$KAlSi_3O_8$ on the basis of experimental data by Seck (1971, 1972) and Johannes (1979) have been made by several authors (Saxena & Ribbe, 1972, Powell & Powell, 1977a, Ghiorso, 1984, Price, 1985, Green & Usdansky, 1984, 1986) with the purpose of creating a two-feldspar geothermobarometer, but due to low solubility of Ca in alkali feldspar and that of K in plagioclase the mixing properties derived have large uncertainties.

It should be noted that application of the ternary feldspar solution model may be restricted to high-temperature rocks, mainly of volcanic origin, and the binary solution

Table 7.1. *Sources of experimental data for evaluation of thermodynamics of some framework silicate solid solutions*

Solid solution	Solvus	Ion exchange	Calorimetry	Volume measurements
Ab^high–San	1–7,50	1,6–14	15–18	1,19–25
Ab^low–Mic	7,13,24	7,13,26,27	Data unknown	28,29
Ab^int–Or	7	7	Data unknown	Data unknown
Ab–An	30,31	12,13,30–32	15,33	33
Ab–San–An	34–37	12	Data unknown	Data unknown
Ne–Ks	7,13,38–40	7,13,41–44	Data unknown	7,39,45,46
Ne–Ks–Qz	47–49	7	Data unknown	Data unknown

Notes:
1 – Orville (1963); 2 – Luth & Tuttle (1966); 3 – Goldsmith & Newton (1974); 4 – Martin (1974); 5 – Smith & Parsons (1974); 6 – Lagache & Weisbrod (1977); 7 – Zyrianov *et al.* (1978); 8 and 9 – Wyart & Sabatier (1956, 1962); 10 – Iiyama *et al.* (1963); 11 and 12 – Iiyama (1965, 1966); 13 – Zyrianov & Perchuk (1975); 14 – Traetteberg & Flood (1972); 15 – Kracek & Neuvonen (1952); 16 – Waldbaum & Robie (1971); 17 – Hovis & Waldbaum (1977); 18 – Haselton *et al.* (1983); 19 – Perchuk & Andrianova (1968); 20 – Hovis (1977); 21 – Wright & Stewart (1968); 22 – Donnay & Donnay (1952); 23 – Luth & Querol-Suñe (1970); 24 – Orville (1967); 25 – Perchuk (1970); 26 – Bachinski & Müller (1971); 27 – Delbove (1975); 28 – Hovis & Peckins (1978); 29 – Kroll *et al.* (1986); 30 – Orville (1972); 31 and 32 – Kotel'nikov *et al.* (1981, 1986); 33 – Newton *et al* (1980); 34 and 35 – Seck (1971a, b); 36 – Seck (1972); 37 – Johannes (1979); 38 – Tuttle & Smith (1958); 39 – Ferry & Blencoe (1978); 40 – Yund *et al.* (1972); 41 – Debron *et al.* (1961); 42 – Debron (1965); 43 – Roux (1971); 44 – Wellman (1970); 45 – Smith & Tuttle (1957); 46 – Donnay *et al.* (1959); 47 – Hamilton & Mackenzie (1960); 48 – Hamilton (1961); 49 – Edgar (1964); 50 – Goldsmith & Newton (1974).

models can be satisfactorily used for thermometry of intrusive granitoids, pegmatites, migmatites and various metamorphic rocks.

Experimental studies of equilibria involving nepheline solid solution are limited to a few works (see Table 7.1). On their basis several attempts were made to create a thermodynamic model of nepheline solid solution and a nepheline–feldspar thermometer (Perchuk, 1965, 1968; Powell & Powell, 1977b). It was not clear for a long time why so sharp a difference in distribution of sodium between coexisting nepheline and alkali feldspar was observed for volcanic and plutonic rocks. Tilley (1954) explained this by a different degree of ordering in alkali feldspars from volcanic and intrusive rocks. Experimental studies (Zyrianov & Perchuk, 1975, Zyrianov *et al.*, 1978) have shown however that the difference is likely to be connected with high silica solubility in nepheline at high temperatures (see Fig. 7.1), which has been observed by Hamilton & MacKenzie (1960), Hamilton (1961) and Edgar (1964) up to 1065 °C when melting of nepheline–feldspar assemblage occurs at 1 kbar water pressure. Zyrianov *et al.* (1978) have fitted the solubility of albite in the nepheline–albite system to the following equation

$$X(\mathrm{Ab, Ne_{ss}}) = 0.03722 + 1.584 \times 10^{-4} t \tag{1}$$

or

$$N(\mathrm{Ab, Ne_{ss}}), \mathrm{wt\%} = 1.8514 + 2.8697 \times 10^{-2} t \tag{2}$$

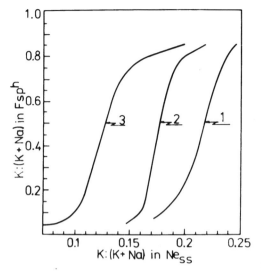

Fig. 7.1. Distribution of K and Na between alkali feldspar and nepheline solid solution of different silica content at temperature 800 °C and pressure 1000 bar (published with permission of the authors after Zyrianov et al., 1978): 1 – Ne$_{ss}$ without normative Qtz; 2 – Ne$_{ss}$ with 4.6 wt. % of normative Qtz; 3 – Ne$_{ss}$ with 13.8 wt. % of normative Qtz (silica saturated solvus). Isolines 1 and 2 reflect metastable equilibria.

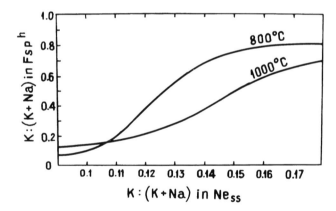

Fig. 7.2. Distribution of the alkalis in the sodium-rich region of the nepheline–alkali feldspar system at 800 °C (P = 6 bars) provided nepheline solid solution is saturated with normative silica (published with permission of the authors after Zyrianov et al., 1978).

Hamilton & MacKenzie (1960) and Hamilton (1961) also observed a decrease in the normative silica content in nepheline solid solution with potassium at constant T and P. Perchuk (1965) and Perchuk & Ryabchikov (1968) showed the insignificant effect of temperature on the nepheline–feldspar exchange equilibrium in the range of 400–700 °C. At higher temperatures, however, increase in the normative silica content in nepheline solid solution and homogenization of alkali feldspars affect exchange equilibrium dramatically making this thermometer applicable to volcanic rocks (Fig. 7.2).

Table 7.2. *Papers with results of evaluation of thermodynamic properties for feldspar and nepheline solid solutions*

| Solid solution | Experimental data used | | | |
	Solvus	Ion exchange	Calorimetry	Volume measurements
Ab^{high}–San	1–10	11–16,18	21,22	17–20,23
Ab^{low}–Mic	24	12,15,24,25	Data unknown	11,17,20,23,26
Ab–An	Data unknown	27–32	33	33
Ab–An–San	10,34–39	34–39	Data unknown	Data unknown
Ne–Ks–Qz	2–4,40	12–15,41,42	Data unknown	Data unknown

Notes:
1 – Ryabchikov (1965); 2–4 – Perchuk (1965, 1968, 1970); 5 and 6 – Thompson & Waldbaum (1968, 1969); 7 – Perchuk & Ryabchikov (1968); 8 – Haselton *et al.* (1983); 9 – Luth & Fenn (1973); 10 – Price (1985); 11 – Perchuk & Andrianova (1968); 12 – Perchuk *et al.* (1978a); 13 and 14 – Delbove (1971, 1975); 15 – Zyrianov *et al.* (1978); 16 – Traetteberg & Flood (1972); 17 – Orville (1967); 18 – Hovis & Waldbaum (1977); 19 and 20 – Hovis (1977, 1987 (personal communication)); 21 – Thompson & Hovis (1979); 22 – Haselton *et al.* (1983); 23 – Kroll *et al.* (1986); 24 – Bachinski & Müller (1971); 25 – Hovis (1979, 1983); 26 – Delbove (1975); 27 – Orville (1972); 28 – Perchuk *et al.* (1978b); 29 – Kotel'nikov (1980); 30 – Kotel'nikov *et al.* (1981, 1986); 31 – Seil & Blencoe (1979); 32 – Kurepin (1981); 33 – Newton *et al.* (1980); 34 – Saxena & Ribbe (1972); 35 – Ryabchikov (1975); 36 – Barron (1976); 37 – Powell & Powell (1977a); 38 – Ghiorso (1984); 39 – Green & Usdansky (1986); 40 – Ferry & Blencoe (1978); 41 – Powell & Powell (1977b); 42 – Roux (1974).

Thermodynamics of solid solutions

Alkali feldspars

High albite–sanidine solid solution. Ryabchikov (1965) and Perchuk (1965) evaluated mixing properties of albite–sanidine solution with the aim of modifying Barth's (1951) geothermometer and of deducing a nepheline–feldspar geothermometer (Perchuk, 1965; Perchuk & Ryabchikov, 1968). Extensive experimental data on solvus and exchange equilibria involving disordered alkali feldspars (Tables 7.1 and 7.2) provided a basis for further studies of thermodynamics of this solid solution.

In addition to volumetric data on albite–sanidine series derived from X-ray measurements (Hovis, 1977), results of the density measurements (Perchuk & Andrianova, 1968, Perchuk, 1970) are also available. The latter correspond to Margules parameters $W^V_{Ab} = 0.134$ and $W^V_{San} = 0.096$ cal/bar. Recently Hovis (1987, pers. comm.) reported $W^V_{Ab} = 0.076$ and $W^V_{San} = 0.111$ cal/bar, which are in contrast to the data by many other authors (see Tables 7.1–7.3).

In comparison to the volumetric data, calorimetric measurements of the mixing properties are relatively scarce (see Table 7.2). Margules parameters for the high albite – sanidine solution were calculated by many authors (Table 7.2). Table 7.3 shows that there are no dramatic differences between the parameters from different sources.

Table 7.3. *The Margules parameters for the system* $KAlSi_3O_8$–$NaAlSi_3O_8$–$CaAl_2Si_3O_8$

No.	Inter-action	W^H	W^S	W^V	Reference from Table 7.2
1	$\{$ Ab^h–San	7474	2.75	0.107 $\}$	17 (W^H and W^S), 18 (W^V)
	$\{$ San–Ab^h	4177	2.59	0.072 $\}$	
2	$\{$ Ab^h–San	5260	2.739	0.093 $\}$	11 and 12 (W^H, W^S and W^V)
	$\{$ San–Ab^h	6727	2.619	0.057 $\}$	
3	$\{$ Ab^h–San	7404	5.12	0.072 $\}$	21 (W^H and W^S), 20 (W^V)
	$\{$ San–Ab^h	4078	0	0.122 $\}$	
4	$\{$ Ab^h–San	4496	2.462	0.0862 $\}$	8 (W^H and W^S), 19 (W^V)
	$\{$ San–Ab^h	6530	2.462	0.0862 $\}$	
5	$\{$ Ab^h–San	4612	2.504	0.101 $\}$	This study
	$\{$ San–Ab^h	6560	2.486	0.074 $\}$	
6	$\{$ Ab^l–Mic	7973	6.476	0.123 $\}$	24 (W^H and W^S), 19 (W^V)
	$\{$ Mic–Ab^l	7491	2.165	0.123 $\}$	
7	$\{$ Ab^l–Mic	7151	5.307	0.142 $\}$	12 (W^H and W^S), 11 (W^V)
	$\{$ Mic–Ab^l	8227	3.227	0.074 $\}$	
8	$\{$ Ab^l–Mic	7594	5.931	0.142 $\}$	This study
	$\{$ Mic–Ab^l	7832	2.657	0.074 $\}$	(W^H and W^S), 11 (W^V)
9	$\{$ Ab–An	6748	—	— $\}$	33 (ΔH_{sol})
	$\{$ An–Ab	2025	—	— $\}$	
10	$\{$ Ab–An	6860	3.877	0 $\}$	This study
	$\{$ An–Ab	1980	1.526	0 $\}$	
11	$\{$ An–San	6688	-4.83	0.010 $\}$	38 (W^H and W^S),
	$\{$ San–An	16125	2.644	0.074 $\}$	this study (W^V)
	$\{$ An–San–Ab	-3314	-3.498	-0.058 $\}$	
13	$\{$ An–San	-15634	-27.27	0.505 $\}$	39 (W^H, W^S and W^V)
	$\{$ San–An	15609	2.997	0.232 $\}$	

Compositional dependences of the excess Gibbs free energy, enthalpy and entropy based on the parameters from Table 7.3 are shown in Figs. 7.3 and 7.4. It can be seen that the asymmetry of curve 1, which reflects calorimetrically determined excess enthalpy (Hovis & Waldbaum, 1977), is in contrast with the enthalpy data derived from the phase equilibria (compare also W^H for No. 1 with W^H for No. 2, No. 4 and No. 5 in Table 7.3).

Haselton *et al.* (1983) have computed the Margules parameters from 128 data points on solvus to deduce equations for mixing enthalpy –

$$H^m = X_{Ab}X_{San}(4496X_{San} + 6530X_{Ab}) \tag{3}$$

– and mixing entropy –

$$S^m = -R(X_{Ab}\ln X_{Ab} + X_{San}\ln X_{San}) + 2.462X_{Ab}X_{San}. \tag{4}$$

These authors have also concluded that calorimetric measurements of the heats of solution (Waldbaum & Robie, 1971; Hovis & Waldbaum, 1977) disagree with the data on the solvus and recommended the Margules parameters based on the phase equilibria data. Compositional dependences of the excess Gibbs free energies calculated with

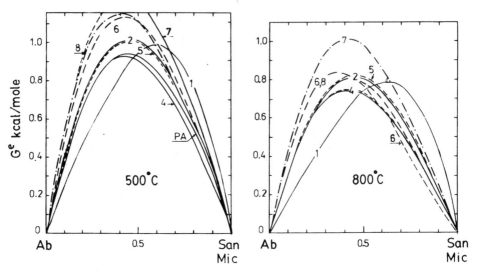

Fig. 7.3. Integral excess Gibbs free energy for disordered (dashed and solid curves) and ordered (curve 8) alkali feldspar solid solutions calculated with the Margules parameters from Table 7.3 at P = 1 bar. Numbers of curves correspond to row numbers in Table 7.3.

equations (I–10) and (I–11)* from extensive experimental data on exchange equilibria of disordered alkali feldspar with aqueous chloride solutions and chloride melts (Perchuk *et al.*, 1978a, b; Zyrianov *et al.*, 1978) practically coincide with the earlier results by Perchuk & Andrianova (1968) and Haselton *et al.* (1983). This is demonstrated in Fig. 7.3. However, the data of Zyrianov *et al.* (1978) have not been taken into account by Haselton *et al.* (1983). Since the Margules parameters derived from the data on the solvus and on the exchange equilibrium between disordered alkali feldspars and aqueous solutions (39 data points† from Zyrianov *et al.* (1978).

$$Ab_{(high)} + KCl_{(aq)} = NaCl_{(aq)} + San \qquad (5)$$

are similar (Table 7.3, No. 2 and No. 4), we have treated them together to get the results shown in Table 7.3, No. 5. The equation

$$G^e_{(Fsp,\,high)} = X_{Ab}X_{San}[X_{Ab}(6548 - 2.42T + 0.0714P)$$
$$+ X_{San}(5014 - 2.269T + 0.0981P)] \qquad (6)$$

fits these data points (167 in total), and curve 5 in Fig. 7.3 illustrates compositional dependence of G^e at 500° and 800 °C and 1 bar. Diagram (a) in Fig. 7.5 shows that the mixing properties of high alkali feldspar series agree with the experimental data on both solvus and exchange equilibria.

---

* here and further on references to formula numbers with 'I' correspond to equation numbers in 'Introduction' to this part of the book.

† in order to avoid possible effect of the different ion-exchange media the data on the equilibrium of the feldspar with chloride melts were not incorporated.

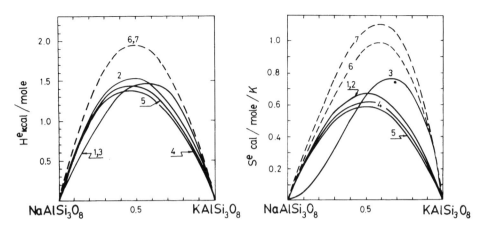

Fig. 7.4. Integral excess enthalpy and entropy for disordered (solid lines) and ordered (dashed lines) alkali feldspar solid solutions calculated with the Margules parameters from Table 7.3 at $P = 1$ bar. Numbers of curves correspond to row numbers in Table 7.3.

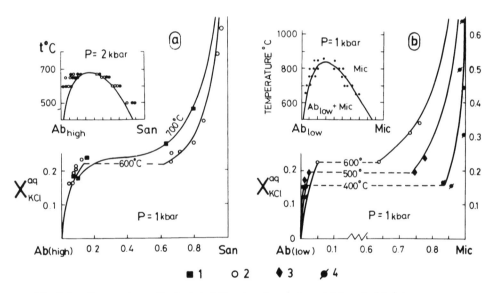

Fig. 7.5. Comparison of calculated (lines) and experimental (symbols) data on phase equilibria of disordered (diagram a) and ordered (diagram b) feldspars. The distribution isotherms: 1 – 700°, 2 – 600°, 3 – 500° and 4 – 400 °C.

Low albite–microcline solid solution. Hovis (1987, pers. comm) has observed results of X-ray measurements for low series of alkali feldspars and fitted them with the Margules parameters $W^V_{\text{Ab(low)}} = 0.141$ and $W^V_{\text{Mic}} = 0.066$ cal/bar, which are comparable to $W^V_{\text{Ab(low)}} = 0.142$ and $W^V_{\text{Mic}} = 0.074$ cal/bar derived from density determinations by Perchuk & Andrianova (1968) and Perchuk (1970).

Calorimetric data on mixing properties of ordered alkali feldspar are lacking, and experimental data on equilibria involving this solid solution are quite limited

(Bachinski & Müller, 1971; Zyrianov & Perchuk, 1975; Zyrianov et al., 1978). Bachinski and Müller (1971) have shown that thermodynamic treatment of their data on the low albite–microcline solvus based on homogenization and ion-exchange experiments at 650–900 °C give a wide range of the Margules parameters depending on the experimental method and treatment procedure. The Margules parameters derived from the 22 exchange data pairs are presented in Table 7.3, No. 6.

Zyrianov & Perchuk (1975) conducted 55 runs on ion exchange of low feldspar series with K–Na chloride melts and aqueous solutions in a wide range of P–T conditions. Perchuk et al. (1978) treated these data on the basis of equations (I–10) and (I–11) and tabulated the excess free energies. Thermodynamic treatment of the 19 data points on exchange equilibria

$$Ab_{(low)} + KCl_{(aq)} = NaCl_{(aq)} + Mic \tag{7}$$

at a temperature of 400–600 °C and pressure of 1 kbar led to the asymmetric solid solution model (equations I–25 and I–27) with the following Margules parameters:

$$\left. \begin{aligned} W^G_{(Ab,\,low)} &= 7151 - 5.307T + 0.142P, \\ W^G_{(Mic)} &= 8227 - 3.227T + 0.074P. \end{aligned} \right\} \tag{8}$$

Within experimental uncertainties these parameters are similar to those derived by Bachinski & Müller (1971) from high temperature data (Table 7.3, No. 6), and we have combined them to get the Margules parameters for a temperature range of 400–900 °C and pressure up to 10 kbar:

$$\left. \begin{aligned} W^G_{(Ab,\,low)} &= 7594 - 5.931T + 0.142P, \\ W^G_{(Mic)} &= 7832 - 2.657T + 0.074P. \end{aligned} \right\} \tag{9}$$

According to equation (I–25)

$$\begin{aligned} G^e_{(Fsp,\,low)} = X_{Ab}X_{Mic}[&X_{Ab}(7832 - 2.657T + 0.074P)] \\ &+ X_{Mic}(7594 - 5.931T + 0.142P). \end{aligned} \tag{10}$$

Compositional dependences of G^e, H^e and S^e calculated with these parameters are compared with those for high feldspar series in Figs. 7.3 and 7.4, and in Table 7.4. Diagram (b) in Fig. 7.5 demonstrates satisfactory agreement of isotherms and solvi calculated using equation (10) with the experimental data by Bachinski & Müller (1971).

Senderov & Yas'kin (1976) have determined the high-temperature stability limit of microcline at 500 °C. According to Zyrianov (1981) microcline/orthoclase transition occurs at 375 ± 25 °C. The transition temperatures calculated from reference data by Robie et al. (1978) and Dorogokupets & Karpov (1984) are 300 °C and 330 °C, respectively.

Table 7.4 and Figs. 7.4 and 7.5 show that thermodynamic properties are functions

Table 7.4. *Thermodynamic properties of K–Na feldspars as a function of molar fraction of* $KAlSi_3O_8$

	No. in Table 7.3						
	1	2	3	4	5	7	8
W^H_{Ab}	7474	5260	7404	4496	4612	7151	7594
W^H_{Or}	4177	6742	4078	6530	6560	8227	7832
X^{Fsp}_{Or}	Mixing enthalpy, H^e						
0.1	405	592	396	569	572	730	702
0.2	773	1029	758	979	978	1282	1245
0.3	1085	1320	1065	1243	1254	1660	1630
0.4	1318	1474	1298	1371	1387	1871	1857
0.5	1456	1498	1435	1378	1396	1922	1928
0.6	1477	1403	1457	1274	1293	1820	1845
0.7	1362	1197	1345	1072	1091	1569	1609
0.8	1090	888	1078	784	800	1179	1223
0.9	643	487	636	423	432	653	685
W^S_{Ab}	2.75	2.739	5.12	2.462	2.504	5.307	5.931
W^S_{Or}	2.59	2.619	0	2.462	2.486	3.227	2.657
X^{Fsp}_{Or}	Excess entropy, S^e						
0.1	0.234	0.237	0.046	0.222	0.225	0.309	0.268
0.2	0.420	0.423	0.164	0.393	0.400	0.583	0.530
0.3	0.554	0.558	0.322	0.517	0.525	0.809	0.764
0.4	0.637	0.640	0.492	0.591	0.600	0.974	0.952
0.5	0.668	0.670	0.640	0.616	0.625	1.067	1.074
0.6	0.644	0.646	0.737	0.591	0.600	1.074	1.109
0.7	0.567	0.567	0.753	0.517	0.525	0.983	1.039
0.8	0.435	0.434	0.655	0.393	0.400	0.783	0.844
0.9	0.246	0.245	0.415	0.223	0.225	0.459	0.504
W^V_{Ab}	0.107	0.093	0.072	0.086	0.101	0.141	0.142
W^V_{Or}	0.072	0.057	0.122	0.086	0.074	0.066	0.074
X^{Fsp}_{Or}	Mixing volume, V^e						
0.1	0.007	0.005	0.011	0.008	0.007	0.007	0.007
0.2	0.013	0.010	0.018	0.014	0.013	0.014	0.014
0.3	0.013	0.010	0.018	0.014	0.013	0.014	0.014
0.4	0.021	0.017	0.024	0.021	0.020	0.024	0.024
0.5	0.022	0.019	0.024	0.022	0.022	0.027	0.027
0.6	0.022	0.019	0.022	0.021	0.022	0.028	0.028
0.7	0.020	0.018	0.018	0.018	0.020	0.026	0.026
0.8	0.016	0.015	0.013	0.014	0.015	0.021	0.021
0.9	0.019	0.008	0.007	0.008	0.009	0.012	0.012

of alkali feldspar ordering. Since similar data on feldspars with intermediate ordering are lacking, a linear combination of the mixing parameters obtained for the high and low series can be proposed as a substitute for the intermediate feldspar parameters in practical calculations. The index Z which has been proposed by Hovis (1986) to describe ordering of Al and Si on tetrahedral sites can be employed to achieve this aim:

$$\left.\begin{array}{l} W_i^H = ZW_i^H{}_{\text{(low)}} + (1-Z)W_i^H{}_{\text{(high)}}, \\ W_i^S = ZW_i^S{}_{\text{(low)}} + (1-Z)W_i^S{}_{\text{(high)}}, \\ W_i^V = ZW_i^V{}_{\text{(low)}} + (1-Z)W_i^V{}_{\text{(high)}}, \end{array}\right\} \tag{11}$$

where Z is defined by the equation

$$Z = 138.575 + 19.135\, c_k \tag{12}$$

and c_k is the unit cell parameter which can be calculated from the observed value c_{obs}, as proposed by Hovis (1986).

Plagioclase

Anorthite–albite solid solution.   According to Newton *et al.* (1980) and Carpenter *et al.* (1985) mixing volume of the anorthite–albite solution is close to zero. After Kracek & Neuvonen (1952) had measured heat of solution for plagioclase series, many authors proposed their description of thermodynamic properties of the solid solution (see Tables 7.1 and 7.2). As a result, the following three main approaches may be distinguished:

(i) Thermodynamic treatment of experimental data on phase equilibria and high-temperature calorimetry using Margules empiric formulation and equations (I–10) and (I–11). In particular, with this approach Orville (1972), Perchuk & Ryabchikov (1976) and Fei *et al.* (1986) have shown that activity of albite component is close to its mole fraction at $X_{\text{Ab}}^{\text{Pl}} > 0.5$;

(ii) Evaluation of the mixing properties based on statistical models for configuration entropy with some crystal-chemical constraints based on the local charge-balance and Al-avoidance models (Kerrick & Darken, 1975; Anderson & Mazo, 1980; Kotel'nikov *et al.*, 1986)

(iii) Calculation of the mixing properties with regard to phase transition of the second (or higher) order from albite ($C\bar{1}$) structure to anorthite ($I\bar{1}$) structure at high temperatures and anorthite-rich compositions (Carpenter *et al.*, 1985).

The first approach does not require knowledge of detailed characteristics of the solid solution structure. With this approach Perchuk (1970) and Zyrianov *et al.* (1978), for example, treated data on equilibria of alkali feldspars of different ordering degree with ion-exchange media and showed that equations (10–14) in Chapter 1 describe mixing properties of these solid solutions with reasonable accuracy.

Diagnostics of Al/Si ordering in the plagioclase solid solution is a problem. Since solution of the configuration entropy equation is related to both Lovenstein's rule and the electro-neutrality principle (Kerrick & Darken, 1975; Kurepin, 1981) we have to know the occupancy of tetrahedral sites by Al atoms as a function of temperature and composition of plagioclase. Most multi-site models of plagioclase solid solution assume equal distribution of Al atoms between four tetrahedral sites. Kerrick & Darken (1975), for example, have deduced the following equation for configuration entropy:

$$S^m = S^{conf} = -R\{X\ln a_{Ab}^{conf} + (1-X)\ln a_{An}^{conf}\}$$
$$= -R\{X\ln(X^2(2-X)) + (1-X)\ln((1-X)(2-X))^2/4\} \qquad (13)$$

where $X = X_{Ab}^{Pl}$. However, preferential occupancy of the T_{10} site by Al atoms revealed by Kroll & Ribbe (1980) should lead to decrease of the configuration entropy with respect to that calculated from the statistical distribution model.

Calculations of the mixing properties based on the $C\bar{1}$–$I\bar{1}$ transition (Carpenter & Ferry, 1984, Carpenter et al., 1985, Hodges & Crowley, 1985) seem uncertain because the composition range of the transition at each given temperature is unknown and both temperature and heat of ordering of pure anorthite are obtained by extrapolation. The assumption of ideal mixing properties of the $C\bar{1}$ plagioclase, on which the extrapolation is based, is likely to be an oversimplification.

As the first approach is accurate enough to describe the data on phase equilibria, we have employed it to calculate Margules parameters.

Coefficients in the equation

$$\Delta H_{sol} = \sum_{n=0}^{3} B_n X_{An}^n$$

based on the subregular solid solution model have been derived from heat of solution of plagioclases of different composition (including pure anorthite and albite) in $2PbO \cdot B_2O_3$ at 970 K measured by Newton et al. (1980):

$$\begin{aligned} B_0 &= 16.57\ (\pm 0.38); & B_2 &= -2.90\ (\pm 0.34); \\ B_1 &= -2.35\ (\pm 2.21); & B_3 &= 4.88\ (\pm 0.46). \end{aligned} \qquad (14)$$

Coefficients B_n are related to the heat of solution and Margules parameters as follows

$$\begin{aligned} \Delta H_{sol}(\mathrm{Ab}) &= B_0; & 2W_{An}^H - W_{Ab}^H &= B_2; \\ \Delta H_{sol}(\mathrm{An}) &= \sum_{n=0}^{3} B_n; & W_{Ab}^H - W_{An}^H &= B_3. \end{aligned} \qquad (15)$$

The Margules parameters can be readily obtained from equations (14) and (15):

$$\begin{aligned} W_{An}^H &= 1980(\pm 800); \\ W_{Ab}^H &= 6860\ (\pm 800). \end{aligned} \qquad (16)$$

Parameters (16) and their uncertainties differ slightly from those given by Newton et al. (1980) (see Table 7.3, No 9), and this is probably due to the fact that we have treated the calorimetric data for both solid solutions and end-members.

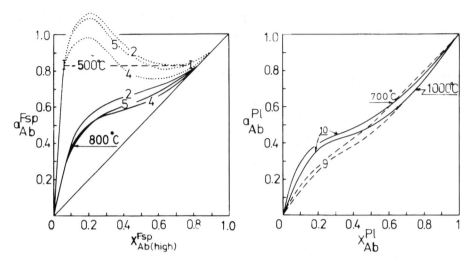

Fig. 7.6. Activities of NaAlSi$_3$O$_8$ in alkali feldspar (diagram a) and plagioclase
(diagram b) calculated with the Margules parameters from Table 7.3 at
pressure 1 bar. Numbers of curves correspond to row numbers in Table
7.3. Dotted curves in diagram a show activities of NaAlSi$_3$O$_8$ in the alkali
feldspar binary solution and a dashed line connects compositions of the
two phases in equilibrium at 500 °C and a(Ab, Fsp^h) = 0.83 ± 0.03.
Activities of albite in plagioclase are calculated after equation (22) with
equation (17) for line 9 and equation (20) for line 10. Bars reflect
uncertainties at temperature 500 °C.

To evaluate the entropy of mixing Newton *et al.* (1980) used a model of Kerrick &
Darken (1975) and obtained the following equations for excess Gibbs free energies of
albite and anorthite in the binary plagioclase solid solution;

$$G^e_{Ab} = X^2_{An}(6746-9442X^{Pl}_{Ab}) + RT\ln[X^{Pl}_{Ab}(2 - X^{Pl}_{Ab})], \qquad (17)$$

$$G^e_{An} = X^2_{Ab}(2025 + 9442X^{Pl}_{An}) + RT\ln[X^{Pl}_{An}(1 + X^{Pl}_{An}{}^2/4], \qquad (18)$$

where the second terms of the right sides represent the configuration activities in the
plagioclase solution. We have used the mixing enthalpy parameter (16) to calculate the
excess entropy parameters from phase equilibrium data referred to in Table 7.1.

$$\left.\begin{array}{l} W^S_{An} = 1.526 \pm 0.478; \\ W^S_{Ab} = 3.874 \pm 0.916. \end{array}\right\} \qquad (19)$$

Relatively large uncertainties of the entropy parameters (19) reflect scatter in activity
coefficients estimated from the experimental data (see also Carpenter & Ferry, 1984).
    With regard to equation (19) the excess Gibbs free energies can be defined as

$$G^e_{Ab} = (1 - X^{Pl}_{Ab})^2[(6860 - 3.874T) - 2X^{Pl}_{Ab}(4880 - 2.348T)], \qquad (20)$$

$$G^e_{An} = (1 - X^{Pl}_{An})^2[(1980 - 1.526T) - 2X^{Pl}_{An}(4880 - 2.348T)]. \qquad (21)$$

Fig. 7.6 presents a comparison of activities of albite in the plagioclase solid solution
based on equations (17) and (20) with regard to the equation

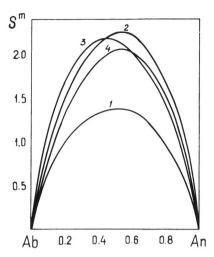

Fig. 7.7. Mixing entropies for plagioclase as a function of its composition. Calculations were made after different models; 1 – ideal one-site solid solution; 2 – after Kerrick & Darken (1975); 3 – after equation (23); 4 – after subregular model with parameters from row 10 in Table 7.3.

$$a_{Ab}^{Pl} = X_{Ab}^{Pl} \exp(G_{Ab}^{e}/RT). \tag{22}$$

The configuration entropies have also been estimated with a statistical approach based on the assumption that the T_{10} sites in the plagioclase structure are occupied exclusively by Al atoms, i.e. the plagioclase formula is $(Ca_xNa_{1-x})Al(Al_xSi_{3-x})O_8$. According to this model the configuration entropy is described as

$$S^m = -R\{X\ln X + (1-X)\ln(1-X) + X[\ln X + 2\ln((3-X)/2)] + 3(1-X)\ln((3-X)/3)\} = S^{ideal} + S^e \tag{23}$$

where $X = X_{An}^{Pl}$.

In Fig. 7.7 the excess entropy calculated with this equation is compared to the results obtained with both the Kerrick–Darken model and the subregular model with the Margules parameters defined by equations (19). Curve 4 which is based on the phase equilibrium data probably corresponds to the real distribution of Al atoms in plagioclase structure.

Nepheline solid solution

Data on structure and chemistry of nepheline solid solution were compiled by Deer *et al.* (1963). Donnay *et al.* (1959) proposed the nepheline formula based on 32 oxygens, and Barth (1963) modified this formula in terms of the following end-members:

nepheline	$Na_4Na_4[Al_8Si_8]O_{32}$
kalsilite	$K_4K_4[Al_8Si_8]O_{32}$
anorthite	$\square_4Ca_4[Al_8Si_8]O_{32}$
'quartz'	$\square_4\square_4[Si_{16}]O_{32}$

where $\square$ denotes a vacant site. According to Barth, the 'ideal' nepheline of Buerger *et al.* (1954) can be described by the formula

$$Na_{5.5}K_{1.3}\square_{0.8}Ca_{0.4}[Al_{7.6}Si_{8.4}]O_{32}$$

with $X_K = 0.191$. Morozewicz (1930) believed that nepheline $Na_3KAl_4Si_4O_{16}$ should be considered as an intermediate compound in the $NaAlSi_3O_8$–$KAlSi_3O_8$ system. This idea has been supported by Tuttle & Smith (1958). In fact, compositional dependence of the nepheline unit-cell parameters (Smith & Tuttle, 1957, Donnay *et al.*, 1959) shows inflection at $X_K \cong 0.25$. Excess enthalpy and entropy of the binary nepheline–kalsilite solution estimated by Perchuk (1965, 1968), Perchuk & Ryabchikov (1968), Perchuk *et al.* (1978) and Zyrianov *et al.* (1978) change sign at compositions close to the Morozewicz nepheline. Perchuk & Ryabchikov (1968) showed that the stability field of the $Ne_{75}Ks_{25}$ phase 'wedges out' at temperatures above 900 °C.

Using experimental data of Debron (see Tables 7.1 and 7.2) Perchuk (1965, 1968) calculated the following Margules parameters for the nepheline–kalsilite binary solution:

$$\left.\begin{array}{l} W_{Ne}^G = 6830 - 2.1T; \\ W_{Ks}^G = 6015 - 2.4T. \end{array}\right\} \tag{24}$$

Ferry & Blencoe (1978) presented experimental data on solvus and thermodynamic properties in terms of $Na_3KAl_4Si_4O_{16}$ and $K_4Al_4Si_4O_{16}$ end-members. M. Powell & R. Powell (1977) have calculated thermodynamic properties of Ne_{ss} in the system nepheline–kalsilite with the $Ne_{75}Ks_{25}$ intermediate compound in terms of the structural model of Tuttle & Smith (1958).

Hamilton & MacKenzie (1960) and Hamilton (1961) found correlation between sodium and excess silica content in nepheline, and, with regard to this, we consider the nepheline solid solution as a ternary mixture of $NaAlSiO_4$(Ne), $KAlSiO_4$(Ks) and $Na_{0.5}Al_{0.5}Si_{1.5}O_4$ (Ab^{Ne}), with mole fractions of the end-members being expressed as

$$\left.\begin{array}{l} X(Ks, Ne_{ss}) = Ks/(Ks + Ne + Ab^{Ne}) = X_1, \\ X(Ne, Ne_{ss}) = Ne/(Ks + Ne + Ab^{Ne}) = X_2, \\ X(Ab, Ne_{ss}) = Ab^{Ne}/(Ks + Ne + Ab^{Ne}) = X_3. \end{array}\right\} \tag{25}$$

This model allows us (i) to avoid problems related to distribution of the alkalis between different sites and (ii) to account for major isomorphism in nepheline.

Experimental data by Zyrianov *et al.* (Table 7.1) on exchange equilibria of nepheline with alkali chloride melts and aqueous solutions

$$NaAlSiO_4 + KCl = NaCl + KAlSiO_4 \tag{26}$$

were treated to obtain Margules parameters for binary nepheline–kalsilite solution;

$$H^e = X_2X_1[3301 + 5306(X_1 - X_2) - 2591(X_1 - X_2)^2], \tag{27}$$

$$S^e = X_2X_1[1.10 + 2.201(X_1 - X_2)]. \tag{28}$$

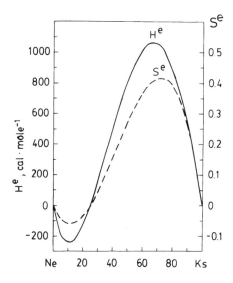

Fig. 7.8. Compositional dependences of enthalpy and entropy for the pseudobinary nepheline solid solution $(Na,K)_{0.94}Al_{0.9}Si_{1.065}O_4$ calculated after equations (27) and (28), respectively, at pressure 1 bar.

Fig. 7.8 shows results of calculations with these formulae. Partial properties can be derived from the following equations:

$$G^e(Ks, Ne_{ss}) = X_2^2[A_0^G + A_1^G(3X_1 - X_2) - A_2^G(X_1 - X_2)(5X_1 - X_2)], \qquad (29)$$

$$G^e(Ne, Ne_{ss}) = X_1^2[A_0^G + A_1^G(3X_2 - X_1) - A_2^G(X_1 - X_2)(5X_1 - X_2)], \qquad (30)$$

where $A_i^G = A_i^H - A_i^S T$; and

$$\left.\begin{array}{l} A_0^G = 3301 - 1.1T, \\ A_1^G = 5306 - 2.201T, \\ A_2^G = -2591. \end{array}\right\} \qquad (31)$$

These parameters correspond to Gibbs free energy of reaction (26)

$$\Delta G_{T(26)}^0 = \Delta H_{T(26)}^0 - T\Delta S_{T(26)}^0 \qquad (32)$$

presented in Table 7.5.

Fig. 7.8 shows that both excess enthalpy and entropy of the nepheline solid solution change sign at $K/(K + Na) \simeq 0.25$ in accordance with previous calculations by Perchuk (1965, 1970) and Perchuk et al. (1978a).

Zyrianov et al. (1978) deduced the compositional dependence of the volume of mixing from data on the unit-cell volumes of *natural* nephelines. The authors revealed negative deviation from ideal mixing volume for the nephelines with $K/(K + Na) \leqslant 0.6$ and the positive mixing volume for the nephelines with $K/(K + Na) > 0.6$. X-ray data by Ferry & Blencoe (1978) on the unit-cell dimensions of *synthetic* nephelines imply a marked difference in compositional dependences of the mixing volume for the same

145

Table 7.5. *Standard thermodynamic parameters for some equilibria discussed*

No in text	Reaction	ΔH_T^0	ΔS_T^0	ΔV^0 cal/ kbar
5	$Ab_{(high)} + KCl_{(aq)} = NaCl_{(aq)} + San$	2961	0.268	3
7	$Ab_{(low)} + KCl_{(aq)} = NaCl_{(aq)} + Mic$	4291	2.502	3
26	$Ne + KCl_{(aq)} = NaCl_{(aq)} + Ks$	-304	-0.214	1
74	$Ne + San = Ab_{(high)} + Ks$	-3265	-0.482	-13
75	$Ne + Mic = Ab_{(low)} + Ks$	-4595	-2.716	-29
35	$_{0.5}Ab$ (in Ne_{ss}) $= _{0.5}Ab$ (in Fsp)	-3028	≈ 0	-93
65[a]	$Ab_{(low)} = Ab_{(high)}$	2600	4.541	9

Notes:

[a] Parameters are calculated at temperature 298 K and pressure 1 bar (Robie *et al.*, 1978)

range of $K/(K + Na)$ but give no indication of the positive deviation of the mixing volume in the potassium-rich region.

We have compiled the volumetric data by Smith & Tuttle (1957), Donnay *et al.* (1959), Ferry & Blencoe (1978) and Zyrianov *et al.* (1978) to derive the compositional dependence of the $NaAlSiO_4$–$KAlSiO_4$ mixing volume. With sharp increase in the nepheline volume at $K/(K + Na) \simeq 0.02$ and insignificant inflection at $K/(K + Na) \simeq 0.25$ being ignored, the dependence at $V^0(Ne) = 1.2945$ cal/bar and $V^0(Ks) = 1.4461$ cal/bar can be expressed in terms of equation (I–12) as

$$V^e = X(1 - X)[-0.077 + 0.115(2X - 1) - 0.013(2X - 1)^2] \qquad (33)$$

where $X = K/(K + Na)$ and $0 \le X \le 1$.

The excess volume dependence for the range $0 < X < 0.6$ ('nepheline' solution) can also be described as

$$V^e = -0.15082X(1 - X) \qquad (34)$$

with $V^0(Ne) = 1.2945$ cal/bar and $V^0(Ks) = 1.4726$ cal/bar. Within a range of $0.6 > X > 1$ ('kalsilite' solution) the solid solution volume can be obtained as a volume of the ideal mixture with $V^0(Ne) = 1.284$ cal/bar and $V^0(Ks) = 1.4461$ cal/bar. Fig. 7.9 illustrates both descriptions discussed.

Experimental data by Greig & Barth (1938), Hamilton & MacKenzie (1960) and Hamilton (1961) on the excess silica content in nepheline in equilibrium with alkali feldspar have been used to describe thermodynamic properties of the ternary solid solution in the system $NaAlSiO_4$–$KAlSiO_4$–$Na_{0.5}Al_{0.5}Si_{1.5}O_4$ using parameters (31) for the binary Ne–Ks solution. Treatment of the above-listed experimental data in terms of the reaction

$$Na_{0.5}Al_{0.5}Si_{1.5}O_4 = \tfrac{1}{2}NaAlSi_3O_8$$
$$\text{Ab in } Ne_{ss} \qquad \text{Ab in Fsp} \qquad (35)$$

resulted in the following parameters:

146

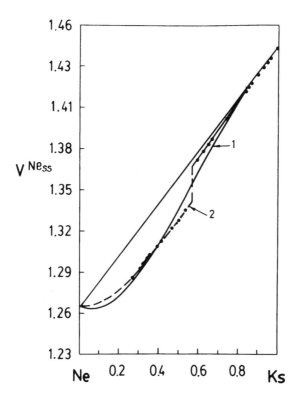

Fig. 7.9. The mixing volume properties of the NaAlSiO$_4$–KAlSiO$_4$ solid solution calculated after two different models: 1 – with equation (33); 2 – with equation (34) for 'nepheline' solution and ideal 'kalsilite' solution.

$$\Delta G^0(35) = -3028, \tag{36}$$

$$W^H_{23} = W^G_{23} = -554, \tag{37}$$

$$W^H_{13} = W^G_{13} = 1013, \tag{38}$$

where 1 stands for Ks, 2 – Ne, and 3 – Ab in Ne$_{ss}$ (Ab^{Ne}). Excess Gibbs free energy for the ternary nepheline solution can be described with the Redlich–Kister equation

$$G^e = X_1 X_2 [A^G_0 + A^G_1(X_1 - X_2) + A^G_2(X_1 - X_2)^2] \\ + 1013 X_1 X_3 - 554 X_2 X_3 \tag{39}$$

where A^G_0, A^G_1 and A^G_2 correspond to equations (31). Equation (39) can be easily modified to an equation of (I–30) type.

Taking into account that

$$Q_1 = X_1 + X_3, \tag{40}$$

$$Q_2 = 4X_1^2 - 3X_1 - 2X_3 + 6X_1 X_3 + 2X_3^2, \tag{41}$$

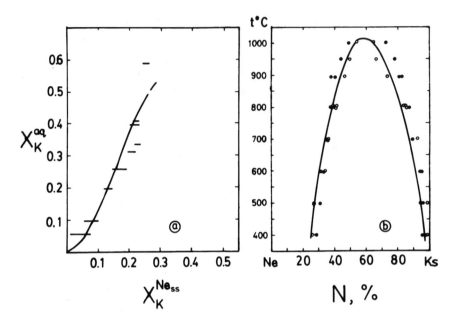

Fig. 7.10. Distribution of the alkalis between aqueous chloride solution (aq) and nepheline solid solution at 800 °C and $P = 1$ kbar (diagram a) and solvus for Ne_{ss} at $P = 0.5$ kbar (diagram b). Lines calculated with the nepheline solid solution properties defined by equations (46)–(51) and the data of Table 7.5. Bars in diagram a reflect uncertainties of experimental results by Zyrianov *et al.* (1978). Circles in diagram b show experimental data of Ferry & Blencoe (1978) for the Ne–Ks solvus at 5 kbar (solid) and 0.5 kbar (open).

$$Q_3 = 12X_1^3 - 16X_1^2 + 5X_1 + 3X_3 - 20X_1X_3 + 15X_1X_3^2$$
$$- 6X_3^2 + 24X_1^2X_3 + 3X_3^2, \tag{42}$$

and

$$q_1 = X_2 + X_3, \tag{43}$$

$$q_2 = X_2(3 - 4X_2) + X_3(2 - 6X_2 - 2X_3), \tag{44}$$

$$q_3 = 4X_2^2(3X_2 - 4) + X_2(5 - 20X_3 + 24X_2X_3)$$
$$+ X_3^2(15X_2 - 6 + 3X_3), \tag{45}$$

we get the following expressions for the partial thermodynamic functions of Ne, Ks and Ab^{Ne} in the nepheline solid solution:

$$H^e(Ne, Ne_{ss}) = X_1(3301Q_1 + 5306Q_2 - 2591Q_3)$$
$$- 1013X_1X_3 - 554Q_1X_3, \tag{46}$$

$$S^e(Ne, Ne_{ss}) = X_1(1.1Q_1 + 2.201Q_2), \tag{47}$$

$$V^e(Ne, Ne_{ss}) = -X(0.077Q_1 - 0.115Q_2 - 0.013Q_3). \tag{48}$$

148

$$H^e(\text{Ks, Ne}_{ss}) = X_2(3301q_1 + 5306q_2 - 2591q_3),$$
$$- 1013q_1X_3 + 554X_2X_3, \tag{49}$$

$$S^e(\text{Ks, Ne}_{ss}) = X_2(1.1q_1 + 2.201q_2), \tag{50}$$

$$V^e(\text{Ks, Ne}_{ss}) = -X_2(0.077q_1 - 0.115q_2 - 0.013q_3), \tag{51}$$

$$H^e(\text{Ab, Ne}_{ss}) = -X_1X_2[3301 + 10612(X_1 - X_2) - 7773(X_1 - X_2^2)]$$
$$+ 1013(X_1^2 + X_1X_2 - 554(X_2^2 + X_1X_2), \tag{52}$$

$$S^e(\text{Ab, Ne}_{ss}) = -X_1X_2[1.1 + 4.402(X_1 - X_2)], \tag{53}$$

$$V^e(\text{Ab, Ne}_{ss}) = X_1X_2[0.077 - 0.23(X_1 - X_2)$$
$$+ 0.039(X_1 - X_2)^2]. \tag{54}$$

On the basis of the volumetric data by Hamilton & MacKenzie (1960) we assumed that

$$V^0(\text{Ab, Ne}_{ss}) = V^0(\text{Ne, Ne}_{ss}). \tag{55}$$

Fig. 7.10 demonstrates a satisfactory agreement of equations (46)–(54) with both (i) experimental data by Zyrianov *et al.* (1978) on exchange equilibria and (ii) data by Ferry & Blencoe (1978) on solvus.

Thermometry

Two-feldspar thermometer

Formulation of a two-feldspar thermometer. The two-feldspar thermometer proposed by Barth (1951) is based on the partition of albite between alkali feldspar and plagioclase,

$$\frac{\text{NaAlSi}_3\text{O}_8}{(\text{Ab in Pl})} = \frac{\text{NaAlSi}_3\text{O}_8}{(\text{Ab in Fsp})} \tag{56}$$

In the general case, the partition coefficient for albite of given structural state is defined by

$$\overline{K}_{\text{Ab}} = \frac{X(\text{Ab, Fsp})}{X(\text{Ab, Pl})} = \exp\frac{G^e(\text{Ab, Pl}) - G^e(\text{Ab, Fsp})}{RT}, \tag{57}$$

i.e. $\overline{K}_{\text{Ab}}$ depends on the difference in excess Gibbs free energies of albite in alkali feldspar and plagioclase. The binary solution models can be used for description of the excess partial energies of albite component if the solubility of KAlSi_3O_8 in plagioclase and that of $\text{CaAl}_2\text{Si}_2\text{O}_8$ in alkali feldspar are neglected.

Graphic versions of the two-feldspar thermometer have been reported in many papers (e.g. Ryabchikov, 1965, Perchuk & Ryabchikov, 1968, Perchuk & Aleksandrov,

1976). The first analytical version of the thermometer was developed by Stormer (1975) and modifed later by a number of the authors listed in Table 7.2.

Ryabchikov (1965) used the Margules parameters based on the disordered alkali feldspar solvus data and assumed ideal mixing of albite and anorthite in plagioclase to derive his graphic version of the two-feldspar geothermometer. A similar model has been accepted by Stormer (1975) to deduce the following equations:

$$t = (\psi/\phi) - 273, \tag{58}$$

where

$$\psi = 6326.7 - 9963.2X + 943.3X^2 + 2690.2X^3$$
$$+ (0.0925 - 0.1458X + 0.0141X^2 + 0.0392X^3)P,$$
$$\phi = 4.621 - 1.987 \ln (X/Y) - 10.815X + 7.7345X^2 - 1.5512X^3$$

with

$$X = X(\text{Ab, Fsp}) \text{ and } Y = X(\text{Ab, Pl}). \tag{59}$$

Perchuk & Aleksandrov (1976) used experimental data by Seck (1971) to deduce the pressure dependence for the albite of the partition coefficient with regard to the excess properties of feldspar and plagioclase solid solutions. They have obtained an analytical expression for the two-feldspar geothermometer and calculated ten isobaric sections of the diagram of partition of the albite component between plagioclase and alkali feldspar.

Haselton *et al.* (1983) used the Margules parameters presented in Table 7.3 (No. 4 and No. 9) and deduced the thermometer equation formulated in a form of equation (58) with

$$\psi = (1 - X)^2(4496 + 4070.6X + 0.087P) - (1 - Y)^2(6747.6 - 9446.18Y)$$

and

$$\phi = 2.46(1 - X)^2 + R\ln\left\{\frac{Y^2(2 - Y)}{X}\right\},$$

where X and Y correspond to (59). It should be noted that Haselton *et al.* (1983) correctly used different standard states for components of the feldspars in equilibrium to derive the mixing entropies from experimental data. The entropy for the alkali feldspar was obtained by treatment of both the solvus data and calorimetric measurements, and that for the plagioclase was calculated using the Kerrick–Darken model plus calorimetric data by Newton *et al.* (1980).

Equation (57) can be easily transformed to the equation

$$T = \frac{G^e(\text{Ab, Pl}) - G^e(\text{Ab, Fsp})}{R\ln (X_{\text{Ab}}^{\text{Fsp}}/X_{\text{Ab}}^{\text{Pl}})}, \tag{60}$$

where

$$G_{Ab}^e = H_{Ab}^e - TS_{Ab}^e + V_{Ab}^e P. \tag{61}$$

With the partial excess functions defined for alkali feldspar from the equations

$$\left.\begin{array}{l} H_{Ab}^e(Fsp) = (1 - X_{Ab}^{Fsp})^2\{W_{Ab}^H(Fsp) + 2X_{Ab}^{Fsp}[W_{Or}^H(Fsp) - W_{Ab}^H(Fsp)]\}, \\ S_{Ab}^e(Fsp) = (1 - X_{Ab}^{Fsp})^2\{W_{Ab}^S(Fsp) + 2X_{Ab}^{Fsp}[W_{Or}^S(Fsp) - W_{Ab}^S(Fsp)]\}, \\ V_{Ab}^e(Fsp) = (1 - X_{Ab}^{Fsp})^2\{W_{Ab}^V(Fsp) + 2X_{Ab}^{Fsp}[W_{Or}^V(Fsp) - W_{Ab}^V(Fsp)]\}, \end{array}\right\} \tag{62}$$

and for plagioclase from the equations

$$\left.\begin{array}{l} H_{Ab}^e(Pl) = (1 - X_{Ab}^{Pl})^2\{W_{Ab}^H(Pl) + 2X_{Ab}^{Pl}[W_{An}^H(Pl) - W_{Ab}^H(Pl)]\}, \\ S_{Ab}^e(Pl) = (1 - X_{Ab}^{Pl})^2\{W_{Ab}^S(Pl) + 2X_{Ab}^{Pl}[W_{An}^S(Pl) - W_{Ab}^S(Pl)]\}, \\ V_{Ab}^e(Pl) = (1 - X_{Ab}^{Pl})^2\{W_{Ab}^V(Pl) + 2X_{Ab}^{Pl}[W_{An}^V(Pl) - W_{Ab}^V(Pl)]\}, \end{array}\right\} \tag{63}$$

a general expression of an isostructural two-feldspar thermometer can be derived from equations (60) and (61):

$$T = \frac{H_{Ab}^e(Pl) - H_{Ab}^e(Fsp^h) - [V_{Ab}^e(Fsp^h) - V_{Ab}^e(Pl)]P}{R\ln[X_{Ab}(Fsp^h)/X_{Ab}(Pl)] - S_{Ab}^e(Fsp^h) + S_{Ab}^e(Pl)}. \tag{64}$$

This is a version of the two-feldspar thermometer for the binary solid solutions of plagioclase and disordered alkali feldspar. The equilibrium temperatures can be obtained with this equation (64) if equations (I–26) and (I–27) with the Margules parameters from Table 7.3 (No. 5 and No. 10) are employed. A graphic version of the geothermometer (64) is presented in Fig. 7.11. Parageneses of disordered alkali feldspar with plagioclase are widely spread in metamorphic rocks and related magmatic rocks (different types of granites, charnockites and some enderbites).

Using Margules parameters from Table 7.3 (No. 6) for alkali feldspar and an ideal solid solution model for plagioclase, Whitney & Stormer (1977) obtained the temperatures for the Brazilian granulite parageneses of ordered alkali feldspar and plagioclase higher than those for sanidine + plagioclase assemblages. These authors did not consider different standard states of albite in plagioclase and low alkali feldspar, i.e. they assumed that μ^0 (low Ab, Fsp) $- \mu^0$ (high Ab, Pl) $= 0$. However, it is clear that for the high–low albite phase transition

$$Ab_{(low)} = Ab_{(high)} \tag{65}$$

ΔG^0 (65) is not zero. For example, according to Robie et al. (1978)

$$\Delta G^0 (65) = \Delta H^0 (65) - T\Delta S^0 (65) + P\Delta V^0(65) = 2600 - 4.54T + 0.0086P, \tag{66}$$

which is not correct enough. With reference to (64) and (65) we can derive the thermometric equation for ordered feldspar and plagioclase in the following general form:

$$T = \frac{H_{Ab}^e(Pl) - H_{Ab}^e(Fsp^l) + \Delta H^0(65) - [V_{Ab}^e(Fsp^l) - V_{Ab}^e(Pl) - \Delta V^0(65)]P}{R\ln[X_{Ab}(Fsp^l)/X_{Ab}(Pl)] - S_{Ab}^e(Fsp^l) + S_{Ab}^e(Pl) + \Delta S^0(65)}. \tag{67}$$

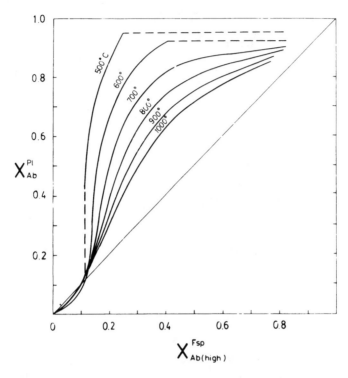

Fig. 7.11. Graphic version of two-feldspar thermometer with albite–sanidine series involved.

Equation (67) can be used for thermometry of microcline-bearing two-feldspar assemblages (in some pegmatites, metasomatites, nepheline syenites and other rocks), if ΔH^0 (65), ΔS^0 (65) and ΔV^0 (65) are correct.

Most feldspar-bearing rocks contain alkali feldspar with intermediate ordering. As proposed above, the index Z can be used in practical calculations to account for the variable ordering, and, with regard to this, the thermometer equation can be modified to

$$T = \frac{H_{Ab}^e(Pl) - H_{Ab}^e(Fsp) + \Delta H^0(65) \times Z - [\Delta V^0(65) \times Z - V_{Ab}^e(Fsp) + V_{Ab}^e(Pl)]P}{R\ln[X_{Ab}^{Fsp}/X_{Ab}^{Pl}] - S_{Ab}^e(Fsp) + S_{Ab}^e(Pl) + \Delta S^0(65) \times Z}, \qquad (68)$$

where excess mixing properties are defined by equations (62) and (63) with Margules parameters according to equations (11). At $Z = 0$, equation (68) becomes identical to equation (64).

Since excess molar volume of albite component in alkali feldspar rich in orthoclase is rather large, the albite distribution between coexisting feldspars appears to be strongly dependent on pressure conditions, and thus may serve for barometric purposes. The barometer equation can be easily derived from equation (68):

$$P = \frac{T \times a - H_{Ab}^e(Pl) + H_{Ab}^e(Fsp) - \Delta H^0(65) \times Z}{\Delta V^0(65) \times Z - V_{Ab}^e(Pl)} \qquad (69)$$

where a is the denominator of equation (68). It is evident that the pressure estimates depend on both the thermodynamic properties involved and feldspar ordering.

Testing the two-feldspar thermometer. The thermodynamic model developed for the two-feldspar equilibrium has been tested with geothermometry of some shallow-depth granites and granulite-facies metamorphic rocks, for which effects of Ca contents in alkali feldspar and K content in plagioclase on the temperature estimates can be neglected.

The temperature estimates obtained with different versions of the two-feldspar thermometer are presented in Table 7.6. The table shows that, for a case of disordered alkali feldspar, the estimates (columns 1–4) are similar. However, a systematic deviation from temperatures calculated from the version of Haselton et al. (1983) is clearly seen.

Metamorphic rocks from the Kanskiy granulite facies complex, Yenisei Ridge, USSR, have been used to test the two-feldspar thermometer. According to Gerya et al. (1986), X(Ab, Fsp) varies within a range of 0.09–0.21 and X(Ab, Pl) = 0.575–0.755. Temperature estimates obtained for these compositions with equation (64) fall within a range of 488–664 °C (Table 7.6, column 4) and give an average temperature 552 °C ($\sigma_n = 49.4$, $n = 15$).

In Table 7.7 these estimates are presented together with temperature and pressure values estimated for gneisses of the Kanskiy complex from data on the composition of Fe–Mg minerals. The temperatures estimated with biotite–garnet and cordierite–garnet thermometers (Perchuk & Lavrent'eva, 1983) show similar average values at pressures estimated with the cordierite–garnet–sillimanite–quartz barometer (Aranovich & Podlesskii, 1983).

Temperature ranges estimated for metamorphic rocks from the Hankay granulite-facies complex, the USSR Far East, and Sharyzhalgay complex, Eastern Siberia, with the two-feldspar thermometer practically coincide with that estimated with the Fe–Mg thermometers (Table 7.7).

The two-feldspar thermometer provides an important tool to restore thermal evolution of granite intrusions. The estimates thus obtained from biotite granites from Kazakhstan and the Baikal foldbelt area demonstrate applicability of the two-feldspar thermometer to these rock types.

Nepheline–feldspar thermometry

Formulation of nepheline–feldspar thermometer. Perchuk (1965) and later Perchuk & Ryabchikov (1968) developed a nepheline–feldspar geothermometer based on the exchange reaction

$$NaAlSiO_4 + KAlSi_3O_8 = KAlSiO_4 + NaAlSi_3O_8.$$

$$\text{Ne in Ne}_{ss} \quad \text{Or in Fsp} \quad \text{Ks in Ne}_{ss} \quad \text{Ab in Fsp} \tag{70}$$

Thermodynamics of this reaction received further development in studies of Powell & Powell (1977) and Zyrianov et al. (1978), who showed that calibration of the nepheline–

Table 7.6. *Comparison of temperatures for feldspar equilibria from metamorphic and granitic rocks with different versions of two-feldspar geothermometer*

Sample no.	P,kb	N_{Ab}		t, °C				
		Fsp	Pl	1[a]	2	3	4	5
Gneisses from Hankay complex, Far East, USSR								
Han-15(1)	4	21	53	732	800	828	775	806
(2)	4	24	57	747	760	864	804	837
(3)	4	15	53	632	650	665	638	682
(4)	4	15	55	620	630	654	629	671
(5)	4	15	58	604	605	638	616	655
(6)	4	17	53	666	700	718	683	722
Han-73(1)	4	8	59	480	400	461	457	522
(2)	4	11	41	630	560	586	569	668
Han-71(1)	4	8	60	477	380	459	455	519
(2)	4	7	42	525	400	445	441	562
Han-81(1)	4	12	58	556	500	566	552	602
(2)	4	12	57	560	520	571	555	601
(3)	4	12	39	669	640	626	605	707
Gneisses from Scharyzhalgay complex, W.S. Baikal Lake								
BL30M2	3	11	78.4	463	450	448	451	511
	3	8	78.4	419	400	391	—	465
	3	12	78.4	475	470	465	467	525
	3	14	63.2	553	580	574	559	604
Gneisses from Kanskiy complex, Yenisei River								
A-115-72	6	11.0	58.3	549	500	555	530	533
	5.8	13.9	57.3	602	610	630	596	651
	5.5	10.5	59.5	532	500	536	515	577
	6.2	10.7	60.2	536	500	540	517	540
	5.9	8.7	59.2	503	470	493	474	596
538	5.6	13.8	72.6	538	510	549	532	590
	4.8	16.2	71.7	571	550	596	573	628
	—	16.2	75.5	557	550	573	556	614
	4.9	9.6	65	500	480	497	480	545
	5.1	10.9	73.7	495	450	490	478	542
	5.3	12.6	67.4	540	510	552	537	588
	4.5	17.6	68.1	604	600	644	613	639
	5.2	11.8	67.1	529	510	537	517	576
	4.6	13.1	74.5	523	510	527	513	573
	4.5	20.7	69.9	684	620	761	650	755
Granites from Baikal foldbelt[b]								
90	1	18.3	75.7	534	540	551	542	594
103	1	33.8	78.5	652	640	720	696	788
109	1	14.9	62.9	544	490	565	547	595
140	1	31.3	78.3	635	620	595	673	753
707	1	13.6	78.4	474	485	466	678	525
708	1	32.8	79.3	640	630	700	678	767
709	1	27.1	76.3	613	580	665	645	707

Table 7.6 (*cont.*)

Sample no.	$P_{,kb}$	N_{Ab}		t, °C				
		Fsp	Pl	1[a]	2	3	4	5
496	1	23.4	81.7	557	560	576	570	633
503	1	42.5	81.9	684	700	760	731	895
508	1	26.3	80.6	585	575	617	606	673
Granites from Kazakhstan, USSR								
6391	3	47.1	73.1	855	780	1056	954	1208
5712	3	49.6	85.0	716	680	787	754	1018
507	3	56.0	92.6	642	700	664	649	994
9	5	43.0	94.3	617	595	628	645	799
2	5	20.1	56.0	681	750	757	749	778

Notes:
[a] 1 – after Stormer (1975); 2 – Perchuk & Aleksandrov (1976); 3 – Haselton *et al.* (1983); 4 – with equation (64) using parameters from rows 5 and 10 in Table 7.3; 5 – Whitney & Stormer (1977).
[b] Perchuk & Aleksandrov (1976).
[c] Letnikov (1975).

Table 7.7. *Comparison of P–T parameters for different mineral equilibria in gneisses from the Kanskiy complex (Precambrian), Yenisei Mountain-Ridge, Eastern Siberia.*

	Sp.A-115-72	Sp.538	Equilibrium
P, kb	6.4–4.0	5.6–4.0	Crd = Grt + Sil + Qtz
t, °C	603–560	628–530	Bt + Grt, Crd + Grt
t, °C	611–488	664–492	Pl + Fsp(high)
t, °C	583	578	
σ_n	14.8°	29°	Bt + Grt, Crd + Grt
n[a]	9	29	(average)

Notes:
[a] number of estimates.

feldspar thermometer based on reaction (70) also requires consideration of the effect of (i) pressure, (ii) phase transitions in nepheline and alkali feldspar including ordering, (iii) excess silica content in nepheline, and (iv) anorthite content in nepheline and feldspar. The last might appear very important for the nepheline–feldspar thermometry of volcanic rocks. However, at present we are unable to analyze it quantitatively because of a lack of appropriate experimental data.

For reaction (70) the thermometric equation can be expressed as

$$T = \frac{\Delta H^0(70) + \Delta H^e(70) + [\Delta V^0(70) + \Delta V^e(70)]P}{\Delta S^0(70) + \Delta S^e(70) - R\ln K_D} \tag{71}$$

where

$$\Delta\Phi(70) = \Phi(\text{Ab, Fsp}) - \Phi(\text{Or, Fsp}) + \Phi(\text{Ks, Ne}_{ss}) - \Phi(\text{Ne, Ne}_{ss}) \qquad (72)$$

and

$$K_D = \left[\frac{X(\text{Ks})}{X(\text{Ne})}\right]_{\text{Ne}_{ss}} \times \left[\frac{X(\text{Ab})}{X(\text{Or})}\right]_{\text{Fsp}} \qquad (73)$$

$\Delta\Phi^0(70)$ values for the reactions

$$\text{Ne} + \text{San} = \text{Ab}_{\text{(high)}} + \text{Ks}, \qquad (74)$$

$$\text{Ne} + \text{Mic} = \text{Ab}_{\text{(low)}} + \text{Ks} \qquad (75)$$

have been obtained by combination of the corresponding standard values for exchange reactions (5), (7) and (26), and are presented in Table 7.5. $\Delta\Phi^e(70)$ values can be readily calculated with excess mixing properties for the alkali feldspar end-members defined by the Margules parameters from Table 7.3 (No. 5 and No. 8) and those for the nepheline solid solution end-members according to equations (46)–(51). The nepheline–feldspar thermometer equations thus expressed for reactions (74) and (75) are, respectively,

$$T = \frac{3265 + \Delta H^e(74) + [0.01286 + \Delta V^e(74)]P}{0.482 + \Delta S^e(74) - R\ln K_D}, \qquad (76)$$

$$T = \frac{4595 + \Delta H^e(75) + [0.02868 + \Delta V^e(75)]P}{2.716 + \Delta S^e(75) - R\ln K_D}. \qquad (77)$$

Similarly to the two-feldspar thermometer, the intermediate structural state of alkali feldspar can be taken into account using the index Z:

$$\Delta\Phi(70) = \Delta\Phi(74)Z + \Delta\Phi(75)(1 - Z). \qquad (78)$$

Reaction (35) which governs the excess silica content in the nepheline solid solution in equilibrium with feldspar can also be employed for thermometric purposes, with the corresponding thermometer equation being expressed as

$$T = \frac{\Delta H^0 + H^e(\text{Ab, Ne}_{ss}) - 0.5H^e(\text{Ab, Fsp})}{\Delta S^0 + S^e(\text{Ab, Ne}_{ss}) - 0.5S^e(\text{Ab, Fsp}) + R\ln K_R}$$

$$+ \frac{[\Delta V^0 + V^e(\text{Ab, Ne}) - 0.5V^e(\text{Ab, Fsp})]P}{\Delta S^0 + S^e(\text{Ab, Ne}_{ss}) - 0.5S^e(\text{Ab, Fsp}) + R\ln K_R}, \qquad (79)$$

where

$$K_R = \frac{\sqrt{X(\text{Ab, Fsp})}}{X(\text{Ab, Ne}_{ss})} \qquad (80)$$

and

$$\Delta\Phi^0 = 0.5Z\Delta\Phi^0(65) - \Delta\Phi^0(35). \qquad (81)$$

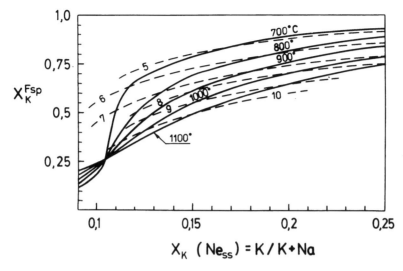

Fig. 7.12.   Graphic version of nepheline–feldspar thermometer with albite–sanidine series involved: sodium-rich portion. Solid lines – isothermic projections of the alkali distribution surface onto the silica saturated plan. Dashed lines show silica content (wt. %) in the nepheline solid solution, which is in equilibrium with feldspar of the albite– sanidine series at given T and P.

Equation (79) can be used to estimate temperature for both nepheline–feldspar and nepheline–plagioclase parageneses.

In an ideal case, if equilibrium is achieved in respect of all the components of coexisting alkali feldspar and nepheline, temperature estimates with equations (71) and (79) should coincide. However, not only due to uncertainties of the thermodynamic description, but also due to different kinetics of equilibration of alkalis and silica in nepheline–feldspar assemblages the coincidence may not be observed.

Testing the nepheline–feldspar thermometer.   Some mineral parageneses from alkaline rocks compiled by Perchuk (1971, 1976) were chosen to test the versions of the nepheline–feldspar thermometer deduced above. At the same time, there are many data on compositions of coexisting nepheline and feldspar from intrusive rocks, while the data for volcanics are quite limited.

Temperature estimates for nepheline–feldspar assemblages from plutonic rocks which have been obtained with equations (76) and (79) at $Z = 0$ and $P = 1$ kbar are presented in Table 7.8. A dramatic difference between estimates with the two equations is evident. Unrealistic temperatures calculated with equation (79) may reflect poor accuracy of determination of excess silica content in the nephelines, possible influence of the above-mentioned equilibration kinetics, and effects of additional components on reaction (35). Equilibrium temperatures for the Igdlerligssalik nepheline syenites (Powell & Powell, 1977b) with equation (76) are some 100 °C higher than those calculated with the Powell's thermometer.

Table 7.8. *Compositions and temperatures of coexisting nephelines and alkali feldspars from plutonic and volcanic rocks*

Sample no.	Mole per cent			Weight per cent				After equation	
	Ab	San	An	Ne	Ks	An	Qtz	(76)	(79)
Plutonic rocks									
71-I-38*	39.4	57.2	3.4	71.1	15.0	4.9	9	899	948
71-I-61	33.6	64.0	2.4	74.3	17.1	4.5	4.1	914	467
71-I-62	31.7	65.9	2.4	74.4	18.0	3.5	4.1	897	470
71-I-63	31.1	68.6	0.2	74.5	18.0	4.0	3.5	900	415
71-I-75	2.7	93.3	0.0	71.9	26.5	0.1	1.5	558	224
71-I-76	38.2	61.8	0.0	80.2	11.8	7.2	0.8	587	111
71-I-80	4.5	93.7	1.6	54.2	44.5	0.7	0.6	757	−121
71-I-86	27.9	67.0	5.0	64.3	14.7	12.2	8.8	900	1139
71-I-87	11.2	88.8	0.0	65.7	23.1	8.9	2.4	775	389
71-I-100	31.0	62.9	5.9	67.6	16.8	12.2	3.3	945	396
76-I-9	8.1	91.7	0.2	74.3	24.4	1.2	0.2	682	213
76-I-10	7.2	92.6	0.2	74.0	24.2	1.8	0.0	662	355
Volcanic rocks									
G-18/1/12	67.0	28.2	4.7	70.57	9.95	6.0	13.46	1305	1225
G-18/2/11	67.1	25.1	7.9	73.42	9.52	5.06	11.98	1086	1058
G-18/3/9	67.3	19.6	13.1	76.79	8.99	4.49	9.71	849	836
G-18/4/15	65.0	20.4	14.6	76.57	9.29	4.17	9.93	869	853
G-3/1	70.15	25.1	4.8	75.36	12.26	6.53	5.83	1307	600
G-3/2	70.15	25.1	4.8	76.44	12.46	5.18	5.90	1314	602
G-3/3	70.15	25.1	4.8	74.95	11.68	6.29	7.08	1290	681

Precision of the nepheline–feldspar thermometer based on reaction (35) probably improves with temperature due to both increase in solubility of albite in the nepheline solid solution and higher equilibration rates. To check this assumption two specimens of nepheline–feldspar bearing basalts from Vogelsberg, Hessen, West Germany (e.g. Tilley, 1958), were studied in detail. All volcanic rocks in this area are tholeutic and alkali basalts which were erupted or intruded near the surface 19 m.y.a. The alkali basalts were differentiated to trachites and nephelinites. In places they bear periodotite nodules.

Bulk chemical composition of alkali basalt from specimen G-18 was determined by microprobe analysis of glass prepared with a technique which is common for the RFA method. An average of 16 probe measurements is the following (wt.%):

SiO_2 – 43.60	MnO – 0.15
TiO_2 – 4.65	CaO – 9.95
Al_2O_3 – 12.14	Na_2O – 3.33
FeO – 10.25	K_2O – 2.13
MgO – 12.41	Total – 98.61

The rock contains phenocrysts of olivine (Fo_{92}), plagioclase (An_{39-57}) and augite. The rock matrix is composed of the same minerals plus nepheline and alkali feldspar.

Results of microprobe analysis of the nepheline and feldspar are given in Table 7.8.

P–T parameters of the *magma generation* have been estimated on the basis of the bulk rock and olivine composition with the method proposed by Perchuk (1987) to yield 1400 °C and 24 kbar, or 1310 °C for olivine–glass equilibration at 1 bar. The highest temperatures 1307–1225 °C obtained for the nepheline–feldspar paragenesis using equations (76) and (79) with the data from Table 7.8 are close to the liquidus temperature calculated above.

Table 7.8 also presents data on compositions of coexisting nepheline and feldspar from a sample of essexite (specimen G–3). The rock which is taken from a dyke in a quarry of the Hauri Company, Oberschaffhausen, West Germany, also contains olivine and clinopyroxene. Significant difference in the temperature estimates from equations (76) and (79) in Table 7.8 probably implies that the nepheline and alkali feldspar have not been equilibrated.

Conclusions

Thermodynamic data on feldspar solid solutions obtained during the last 40 years have been reviewed with the aim of revising the two-feldspar geothermometer. Compilation of the data on the phase equilibria and their thermodynamic treatment allowed us to derive the internally consistent Margules parameters. Based on these results two-feldspar geothermometers which account for the alkali feldspar ordering degree have been deduced. The equations were tested by estimating the P–T parameters of mineral equilibration in granites and some metamorphic rocks and comparing them to parameters independently calculated from Fe–Mg thermobarometry. Contribution of the Si–Al ordering degree in energetics of the albite partition between alkali and plagioclase feldspar has also been demonstrated. For the majority of granites and metamorphic rocks a binary thermodynamic model for both coexisting feldspars provides reasonable thermobarometric results.

Mixing properties of the nepheline solid solutions were evaluated on the basis of existing phase equilibrium experimental data. Results of these calculations were used to deduce geothermometric equations for nepheline–feldspar paragenesis. The equations were tested with Ne–Fsp pairs from plutonic and volcanic rocks. Despite the low resolution, the nepheline–feldspar geothermometer can reasonably reflect temperatures of rock emplacements and the degree of mineral equilibration quenched in the lava flows or other magmatic bodies.

Acknowledgements   We thank G.L. Hovis for providing us with unpublished data on alkali feldspar molar volumes; T.V. Gerya for probe analyses of coexisting phases from gneisses of the Yenisei River, and E.V. Zabotina for her help in typing the manuscript.

References

Andersen, G.R. & Mazo, R.M. (1980). Statistical mechanical models for aluminium–silicon disorder in plagioclase. *Amer. Mineral.*, **65**, 75–80.
Bachinski, S.W. & Müller, G. (1971). Experimental determination of the microcline – low albite solvus. *J. Petrol.*, **12**, 329–56.

Barron, L.M. (1976). A comparison of two models of ternary feldspars' excess free energy. *Contrib. Mineral. Petrol.*, **57**, 71–81.

Barth, T.F.W. (1951). The feldspar geologic geothermometer. *Neues Jb. Mineral. Abh.*, **82**, 143—50.

— (1963). The composition of nepheline. *Schweiz. Mineral. Petrol. Mitt.*, **43**, 153–64.

Brown, W.L. & Parsons, I. (1981). Towards a more practical two-feldspar geothermometer. *Contrib. Mineral. Petrol.*, **76**, 369–77.

—— (1985). Calorimetric and phase diagram approaches to two-feldspar geothermometry: a critique. *Amer. Mineral.*, **70**, 356–61.

Buerger, M.J., Klein, G.E. & Donnay, G. (1954). Determination of the crystal structure of nepheline. *Amer. Mineral.*, **39**, 805–18.

Carpenter, M.A. & Ferry, J.M. (1984). Constraints on the thermodynamic mixing properties of plagioclase feldspars. *Contrib. Mineral. Petrol.*, **87**, 138–48.

Carpenter, M.A. & McConnell, J.D.C. (1984). Experimental delineation of the $C\bar{1}\text{-} = I\bar{1}$ transformation in intermediate plagioclase feldspars. *Amer. Mineral.*, **69**, 112–21.

Carpenter, M.A., Navrotsky, A. & McConnell, J.D.C. (1985). Enthalpies of ordering in the plagioclase feldspar solid solution. *Geochim. Cosmochim. Acta*, **49**, 947–66.

Debron, G. (1965). Contribution à l'étude des réactions d'échange des ions alcalino-terreux dans les feldspathoïdes. *Bull. Soc. franç. Minéral. Cristall.*, **88**, 69–96.

Debron, G., Donnay, G., Wyart, J. & Sabatier, G. (1961). Réaction d'échange des ions sodium par les ions potassium dans la néphéline. Application à l'étude du système néphéline–kalsilite. *Compt. rendus des séances de l'Acad. des Sci.*, **252**, 1255–7.

Deer, W.A., Howie, R.A. & Zussman, J. (1963). *Rock-forming, Vol. 4: Framework silicates.* London: Longmans.

Delbove, F. (1971). Equilibria d'échange d'ions entre feldspaths alcalins et halogénures sodi-potassiques fondus. Application au calcul des propriétés thermodynamiques de la série des feldspaths alcalins. *Bull. Soc. franç. Minéral. Cristall.*, **94**, 456–66.

— (1975). Excess Gibbs energy of microcline – low albite alkali feldspars at 800 °C and 1 bar, based on fused alkali bromide ion exchange experiments. *Amer. Mineral.*, **60**, 972–84.

Donnay, G. & Donnay, D.H. (1952). The symmetry change in the high-temperature alkali feldspar series. *Amer. J. Sci.*, **250A**, 115–32.

Donnay, G., Schairer, J.F. & Donney, J.D.H. (1959). Nepheline solid solutions. *Mineral. Mag.*, **32**, 93–109.

Dorogokupets, P.I. & Karpov, I.K. (1984). *Thermodynamics of minerals and mineral equilibria.* Novosibirsk: Nauka.

Edgar, A.D. (1964). Phase-equilibrium relations in the system nepheline–albite–water at 1000 kg/sm². *J. Geol.*, **72**, 448.

Fei, Y., Saxena, S.K. & Eriksson, G. (1986). Some binary and ternary silicate solution models. *Contrib. Mineral. Petrol.*, **94**, 221–3.

Ferry, I.M. & Blencoe, G.J. (1978). Subsolidus phase relations in the nepheline–kalsilite system at 0.5, 2.0, and 5.0 kbar. *Amer. Mineral.*, **63**, 1225–40.

Gerya, T.V., Dotsenko, V.M., Zabolotskiy, K.A. *et al.* (1986). *Precambrian crystalline complexes of the Enisey mountain-ridge: guide-book* [in Russian]. Novosibirsk: Nauka.

Ghiorso, M.S. (1984). Activity/composition relations in the ternary feldspars. *Contrib. Mineral. Petrol.*, **87**, 282–96.

Goldsmith, J.R. & Newton, R.C. (1974). An experimental determination of the alkali feldspar solvus. In *The feldspars*, ed. W.S. MacKenzie & J. Zussman, pp. 337–59. Manchester University Press.

Green, N.L. & Usdansky, S.I. (1984). Feldspar–mica equilibria: 1. Plagioclase-muscovite geothermometry. *Geol. Soc. Amer. Abstr. with Programs*, **16**, 222.

—— (1986). Ternary-feldspar mixing relations and thermobarometry. *Amer. Mineral.*, **71**, 1100–8.

Greig, J.W. & Barth, T.F.W. (1938). The system $Na_2O–Al_2O_3–2SiO_2$ (nepheline, carnegieite) – $Na_2O \times Al_2O_3 \times 6SiO_2$ (albite). *Amer. J. Sci. 5th ser.*, **35A**, 93–108.

Hamilton, D.L. (1961). Nepheline as crystallization temperature indicator, *J. Geol.*, **69**, 321–9.

Hamilton, D.L. & MacKenzie, W.S. (1960). Nepheline solid solution in the system $NaAlSiO_4–KAlSiO_4–SiO_2$. *J. Petrol.*, **1**, 56–72.

Haselton, H.T., Hovis, G.L., Hemingway, B.S. & Robie, R.A. (1983). Calorimetric investigation of the excess entropy of mixing in albite–sanidine solid solutions: lack of evidence for two-feldspar thermometry. *Amer. Mineral.*, **68**, 398–413.

Hodges, K.V. & Crowley, P.D. (1985). Error estimation and empirical geothermobarometry for pelitic systems. *Amer. Mineral.*, **70**, 702–9.

Hovis, G.L. (1977). Unit-cell dimensions and molar volumes for a sanidine–albite ion-exchange series. *Amer. Mineral.*, **62**, 672–9.

— (1979). A solution calorimetric investigation of K–Na mixing in a sanidine–analbite ion-exchange series: corrections. *Amer. Mineral.*, **64**, 925.

— (1983). Thermodynamic mixing properties and a calculated solvus for low albite–microcline crystalline solutions. *Geological Society of America, Abstracts with Programs*, **15**, 519.

— (1986). Behavior of alkali feldspars: crystallographic properties and characterization of composition and Al–Si distribution. *Amer. Mineral.*, **71**, 869–90.

Hovis, G.L. & Peckins, E. (1978). A new X-ray investigation of maximum microcline crystalline solutions. *Contrib. Mineral. Petrol.*, **66**, 345–9.

Hovis, G.L. & Waldbaum, D.R. (1977). A solution calorimetric investigation of K–Na mixing in a sanidine–albite ion-exchange series. *Amer. Mineral.*, **62**, 680–6.

Iiyama, J.T. (1965). Influence des anions sur les équilibres d'échange d'ions Na–K dans les feldspaths alcalins à 6000 °C sous un pression de 1000 bars. *Bull. Soc. franç. Minéral. Cristall.*, **88**, 618–22.

— (1966). Contribution à l'étude des équilibres sub-solidus du système ternaire orthoclase–albite–anorthite à l'aide de réactions d'échanges d'ions Na–K au contact d'une solution hydrothermale. *Bull. Soc. franç. Minéral. Cristall.*, **89**, 442–54.

Iiyama, J.T., Wyart, J. & Sabatier, G. (1963). Equilibre des feldspaths alcalins et des plagioclases à 500, 600, 700 et 800 °C sous une pression d'eau de 1000 bars. *Compt. rendus des séances de l'Acad. des Sci.*, **256**, 5016.

Johannes, W. (1979). Ternary feldspars: kinetics and possible equilibria at 800 °C. *Contrib. Mineral. Petrol.*, **68**, 221–30.

Kerrick, D.M. & Darken, L.S. (1975). Statistical thermodynamic models for ideal oxide and silicate solid solutions, with application to plagioclase. *Geochim. Cosmochim. Acta.*, **39**, 1431–42.

Kotel'nikov, A.R. (1980). Calculation of the mixing functions for the plagioclase solid solution. *Geochimia*, **2**, 226–30.

Kotel'nikov, A.R., Bychkov, A.M. & Chernavina, N.A. (1981). The distribution of calcium between plagioclase and water–salt fluid at 700 °C and $P_{\text{fl}} = 1000$ kg/cm². *Geochem. Intern.*, **18**, 61–75.

Kotel'nikov, A.R., Vinograd, V.L. & Petukhov, P.A., (1986). Experimental study of equilibria of scapolite and plagioclase in hydrothermal conditions. In *Experiment in geology*, ed. V.A. Zharikov & V.V. Fed'kin, pp. 399–418. Moscow: Nauka.

Kracek, F.C. & Neuvonen, K.J. (1952). Thermochemistry of plagioclase and alkali feldspars. *Amer. J. Sci., The Bowen Volume, Part 1*, pp. 293–318.

Kroll, H. & Ribbe, P.H. (1980). Determinative diagrams for Al–Si order in plagioclases. *Amer. Mineral.*, **65**, 449–57.

Kroll, H., Schmiemann, I. & Cölln, G. (1986). Feldspar solid solution. *Amer. Mineral.*, **71**, 1–17.

Kurepin, V.A. (1981). Thermodynamics of minerals of variable composition and geothermometry. Kiev: Naukova Dumka.

Lagache, M. & Weisbrod, A. (1977). The system: two alkali feldspar–KCl–NaCl–H_2O. *Contrib. Mineral. Petrol.*, **62**, 77–101.

Letnikov, F.A. (1975). *Granitoids of the Central Massifs.* Novosibirsk: Nauka.

Luth, W.C. & Fenn, P.M. (1973). Calculation of binary solvi with special reference to the sanidine–high albite solvus. *Amer. Mineral.*, **58**, 1009–15.

Luth, W.C. & Querol-Suñe, F. (1970). An alkali feldspar series. *Contrib. Mineral. Petrol.*, **25**, 25–40.

Luth, W.C. & Tuttle, O.F. (1966). The alkali feldspar solvus in the system Na_2O–K_2O–Al_2O_3–SiO_2H_2O. *Amer. Mineral.*, **51**, N 9–10.

Martin, R.F. (1974). The alkali feldspar solvus: the case for a first-order break on the K-limb. *Bull. Soc. franç Minéral. Cristall.*, **97**, 346–55.

Morozewicz, J. (1930). Der Mariupolit und seine Blutsverwandten. *Mineral. Petrol. Mitt.*, **40**, N 5–6.

Newton, R.C., Charlu, T.V. & Kleppa, O.J. (1980). Thermochemistry of high structural state plagioclases. *Geochim. Cosmochim. Acta*, **44**, 933–41.

Orville, P.M. (1963). Alkali ion exchange between vapor and feldspar phase. *Amer. J. Sci.*, **261**, 201–37.

— (1967). Unit-cell parameters of the microcline–low albite and sanidine–high albite solid solution series. *Amer. Mineral.*, **52**, 55–86.

— (1972). Plagioclase cation exchange equilibria with aqueous chloride solution. Results at 700 °C and 2000 bars in the presence of quartz. *Amer. J. Sci.*, **272**, 234–72.

Perchuk, L.L. (1965). Paragenesis of nepheline with feldspar as an indicator of thermodynamic conditions of mineral equilibrium. *Doklady Akademii Nauk SSSR*, **161**, 132–5.

— (1968). The phase relationships in the system nepheline–alkali feldspar – aqueous solutions [in Russian]. In *Physico-chemical and other problems of metasomatism and metamorphism*, pp. 53–95. Moscow: Nauka.

— (1987). Studies of volcanic series related to the origin of some marginal sea floors. In *Magmatic processes: physicochemical principles*, pp. 209–30. Washington: The Geochemical Society, Special Publication 1, 209–30.

Perchuk, L.L. & Alexsandrov, A.L. (1976). Evaluation of isobaric diagrams for two-feldspar equilibria. In *Modern techniques of petrological investigations*, ed. A.A. Marakushev, pp. 5–10. Moscow: Nauka.

Perchuk, L.L. & Andrianova, L.L. (1968). Thermodynamics of equilibrium of alkali feldspar (K, Na)$AlSi_3O_8$ with aqueous solution (K, Na)Cl at 500–800 °C and 2000–1000 bar pressure. In *Theoretical and experimental studies of mineral equilibria*, ed. I.Ya. Nekrasov, pp. 37–72. Moscow: Nauka.

Perchuk, L.L. & Lavrent'eva, I.V. (1983). Experimental investigation of exchange equilibria in the system cordierite–garnet–biotite. *Advances in Physical Geochemistry*, **3**, 199–239.

Perchuk, L.L., Podlesskii, K.K. & Zyrianov, V.N. (1978a). Thermodynamic mixing functions for the nepheline and feldspar solid solutions at $1000 > T > 400$ °C. *Intern. Geochem.*, **20**, 116–24.

Perchuk, L.L. & Ryabchikov, I.D. (1968). Mineral equilibria in the system nepheline–alkali feldspar–plagioclase and their petrological significance. *J. Petrol.*, **9**, 123–67.

—— (1976). *Phase correspondence in the mineral systems.* Moscow: Nedra.

Perchuk, L.L., Zyrianov, V.N., Podlesskii, K.K. *et al.* (1978). Mixing energies for some minerals of varying composition. *Phys. and Chem. Mineral.*, **3**, 301–7.

Powell, R. & Powell, M. (1977a). Plagioclase–alkali feldspar geothermometry revisited. *Mineral. Magazine*, **41**, 253–6.

Powell, M. & Powell, R.A. (1977b). A nepheline–alkali feldspar geothermometer. *Contrib. Mineral. Petrol.*, **62**, 193–204.

Price, J.G. (1985). Ideal site mixing in solid solutions, with application to two-feldspar geothermometry. *Amer. Mineral.*, **70**, 696–701.

Robie, R.A., Hemingway, B.S., & Fisher, J.R. (1978). *Thermodynamic properties and related substances at 298 °K and 1 bar pressure and at higher temperatures*. Washington DC: US Government Printing Office.

Roux, J. (1971). Sur l'équilibre de la réaction d'échange d'ions Na–K dans la néphéline à 600 °C–2000 bars. *Compt. rendus Acad. Sci. colon.*, **272**, N 26.

— (1974). Etude des solutions solides des néphélines (Na, K)AlSiO₄, *Geochim. Cosmochim. Acta*, **38**, 1213–24.

Ryabchikov, I.D. (1965). New diagram for two-feldspar geological thermometer deduced with thermodynamic treatment of experimental data. *Doklady Akademii Nauk SSSR*, **165**, N 3.

— (1975). *Thermodynamics of fluid related to granite magmas*. Moscow: Nauka.

Saxena, S.K. & Ribbe, P.H. (1972). Activity–composition relations in feldspars. *Contrib. Mineral. Petrol.*, **37**, 131–8.

Seck, H.A. (1971). Der Einfluß des Drucks auf die Zusammensetzung koexistierender Akalifeldspäte und Plagioklase im System $NaAlSi_3O_8$–$KAlSi_3O_8$–$CaAl_2Si_2O_8$–H_2O. *Contrib. Mineral. Petrol.*, **31**, 67–86.

— (1971). Koexistierende Akalifeldspäte und Plagioklase im System $NaAlSi_3O_8$–$KAlSi_3O_8H_2O$ bei Temperaturen von 650 °C bis 900 °C. *N. Jb. Mineral. Abh.*, **115**, 315—45.

— (1972). The influence of pressure on the alkali feldspar solvus from peraluminous and persilicic materials. *Fortschritte der Mineralogie*, **49**, 31–49.

Seil, M.K. & Blencoe, J.G. (1979). Activity–composition relations of $NaAlSi_3O_8$–$CaAl_2Si_2O_8$ feldspar at 2 kb, 600–800 °C. *Geol. Soc. Amer. Abstracts with Programs*, **11**, 513.

Senderov, E.E. & Yas'kin, G.M. (1976). Stability of the monoclinic alkali feldspar. *Geochimia*, 7.

Smith, J.V. & Parsons, I. (1974). The alkali-feldspar solvus at 1 kbar water-vapor pressure. *Mineral. Magazine*, **39**, 747–67.

Smith, J.V. & Tuttle, O.P. (1957). The nepheline–kalsilite system: 1. X-ray data for the crystalline phases. *Amer. J. Sci.*, **255**, 282–305.

Stormer, J.C. (1975). A practical two-feldspar geothermometer. *Amer. Mineral.*, **60**, 667–74.

Thompson, J.B. & Hovis, G.L. (1979). Entropy of mixing in sanidine. *Amer. Mineral.*, **64**, 57–65.

Thompson, J.B. & Waldbaum, D.K. (1968). Mixing properties of sanidine solutions: 1. Calculations based on ion exchange data. *Amer. Mineral.*, **53**, 1965–99.

Thompson, J.B. & Waldbaum, D.R. (1969). Mixing properties of sanidine crystalline solutions: 3. Calculations based on two-phase data. *Amer. Mineral.*, **54**, 811–38.

Tilley, C.E. (1954). Nepheline–alkali feldspar paragenesis. *Amer. J. Sci.*, **252**, 65–75.

—— (1958). The leucite nepheline dolerite of Meiches, Vogelsberg. *Amer. Mineral.*, **43**, No 7–8.

Traetteberg, A. & Flood, H. (1972). Alkali ion exchange equilibria between feldspar phases and molten mixtures of potassium and sodium chloride. *Kungliga Tekniska Hogskolans Handlinga, Norsk.*, **296**, 609–18.

Tuttle, O.F. & Smith, I.V. (1958). The nepheline–kalsilite system. 2. Phase relations. *Amer. J. Sci.*, **256**, N 8.

Waldbaum, D.R. & Robie, R.A. (1971). Calorimetric investigation of Na–K mixing and polymorphism in the alkali feldspars. *Zeitschrift für Kristallographie*, **134**, 381–420.

Wellman, T.R. (1970). The stability of sodalite in a synthetic syenite and aqueous chloride fluid system. *J. Petrol.*, **11**, 48–71.

Whitney, J.A. & Stormer, J.C. (1977). The distribution of $NaAlSi_3O_8$ between coexisting microcline and plagioclase and its effects on geothermometric calculations. *Amer. Mineral.*, **62**, 687–91.

Wright, T.L. & Stewart, D.B. (1968). X-ray and optical study of alkali feldspar. 1. Determination of composition and structural state from refined unit cell parameters and $2V$. *Amer. Mineral.*, **53**, 38–87.

Wyart, J. & Sabatier, G. (1956). Transformations mutuelles des feldspaths alcalins; reproduction du microcline et de l'albite. *Bull. Soc. franç Minéral Cristall.*, **79**, 574–81.

——— (1962). Sur le problème de l'équilibre des feldspaths alcalins et des plagioclases. *Compt. rendus des séances de l'Acad. des Sci.*, **255**, 1551–6.

Yund, R.A., McCallister, R.A. & Savin, S.M. (1972). An experimental study of nepheline–kalsilite exsolution. *J. Petrol.*, **13**, 255–72.

Zyrianov, V.N. (1981). *Phase correspondence in the systems of alkali feldspars and feldspathoids.* Moscow: Nauka.

Zyrianov, V.N. & Perchuk, L.L. (1975). Experimental investigation of phase correspondence in the system nepheline-alkali feldspar. In *Contributions to physico-chemical petrology*, **5**, pp. 51–77. Moscow: Nauka.

Zyrianov, V.N., Perchuk, L.L. & Podlesskii, K.K. (1978). Nepheline–alkali feldspar equilibria: 1. Experimental data and thermodynamic calculations, *J. Petrol.*, **19**, 1–44.

PART II

Metamorphic and metasomatic processes

8

Macrokinetic model of origin and development of a monomineralic bimetasomatic zone

V.N. BALASHOV and M.I. LEBEDEVA

Introduction

Contemporary understanding of metasomatic zoning is based on fundamental works by Korzhinskii (1959, 1970), who considered a metasomatic process as a transfer of matter with chemical interaction in natural heterophase systems. Korzhinskii summarized geological data on metasomatic processes and formulated the principle of mosaic or local chemical equilibrium (LCE). In particular, thermodynamic ordering of metasomatic zones follows from the principle of LCE and corresponds to their ordering on equilibrium phase diagrams.

The studies of Korzhinskii have given impetus to a great number of geological, theoretical and experimental investigations in metasomatism. Extensive experimental studies conducted by Zaraisky et al. (1981, 1986) under hydrothermal conditions in a temperature range of 300°–700 °C have confirmed the equilibrium metasomatic theory. Some principal results of the experimental investigations are given by Zharikov and Zaraisky (Chapter 9 in this volume).

In the seventies the LCE-theory of metasomatism was further developed by Frantz & Mao (1975, 1976, 1979). They have dealt with a joint system of differential mass transfer equations and algebraic equations corresponding to the law of mass action under conditions of constant porosity and small ratio of component concentration in pore solution to its concentration in solid phases. Solid phases were considered as constant composition minerals. In these works possibilities of inner-zone mineral production and inhomogeneity of mineral composition in a polymineralic zone were discussed. Frantz & Mao (1975) have also formulated some rules of boundary zone movement and developed (Frantz & Mao, 1976, 1979) a quantitative model for the bimetasomatic systems $MgO–SiO_2–H_2O–HCl$ and $CaO–MgO–SiO_2–CO_2–H_2O$ for small constant porosity. The assumption of constant porosity under conditions of mineral production requires the corresponding transfer of solid phases.

A simplified model of metasomatic zoning with chemical equilibrium only at zone boundaries was considered by Balashov (1985). The numerical solutions of this model for the system MgO–SiO$_2$–H$_2$O–HCl are given by Zaraisky et al. (1986).

A new level of theoretical consideration of transfer phenomena in heterogeneous hydrothermal systems with chemical interaction has been achieved by Lichtner (1985) using a time-space continuum model. Lichtner formulated the main definitions required for a continuous field description, considered different possibilities of partitioning the reacting species into primary (independent) and secondary (dependent) species and deduced the relations at zone boundaries (discontinuities) from equations of the inner-zone transfer through limit transition.

In further studies Lichtner et al., (1986a, b) worked out the method of numerically solving the LCE-system on the basis of the weak formulation of the moving boundary problem. The calculations are in good agreement with the steady-state solution for the precipitation zone resulting from counter-diffusion (Helfferich & Katchalsky, 1970). Some numerical results are given by Lichtner (1986b) for the system MgO–SiO$_2$–H$_2$O–HCl at 390 °C under conditions of constant porosity which is taken as 1% and the total chloride concentration is taken to be 1 molal throughout the system.

The problem of metasomatism and metasomatic zoning has been investigated not only from the viewpoint of LCE-theory. Golubev & Sharapov (1974) and Golubev (1981) have proposed that kinetic constants of irreversible reactions play a decisive role for the zoning sequence and other features of metasomatism. These authors have studied a wide set of relatively simple cases with analytical solutions from schemes of parallel and sequential irreversible reactions of first order coupled with mass transfer in model heterogeneous systems.

The detailed analysis of diffusion metasomatism with one irreversible reaction is given by Volkova & Sheplev (1980) who considered the ratio of component concentration in the intergranular fluid to mineral density to be a small parameter.

In our view, the development of macrokinetic theory as based on reversible chemical kinetics has more perspective. Because this theory can describe regions which are both near and relatively far from equilibrium, it satisfies experimental and natural data and takes into account the finite rates of chemical transformation.

This approach was developed by Lebedeva et al. (1985, 1987), who considered systems with diffusional mass transfer of a component C in a pore solution and the reversible reaction $Q_2 + C = Q_1$, where Q_1 and Q_2 are product and reactant minerals, respectively. It was found that the column $Q_1/Q_1 + Q_2/Q_2$ corresponding to a kinetic region approaches a LCE-column with sharp phase boundary Q_1/Q_2 as $k^{\pm}$ tends to ∞ with $k^+/k^- = $ const. The concentration of component C at the zone boundary is defined by the equilibrium relation $y^e = k^-/k^+$. Practically simultaneously with Lichtner (1985), Lebedeva et al. (1985) deduced the relation on jump discontinuities at reaction zone boundaries from differential process equations without additional physical assumptions.

168

In this work we concentrate on the numerical investigation of a system with diffusional mass transfer of two components corresponding to the case of simple bimetasomatism.

In this system, in the limit from kinetic to LCE- description, the equilibrium relations must be fulfilled in all space intervals corresponding to zones of mineral alteration. This paper presents numerical results for the case of variable porosity both in kinetic and in LCE-regions. The LCE-algorithm used in this work is based on the method of front rectification in the Stephan problem (Budak *et al.*, 1966). This algorithm allows for the possibility of an exact definition of moving boundary coordinates.

Some explanation of the terminology used in this paper is necessary. The term 'macro' is used for a continuous phenomenological description based on a representative elemental volume (REV). Kinetic and LCE-descriptions of heterogeneous processes with mass transfer is referred to by the term 'macrokinetics', which implies a definite macrostructure described by model equations. In applications to metasomatism, the main element of this macrostructure is metasomatic zoning.

Macroexpressions for rate of chemical reaction (the kinetic law)

For a kinetic description of the regions which are close to equilibrium, we must firstly satisfy a set of restrictions on the kinetic rate law. Considerations in this part of the paper are based on Lasaga's (1981) approach.

Consider a porous medium consisting of minerals of constant composition at a given P, T approximated by a lumped-parameter system. The rate of rth heterogeneous reaction is described by the general relation

$$\frac{\mathrm{d}\xi_r}{\mathrm{d}t} = f(k_r^{\pm}, m_i, s_j), \tag{1}$$

where ξ_r denotes the reaction progress variable of the rth reaction, $k_r^{\pm}$ refers to kinetic rate constants of the rth reaction, m_i signifies molalities of the ith aqueous species in the pore solution and s_j denotes the value of the specific surface of the jth mineral phase taking part in reaction. The rate is defined per volume unit of porous medium.

As m_i approaches the equilibrium values m_i^e, $\mathrm{d}\xi_r/\mathrm{d}t$ must approach zero according to

$$\lim_{m_i \to m_i^e}(\mathrm{d}\xi_r/\mathrm{d}t) = 0 \tag{2}$$

Here the quantities m are constrained by the equilibrium relations corresponding to the law of mass action:

$$\sum_i \nu_{ri}\ln(\gamma_i m_i^e) = \ln K_r \tag{3}$$

169

where γ_i refers to the activity coefficients of the ith species activity, ν_{ri} denotes the stoichiometric coefficients in the rth reaction and K_r refers to the equilibrium constant of the rth reaction.

According to Lasaga (1981), for any state of the system the quantity Ω_r can be used to characterise the relative deviation of reaction from the equilibrium state:

$$\Omega_r = \frac{\prod_i (\gamma_i m_i)^{\nu_{ri}}}{K_r} \tag{4}$$

where $\lim_{m_i \to m_i^e} \Omega_r = 1$. Thus any function of the form

$$f(k_r^{\pm}, m_i, s_j) = \varphi(k_r^{\pm}, m_i, s_j)(1 - \Omega_r)^n, \, n > 0, \tag{5}$$

where φ is a limited function, satisfies relations (1) and (2). Thus one of many possible forms for the kinetic rate law is given by

$$\frac{d\xi_r}{dt} = \varphi(k_r^{\pm}, m_i, s_j)(1 - \Omega_r). \tag{6}$$

Time evolution equations

We shall consider the case of constant lithostatic and pore fluid pressure under isothermal conditions. All numerical results obtained in this study refer to the system $MgO-SiO_2-H_2O-HCl$ at $P = 1000$ bar and $T = 600$ °C. Possible stable mineral phases in the system under these P, T conditions are Q, Tlc, Fo and Brc (Hemley $et\ al.$, 1977a, b, Helgeson $et\ al$, 1978, Zaraisky $et\ al.$, 1986). As water is a main component of the pore solution, to a good approximation the chemical potential of H_2O is defined by P and T. Computations by Zaraisky $et\ al.$ (1986) based on thermodynamic and experimental data (Helgeson & Kirkham, 1974, 1981, Frantz & Marshall, 1982, 1983), suggest that the major species in solution are $MgCl_2^0$, SiO_2^0 and HCl^0. It is assumed that the activity coefficients are equal to 1.

Consider the simplest scheme of linearly independent reversible dissolution – precipitation reactions:

$$\left.\begin{array}{l} Q = SiO_2^0 \\ Tlc + 6HCl^0 = 3MgCl_2^0 + 4SiO_2^0 + 4H_2O, \\ Fo + 4HCl^0 = 2MgCl_2^0 + SiO_2^0 + 2H_2O, \\ Brc + 2HCl^0 = MgCl_2^0 + 2H_2O. \end{array}\right\} \tag{7}$$

The form of the reactions in (7) allows the systems to be described as a two-component system for constant HCL^0 activity (molality). In this way the change of m_{HClO} corresponds to changes in the solubility constants of Tlc, Fo, Brc and concentration of $MgCl_2^0$ in the pore solution. However, the assignment of constant total molality of Cl is more correct (for equal component diffusion coefficients):

$$m_{Cl} = 2m_{MgCl_2^0} + m_{HCl0} = \text{const.}$$

In the case where $m_{HCl0} \gg m_{MgCl_2^0}$ both assumptions are equivalent. In this study the more approximate variant of constant HCl^0 molality is assumed.

Kinetic system of equations

The main goal of this study is to simulate development of monomineralic zones from the starting contacts: Q/Fo or Tlc/Brc. Obviously, in the former case a talc zone arises, in the latter a forsterite zone. The chemical reaction rate corresponds to kinetic law (6). For dissolution–precipitation of talc, the rate of reaction is given by

$$\frac{\partial \rho_{Tlc}}{\partial t} = -\frac{\partial \xi_{Tlc}}{\partial t} = -(k_{Tlc}^+ m_{HCl0}^6 - k_{Tlc}^- m_{MgCl_2^0}^3 m_{SiO_2^0}^4), \tag{8}$$

where ρ_{Tlc} denotes the concentration of Tlc per volume unit of porous medium, $\{mol/cm^3\}$. The equilibrium constant K_{Tlc} for this reaction and Ω_{Tlc} are defined by

$$k_{Tlc} = \frac{k_{Tlc}^+}{k_{Tlc}^-}, \quad \Omega_{Tlc} = \frac{m_{MgCl_2^0}^3 m_{SiO_2^0}^4}{m_{HCl0}^6 k_{Tlc}}; \tag{9}$$

consequently,

$$\frac{\partial \rho_{Tlc}}{\partial t} = -k_{Tlc}^+ m_{HCl0}^6 (1 - \Omega_{Tlc}). \tag{10}$$

Similar relations apply to other minerals:

$$\begin{aligned}
\frac{\partial \rho_Q}{\partial t} &= -k_Q^+ (1 - \Omega_Q), \\
\frac{\partial \rho_{Fo}}{\partial t} &= -k_{Fo}^+ (1 - \Omega_{Fo}) m_{HCl0}^4, \\
\frac{\partial \rho_{Brc}}{\partial t} &= -k_{Brc}^+ (1 - \Omega_{Brc}) m_{HCl0}^2.
\end{aligned} \tag{11}$$

In the general case the constants k_m^+ depend on the specific mineral surface area. In the model calculation, variation in surface area is neglected i.e. $k_m^+ = \text{const.}$ The mass transfer equations for the system with components Mg and Si, index 1 assigned to SiO_2^0, 2 to $MgCl_2^0$ are as follows:

$$\left.\begin{aligned}
\beta \frac{\partial a m_1}{\partial t} + \frac{\partial \rho_Q}{\partial t} + 4\frac{\partial \rho_{Tlc}}{\partial t} + \frac{\partial \rho_{Fo}}{\partial t} &= \beta D_1 \frac{\partial}{\partial x}\left(a\frac{\partial m_1}{\partial x}\right), \\
\beta \frac{\partial a m_2}{\partial t} + 3\frac{\partial \rho_{Tlc}}{\partial t} + 2\frac{\partial \rho_{Fo}}{\partial t} + \frac{\partial \rho_{Br}}{\partial t} &= \beta D_2 \frac{\partial}{\partial x}\left(a\frac{\partial m_2}{\partial x}\right),
\end{aligned}\right\} \tag{12}$$

where $\beta = \rho_{H_2O}/1000$ denotes the coefficient for converting from molality to concentration units $\{mol/cm^3\}$ per 1 cm^3 of porous solution, a denotes the porosity, which may be computed according to: $a = 1 - \sum_m V_m^0 \rho_m$, where $m = $ Q, Tlc, Fo, Brc, V_m^0 is molar

171

volume of mineral m.

It is necessary to complement equations (12) by the four kinetic equations (11) describing the mineral phases. As the integration of (11) for $\Omega_m < 1$ and $\rho_m = 0$, at $t = t_0$, the initial moment of time, results in $\rho_m < 0$ for $t > t_0$, some supplementary constraints on kinetic functions must be taken into account according to

$$\frac{\partial \rho_m}{\partial t} = -k_m^+ m_{HClO}^{2v_{rm}}(1 - \Omega_m)\delta_m(\rho_m, \Omega_m)$$

(13)

where

$$\delta_m(\rho_m, \Omega_m) = \begin{cases} 1, & \rho_m > 0, \\ 0, & \rho_m = 0 \text{ and } \Omega_m < 1, \end{cases}$$
$$m = Q, \text{ Tlc, Fo, Brc.}$$

This condition ensures that physically it is not possible to dissolve a mineral that is not present. If initial and boundary conditions are specified the system of differential equations (12, 13) completely defines mineral formation in the distributed parameter system of interest. Equations (12) are valid only for the case of immobile mineral phases with flux of water through chemical reactions and changes in porosity being ignored.

Though we are interested in the solution to the problem (12, 13) in an interval of space coordinate $x(-\infty, +\infty)$, we shall numerically solve the problem in the finite interval of $(-L, L)$ with uniform boundary conditions of second kind and sufficiently large L. At the initial time, $t = t_0$, mineral and solution concentrations on both sides of the contact of two incompatible minerals are specified.

The following dimensionless variables and parameters are introduced:

$$\left. \begin{array}{l} \epsilon = \beta m_1^0 V_Q^0, \ C_i = m_i/m^0, \ m_1^0 = K_Q, \ m_2^0 = K_{Br} m_{HClO}^2, \\[4pt] q = m_1^0/m_2^0, \ \eta_m = \rho_m/\rho_m^0 = \rho_m V_m^0, \ R_m = V_m^0/V_Q^0 = \rho_Q^0/\rho_m^0, \\[4pt] \xi = \frac{x}{L}, \ \tau = \frac{\epsilon D_1 t}{L^2}, \ D = \frac{D_2}{D_1}, \ \kappa = \frac{L^2}{D_1 \beta m_1^0}, \\[4pt] \bar{\Omega}_m = \Omega_m, \ \bar{k}_m^+ = k_m^+ \kappa, \ \bar{K}_m = \frac{K_m}{(m_1^0)^{v_{1m}}(m_2^0)^{v_{2m}}}. \end{array} \right\}$$

(14)

Here m_1^0 and m_2^0 denote the concentrations of SiO_2^0 and $MgCl_2^0$ in equilibrium with quartz and brucite, $\epsilon \ll 1$ is a parameter which reflects the ratio of concentrations in solution to mineral densities, $V_m^0 = 1/\rho_m^0$ are molar volumes of minerals, C_i are relative concentrations of species in porous solution, η_m are mineral volume fractions, τ and ξ are the dimensionless time and space coordinates, respectively.

With these variables the initial system of equations reduces to the form

$$\left. \begin{array}{l} \epsilon \frac{\partial a C_1}{\partial \tau} + R_Q^{-1}\frac{\partial \eta_Q}{\partial \tau} + 4R_{Tlc}^{-1}\frac{\partial \eta_{Tlc}}{\partial \tau} + R_{Fo}^{-1}\frac{\partial \eta_{Fo}}{\partial \tau} = \frac{\partial}{\partial \xi}(a\frac{\partial C_1}{\partial \xi}), \\[8pt] \epsilon \frac{\partial a C_2}{\partial \tau} + q(3R_{Tlc}^{-1}\frac{\partial \eta_{Tlc}}{\partial \tau} + 2R_{Fo}^{-1}\frac{\partial \eta_{Fo}}{\partial \tau} + R_{Brc}^{-1}\frac{\partial \eta_{Brc}}{\partial \tau}) = D\frac{\partial}{\partial \xi}(a\frac{\partial C_2}{\partial \xi}), \\[8pt] \frac{\partial \eta_m}{\partial \tau} = -R_m \bar{k}_m^+ m_{HClO}^{2v_{2m}}(1 - \Omega_m)\delta_m(\eta_m, \Omega_m), \end{array} \right\}$$

(15)

where $m = Q$, Tlc, Fo, Brc.

The variables x, t and constant k_m^+ can be expressed in terms of ξ, τ, k_m according to

$$\left.\begin{array}{l} t = \dfrac{L^2}{\epsilon D_1}\,\tau, \\[2ex] x = L\xi, \\[2ex] k_m^+ = \dfrac{D_1 \beta m_1^0}{L^2}\,\bar{k}_m^+. \end{array}\right\} \tag{16}$$

The LCE-system of equations

The last four equations of system (15) are rewritten introducing the small parameter $\mu = 1/\bar{k}_m^+$:

$$\mu \frac{\partial \eta_m}{\partial \tau} = - R_m m_{\mathrm{HClO}}^{2\nu_{2m}}(1 - \Omega_m)\delta_m(\eta_m, \Omega_m), \tag{17}$$

where $m = Q$, Tlc, Fo, Brc.

It follows for μ from (14), (16) that

$$\mu = \frac{D_1 \beta m_1^0}{L^2 k_m^+} = \frac{\tau}{V_Q^0 k_m^+ t}.$$

μ becomes a small parameter at large values of k_m^+, which, in a physical sense, corresponds to large specific surface area of mineral and high temperatures. On the other hand, for given dimensionless time τ and chemical transport in nonrestricted space $x(-\infty, +\infty)$: $\mu \to 0$ as $t \to \infty$. In the limit as $\mu \to 0$ and with $|\partial \eta_m/\partial \tau|$ a limited function one gets

$$(1 - \Omega_m)\delta_m = 0, \quad m = Q, \text{ Tlc, Fo, Brc} \tag{18}$$

At $\delta_m \neq 0$ it follows that

$$\begin{array}{l} \nu_{1m}\ln C_1 + \nu_{2m}\ln C_2 = \ln \overline{K}_m + 2\nu_{2m}\ln m_{\mathrm{HClO}}, \\ m = Q, \text{ Tlc, Fo, Brc} \end{array} \tag{19}$$

and therefore at given $\overline{m}_{\mathrm{HClO}}$ a linear system of four equations is obtained with two unknowns. Thus the system is overdefined and a unique solution can be given only for mineral pairs. The following thermodynamically stable mineral pairs are possible: Q/Tlc, Tlc/Fo, Fo/Brc. However, volume fractions of minerals corresponding to unique definition of C_i which follows from the two first equations (8) are not defined. Thus, there is nothing but to assume that, in the limit case, the system is separated into monomineralic zones. In this case, a system of two differential equations and one algebraic equation uniquely defines each zone under corresponding initial and boundary conditions:

$$\left.\begin{array}{l} \epsilon\dfrac{\partial a C_1}{\partial \tau} + \dfrac{\nu_{1m}}{R_m}\dfrac{\partial \eta_m}{\partial \tau} = \dfrac{\partial}{\partial \xi}(a\dfrac{\partial C_1}{\partial \xi}), \\[2ex] \epsilon\dfrac{\partial a C_2}{\partial \tau} + q\dfrac{\nu_{2m}}{R_m}\dfrac{\partial \eta_m}{\partial \tau} = \dfrac{\partial}{\partial \xi}(a\dfrac{\partial C_2}{\partial \xi}), \\[2ex] \nu_{1m}\ln C_1 + \nu_{2m}\ln C_2 = \ln \overline{K}_m + 2\nu_{rm}\ln m_{\mathrm{HClO}}, \end{array}\right\} \tag{20}$$

where $m = $ Q, Tlc, Fo or Brc.

The zone separation sequence can be deduced from the array of thermodynamically stable pairs: Q/Tlc/Fo/Brc. It is evident from the example given that with regard to definition of the equilibrium zone sequence the limit transition to LCE does not give rise to any difficulties. In the general case of systems of three or more components, the problem of the LCE zone sequence becomes more complicated in character and has not been completely solved.

Methods of solving the system of equations

The system of equations (15, 20) were solved by methods of finite differences. In this part we give the initial and boundary conditions and the main elements of the algorithms used.

The kinetic system of equations

The origin and development of talc or forsterite zones is described by system (15) which may be expressed according to

$$\left. \begin{aligned} \epsilon\frac{\partial aC_k}{\partial \tau} + ((q-1)k + 2 - q)\sum_m \frac{\nu_{km}}{R_m}\frac{\partial \eta_m}{\partial \tau} &= ((D-1)k + 2 - D)\frac{\partial}{\partial \xi}(a\frac{\partial C_k}{\partial \xi}), k = 1, 2, \\ \frac{\partial \eta_m}{\partial \tau} &= -R_m \bar{k}_m^+ m_{\text{HClO}}^{2\nu 2m}(1 - \bar{\Omega}_m)\delta_m, \end{aligned} \right\} \quad (21)$$

where $m = 1, 2, 3$ implying the sequence Q, Tlc, Fo or Tlc, Fo, Brc.

Initial and boundary conditions of the problem for $-1 \leqslant \xi \leqslant 1$ are as follows:

$$t = 0: \qquad \eta_1 = \begin{cases} \eta_{10}, & -1 \leqslant \xi \leqslant 0, \\ 0, & 0 < \xi \leqslant 1; \end{cases} \quad \eta_2 = 0, \quad -1 \leqslant \xi \leqslant 1,$$

$$\left. \begin{aligned} \eta_3 = \begin{cases} 0, & -1 \leqslant \xi \leqslant 0, \\ \eta_{30}, & 0 < \xi \leqslant 1, \end{cases} \quad C_1 = \begin{cases} C_{10}^-, & -1 \leqslant \xi \leqslant 0, \\ C_{10}^+, & 0 < \xi \leqslant 1, \end{cases} \quad C_2 = \begin{cases} C_{20}^-, & -1 \leqslant \xi \leqslant 0, \\ C_{20}^+, & 0 < \xi \leqslant 1, \end{cases} \end{aligned} \right\} \quad (22)$$

$$t > 0: \qquad \frac{\partial C_k}{\partial \xi}\bigg|_{\xi = \pm 1} = 0, \, k = 1, 2.$$

In the case of the starting contact Q/Fo: $C_{10}^- = 1$, $C_{20}^- = 0$; C_{10}^+ and C_{20}^+ are defined by the equilibrium Tlc/Fo. For the starting contact Tlc/Brc: C_{10}^- and C_{20}^- are defined by the equilibrium condition Tlc/Fo; $C_{10}^+ = 0$, $C_{20}^+ = 1$.

Substituting expressions for $\partial \eta_m / \partial \tau$ into the component mass transfer equations and transforming

$$\bar{\Omega}_m = \frac{C_1^{\nu 1m} C_2^{\nu 2m}}{m_{\text{HClO}}^{2\nu 2m}\bar{K}_m}, \quad \frac{\partial aC_k}{\partial \tau} = a\frac{\partial C_k}{\partial \tau} + C_k\frac{\partial a}{\partial \tau} = a\frac{\partial C_k}{\partial \tau} - C_k\sum_m\frac{\partial \eta_m}{\partial \tau}$$

we obtained

$$\epsilon a \frac{\partial C_k}{\partial \tau} = \sum_m \{(((q-1)k + 2 - q)\frac{\nu_{km}}{R_m} - C_k)R_m \bar{k}_m^+ m_{\mathrm{HClO}}^{2\nu 2m}$$

$$\times (1 - \frac{C_1^{\nu 1m} C_2^{\nu 2m}}{m_{\mathrm{HClO}}^{2\nu 2m} \bar{K}_m})\delta(\eta_m, \Omega_m)\} + ((D-1)k + 2 - D)\frac{\partial}{\partial \xi}(a\frac{\partial C_k}{\partial \xi}). \tag{23}$$

Transforming the system of equations (23) to the finite difference form according to the implicit scheme (Samarsky, 1977) in a uniform space grid of N nodes in an interval of $-1 \leqslant \xi \leqslant 1$ we get

$$\epsilon a_i^j \frac{C_{ki}^{j+1} - C_{ki}^j}{\tau} = \sum_m \{(((q-1)k + 2 - q)\frac{\nu_{km}}{R_m} - C_{ki}^{j+1})R_m \bar{k}_m^+ m_{\mathrm{HClO}}^{2\nu 2m}$$

$$\times (1 - \frac{(C_{1i}^{j+1})^{\nu 1m}(C_{2i}^{j+1})^{\nu 2m}}{m_{\mathrm{HClO}}^{2\nu 2m} \bar{K}_m})\delta(\eta_{lm}^j, \Omega_m^j)\} + ((D-1)k + 2 - D)\frac{1}{h}$$

$$\times (\alpha_{i+1}^j \frac{C_{k, i+1}^{j+1} - C_{ki}^{j+1}}{h} - \alpha_i^j \frac{C_{ki}^{j+1} - C_{ki-1}^{j+1}}{h}), \tag{24}$$

where $k = 1, 2$, $i = 2, \ldots, n - 1$, the index j relates to time steps, τ is time step, h is space step. Taking into account uniform conditions of the second kind on $\xi = \pm 1$ it follows that $C_{k1}^{j+1} = C_{k2}^{j+1}$ and $C_{kN-1}^{j+1} = C_{kN}^{j+1}$. To compute the concentration at the $(j+1)$th time step C_{ki}^{j+1} from the jth, it is necessary to solve the system of $2(N-2)$ non-linear algebraic equations (24). This system has been solved by Newton's method. Let $r = 1, 2, \ldots$ by the sequence of iteration numbers in Newton's method. Dropping the index $j + 1$, denote each equation (12) by $F_{ki}(C_{ki-1}, C_{ki}, C_{ki+1}) = 0$, and the matrix of first derivatives by $|\partial F_{ki}/\partial C_{ki}|$. Then

$$C_k^{r+1} = C_k^r - \left|\left(\frac{\partial F_{ki}}{\partial C_{ki}}\right)^r\right|^{-1} F_{ki}(C_{k, i-1}^r, C_{k, i}^r, C_{k, i+1}^r). \tag{25}$$

The system is solved if $|(C_{ki}^{r+1} - C_{ki}^r)/C_{ki}^{r+1}| < \epsilon^0$, where ϵ^0 is a given small number. Thus in each step of the Newton iteration sequence it is necessary to solve the system of linear equations corresponding to the matrix $|\partial F_{ki}/\partial C_{ki}|$. This matrix has the tri-diagonal block structure, with each block being a square matrix of dimension (2×2).

After solving system (24) the value of η_m at the $(j + 1)$ time step is obtained from the following explicit relations:

$$\eta_{mi}^{j+1} = \eta_{mi}^j - \tau R_m \bar{k}_m^+ m_{\mathrm{HClO}}^{2\nu 2m}(1 - \Omega_m^{j+1})\delta(\eta_m^j, \Omega_m^{j+1}). \tag{26}$$

The limit LCE-system of equations

We shall solve the system of equations (20) using the front rectification method given by Budak et al. (1966) for Stephan multifront problems. Consider a monomineralic zone consisting of mineral X_2 which has developed at the starting mineral contact X_1/X_3.

Designate the coordinates of phase fronts X_1/X_2 and X_2/X_3 according to $\xi_0(\tau)$ and $\xi_1(\tau)$.

Introduce new variables

$$\zeta(\xi, \tau) = (\xi - \xi_0(\tau))/(\xi_1(\tau) - \xi_0(\tau)), \quad \bar{\tau} = \tau,$$

defined on the regions (ζ, τ): $0 \leqslant \zeta \leqslant 1$, $\tau > 0$.

Designate $\bar{\rho}_k = \epsilon a C_k + ((q-1)k + 2 - q)\dfrac{\nu_{km}}{R_m}\eta_m$.

Expressing equations (20) in terms of the new variables gives ($\xi_i' = \mathrm{d}\xi_i/\mathrm{d}\tau$, $i = 0, 1$)

$$\left.\begin{aligned}
&\frac{\partial \bar{\rho}_k}{\partial \tau} = \frac{\xi_0' + (\xi_1' - \xi_0')}{\xi_1 - \xi_0}\frac{\partial \rho_k}{\partial \xi} + \frac{((D-1)k + 2 - D)}{(\xi_1 - \xi_0)^2}\frac{\partial}{\partial \xi}\left(a\frac{\partial C_k}{\partial \xi}\right), \\
&\sum \nu_{km}\ln C_k = \ln \bar{K}_m + 2\nu_{2m}\ln(m_{\mathrm{HClO}}), \\
&a = 1 - \eta_m, \quad m = 2.
\end{aligned}\right\} \tag{27}$$

As a consequence of conservation of mass at the region boundaries (discontinuities) the following conditions must be satisifed:

$$\left.\begin{aligned}
&\xi_0((D-1)k + 2 - D)\{(a\frac{\partial C_k}{\partial \xi})_{\xi_0^{+0}} - (a\frac{\partial C_k}{\partial \xi})_{\xi_0^{-0}}\} = (\bar{\rho}_k|_{\xi_0^{-0}} - \bar{\rho}_k|_{\xi_0^{+0}})\xi_0', \\
&\xi_1((D-1)k + 2 - D)\{(a\frac{\partial C_k}{\partial \xi})_{\xi_1^{+0}} - (a\frac{\partial C_k}{\partial \xi})_{\xi_1^{-0}}\} = (\bar{\rho}_k|_{\xi_1^{-0}} - \bar{\rho}_k|_{\xi_1^{+0}})\xi_1'.
\end{aligned}\right\} \tag{28}$$

The diffusion character of the process requires the following constraints on the concentration at the boundaries:

$$C_k|_{\xi_i}^{+0} = C_k|_{\xi_i}^{-0} = C_k^i, \quad i = 0, 1; \ k = 1, 2. \tag{29}$$

The quantities are defined from equilibrium relations for the pairs X_1/X_2, X_2/X_3. The initial and boundary conditions of problem (20), in which we are interested, are similar to those of kinetic problems (15). However, to solve the system of equations (27) with the conditions of discontinuity (28), it is necessary to determine at the initial time, $\tau = \tau_0$, the concentrations of minerals and species for the interval $0 \leqslant \xi \leqslant 1$ and positions of boundaries $\xi_0(\tau_0)$ and $\xi_1(\tau_0)$.

Define the initial and boundary conditions of the problem:

$$\left.\begin{aligned}
\tau = 0: \quad &\eta_1 = \begin{cases} \eta_{10}, & \zeta = 0, \\ 0, & 0 < \zeta \leqslant 1, \end{cases} \quad \eta_2 = \eta_{20}, \ 0 \leqslant \xi \leqslant 1, \\
&\eta_3 = \begin{cases} \eta_{30}, & \zeta = 1, \\ 0, & 0 \leqslant \zeta < 1, \end{cases} \quad C_k = (C_k^1 - C_k^0)\zeta + C_k^0, \ k = 1, 2; \\
&\xi_0 = -l_0, \ \xi_1 = l_0; \\
\tau > 0: \quad &\eta_1 = \begin{cases} \eta_{10}, & \zeta = 0, \\ 0, & 0 < \zeta \leqslant 1, \end{cases} \eta_3 = \begin{cases} \eta_{30}, & \zeta = 1, \\ 0, & 0 \leqslant \zeta < 1, \end{cases} C_k = \begin{cases} C_k^0, & \zeta = 0, \\ C_k^1, & \zeta = 1, \end{cases}
\end{aligned}\right\} \tag{30}$$

The relations on discontinuities in this case are the following.

$\xi_0(\zeta=0):$ $\qquad ((D-1)k+2-D)\left(\dfrac{a}{(\xi_1-\xi_0)}\dfrac{\partial C_k}{\partial\zeta}\right)_{\zeta=0}=\left(\bar{\rho}_k\big|_{\xi_0^- 0}-\bar{\rho}_k\big|_{\xi_0^+ 0}\right)\xi_0',$

$\xi_1(\zeta=1):$ $\qquad ((D-1)k+2-D)\left(\dfrac{a}{(\xi_1-\xi_0)}\dfrac{\partial C_k}{\partial\zeta}\right)_{\zeta=1}=\left(\bar{\rho}_k\big|_{\xi_1^- 0}-\bar{\rho}_k\big|_{\xi_1^+ 0}\right)\xi_1',$ $\qquad (31)$

where

$$\bar{\rho}_k\big|_{\xi_0^- 0}=\epsilon(1-\eta_{10})C_k^0+((q-1)k+2-q)\frac{v_{k1}}{R_1}\eta_{10},$$

$$\bar{\rho}_k\big|_{\xi_1^+ 0}=\epsilon(1-\eta_{30})C_k^1+((q-1)k+2-q)\frac{v_{k3}}{R_3}\eta_{30}. \qquad (32)$$

Solution of the system of equations (27, 28) allows us to characterize development of the monomineralic zone from the starting state with time. For times sufficiently large, when the monomineralic zone width is considerably larger than $2l_0$, the solution obtained must be in good agreement with that corresponding to the initial and boundary conditions (22).

The equation system (27) is transformed according to the implicit scheme (Samarsky, 1977) to finite differences relative to C_k and η_2 on a uniform space grid of N nodes in the interval $0\leqslant\xi\leqslant1$. In order to account for the changes in positions and rates of boundaries ξ_0, ξ_1, ξ_0' and ξ_1', which are defined together with boundary densities η_{21} and η_{2N} from relations (31), it is necessary to use the revising iteration cycle. Designate

$$\varphi_s=\frac{\xi_0'^s-(\xi_1'^s-\xi_0'^s)\zeta}{\xi_1^s-\xi_0^s},\quad \bar{\rho}_{ki}^j=\epsilon a^j C_{ki}^j+((q-1)k+2-q)\frac{v_{k2}}{R_2}\eta_{2i}^j,$$

$$\bar{\rho}_{ki}^{j+1,s+1}=\epsilon a^s C_{ki}^{j+1,s+1}+((q-1)k+2-q)\frac{v_{k2}}{R_2}\eta_{2i}^{j+1,s+1}.$$

For the ith node the finite difference scheme is written as

$$\frac{\bar{\rho}_{ki}^{j+1,s+1}-\bar{\rho}_{ki}^j}{\tau}=\frac{\varphi_s}{h}\left(-\frac{1+\mathrm{sign}(\varphi_s)}{2}\bar{\rho}_{ki+1}^{j+1,s+1}-\mathrm{sign}(\varphi_s)\bar{\rho}_{ki}^{j+1,s+1}-\frac{1-\mathrm{sign}(\varphi_s)}{2}\bar{\rho}_{ki-1}^{j+1,s+1}\right)$$

$$+\frac{((D-1)k+2-D)}{(\xi_1^s-\xi_0^s)^2 2h^2}\left((\alpha_{i-1}^{j+1,s}+\alpha_i^{j+1,s})C_{ki-1}^{j+1,s+1}-(\alpha_{i-1}^{j+1,s}+2\alpha_i^{j+1,s}+\alpha_{i+1}^{j+1,s})C_{ki}^{j+1,s+1}\right. \qquad (33)$$

$$\left.+(\alpha_i^{j+1,s}+\alpha_{i+1}^{j+1,s})C_{ki+1}^{j+1,s+1}\right),$$

$$\sum_{k=1}^{2}v_{k2}\ln C_k^{j+1,s+1}=\ln\bar{K}_2+2v_{22}\ln m_{HClO},$$

where τ is time step.

The quantities of component concentrations in porous solution for $i=1$ and for $i=N$ are defined in the following way: $C_{k1}=C_k^0$, $C_{kN}=C_k^1$. The nonlinear system of $3(N-2)$ equations (33) was solved by Newton's method. The matrix of first derivatives has the block tri-diagonal structure, with each block being a square matrix (3×3) according to the number of unknowns at the ith node: C_{ki}, $k=1, 2$ and η_{2i}. The nonlinear system (33) is considered to be solved if $\sum\limits_{n=1}^{3(N-2)} |\Delta_n| \leqslant \epsilon^0$, where Δ_n is the discrepancy of the nth

equation (33), ϵ^0 is a given small number. Unknowns C_{ki} have been found in logarithm form. After solving equations (33), equations (31) are transformed to

$$\xi_0(\zeta=0): \quad ((D-1)k+2-D)\frac{1-\eta_{21}^{j+1,s+1}}{\xi_1^{j+1,s}-\xi_0^{j+1,s}}\frac{C_{k2}^{j+1,s+1}-C_{k1}}{h} \tag{34}$$

$$= (\bar{\rho_k}\big|_{\xi_0^{-0}} - \bar{\rho_k}^{j+1,s+1}\big|_{\xi_0^{+0}})\xi_0'^{s+1}$$

where $k=1, 2$,

$$\bar{\rho_k}^{j+1,s+1}\big|_{\xi_0^{+0}} = \epsilon(1-\eta_{21}^{j+1,s+1})C_k^0 + ((q-1)k+2-q)\frac{\nu_{k2}}{R_2}\eta_{21}^{j+1,s+1} \tag{35}$$

has been solved to refine $\eta_{21}^{j+1,s+1}$ and $\xi'^{j+1,s+1}$. Similarly, in the case of

$$\xi_1(\zeta=1): \quad ((D-1)k+2-D)\frac{1-\eta_{2N}^{j+1,s+1}}{(\xi_1^{j+1,s}-\xi_0^{j+1,s})}\frac{C_{kN}-C_{k,N-1}^{j+1,s+1}}{h} \tag{36}$$

$$= -(\bar{\rho_k}^{j+1,s+1}\big|_{\xi_1^{-0}} - \bar{\rho_k}\big|_{\xi_1^{+0}})\xi_1'^{s+1},$$

$$\bar{\rho_k}^{j+1,s+1}\big|_{\xi_1^{-0}} = \epsilon(1-\eta_{2N}^{j+1,s+1})C_k^1 + ((q-1)(k+2-q)\frac{\nu_{k2}}{R_2}\eta_{2N}^{j+1,s+1}), \tag{37}$$

$\eta_{2N}^{j+1,s+1}$ and $\xi_1'^{j+1,s+1}$ have been refined. Via relations

$$\xi_1^{j+1,s+1} = \xi_i^j + \tau(\xi_i')^{j+1,s+1} \tag{38}$$

all the values have been defined to resolving system (33). The iteration cycle is completed and, consequently, the time step (from j to $j+1$) is completed, if the following inequalities are satisfied:

$$\frac{|\eta_{2p}^{j+1,s+1}-\eta_{2p}^{j+1,s}|}{\eta_{2p}^{j+1,s+1}} \leqslant \epsilon_\eta^0, \quad p=1,N; \quad \left|\frac{\xi_p'^{j+1,s+1}-\xi_p'^{j+1,s}}{\xi_p'^{j+1,s+1}}\right| \leqslant \epsilon_\xi^0, \quad p=0,1, \tag{39}$$

where ϵ_η^0 and ϵ_ξ^0 are given small quantities. The numerical solutions in the cases of the zone sequences Q/Tlc/Fo and Tlc/Fo/Brc have been found by the above described method.

Thermodynamic and kinetic data for computations

The equilibrium solubility constants of minerals and species concentrations for specified conditions are shown in Table 8.1. The table is compiled on the basis of thermodynamic and calculated data (Hemley *et al.*, 1977a, 1977b, Helgeson *et al.*, 1978, Helgeson & Kirkham, 1974, 1981, Frantz & Marshall, 1982, 1983, Zaraisky *et al.*, 1986).

To solve kinetic equations (15), (21) the kinetic constants and surface area must be

Table 8.1. *Equilibrium thermodynamic data*

Solubility constants in molal scale and molar volumes of minerals, cm^3/mole

Minerals	Q	Tlc	Fo	Brc
Constants	$4.94 \cdot 10^{-2}$	$2.37 \cdot 10^{-7}$	$2.61 \cdot 10^{-2}$	45.25
Molar volumes	22.688	136.25	43.79	24.63

Relative concentrations in porous solution on boundaries at $m_{HClO} = $ const.

Species	Type of boundary		
	Q/Tlc	Tlc/Fo	Fo/Brc
SiO_2^0	1.0	$4.03 \cdot 10^{-1}$	$2.57 \cdot 10^{-4}$
$MgCl_2^0$	$7.52 \cdot 10^{-3}$	$2.52 \cdot 10^{-2}$	1.0

Molality of SiO_2^0 in equilibrium with Q
$m_{SiO_2^0} = 4.94 \cdot 10^{-2}$

Molalities of $MgCl_0^2$ in equilibrium with Brc at the different m_{HClO}

m_{HClO}	10^{-3}	$5 \cdot 10^{-3}$	10^{-2}	$2 \cdot 10^{-2}$	$5 \cdot 10^{-2}$	10^{-1}
$m_{MgCl_0^2}$	$4.54 \cdot 10^{-5}$	$1.14 \cdot 10^{-3}$	$4.54 \cdot 10^{-3}$	$1.82 \cdot 10^{-2}$	$1.14 \cdot 10^{-1}$	$4.54 \cdot 10^{-1}$

defined. For quartz the experimentally determined kinetic rate constants from Rimstidt & Barnes (1980) consistent with thermodynamic data have been used.

The rate of quartz dissolution in water satisfies the relation

$$\frac{\partial m_{SiO_2^0}}{\partial t} = \frac{A}{M_{H_2O}}(\hat{k}_Q^+ - \hat{k}_Q^- m_{SiO_2^0}), \quad K_Q = \frac{\hat{k}_Q^+}{\hat{k}_Q^-}, \tag{40}$$

where $m_{SiO_2^0}$ is molality, t is time in seconds, A/M_{H_2O} is surface area of mineral per unit of water mass, $\{m^2/kgH_2O\}$.

It is necessary to normalize the constants per unit volume of the porous medium. To achieve this goal we assumed cubic grains as a hypothetical model of the porous medium. In this case

$$\frac{A}{M_{H_2O}}\left[\frac{m^2}{kgH_2O}\right] = \frac{0.6(1-a)}{a\rho_{H_2O}d} \tag{41}$$

where a is porosity, ρ_{H_2O} is density of water (or hydrothermal solution) in g/cm^3, d is length of edge (diameter) of cubic grain in cm.

Then the required constants are defined by the relation

$$k_Q^\pm = \hat{k}_Q^\pm \frac{A}{M_{H_2O}} \cdot \frac{a\rho_{H_2O}}{1000} = \hat{k}_Q^\pm \cdot 6 \cdot 10^{-4}(1-a)/d. \tag{42}$$

Temperature dependence of $\hat{k}_Q^-$ is, according to Rimstidt & Barnes (1980),

179

$$\lg \hat{k}_Q^- = -0.707 - 2598/TK. \tag{43}$$

With thermodynamic quartz solubility constant, the extrapolation based on the latter dependence gives a crude estimate of $\hat{k}_Q^+(P=1000, T=600\ °C)=1.027\cdot10^{-5}$. Then relation (42) gives

$$k_Q^+ = 1.027\cdot10^{-5}\cdot6\cdot10^{-4}(1-a)/d. \tag{44}$$

Substituting $a=0.4$ and $d=10^{-4}$ cm we get $k_Q^+ = k_Q^{st} = 3.7\cdot10^{-5}$ {mol/cm^3/s}, which can be considered a reference ('standard') value.

The apparent kinetic constants of magnesium minerals are dependent on (according to equation (13))

$$\tilde{k}_m^+ = k_m^+(m_{HClO})^{2v2m} \tag{45}$$

where $m=$ Q, Tlc, Fo, Brc. Then $\tilde{k}_Q^+ = k_Q^+$.

To characterize the kinetic constants relative to the conventional 'standard' value k^{st} it is convenient to use relative quantities:

$$\tilde{k}_m^* = \tilde{k}_m^+/k_Q^{st}. \tag{46}$$

The diffusion coefficients for all species in porous solution at $P=1000$ bar, $T=600\ °C$ have been taken equal to $3.2\cdot10^{-4}$ {cm^2/s} in accordance with the data by Ildefonse & Gabis (1976), Balashov et al. (1983), and Zaraisky et al. (1986).

The initial porosities of the starting mineral contacts have been assumed close to that obtained from experiments (Zaraisky et al., 1981, Zaraisky et al., 1986, Zharikov & Zaraisky, Chapter 9 in this volume). For starting Q/Fo contact relative densities are $\eta_{10}=0.6$, $\eta_{30}=0.4$, corresponding to $a_Q^0=0.4$, $a_{Fo}^0=0.6$, respectively. For the starting contact Tlc/Brc $- \eta_{10}=0.6$ and $\eta_{30}=0.5$.

Results and discussion

The metasomatic reactions for both the starting contacts are

$$\left.\begin{array}{ll} \text{Q/Fo:} & 5Q+3Fo+2H_2O=2Tlc, \\ \text{Tlc/Brc:} & Tlc+5Brc=4Fo+6H_2O. \end{array}\right\} \tag{47}$$

According to the data of Table 8.1 the mineral volume change for the first reaction is 27.69 cm^3, and for the second -84.24 cm^3. Thus, for free water migration (at $P=$const. and stoichiometric mineral composition) one can assume that an immobile solid phase regime of the first reaction is possible only for initial porosity greater than 0.1131, but in the second case the formation of additional porosity equal to 0.3427 is possible. These estimates of porosity are valid only for the case of uniform reaction progress (for lumped-parameter systems). These estimates should be considered as a rough guide to conditions of mineral formation in a zone of mutual diffusion. For example, in the case of complete immobility of magnesium the following relation of volume fractions of magnesium minerals is correct:

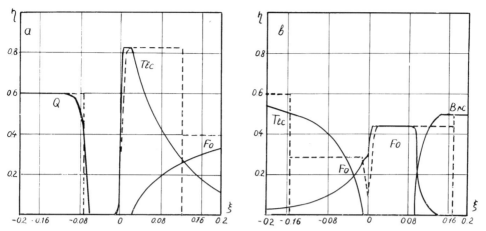

Fig. 8.1. Distribution of mineral volume fractions at $m_{HClO} = 10^{-3}$. In all figures solid lines correspond to kinetic solutions, dotted lines LCE-solutions. (a) zoning Q/Tlc/Fo, $\tau_{kin} = 2.56 \cdot 10^{-2}$, $\tau_{LCE} = 2.55 \cdot 10^{-2}$; (b) zoning Tlc/Fo/Brc, $\tau_{kin} = 2.46 \cdot 10^{-2}$, $\tau_{LCE} = 2.46 \cdot 10^{-2}$.

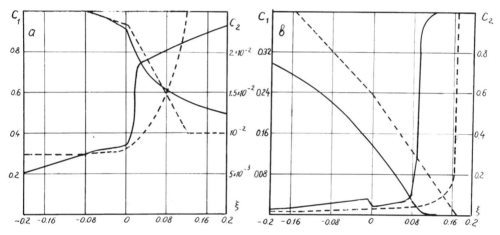

Fig. 8.2. The concentration profiles in porous solution at $m_{HClO} = 10^{-3}$. (a) zoning Q/Tlc/Fo, $\tau_{kin} = 2.56 \cdot 10^{-2}$, $\tau_{LCE} = 2.55 \cdot 10^{-2}$; (b) zoning Tlc/Fo/Brc, $\tau_{kin} = 2.46 \cdot 10^{-2}$, $\tau_{LCE} = 2.46 \cdot 10^{-2}$.

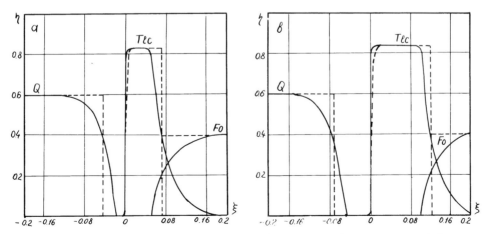

Fig. 8.3. The mineral volume fraction profiles for zoning Q/Tlc/Fo at $m_{HClO} = 10^{-2}$. (a) $\tau_{kin} = 8.8 \cdot 10^{-3}$, $\tau_{LCE} = 8.8 \cdot 10^{-3}$; (b) $\tau_{kin} = 2.56 \cdot 10^{-2}$, $\tau_{LCE} = 2.52 \cdot 10^{-2}$.

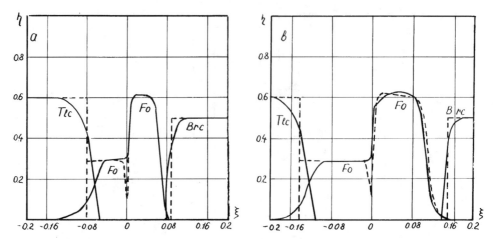

Fig. 8.4.   The mineral volume fraction profiles for zoning Tlc/Fo/Brc at
$m_{HCL0} = 10^{-2}$. (a) $\tau_{kin} = 6.84 \cdot 10^{-2}$, $\tau_{LCE} = 6.56 \cdot 10^{-2}$; (b) $\tau_{kin} = 1.96 \cdot 10^{-2}$,
$\tau_{LCE} = 1.95 \cdot 10^{-2}$.

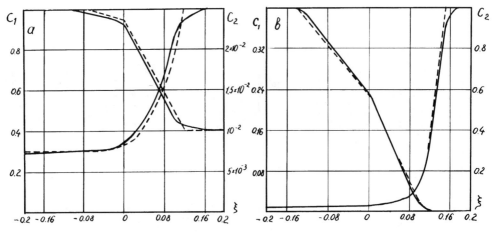

Fig. 8.5.   The concentration profiles in porous solution at $m_{HCl0} = 10^{-2}$. (a) zoning
Q/Tlc/Fo, $\tau_{kin} = 2.56 \cdot 10^2$, $\tau_{LCE} = 2.52 \cdot 10^{-2}$; (b) zoning Tlc/Fo/Brc,
$\tau_{kin} = 1.96 \cdot 10^{-2}$, $\tau_{LCE} = 1.95 \cdot 10^{-2}$.

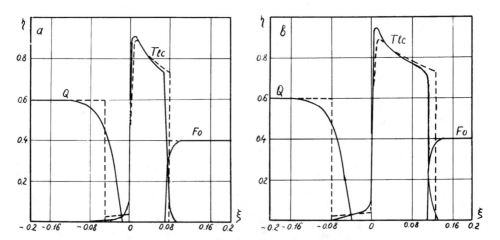

Fig. 8.6.   The mineral volume fraction profiles for zoning Q/Tlc/Fo at $m = 5 \cdot 10^{-2}$.
(a) $\tau_{kin} = 9.42 \cdot 10^{-3}$, $\tau_{LCE} = 9.64 \cdot 10^{-3}$; (b) $\tau_{kin} = 2.38 \cdot 10^{-2}$, $\tau_{LCE} = 2.37 \cdot 10^{-2}$.

$$\eta_{\text{Brc}} = 1.125\eta_{\text{Fo}} = 0.542\eta_{\text{Tlc}}. \qquad (48)$$

Thus, the relative concentration of forsterite replaced by talc of unit relative concentration is 0.482 ($a_{\text{Fo}} = 0.518$). In the case of lower initial forsterite porosity in contact Q/Fo a critical situation of the volume deficiency arises.

Computations for the limit LCE-case have been made on a digital computer for six different molalities m (Table 8.1). In LCE-simulation on the non-dimensional width of the initial zone was $2l_0 = 4 \cdot 10^{-3}$, the initial relative concentration of the forming mineral (talc or forsterite) being $\eta_{20} = 0.01$, ($a_{20} = 0.99$). Thus the initial time zones were in practice narrow slots.

Dashed lines in Figs. 8.1–8.6 show numerical results for profiles of mineral volume fractions (Figs. 8.1, 8.3, 8.4, 8.6) and SiO_2^0 and $MgCl_2^0$ concentrations in porous solution (Figs. 8.2, 8.5). In all cases a state corresponding to invariant inner structure of developing monomineralic zone is achieved (Figs. 8.1, 8.3, 8.4, 8.6). The resulting stable structure extends in time – Figs. 8.3, 8.4, 8.6. The concentration profiles in porous solution have similar properties – Figs. 8.2, 8.5. The minimum of volume fraction near the starting contact (Fig. 8.1, 8.4) corresponds to the initial zone (hole). It becomes smoothed out in time.

The forming macrostructure of monomineralic zone depends on value of m_{HCl0}. Increasing m_{HCl0} implies an increase in magnesium mineral solubility and $MgCl_2$ concentration in porous solution (Table 8.1).

The transfer of Mg in porous solution at $m_{\text{HCl0}} = 10^{-3}$ is insignificant and the monomineralic zone structure corresponds to that under conditions of full conservation, i.e. at constant mineral concentration of Mg. The talc zone corresponding to initial contact Q/Fo is divided structurally into two parts: the first subzone of practically zero volume fraction ($\eta_{\text{Tlc}} \approx 10^{-4}$–$10^{-5}$) to the left of contact and the second subzone of constant volume fraction, $\eta_{\text{Tlc}} = 0.83$ (Fig. 8.1(a)), to the right. The forsterite zone (the starting contact Tlc/Brc) consists of two subzones with constant volume fractions: to the left of contact $\eta_{\text{Fo}} = 0.289$, to the right $\eta_{\text{Fo}} = 0.445$. All computed constant values of mineral volume fractions correspond to equation (48). Thus at low levels of m_{HCl0} internal mineral production practically does not occur, and the macromechanism of the process corresponds to free transfer of SiO_2^0 in porous solution and mineral replacement at zone boundaries at immobility of Mg. Two interzone replacements take place during forsterite growth. On the left-hand side talc is replaced by Fo with loss of silica, on the right-hand side brucite is replaced by Fo with gain of Si.

The rising level of m_{HCl0} causes an increase in inner-zone mineral production. At $m_{\text{HCl0}} = 10^{-2}$ (Tlc/Brc) and $m_{\text{HCl0}} = 5 \cdot 10^{-2}$ (Q/Fo) the volume fraction of forming mineral decreases at boundaries Fo/Brc and Tlc/Fo in spite of prevailing transfer of Si in porous solution. The macromechanism of forsterite zone development to the right of initial contact corresponds to 'dissolution with following precipitation'. Mg passes into the solution and is precipitated at some distance from the dissolution boundary. The left part of the forsterite zone is formed by the former macromechanism of replacement (Fig. 8.4.).

Table 8.2. *Parameters of kinetic computations*

Zoning: Q/Tlc/Fo

m_{HClO}	$\bar{k}_{\dot{Q}}^+$	$\bar{k}_{Tlc}^+$	$\bar{k}_{Fo}^+$	τ_1	τ_2	Fig. no.	$L/2.5$ cm	t_1 (s)	t_2 (s)	$k_{\dot{Q}}^+$	$k_{\dot{Q}}^*$	k_{Tlc}^+	k_{Tlc}^*	k_{Fo}^+	k_{Fo}^*
10^{-3}	$1.055\cdot10^4$	$1.077\cdot10^{18}$	$1.077\cdot10^{15}$		$2.56\cdot10^{-2}$	1(a), 2(a)	10^{-2}		119.5	$9.8\cdot10^{-3}$	$2.65\cdot10^3$	10^{13}	$2.7\cdot10^{-1}$	10^{10}	$2.7\cdot10^2$
"	"	"	"			9, 10(a)	10^{-1}		11950	$9.8\cdot10^{-4}$	26.5	10^{11}	$2.7\cdot10^{-3}$	10^8	2.7
10^{-2}	$1.055\cdot10^3$	$1.077\cdot10^{15}$	$1.077\cdot10^{10}$	$8.8\cdot10^{-3}$	$2.56\cdot10^{-2}$	3, 5(a)	10^{-2}	41	119.5	$9.8\cdot10^{-3}$	265	10^{10}	270	10^5	27
"	"	"	"			7(a), 10(a)	10^{-1}	4100	11950	$9.8\cdot10^{-5}$	2.65	10^8	2.7	10^3	0.27
"	$1.055\cdot10^5$	$1.077\cdot10^{17}$	$1.077\cdot10^{12}$		$2.40\cdot10^{-2}$	7(b)	10^{-2}		112	$9.8\cdot10^{-1}$	26500	10^{12}	$2.7\cdot10^{-4}$	10^7	$2.7\cdot10^3$
"	"	"	"				10^{-1}		1200	$9.8\cdot10^{-3}$	265	10^{10}	270	10^5	27
$2\cdot10^{-2}$	$1.055\cdot10^3$	$1.077\cdot10^{13}$	$1.077\cdot10^{8}$	$9.4\cdot10^{-3}$	$2.54\cdot10^{-2}$	10(a)	10^{-2}	44	118.7	$9.8\cdot10^{-3}$	265	10^8	173	10^3	4.32
"	"	"	"				10^{-1}	4400	11870	$9.8\cdot10^{-5}$	2.65	10^6	1.73	10	$4.32\cdot10^{-2}$
$5\cdot10^{-2}$	$1.055\cdot10^3$	$1.077\cdot10^{11}$	$1.077\cdot10^{10}$	$9.4\cdot10^{-3}$	$2.38\cdot10^{-2}$	6(a), 6(b)	10^{-2}	44	111.2	$9.8\cdot10^{-3}$	265	10^6	422	10^5	$1.69\cdot10^4$
"	"	"	"				10^{-1}	4400	11120	$9.8\cdot10^{-5}$	2.65	10^4	4.22	10^3	169

Zoning: Tlc/Fo/Brc

m_{HClO}	$\bar{k}_{Tlc}^+$	$\bar{k}_{Fo}^+$	$\bar{k}_{Brc}^+$	τ_1	τ_2	Fig. no.	$L/2.5$ cm	t_1 (s)	t_2 (s)	k_{Tlc}^+	k_{Tlc}^*	k_{Fo}^+	k_{Fo}^*	k_{Brc}^+	k_{Brc}^*
10^{-3}	$1.077\cdot10^{19}$	$1.077\cdot10^{12}$	$1.077\cdot10^9$		$2.46\cdot10^{-2}$	1(b), 2(b)	10^{-2}		115	10^{14}	2.7	10^7	$2.7\cdot10^{-1}$	10^4	270
"	"	"	"			9, 10(b)	10^{-1}		11500	10^{12}	$2.7\cdot10^{-2}$	10^5	$2.7\cdot10^{-3}$	10^2	2.7
$5\cdot10^{-3}$	$1.077\cdot10^{15}$	$1.077\cdot10^{10}$	$1.077\cdot10^8$	$9.6\cdot10^{-3}$	$3.32\cdot10^{-2}$	9	10^{-2}	44.8	108	10^{10}	4.22	10^5	1.69	10^3	676
"	"	"	"				10^{-1}	4480	10800	10^8	$4.22\cdot10^{-2}$	10^3	0.0169	10	6.76
10^{-2}	$1.077\cdot10^{14}$	$1.077\cdot10^{11}$	$1.077\cdot10^7$	$6.8\cdot10^{-3}$	$1.96\cdot10^{-2}$	4, 5(b)	10^{-2}	32	92	10^9	27	10^6	270	10^2	270
"	"	"	"			10(b)	10^{-1}	3200	9200	10^7	0.27	10^4	2.7	1.0	2.7

Fig. 8.6 shows that at $m_{HClo} = 5 \cdot 10^{-2}$ the macromechanism of talc zone formation to the right of the initial contact position is of mixed character. The internal precipitation of talc is visible to the left of contact. Similar patterns are observed in porous solution (Figs. 8.2, 8.5). For forsterite formation (Figs. 8.2(b), 8.5(b)) the maximal mineral precipitation region is marked by a sharp change of $MgCl_2^0$ concentration profile in porous solution.

The above described picture corresponds to the limit LCE-case of infinitely high rates of reversible reactions. Is a similar picture possible in the case of finite rates and what are the differences from the limit case? Answers to these and other questions are given by numerical computations for the kinetic system of equations (15, 21).

The computations were made for five HCl molalities in the case of the starting contact Q/Fo and for three HCl molalities in the case of the starting contact Tlc/Brc. Summaries of rate constants and time parameters are shown in Table 8.2. The general feature of experimental bimetasomatic columns is clearly sharp zone boundaries in the limits of zone width more than $5 \cdot 10^{-3}$ cm (Zaraisky et al., 1986). In Table 8.2 the results of computation for dimensionless equations are related, with regard to relations (16), to two variants of dimension quantities, which correspond to $L = 2.5 \cdot 10^{-2}$ and $L = 2.5 \cdot 10^{-1}$ cm. In the first case the length of abscissa in Figs. 8.1–8.6, 8.7 is 10^{-2} cm, in the second 10^{-1} cm.

Consider the solid lines in Figs. 8.1–8.7, which correspond to solutions of the kinetic equation system. It is convenient to analyse kinetic data using quantities which characterize the rates of chemical reactions in relation to the estimate 'standard'. The largest difference between kinetic and LCE-dependences is observed at $m_{HClo} = 10^{-3}$ (Figs. 8.1, 8.2). This can be explained by the smallest values of $\tilde{k}_{Tlc}^*$ and $\tilde{k}_{Fo}^*$ (Table 8.2), which in these cases are equal to $2.7 \cdot 10^{-1}$ and $2.7 \cdot 10^{-3}$, corresponding to L values $2.5 \cdot 10^{-2}$ and $2.5 \cdot 10^{-1}$ cm, respectively. However, it can be seen from Fig. 8.1. that kinetic curves correspond qualitatively to the limit LCE-type of dependences.

The 1000 times increase of effective kinetic constants of the forming magnesium minerals gives the coincidence of the kinetic and LCE-profiles (Figs. 8.3–8.6). It is to be noted that for $L = 2.5 \cdot 10^{-1}$ cm the constant values are only a few times greater than the 'standard' (Table 8.2). The kinetic solution for the data presented gives the best approximation to LCE-solution in the middle part of a forming monomineralic zone. On boundaries the kinetic solution is characterized by a certain smearing of a sharp LCE-front (Figs. 8.3, 8.4, 8.6). Thus the development of a monomineralic zone of ~ 0.1 cm thick can be described in the two-component system considered, in terms of LCE, if the values of effective kinetic constants are more than $5 \cdot 10^{-5}$ {mol/cm^3/s} (Figs. 8.3–8.6). Similar dependences can be obtained for a zone of 0.01 cm thick by increasing the kinetic constants 100 times.

The character of dimensionless parameters ξ, τ, $\bar{k}_m^+$ (14, 16) of equation systems (15) shows that asymptotic of kinetic system on large times is equivalent in a sense to asymptotic on large kinetic constants and corresponds to the limit LCE-case, as illustrated in Fig. 8.7. The results are characterized by the smearing of front

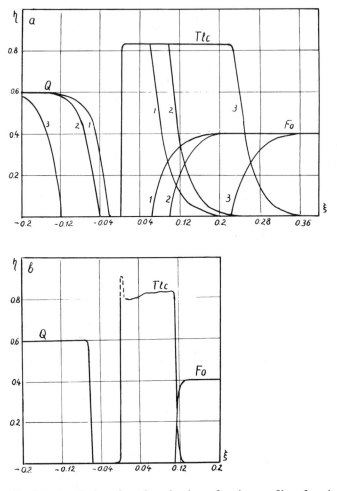

Fig. 8.7. The kinetic solutions for mineral volume fraction profiles of zoning Q/Tlc/
Fo at $m_{HClO} = 10^{-2}$ for different times: (a) $\tau_1 = 1.2 \cdot 10^{-2}$, $\tau_2 = 2.4 \cdot 10^{-2}$,
$\tau_3 = 0.112$; (b) $\tau = 2.4 \cdot 10^{-2}$, $\bar{k}_{Tlc}^+ = 1.077 \cdot 10^{17}$.

approximately constant in time. Thus with time the zone boundaries are getting more
sharp in relative space scale. Results for $L = 2.5 \cdot 10^{-1}$ cm shown in Fig. 8.7(b) can be
considered as a continuation of the process depicted in Fig. 8.7(a) in time for
$L = 2.5 \cdot 10^{-2}$ cm (Table 8.2).

Small oscillation of the upper part of the curve in Fig. 8.7(b) is a consequence of the
relative crudity of the space-time grid for these large values of dimensionless constants
(Table 8.2).

In general, any values of kinetic constants correspond to definite moments of time in
development of metasomatic zoning on the space axis $(-\infty, \infty)$ after which the
observed pattern practically coincides in relative space scale with LCE-zoning. This
may be important for consideration of geological processes which operate in nature
during long periods of time.

186

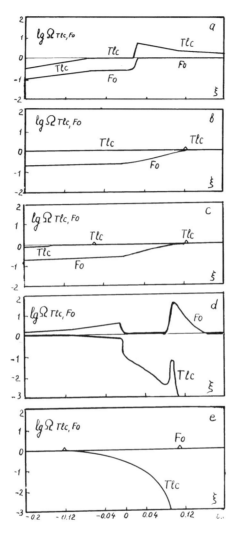

Fig. 8.8. The relative supersaturation levels of talc and forsterite during development of zoning. Zoning Q/Tlc/Fo: (a) $m_{HCl} = 10^{-3}$, $\tau = 2.56 \cdot 10^{-2}$; (b) $m_{HCl0} = 10^{-2}$, $\tau = 1.96 \cdot 10^{-2}$; (c) $m_{HCl} = 5 \cdot 10^{-2}$, $\tau = 2.38 \cdot 10^{-2}$. Zoning Tlc/Fo/Brc: (d) $m_{HCl} = 10^{-3}$, $\tau = 2.46 \cdot 10^{-2}$; (e) $m_{HCl0} = 10^{-2}$, $\tau = 1.96 \cdot 10^{-2}$. Triangles correspond to local maximal supersaturation regions invisible in given scale.

Taking into account that, for $L = 2.5 \cdot 10^{-1}$ cm, $k_{Tlc}^{+} = 10^{11}$ at $m_{HCl0} = 10^{-3}$ and $k_{Tlc}^{+} = 10^{4}$ at $m_{HCl0} = 5 \cdot 10^{-2}$ results in $\tilde{k}_{Tlc}^{*} = 2.7.10^{-3}$ and $\tilde{k}_{Tlc}^{*} = 4.22$, respectively. A similar situation is observed for the forsterite zone formation, given in Table 8.2. Thus to retain a system close to LCE during relatively small times at a low level of m_{HCl0} very high values of 'true' constants k_m^{+} are required. At the same time experimental results (Zaraisky *et al.*, 1986, Ildefonse & Gabis, 1976) show sharp boundaries under conditions of a low level of m_{HCl0}. We believe that replacement reactions independent

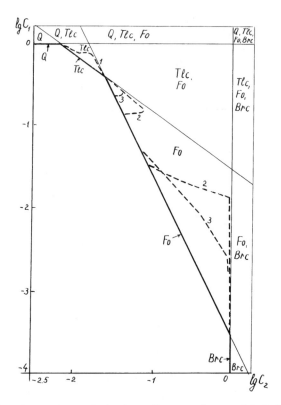

Fig. 8.9. The kinetic reaction paths in phase diagram of system MgO–SiO$_2$–H$_2$O–HCl. 1 – the reaction path of zoning Q/Tlc/Fo at $m_{HClO} = 10^{-3}$, $\tau = 2.56 \cdot 10^{-2}$; 2 and 3 – the reaction paths of zoning Tlc/Fo/Brc; 2 – $m_{HClO} = 10^{-3}$, $\tau = 2.46 \cdot 10^{-2}$; 3 – $m_{HClO} = 5 \cdot 10^{-3}$, $\tau = 2.32 \cdot 10^{-2}$. The reaction paths corresponding to other kinetic and LCE-solutions (Figs. 8.1–8.7) practically coincide with stable equilibrium lines of the diagram. The diagram is divided by metastable extensions of mineral-solution equilibria on supersaturation fields of mineral associations.

of m_{HClO} should be included in the reversible reaction scheme in order to make simulation of zoning at low m_{HClO} more real.

In the region of larger m_{HClO} ($m_{HClO} \geqslant 10^{-2}$) the chosen simple kinetic scheme of reversible reactions of dissolution–precipitation is a good approximation to LCE-case for relatively small k_m^+ that corresponds to LCE-macromechanisms discussed above.

Reaction regions are defined by dependence of supersaturation quantities of minerals on relative distance presented in Fig. 8.8. In the case of talc zone formation (Figs. 8.8(a), 8.8(b), at $m_{HClO} = 10^{-3}$–10^{-2} only one maximum was observed on the curve corresponding to talc forming over forsterite. Comparison of the curves at $m_{HClO} = 10^{-3}$ and $m_{HClO} = 10^{-2}$ is a characteristic of the approximation to LCE for $\Omega_{Tlc} = 1$. For $m_{HClO} = 5 \cdot 10^{-2}$ (Fig. 8.8(c)) the positions of the Ω_{Tlc} inner-zone peaks correspond to talc precipitation. The bimodal type of the curve is observed for the forsterite zone, Figs. 8.8(d), 8.8(e). The first and second maxima correspond to

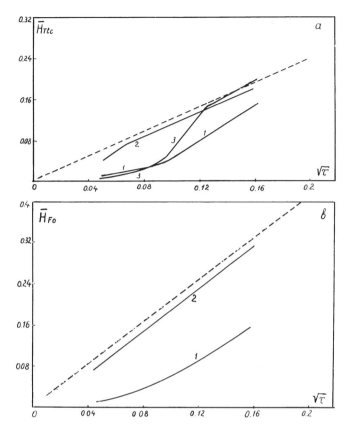

Fig. 8.10. The dynamics of growth of talc (a) and forsterite (b) zones according to LCE and kinetic solutions. (a): $1 - m_{HClO} = 10^{-3}$; $2 - m_{HClO} = 10^{-2}$; $3 - m_{HClO} = 2 \cdot 10^{-2}$. (b): $1 - m_{HClO} = 10^{-3}$; $2 - m_{HClO} = 10^{-2}$.

formation of forsterite on talc and brucite, respectively. With increasing m_{HClO} the second maximum shifts towards the inner part of the zone.

The concentration paths of reactions in the phase diagram (Fig. 8.9) resemble the curves of inner-zone supersaturations. Under conditions of LCE the reaction paths on this diagram must coincide with phase lines, the deviation from LCE corresponding to deviations of reactions paths from phase lines. Fig. 8.9 shows that the reaction paths of given mineral formation are within the limits of the supersaturation region of the forming mineral.

The computations show that concentrations in porous solution and mineral volume fractions tend to LCE-values in different ways. As follows from results presented in Figs. 8.8, 8.9 the component concentrations in porous solution tend to the limit LCE-values in the whole zone interval. It is not so in the case of mineral volume fraction distribution, where, as mentioned above, the smearing of fronts always takes place in absolute scale.

Growth of the talc and forsterite zones with time is illustrated in Fig. 8.10. The limit

Table 8.3. *The zone growth constants for LCE-case*

	b	
m_{HClO}	Tlc	Fo
10^{-3}	$1.22 \cdot 10^{-2}$	$2.11 \cdot 10^{-2}$
$5 \cdot 10^{-3}$	$1.21 \cdot 10^{-2}$	$2.22 \cdot 10^{-2}$
10^{-2}	$1.21 \cdot 10^{-2}$	$2.13 \cdot 10^{-2}$
$2 \cdot 10^{-2}$	$1.25 \cdot 10^{-2}$	$1.37 \cdot 10^{-2}$
$5 \cdot 10^{-2}$	$1.33 \cdot 10^{-2}$	$1.87 \cdot 10^{-2}$
10^{-1}	$9.97 \cdot 10^{-3}$	$2.50 \cdot 10^{-2}$

case of LCE corresponds to a boundary moving in direct proportion to $\sqrt{t}$. This front propagation is characteristic of diffusion in one-dimensional unrestricted space.

As follows from Fig. 8.10(a, b) in the case of variable mineral concentrations and porosity the law of front propagation does not change. The dimensionless constants of zone growth are calculated from the relation $b = H/2\sqrt{(D_1 t)}$, where H, the width of the zone, and t, are measured in cm and s, respectively, and their values are given in Table 8.3.

The growth curves in Figs. 8.10(a): 2, 3 and (b): 2 corresponding to kinetic scheme for calculated time intervals characterize the asymptotic approach to LCE-curves with time (Fig. 8.10).

In the limit LCE-case the numerical computations show self-similarity solutions, i.e. solutions which undergo only the transformation of similarity with time.

The correct self-similarity of a numerical solution is disturbed by specification of the non-zero initial zone with mineral density being equal to η_{20}. However, this disturbance is smoothed with time (Fig. 8.4).

Thus for numerical LCE-solutions of equation (27) we have

$$\left. \frac{\partial \bar{\rho}_k}{\partial \tau} \right|_\zeta = 0. \tag{49}$$

In the given case the self-similarity variable is a variable of type $\xi/\sqrt{\tau}$. Taking $\xi_i = 2b_i\sqrt{(D\tau)}$ we transform the transfer equations in (27) to

$$\frac{b_0 + (b_1 - b_0)\zeta}{2(b_1 - b_0)} \frac{d\bar{\rho}_k}{d\zeta} + \frac{(D-1)k + 2 - D}{4(b_1 - b_0)^2 D} \frac{d}{d\zeta}(a \frac{dC_k}{d\zeta}) = 0, \, k = 1, 2. \tag{50}$$

Now instead of the equation system with partial derivatives on ζ and τ we have a system of ordinary differential equations of second order and the algebraic equation of mass action law. Functions of one variable ζ: $C_k(\zeta)$, $\eta_2(\zeta)$, and constants of moving boundary b_i are the required solutions of this system.

Analysis of the dimensionless equations shows that the self-similarity solutions of the type considered† are valid for the general case of diffusion metasomatism with

† In the case of immobile solid phases.

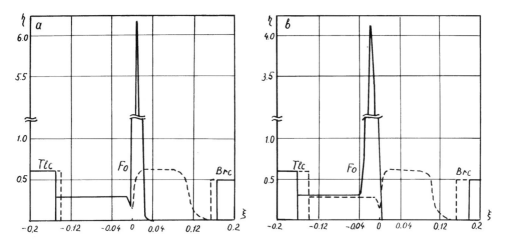

Fig. 8.11. The mineral volume fraction LCE-profiles for zoning Tlc/Fo/Brc at the high levels of m_{HClo}. (a) $m_{\text{HClo}} = 2\cdot10^{-2}$, $\tau = 5.67\cdot10^{-2}$; (b) $m_{\text{HClo}} = 5\cdot10^{-2}$, $\tau = 3.43\cdot10^{-2}$. The dotted lines correspond to $m_{\text{HClo}} = 10^{-2}$, $\tau = 1.95\cdot10^{-2}$.

variable porosity and mineral volume fractions in the space interval $(-\infty, \infty)$. It was D.S. Korzhinskii (1959, 1979) who pointed out a possibility of these solutions in the case of diffusion metasomatism. In Korzhinskii (1979) they were characterized as conventionally steady-state solutions.

For $m_{\text{HClo}} > 5\cdot10^{-2}$ (Q/Tlc/Fo) and $m_{\text{HClo}} > 10^{-2}$ (Tlc/Fo/Brc) the self-similarity LCE-solutions indicate complete closure of pores (due to overgrowing of pores) in the case of immobile solid phases. We assume constant porosity ($a^* = 0.01$) in points of complete closure of pores, i.e. take $a_2 = 0.01$ at $\eta_2 \geqslant 0.99$.

The overgrowing of pores was also observed by Lichtner et al. (1986b) in the case of constant porosity for zoning Q/Fo→Q/Tlc/Serpentine/For at 390 °C. As follows from our results the incorporation of changes in porosity into the model does not reduce the problem of complete rapid filling of the pore space due to internal precipitation.

The results are shown in Figs. 8.11, 8.12. It can be clearly seen in the case of development of the forsterite zone that increasing m_{HClo} moves the forsterite precipitation region to the left from starting contact (Figs. 8.11, 8.12). A sharp local volume fraction maximum of the forming mineral is also a distinctive feature particularly manifested for forsterite at $m_{\text{HClo}} = 2\cdot10^{-2}$ and $5\cdot10^{-2}$. The results given in Figs. 8.11, 8.12 and Table 8.1 allow us to assume that the maximal volume fraction corresponds to stoichiometric ratio of zone concentration differences of Mg and Si in porous solution, when

$$\frac{\Delta\text{Mg}}{\Delta\text{Si}} = \left|\frac{C_2^0 - C_2^1}{C_1^0 - C_1^1}\right| \approx \frac{\nu_{22}}{\nu_{12}}. \tag{51}$$

As follows from Table 8.1, for forsterite zone at $m_{\text{HClo}} = 2\cdot10^{-2}$, $\Delta\text{Mg}/\Delta\text{Si} = 0.89$, at $m_{\text{HClo}} = 5\cdot10^{-2}$, $\Delta\text{Mg}/\Delta\text{Si} = 5.56$.

At $a^* = 0.01$ no serious decrease of zoning growth rate is observed (Table 8.3). In the

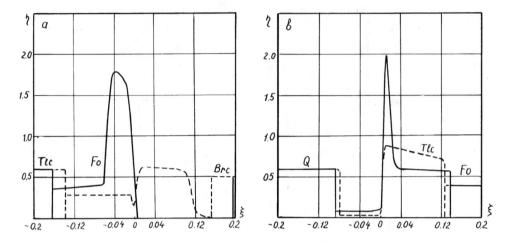

Fig. 8.12. The mineral volume fraction LCE-profiles at $m_{HClO} = 0.1$. (a) zoning Tlc/ Fo/Brc, $\tau = 2.14 \cdot 10^{-2}$, (b) zoning Q/Tlc/Fo, $\tau = 5.36 \cdot 10^{-2}$. The dotted lines correspond to: (a) $m_{HClO} = 10^{-2}$, $\tau = 1.95 \cdot 10^{-2}$; (b) $m_{HClO} = 5.10^{-2}$, $\tau = 2.37 \cdot 10^{-2}$.

case of $m_{HClO} = 0.1$ velocity of forsterite zone growth even increases due to increasing MgCl$_2^0$ concentration. The computation for $m_{HClO} = 2 \cdot 10^{-2}$ and $a^* = 0.001$ detected deceleration of forsterite zone growth, $b = 4.85 \cdot 10^{-3}$, with inner-zone structure remaining practically the same.

It is to be noted that closing of pores is caused by mineral precipitation under conditions of LCE as it is described by equation system (27). The integral volume relations of summary metasomatic reactions are favourable in the cases considered.

In the general case we can assume a relation between local stress + structure of heterogeneous mineral system (crystallization stress, pore size, etc.) and thermodynamic constants of mineral solubility, with the latter increasing in this situation of pore closing.

However from numerical data represented in Figs. 8.11, 8.12 it is clear that relatively small shifts of solid phases from their intensive precipitation regions, for example under the influence of crystallisation stress, relieves the critical situation. Physico-mathematical description of this movement requires modification of the differential transport equations by addition of the corresponding terms, which are responsible for the solid-phase movement.

Conclusions

Isothermal-isobaric diffusion metasomatism with variable porosity and distribution of mineral volume fractions was investigated under conditions of immobile solid phases. The macrokinetic model is suggested to describe development of monomineralic zone in hydrothermal systems with binary counterdiffusion.

The following results are obtained.

– LCE-solution is shown numerically to be asymptotic to kinetic solution at the large kinetic constants of reversible reactions.

– The limit LCE-solutions are self-similar and correspond to invariant inner-zone macrostructure with regard to origin and development of a monomineralic zone from contact of two incompatible minerals under conditions of one-dimensional diffusion in the space interval $(-\infty, \infty)$.

– Generally, numerical investigations based on the kinetic scheme give the following picture of development of diffusion metasomatic zoning with reversible reactions of dissolution–precipitation. At the initial time the zoning profiles of mineral volume fractions and concentrations in porous solution differ from those corresponding to LCE. With time the zoning approaches asymptotically the LCE-solution; the greatest deviations from LCE are observed on boundaries as smearing of fronts. At the same time, the smearing of fronts does not change their absolute widths and tends to zero in relative space scale. In this way for any values of kinetic constants of reversible reactions there exists a moment of time when the zoning becomes practically of LCE-type.

– For the system MgO–SiO_2–H_2O–HCl two macromechanisms of mineral formation have been found: replacement and dissolution followed by precipitation. The latter macro-mechanism becomes significant with increase of m_{HClO}, for formation of talc and forsterite zones at $m_{HClO} \gtrsim 5 \cdot 10^{-2}$ and $m_{HClO} \gtrsim 5 \cdot 10^{-3}$, respectively.

– At $m_{HClO} > 2 \cdot 10^{-2}$ for forsterite and $m_{HClO} \gtrsim 5 \cdot 10^{-2}$ for talc zones the critical situation of complete closure of pores is because of the inner-zone mineral precipitation. It was also noted by Lichtner *et al.* (1986b). This raises a problem of the mechanism of movement of mineral phase and corresponding modifications of transfer equations.

Acknowledgements The authors are grateful to Prof. L.L. Perchuk for permanent support, to Dr G.P. Zaraisky and Dr J.A. Alekhin for useful discussions, to V.S. Khudyaev for help in preparation of the manuscript. The authors particularly thank Dr P.C. Lichtner for critical review of the manuscript and helpful suggestions.

References

Balashov, V.N. (1985). On mathematical description of a model of metasomatic zoning with multicomponent minerals. *Doklady Akad. Nauk SSSR*, **280**, 746–50.

Balashov, V.N., Zaraisky, G.P., Tikhomirova, V.I. & Postnova, L.E. (1983). An experimental study of diffusion of rock forming components in pore solutions at $T = 250\,°C$, $P = 100$ MPa. *Geokhimya*, **1**, 30–42.

Budak, B.M., Gol'dman, N.L. & Uspenskii, A.B. (1966). Difference schemes involving the rectification of fronts for the solution of Stephan type multifront problems. *Doklady Akad. Nauk SSSR*, **167**, 735–8.

Frantz, J.D. & Mao, H.K. (1975). Bimetasomatism resulting from intergranular diffusion: multimineralic zone sequences. In *Annual Report of the Directors of the Carnegie Institution 74*, pp. 417–24.

—— (1976). Bimetasomatism resulting from intergranular diffusion. I. A theoretical model for monomineralic reaction zone sequences. *Amer. J. Sci.*, **276**, 817–40.

—— (1979). Bimetasomatism resulting from intergranular diffusion. II. Prediction of multimineralic zone sequences. *Amer. J. Sci.*, **279**, 302–23.

Frantz, J.D. & Marshall, W.L. (1982). Electrical conductances and ionisation constants of calcium chloride and pressure to 4000 bars. *Amer. J. Sci.*, **282**, 1666–93.

—— (1983). Electrical conductances and ionisation constants of acids and bases in super-critical aqueous fluids: hydrochloric acid from 100 to 700 °C and pressures to 4000 bars. In *Annual Report of the Directors of the Carnegie Institution 82*, 372–7.

Golubev, V.S. & Sharapov, V.N. (1974). *Dynamics of endogenic ore formation*. Moscow: Nedra.

Golubev, V.S. (1981). *Dynamics of geochemical processes*. Moscow: Nedra.

Helfferich, F. & Katchalsky, A. (1970). Simple model of interdiffusion with precipitation. *J. Ph. Chem.*, **74**, 308–14.

Helgeson, H.C., Delany, J.M., Nessbitt, H.W. & Bird, D.K. (1978). Summary and critique of the thermodynamic properties of rock-forming minerals. *Amer. J. Sci.*, **278-a**, 1–229.

Helgeson, H.C. & Kirkham, D.H. (1974). Theoretical prediction of the thermodynamic behavior of aqueous electrolytes at high pressures and temperatures. I. Summary of the thermodynamic electrostatic properties of the solvent. *Amer. J. Sci.*, **274**, 1089–198.

Helgeson H.C., Kirkham D.H. & Flowers G.C. (1981). Theoretical prediction of the thermodynamic behavior of aqueous electrolytes at high pressures and temperatures. IV. Calculation of activity coefficients, osmotic coefficients and apparent molal standard and relative partial molal properties to 600 °C and 5 kb. *Amer. J. Sci.*, **281**, 1249–516.

Hemley, J.J., Montoya, J.W., Christ, C.L. & Hosteller, P.B. (1977a). Mineral equilibria in the $MgO–SiO_2–H_2O$ system. I. Talc–chrysotile–forsterite–brucite stability relations. *Amer. J. Sci.*, **277**, 322–51.

Hemley, J.J., Montoya, J.W., Shaw, D.R., Luce, R.W. (1977b). Mineral equilibria in the $MgO–SiO_2–H_2O$ system. II. Talc–antigorite–forsterite–anthophyllite–enstatite stability relations and some geological implications in the system. *Amer. J. Sci.*, **277**, 353–83.

Ildefonse, J.P. & Gabis, V. (1976). Experimental study of silicon diffusion during metasomatic reactions in the presence of water at 550 °C and 1000 bars. *Geochim. et Cosmochim. Acta*, **40**, 297–303.

Korzhinskii, D.S. (1959). *Physicochemical basis of the analysis of the paragenesis of minerals*. New York: Consultants Bur.

— (1970). *Theory of metasomatic zoning*. Clarendon Press.

— (1979). Conventionally steady state systems. *Zap. VMO*, **108**, 522–3.

Lasaga, A.C. (1981). Rate laws of chemical reactions. In *Reviews in mineralogy, 8, Kinetics of geochemical processes*, ed. A.C. Lasaga & R.J. Kirkpatrick, pp. 1–68.

Lebedeva, M.I., Zaraisky, G.P. & Balashov, V.N. (1985). The application of small parameter method for investigation of diffusion metasomatism model in the case of reversible reactions. Chernogolovka: O.I.Kh.F. AN SSSR.

——— (1987). The application of small parameter method for investigation of diffusion metasomatism model in the case of reversible reactions. *Geokhimia*, **3**, 459–64.

Lichtner, P.C. (1985). Continuum model for simultaneous chemical reactions and mass transport in hydrothermal systems. *Geochim. et Cosmochim. Acta*, **49**, 779–800.

Lichtner, P.C., Oelkers, E.H. & Helgeson, H.C. (1986a). Exact and numerical solutions to the moving boundary problem resulting from reversible heterogeneous reactions and aqueous diffusion in a porous medium. *J. of Geoph. Res.*, **91**, 7531–44.

——— (1986b). Interdiffusion with multiple precipitation/dissolution reactions: transient model and steady-state limit. *Geochim. et Cosmochim. Acta*, **50**, 1951–66.

Rimstidt, J.D. & Barnes, H.L. (1980). The kinetics of silica-water reactions. *Geochim. et Cosmochim. Acta*, **44**, 1683–99.

Samarsky, A.A. (1977). Theory of difference schemes. Moscow: Nauka.

Volkova, N.I. & Sheplev, V.S. (1980). A model of diffusion metasomatism in quasi-stationary approximation. *Doklady Akad. Nauk SSSR*, **252**, 1224–1227.

Zaraisky, G.P., Shapovalov, Yu.B. & Belyaevskaya, O.N. (1981). An experimental study of acid metasomatism. Moscow: Nauka.

Zaraisky, G.P., Zharikov, V.A., Stoyanovskaya, F.M. & Balashov, V.N. (1986). An experimental investigation of skarn formation. Moscow: Nauka.

Zharikov, V.A. & Zaraisky, G.P. (1990). Experimental modelling of wall-rock metasomatism. Chapter 9 in this volume.

9

Experimental modelling of wall-rock metasomatism

V.A. ZHARIKOV and G.P. ZARAISKY

Mineral symbols

Aam	alkaline (sodium) amphibole
Ab	albite
Act	actinolite
Adr	andradite
Aeg	aegirine, aegirine–diopside
Am	amphibole of the actinolite or hornblende series
And	andalusite, x-andalusite
Ank	ankerite and other carbonates of the dolomite–ferrodolomite series
Anl	analcite
Ap	apatite
Bi	biotite
Brc	brucite
Brn	breunnerite and other carbonates of the magnesite–siderite series
Can	cancrinite
Cc	calcite
Chl	chlorite
Cpx	clinopyroxene
Di	diopside
Dol	dolomite
Fl	fluorite
Fo	forsterite
Gr	garnet of the grossularite–

	andradite series
Ilm	ilmenite
Kl	kaolinite and other clay minerals
Ksp	K-feldspar (microcline, orthoclase, sanidine)
Mel	melilite
Mer	merwinite
Mgs	magnesite
Mnt	monticellite
Ms	muscovite
Mt	magnetite
Ne	nepheline
Ol	olivine
Per	periclase
Phl	phlogopite
Pl	plagioclase
Pph	pyrophyllite
Py	pyrite
Q	quartz
Scp	scapolite
Ser	sericite
Sph	sphene
Spl	spinel
Srp	serpentine
Tlc	talc

Tr	tremolite	Wo	wollastonite
Tu	tourmaline	Zo	zoisite, clinozoisite
Tz	topaz	Zrc	zircon

Numbers by the symbols of plagioclase and melilite refer to the content of, respectively, anorthite and gehlenite components (mol. %); numbers by other mineral symbols are for the iron component.

I Introduction

Metasomatic processes involve replacement of some minerals by others with associated changes in the rock chemistry while the rock remains solid. Metasomatic replacement takes place in the presence of a fluid phase. Lindgren (1925) seems to be the first to have given such a definition to metasomatism. He recognized certain characteristic types of hydrothermal metasomatic rocks, showed their relation to ore formation and put forward the rule of 'constant volume during metasomatism'.

A decisive factor in the metasomatic replacement is the dynamic interaction between a rock and the solution with which the rock is out of equilibrium, leading to some chemical species being added to and others removed from the system.

Current concepts of metasomatic processes are based on the fundamental studies by Korzhinskii (1936, 1945, 1946, 1950, 1959, 1970). We assume that readers are acquainted with his theory of metasomatic zoning and therefore only some of the conclusions relevant to this study will be mentioned. The most important is the concept of the metasomatic column as a set of contemporaneously formed metasomatic zones which are distinctly defined and have different mineral and chemical composition. Each zone has its own set of factors of state: the number and mass of inert components, the number and magnitude of the chemical potentials (activities) of the perfectly mobile components, etc., the other intensive parameters being the same throughout the column: temperature, pressure, chemical potentials (activities) of some (virtual) perfectly mobile components. These intensive parameters characterize the external conditions which control a metasomatic system. It has been shown that the development of zoning within a metasomatic column is related to the differential mobility of components caused by successive passage of the components from the inert to the perfectly mobile state as metasomatic alteration advances. This passage, while fixing the changes in the component thermodynamic regime, involves mineral replacement reactions at the zone boundaries which tend to decrease the number of minerals as the intensity of metasomatism increases. If several mineral replacement reactions can occur at the same replacement front, the column structure will be complicated owing to the development of additional zones containing the same number of the reaction minerals.

Inasmuch as metasomatism arises from the interaction of solutions and a rock that are out of equilibrium, the mode of matter transfer in the solutions is critical; the two limiting cases to be distinguished are infiltration and diffusion metasomatism. Infiltration and diffusion metasomatic columns while having several common features differ in their component and mineral behaviour within the zones. In diffusion columns, changes in solution concentration and, accordingly, in the composition of minerals are continuous within the zones. In infiltration columns, the solution and rock compositions remain constant within every zone but undergo abrupt changes at the zone boundaries.

1 Metasomatism and possibilities of experiment

Korzhinskii's ideas have given impetus to a great number of studies by Soviet geologists on the wall-rock alterations in the most varied deposits. However, experimental work on metasomatic processes has so far been scarce. The complexity of the problem deters the activity of experimenters.

Among the studies pertinent to the modelling of metasomatic processes are those of Morey & Ingerson (1937), Gruner (1944), Ames (1961), Ivanov (1961, 1962), Khitarov *et al.* (1962), Garrels & Dreyer (1962), Winkler & Johannes (1963), Ellis & Mahon (1964), Kazitsin *et al.* (1967), Hofmann (1967), Syromyatnikov & Vorobiev (1969), Vidale (1969), Sharapov *et al.* (1970), Lapukhov (1971), Ildefonse & Gabis (1976) and others. These references are selected examples which give the state of the problem.

All these works dealt with various aspects of experimental investigation in metasomatism: rock and mineral alteration under the action of hydrothermal solutions in different physico-chemical conditions, constant interaction of rocks, mechanisms and kinetics of separate alteration processes, ion exchange, development of metasomatic zoning, the effect of temperature gradients, material transport by diffusion in pore solutions, ore formation. Although few papers exhaustively treat the topic, they demonstrate, more or less successfully, the potentialities of the direct modelling of metasomatic zoning.

In 1964, the Institute of Experimental Mineralogy, USSR Academy of Sciences, launched systematic experimental investigations of metasomatism. The research has been directed along two main lines, according to the basic features of metasomatisms stated above.

The first line of research concentrates on mineral equilibria in systems with perfectly mobile components under the controlled factors of metasomatism: the masses of the inert components and the intensive parameters such as temperature, pressure, and effective chemical potentials (activities) of the perfectly mobile components. By varying the extensive and intensive parameters from run to run, it is possible to cover the whole set of parageneses to be found in one or several metasomatic formations. Considering that the stable mineral assemblages obtained in each of the runs characterize the parageneses in only one zone of the metasomatic column, it would require an enormous amount of experimentation to derive a more or less complete picture of the conditions in which a certain group of metasomatic formations developed. The task is all the more difficult owing to the fact that the activities of the most varied components in the system have to be buffered to sustain their perfectly mobile regime. This is why only one system, Al_2O_3–SiO_2–H_2O–HCl–KCl–NaCl, characteristic of the inner zones in some of the most important wall-rock metasomatites has been studied comprehensively with this method (Zharikov *et al.*, 1972, Hemley, 1959, Hemley *et al.*, 1961).

The second line of the experimental metasomatic research has been to model directly metasomatic columns, under the planned factors of metasomatism, by means of diffusion or infiltration interaction between the solutions and rocks of known composition. The feature of this approach is that under the chosen 'external' intensive parameters (temperature, pressure, solution composition) the entire metasomatic column is reproduced, from the unaltered original rock to the rear zones, very heavily altered. To each set of the factors of metasomatism there corresponds a definite metasomatic column (metasomatic formation) having its own mineral assemblage, metasomatic zoning etc. If in the process the local equilibrium is attained then by varying the factors of metasomatism it will be possible to determine reliably the physico-chemical boundary conditions under which metasomatic formations develop.

In natural metasomatic processes, mass transport occurs by both infiltration and diffusion, with one or the other mechanism prevailing, but in experimental modelling they should be studied separately. Experimentation on infiltration metasomatism requires sophisticated apparatus and runs into considerable difficulty in maintaining a uniform solution flow at a definite rate for considerable periods of time so as to suppress the material transfer by diffusion. Such apparatus has been designed and tested. Diffusive metasomatism is technically easier to study and these experiments produce much useful information.

It will be noted that diffusive metasomatism is fairly widespread in nature. Commonly, diffusive metasomatism occurs near veins of fractures when a fluid from a magmatic or some other source flows through open space or in a highly permeable environment for a long period of time. The interaction between the fluid and wall rocks of low permeability occurs by diffusion of the fluid components into the rock through a system of fine intergranular pores and microfissures while the rock components diffuse in the opposite direction (Fig. 9.1(a)). The original composition of the fluid is maintained due to the continuous flow of the solution.

An important variety of diffusive metasomatism is bimetasomatism which occurs when two compositionally contrasting rocks that are out of chemical equilibrium such as granites and marbles are in contact. Here one rock is the source of chemical species for the other. The interaction is by counter-diffusion of the components from the two rocks through stagnant pore solutions (Fig. 9.1(b)).

Near-fissure metasomatism occurs widely in all types of vein hydrothermal deposits while bimetasomatism is typical of skarn deposits. In both cases alteration aureoles commonly display distinct zoning most clearly seen where the neighbouring zones contrast in coloration or structure (Fig. 9.2). The single-stage diffusional mini-zoning

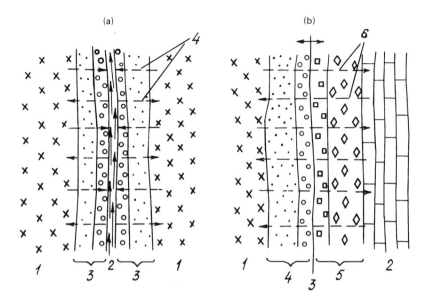

Fig. 9.1. Basic schemes to characterize the occurrence of diffusion metasomatism in nature. (a) Near-fissure metasomatism: (1) original unaltered rock, (2) fissure or high-permeability zone with a hydro-dynamic fluid flow from a deep-seated source, (3) symmetrical zones of diffusion metasomatism, (4) diffusion transport of components in stagnant pore solutions. (b) Bimetasomatism: (1) original unaltered aluminosilicate rock, (2) original carbonate rock, (3) position of the original contact, (4) endocontact metasomatic zones, (5) exocontact metasomatic zones, (6) diffusion transport of components in stagnant pore solutions.

exhibited by a hand-sample often reflects the basic structure of thick altered rock aureoles which develop around ore bodies. The study of such mini-columns is helpful in deriving general schemes of metasomatic zoning developed in the deposit because the regular structure of the metamorphic alteration aureoles within thick productive zones is often upset under the cumulative effect of inhomogeneities in the tectonic structure, complicated combination of infiltration and diffusion paths as well as spatial overlap of different-age mineral assemblages. It will be noted that the mineralogical types of wall-rock metasomatites depend little on the mechanism of mass transfers because the mineral parageneses, zone sequences in the metasomatic column as well as component mobility regimes during the process are largely defined by the initial compositions of the rocks and solutions and the P–T conditions of the process.

Over the last 15 years, all main types of wall-rock metasomatites have been reproduced and studied by the modelling of diffusive metasomatic zoning. Not only were their physico-chemical formation conditions determined but a number of important dynamic characteristics of diffusion metasomatism came to light in these studies.

Fig. 9.2.

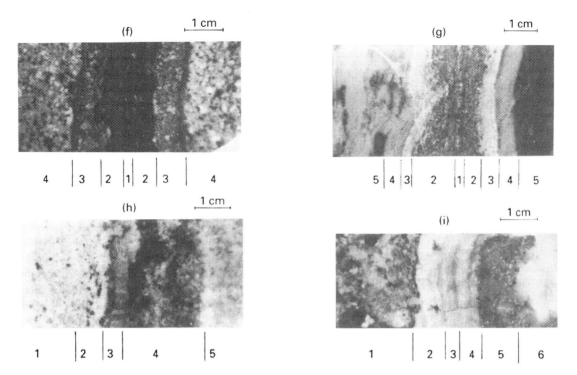

Fig. 9.2. Some examples of natural diffusion metasomatic zoning. (a) Symmetrical
(*cont*) near-vein greisen zoning (Central Kazakhstan, Akchatau): (1) quartz vein,
(2) quartz–muscovite greisen, (3) granite–aplite. (b) Zoning in greisens of
the topaz facies (the same location): (1) quartz–topaz greisen, (2) quartz–
muscovite greisen, (3) muscovitized granite, (4) biotite granite. (c) Beresite
zoning (middle Urals, Shartash): (1) quartz–carbonate vein, (2) beresite
$(Q + Ser + Ank + Py)$, (3) beresitized granodiorite–porphyry, (4) weakly
altered granodiorite–porphyry. (d) Symmetrical gumbeite zoning in
granosyenites (South Urals): (1) quartz–pyrite veinlet, (2) gumbeite
$(Ksp + Ank + Q + Py)$, (3) altered granosyenite (with Ksp, Ank, Phl).
Symmetrical serpentinization zoning in hyperbasite (Middle Urals, the
Bazhenovskoye mine): (1) vein of fibrous chrysolite–asbest $(Srp + Mt)$, (3)
serpentinized hyperbasite $(Ol + Srp + Brc)$. (f) Symmetrical skarn zoning in
granites (South Urals, Magnitogorsk): (1) garnet veinlet, (2)
clinopyroxene–garnet zone, (3) clinopyroxene–plagioclase zone, (4)
granite. (g) Symmetrical skarn zoning in limestones (North Caucasus,
Tyrnyauz): (1) garnet veinlet, (2) clinopyroxene zone, (3) wollastonite
zone, (4) clarified marbleized limestone (calcite marble), (5) original dark
grey limestone. (h) Bimetasomatic skarn zoning at the contact of granite–
aplite and dolomite (Central Asia, Tien-Shan): (1) granite–aplite, (2) zone
$Pl + Scp + Di$, (3) zone $Di + Cc$, (4) zone $Di + Fo + Cc$, (5) zone $Cc + Fo$. (i)
Bimetasomatic skarn zoning at the contact of syenite–porphyry and
dolomite (Central Asia): (1) syenite–porphyry, (2) zone $Cpx + Pl$, (3) zone
$Cpx + Spl$, (4) zone $Fo(Srp)$, (5) zone $Cc + Fo (Srp)$, (6) dolomite marble.

203

2 Experimental technique

A problem with the experimentation on metasomatic systems is the necessity to maintain the activity of 'perfectly mobile components', by Korzhinskii's definition, at constant level while their regimes are independent of the reactions proceeding in the system and are defined by the external conditions such as temperature and pressure. To this end, the volume of the reaction solution was large relative to the mass of solids in order to maintain the planned concentration approximately at the same level despite some loss of the component during the run. Though the experimental procedure varied a little with the type of the metasomatic process or run conditions the general procedure was the same throughout.

Starting materials. Starting materials in most runs were ground rocks (particle size $\leqslant 0.1$ mm): granite, granodiorite, quartz diorite, calcite and dolomite marble and others. Care was taken to use fresh, compositionally uniform samples, from the unaltered zones. The chemical and mineral compositions of the starting materials are given in Table 9.1.

Equipment, experimental procedure. Ground rocks were moistened with distilled water and packed into open gold or platinum tubes ($d = 5$ mm, $l = 50$ mm) which were placed vertically into the 150 cm^3 titanium container (Fig. 9.3 (a, b)). In bimetasomatic runs the contact between the carbonate and aluminosilicate rocks was at half the height of the tube and the place of the contact was marked with a ring from a thin platinum (0.1 mm) wire (Fig. 9.3(c)). The containers were filled with the reaction solution and, depending on the run conditions, then were added mineral buffers, ground quartz to saturate the solution with SiO_2, weighted amounts of solid carbon dioxide, sulphur and other ingredients. The oxygen buffers were normally NNO mixtures and, more rarely, IM or MH. The solution volume in the container was about 100 times that of the solid phase (the rock in the tube) so that the solution composition was held relatively constant. The walls of the titanium container are rather inert and did not affect the composition or properties of the solution.

The sealed containers were placed in autoclaves and to compensate the pressure, distilled water was added in the amounts depending on the loading factor of the container. The experiments were run at 0.5, 1.0, 2.0, 4 kb (accurate to $\pm 5\%$) and up to 600 °C (accurate ± 5 °C). Several high-temperature experiments on skarn formation ($T = 700$–900 °C, $P =$ to 5 kb) were run in gas bombs. The run durations were normally two weeks. Run times for special kinetic series ranged from 1 h to 4 months.

During the run, the solution interacted with the rock through the opening in the tube, penetrating by diffusion. In the oncoming direction there was the diffusion flux of the rock species. Small size of the particles and high porosity (about 40 vol. %) promoted the uniform metasomatic alteration of the rock. As a result, a zoned metasomatic column evolved in place of the initially uniform rock and the alteration was greatest at the open top end of the tube while the zones at the bottom end were close in composition to the original rock.

Table 9.1. *Chemical and mineral composition of starting rocks used in the experimentation (wt. %)*

	SiO_2	TiO_2	Al_2O_3	Fe_2O_3	FeO	MgO	CaO	Na_2O	K_2O	CO_2	H_2O	Total	Q	Ksp	Pl	Am	Bi	Cc	Dol	Other minerals
1	71.06	0.53	14.59	0.30	1.75	1.20	1.74	4.14	4.27	—	0.80	100.38	27	20	44($X_{Ca}=18$)	—	8($X_{Fe}=48$)	—	—	≤1: Ap, Zrc, Ilm, Ms
2	65.90	0.80	15.60	—	2.80	2.20	4.00	3.78	3.50	—	1.16	99.74	24	15	46($X_{Ca}=29$)	5($X_{Fe}=36$)	9($X_{Fe}=48$)	<1	—	<1: Chl, Zo, Ms, Ap, Sph
3	57.58	0.40	17.47	2.46	4.13	3.64	6.36	3.56	1.14	0.70	1.30	98.96	15	—	56($X_{Ca}=45$)	15($X_{Fe}=40$)	13($X_{Fe}=50$)	<1	—	<1: Chl, Mt, Ap, Sph
4	—	—	—	0.03	—	—	56.00	0.07	0.04	44.00	—	100.13	—	—	—	—	—	100	—	—
5	1.28	—	0.01	0.08	0.33	20.93	29.25	0.50	0.10	47.30	—	99.78	1.3	—	—	—	—	—	98.7	—

1 – biotite granite, Western Altai; 2 – granodiorite, Central Asia; 3 – quartz diorite, Northern Kazakhstan; 4 – crystalline limestone (white marble), Central Asia; 5 – crystalline dolomite, Baikal.

X_{Ca} – Ca fraction in plagioclase, at. %: $X_{Ca} = [Ca/(Ca + Na)] \cdot 100$

X_{Fe} – Fe fraction in amphibole and biotite, at. %: $X_{Fe} = [Fe/(Fe + Mg)] \cdot 100$

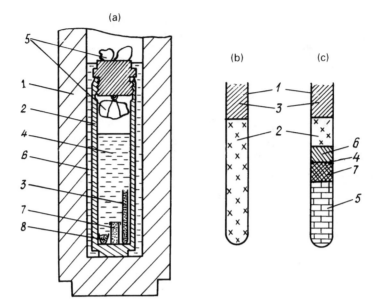

Fig. 9.3. Schematic representation of the apparatus and technique for modelling diffusion metasomatism. (a) Autoclave equipment: (1) autoclave, (2) sealed container of titanium alloy, (3) open gold or platinum tube with compact powder of the original rock, (4) starting reaction solution, (5) solid carbon dioxide, (6) distilled water, (7) oxygen buffer, (8) ground quartz or other solids in an open platinum cup. (b) The tube after the run modelling near-fissure metasomatism: (1) tube, (2) unaltered starting aluminosilicate rock, (3) a column of diffusion metasomatic zones at the open end of the tube. (c) The tube after the run modelling bimetasomatism: (1–3) see above, (4) platinum ring mark, (5) unaltered original carbonate rock, (6) 'endocontact' bimetasomatic zones, (7) 'exocontact' bimetasomatic zones.

The characteristic feature of this process is that, as in nature, metasomatic alteration takes place not continuously but rather in a stepwise fashion, with the sharp replacement fronts developing at the boundaries of metasomatic zones (Fig. 9.4). In bimetasomatic runs zones grow on either side of the contact.

Analysis of experimental columns.   After the runs the experimental columns were consolidated with ciacrine ($C_6H_7NO_2$) which easily penetrates fine pores and then sliced with a diamond saw. One half was studied by optical microscopy (immersion, transparent thin sections), and X-ray examination (powder diffraction) the other was analysed on a microprobe. In addition to the conventional local analyses of mineral compositions, concentration profiles perpendicular to the column axis were scanned to determine the bulk chemical composition. The scanning was performed by moving the sample at a slow uniform speed in the fixed electron beam. Normally, the profiles were 1–3 mm long and spaced apart from 5–10 μm to 0.5–1 mm, depending on the width of the metasomatic zones. The number of analyses within each zone varied from 3 to 10.

II Discussion of experimental results on modelling the main types of wall-rock metasomatites

1 General discussion

By varying the factors of state for an experimental metasomatic system, such as starting rock compositions, the composition and concentration of salic species in the solution, acidity, CO_2 molar fraction, temperature, pressure, all main types (formations) of wall-rock metasomatites have been reproduced: greisens, quartz–feldspar, quartz–tourmaline, quartz–sericite metasomatites, beresites, propylites, secondary quartzites, argillisites, alkali metasomatites, calcareous and magnesian skarns (Zaraisky, 1969, 1979, Zharikov & Zaraisky, 1973, Zaraisky & Zyrianov, 1973, Shapovalov & Zaraisky, 1974, 1978, Zaraisky et al., 1972, 1978, 1981, 1984, 1986, Zaraisky & Stoyanovskaya, 1984, Zharikov et al., 1984, Zaraisky & Balashov, 1987). These works contain comprehensive descriptions of the methods and procedures, give tables stating the run conditions and results, characterize all newly formed minerals, discuss the patterns of the experimental column structures, present tabulated analyses of mineral chemistry and bulk compositions of the metasomatic zones, derive regularities in element migration and component mobility sequences.

In all, about 1000 experimental zoned columns corresponding to various types of wall-rock metasomatites have been reproduced and studied. It will be beyond the scope of this paper to present the prodigious material obtained and we refer the reader to the relevant publications. Here we will only give the more important results from the systematic studies of wall-rock diffusion metasomatism by experimental modelling, with the emphasis on the three main groups of wall-rock metasomatites: acidic, alkaline and skarn.

Interaction between the rock loaded into an open tube and a solution with which it is out of equilibrium, as well as contact interaction between two rocks of contrasting composition, will invariably produce a metasomatic column as a series of contemporaneously developing zones arranged in some fixed sequence. As in nature, the forward zones have mineralogies and chemistries closely approaching original rock whereas the rear zones often have little in common with it, their composition being controlled by the reaction solution. The boundaries between the neighbouring zones are usually distinct, clearly visible to the naked eye. They are fixed by the changes in the mineral composition involving the changes in the coloration and structure (Fig. 9.4). In columns of acidic and alkaline metasomatism, the number of minerals in the neighbouring zones usually differs by one, a decrease taking place from forward zones to those behind. In bimetasomatic columns, two or three neighbouring zones have usually the same number of minerals.

These simple relationships are overshadowed by the presence of the relict minerals from the original rocks, especially in the forward zones, under the low-temperature

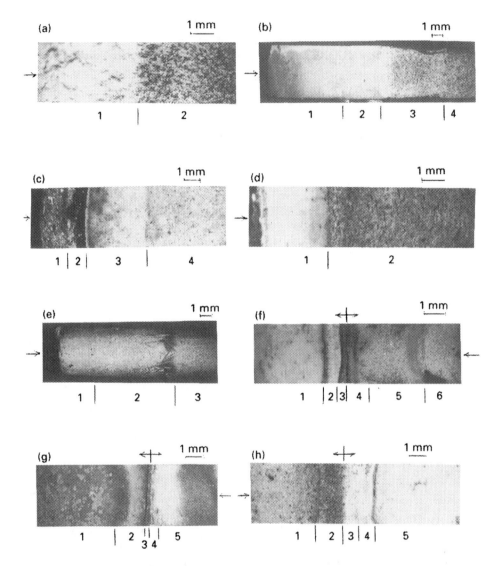

Fig. 9.4. Several examples of experimental diffusion metasomatic zoning. (a) High-temperature acidic metasomatism. A column of andalusitic greisens formed in biotite granite ($T = 500$ °C, $P = 1$ kb, solution 0.1 M KCl + 0.1 M HCl, excess quartz, $t = 686$ h). The sharp boundary divides zones (1) Q + Ms and (2) Q + Ab + Bi + Ms. (b) Medium-temperature acidic metasomatism. A column of quartz–pyrophyllite metasomatites formed in granodiorite ($T = 350$ °C, $P = 1$ kb, solution 0.1 M HCl, excess quartz, $t = 334$ h): (1) zone Q + Pph, (2) zone Q + Ms, (3) zone Q + Ksp + Ms + Chl, (4) zone Q + Pl + Ksp + Bi. (c) Medium–low temperature acidic metasomatism. Dunite listvenitization column ($T = 300$ °C, $P = 1$ kb, $X_{CO_2} = 0.1$, solution 0.025 M K$_2$SiO$_3$ + 0.08 M HCl, $t = 720$ h): (1) quartz zone, (2) zone Q + Mgs, (3) zone Tlc + Mgs, (4) zone Srp + Mgs. (d) Medium-temperature potassium metasomatism. Quartz diorite gumbeization (Q + Ksp + Ank metasomatism) column ($T = 400$ °C, $P = 1$ kb, $X_{CO_2} = 0.2$, solution 1.0 M KCl + 5·10^{-2} M HCl, excess quartz, $t = 333$ h): (1) zone Q + Ksp, (2) zone Q + Ksp + Ank + Bi. (e) High-temperature

conditions and in runs with weak solutions. Normally, identifying both stable (equilibrium) and relict minerals can be made optically with fair accuracy by numerous indications of the phase growth or loss; the development of new crystals, facets, rims, the replacement proceeding from the periphery and along cleavages to produce fine-granular pseudomorphs, dissolution, corroding etc. The relict minerals from the original rack that persist, though only in part, longest are plagioclase and amphibole, occasionally quartz, biotite, potassium feldspar. The production or loss of new minerals at the zone boundaries takes place abruptly, apparently owing to the small size of the grains (0.005–0.05 mm).

The metasomatic columns dealt with below schematically are somewhat idealized: for representing the nature of the process better, only stable mineral parageneses are shown, all the relict minerals being excluded. However, the microprobe bulk chemical analyses of the zones do include the relict minerals with the result that the changes in the rock chemistry at the zone boundaries are smoothed. Bimetasomatic columns did not normally contain the relict phases.

2 Acidic metasomatism

The processes of acidic metasomatism are most apparent in ore deposits. Greisens, beresites, argillisites, secondary quartzites, quartz–feldspar, quartz–sericite metasoma-tites and others belong with this group of wall-rock alterations in country rocks. Their common feature is the tendency to leach all bases from the rock with the result that the final rear zone is solely composed of silica and alumina minerals or else quartz alone. Korzhinskii (1955) called these processes 'acidic leaching', Hemley & Jones (1964) proposed the term 'hydrogen metasomatism'.

High-temperature acidic metasomatism. In one series of experiments, biotite granites (Table 9.1) were exposed to acid potash solutions in which the m_{KCl}/m_{HCl} ratio varied

Fig. 9.4. (*cont.*)

alkaline sodium metasomatism of quartz–biotite schists ($T = 500$ °C, $P = 1$ kb, solution 0.14 M $Na_2CO_3 + 0.36$ M NaOH, solution is not saturated with SiO_2, $t = 166$ h): (1) zone Aeg + Aam, (2) zone Can + Aeg + Aam, (3) zone Ab + Ksp + Aeg + Aam. In the photograph, prismatic cancrinite crystals are clearly seen in zone (2). (f) Bimetasomatic interaction of quartz and dolomite ($T = 600$ °C, $P = 1$ kb, solution 1.0 M NaCl, $t = 2512$ h): (1) quartz, (2) zone Wo, (3) zone Di, (4) zone Fo, (5) zone Cc + Fo, (6) Cc + Brc. (g) Bimetasomatic interaction of granodiorite and limestone column of calcareous skarns ($T = 600$ °C, $P = 1$ kb, solution 1.0 M NaCl, $t = 334$ h): (1) zone Pl + Q + Am, (2) zone Pl + Cpx, (3) zone Cpx, (4) zone Wo + Adr, (5) Cc (white marble). (h) Bimetasomatic interaction of granodiorite and dolomite. Column of magnesium skarns ($T = 600$ °C, $P = 2$ kb, solution 1.0 M NaCl, $t = 386$ h): (1) zone Pl + Q + Am, (2) zone Pl + Cpx, (3) zone Fo, (4) zone Cc + Fo, (5) zone Cc + Brc (altered dolomite). The arrows on the side show the direction of the solution component movement from the reservoir. The double arrows on top mark the original location of the contact in bimetasomatic columns.

from 0.1 to 100 (Shapovalov & Zaraisky, 1974). To ensure the necessary value of this ratio, the HCl concentration was set constant at 0.1 M while the KCl concentration was varied to reach the required ratio. The temperature and pressure were constant (500 °C, 1 kb), run durations were 336 h. Several experiments were run in pure HCl solutions with the concentration 0.1, 0.3 and 1.0 M. One gram of ground quartz was placed on the bottom of the container to saturate the solution with silica. The solution acidity was found to have changed only slightly during the run, not more than 0.5 pH units, and remained rather high (pH $\leqslant$ 1.5).

It has been found that the structure of experimental columns is a function of the m_{KCl}/m_{HCl} in the reaction solution. The columns produced at low ratios ($\log(m_{KCl}/m_{HCl}) < 0.7$) have the rear zone composed of quartz-andalusite* which is replaced by the quartz–muscovite zone and then fades into zones containing feldspars and biotite:

Solution→	Q + And	Q + Ms	Q + Ab + Bi + Ms	Q + Ksp + Ab + Bi + Ms	granite

The zone boundaries are very distinct, relict minerals are very few and there is a sharp change in coloration at the muscovite–biotite replacement front, as the photomicrograph (Fig. 9.4(a)) shows. Albite is replaced by the assemblage Ms + Q simultaneously with biotite. At the boundary between these zones the following reaction occurs:

$$K(Mg, Fe)_3AlSi_3O_{10}(OH)_2 + 2NaAlSi_3O_8 + 8H^+$$

      biotite   albite

$$\rightarrow KAl_3Si_3O_{10}(OH)_2 + 6SiO_2 + 3(Fe, Mg)^{2+} + 2Na^+ + 4H_2O.$$

         muscovite   quartz

This is a relatively rare case when two minerals disappear simultaneously. Normally, the reactions are simpler and the number of minerals decreases by one, as is seen at the next boundary where potash feldspar is being replaced:

$$3KAlSi_3O_8 + 2H^+ \rightarrow KAl_3Si_3O_{10}(OH)_2 + 6SiO_2 + 2K^+.$$

      microcline   muscovite   quartz

The granite alteration process begins with the replacement of plagioclase by the assemblage Ab + Ms:

$$15(Na_{0.8}Ca_{0.2})(Al_{1.2}Si_{2.8})O_8 + 2K^+ + 4H^+$$

           plagioclase

$$\rightarrow 12NaAlSi_3O_8 + 2KAl_3Si_3O_{10}(OH)_2 + 3Ca^{2+}.$$

       albite   muscovite

---

* As with other researchers, the andalusite formed in the present runs was an x-andalusite, which is richer in alumina; its approximate composition as given by the microprobe analysis is $SiO_2.2Al_2O_3.H_2O$ (Zaraisky et al., 1984).

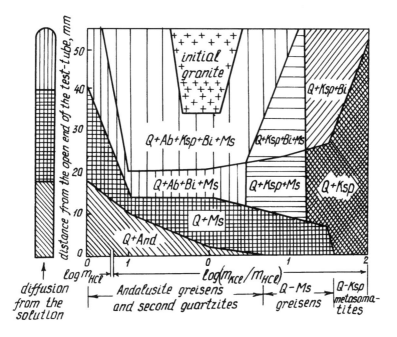

Fig. 9.5. Comprehensive diagram for the dependence of the structure of experimental diffusion columns of high-temperature acidic metasomatism on the composition of operating solution ($T = 500$ °C, $P_{H_2O} = 1$ kb, $t = 336$ h, excess quartz).

The zone assemblages in the column change with increasing potassium content in the solution. At $\log(m_{KCl}/m_{HCl})$ in the range 0.7–1.6 the rear zone in the experimental columns has the quartz–muscovite composition. In the process, the mineral composition of other zones also undergoes changes; feldspar becomes more stable than albite and biotite:

Solution→	Q + Ms	Q + Ksp + Ms	Q + Ksp + Bi + Ms	Q + Ksp + Ab + Bi + Ms

Further increase in m_{KCl}/m_{HCl} leads to the loss of muscovite, and the quartz–potash feldspar appears in the rear zone instead:

Solution→	Q + Ksp	Q + Ksp + Bi

The plagioclase in these columns becomes completely K-feldspathized.

The relationships between experimental metasomatic columns are shown in Fig. 9.5, where the ordinate is the length of a column (mm) from the open end (zero section) inward. Each solution composition gives rise to a characteristic metasomatic zoning. This and other figures state the initial compositions of the solutions. The schematic diagram thus obtained reflects all changes that take place in the structure of experimental columns as the solution composition changes.

A comparison of these experimental columns with natural metasomatites helps one recognize and understand the sequences of zonation leading to assemblages such as secondary quartzites, greisens and quartz–potash feldspar metasomatites. According to experimental results, all these types of wall-rock alteration can be formed at the same temperatures and pressures. They change in regular succession with increasing m_{KCl}/m_{HCl} in the solution. This sequence reflects the equilibrium phase relations between And, Ms and Ksp found by extensive experimentation in the system Al_2O_3–SiO_2–H_2O–HCl–KCl (Hemley, 1959, Zharikov *et al.*, 1972, Ivanov *et al.*, 1974). Comparison with the present results showed a good fit for the limiting stability conditions for the above minerals and thus suggests that equilibrium had been attained in our modelling experiments.

In the runs where the starting solution was not saturated with SiO_2, quartz disappeared from the rear zones to give way to one of the alumina-bearing minerals–andalusite, muscovite or potash feldspar, depending on the m_{KCl}/m_{HCl} ratio in the solution. Although such conditions are less typical for the acidic leaching stage, rear zones in some secondary quartzite facies become relatively depleted in silica with the result that monomineralic andalusite, prophyllite or kaolinite rocks are formed.

By varying the solution, we managed to obtain some of the rarer types of acidic metasomatites. Runs with fluoride solutions produced a column with the topaz greisen zoning ($T = 500\,°C$, $P = 1$ kb, solution 0.1 HF, excess SiO_2 (solid), Al_2O_3 (solid)):

Solution→	Q + Tz	Q + Tz + Ms	Q + Ms + Ksp + Pl	granite

The zones of this column contain some fluorite.

Tourmaline metasomatites could not be formed with only boric acid added to the solution; it was also necessary to enhance the activity of iron and magnesium. Depending on the activity ratio of K, Na and the acidity of the solution, tourmaline was formed in paragenesis with albite, potash feldspar or muscovite. At $T = 500\,°C$, $P = 1$ kb and the solution 1.0 M H_3BO_3 + 0.1 M NaCl + 0.1 M $FeCl_2$ + quartz (excess), the following column was produced:

Solution→	Q + Tu	Q + Tu + Ab	Q + Tu + Pl + Chl	granodiorite

Moderate- to low-temperature acidic metasomatism. Most wall-rock alterations in ore deposits belong with this group of processes. Some of the alterations were given special names, such as 'beresites', 'argillisites', others are identified by the typical mineral assemblage: 'quartz–sericite metasomatites', 'quartz–potash feldspar metasomatites'. In the western literature the following names are common: 'argillic alteration', 'phyllitic alteration' (Creasey, 1959, Meyer & Hemley, 1967). Classification of these metasomatites is based largely on the presence or absence of minerals such as micas, hydromicas, clay minerals, feldspars, carbonates, pyrite. Carbonate minerals other than calcite such as ankerite $Ca(Mg, Fe)(CO_3)_2$ and breunnerite $(Mg, Fe)CO_3$ can be useful identifiers.

The most important type of wall-rock alteration in gold, base metal and uranium deposits is characterized by the paragenesis of quartz, sericite, ankerite and pyrite. In the Soviet literature this type of wall-rock alteration is known as 'beresites' and owes its name to the Beresovskoye gold deposit in the Middle Urals. Quartz–sericite metasomatites commonly occurring in pyrite, base-metal porphyry copper and other deposits also contain quartz, sericite and pyrite but unlike beresites they have no ankerite. A feature of hydrothermal argillisites is the presence of kaolinite, hydromicas or other clay minerals in the rear zones of columns. The stable minerals in the middle and forward zones can include sericite, chlorite, hydromica and various carbonates (ankerite, breunnerite, calcite). Gumbeites normally contain quartz and potash feldspar in parageneses with ankerite, dolomite and phlogopite. The name 'gumbeites' was suggested by Korzhinskii (1955) for the quartz–orthoclase–ankerite wall-rock alteration near scheelite-bearing quartz veins in the granosyenites on the river Gumbeika (South Urals). By contrast, quartz–potash feldspar metasomatites do not contain ankerite; calcite, though uncommon, can occur in middle and front zones.

In our experiments we were able to reproduce all main types of the moderate- to low-temperature metasomatites and to evaluate the boundary (limiting) conditions of their formation (Shapovalov & Zaraisky, 1978, Zaraisky *et al.*, 1981). The starting material was quartz diorite (Table 9.1); the reaction solution contained KCl, HCl, SiO_2 and carbon dioxide and elemental sulphur. Experiments were run at 200–400 °C, 0.5–4.0 kb. In this series of experiments, the H^+ concentration was determined after runs from the pH of the quench solution. The experimentally derived columns corresponded in composition and structure to beresites, argillisites, gumbeites, quartz–sericite and quartz–potash feldspar metasomatites.

In Figs. 9.6 and 9.7 the structure of the experimental columns is given as a function of temperature and solution composition, at relatively high CO_2 activity ($X_{CO_2} = 0.1$; $P_{H_2O + CO_2} = 1$ kb).

The results show that the replacement of gumbeites by beresites and then by argillisites may be caused by both a decrease in the temperature at the constant solution composition and an increase in the solution acidity at the constant temperature. However, upon close examination of the experimental column structures it is possible to determine which factors prevailed in any particular case. For instance, at relatively high temperatures ($\geqslant 350$ °C), iron-free dolomite formed in the middle zones of the columns; as the temperature decreased it was replaced by ankerite and later by the magnesian–iron breunnerite. An increase in the solution acidity, or a decrease in the potassium activity under isothermal conditions, affects negligibly the carbonate composition but favours the early loss of potash feldspar and biotite from the front zones of columns.

V.A. Zharikov and G.P. Zaraisky

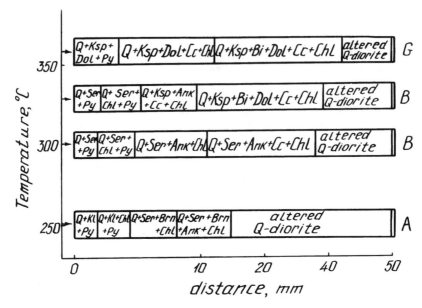

Fig. 9.6. Structure of experimental columns of medium–low temperature acidic metasomatism as a function of temperature. Types of columns: A – argillisite, B – beresite, G – gumbeite [$P_{\mathrm{H_2O+CO_2}}$ = 1 kb, $X_{\mathrm{CO_2}}$ = 0.1, log ($m_{\mathrm{KCl}}/m_{\mathrm{HCl}}$) = 3.0, m_{KCl} = 1.0, $\sum S = 10^{-2}$–$2\cdot10^{-2}$ M, excess quartz, t = 336 h). The H$^+$ molality was determined by the quench pH of the solution which was measured as soon as the autoclave was opened. For clarity, the rear portion of the columns is at double scale.

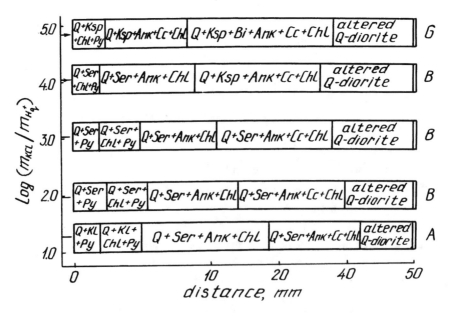

Fig. 9.7. Structure of experimental columns of medium–low temperature acidic metasomatism as a function of solution composition (T = 300 °C, $P_{\mathrm{H_2O+CO_2}}$ = 1 kb, $X_{\mathrm{CO_2}}$ = 0.1, m_{KCl} = 0.1–1.0 M, $\sum S = 10^{-3}$–10^{-1} M, excess quartz, t = 336 h). For clarity, the rear portion of the column is at double scale.

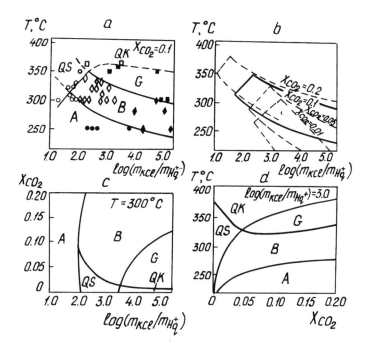

Fig. 9.8. Physico-chemical conditions of formation of medium–low temperature acidic metasomatites as based on experimental modelling data: A – argillisite field; B – beresite field; G – gumbeite field; QS – quartz–sericite metasomatite field; QK – quartz-K-feldspar metasomatite field. (a) T–$\log(m_{KCl}/m_{H_q^+})$ diagram at $X_{CO_2} = 0.1$. (b) T–$\log(m_{KCl}/m_{H_q^+})$ diagram for beresite field at different CO_2 content in the fluid $X_{CO_2} = 0.01, 0.05, 0.1, 0.2$. (c) X_{CO_2}–$\log(m_{KCl}/m_{H_q^+})$ diagram at $T = 300$ °C. (d) T–X_{CO_2} diagram at $\log(m_{KCl}/m_{H_q^+}) = 3.0$. Pressure $P = P_{H_2O+CO_2} = 1$ kb for all diagrams.

The diagrams in Fig. 9.8 clearly show the dependence of the stability fields in the metasomatites under discussion on the main factors of metasomatism: temperature, carbon dioxide and potassium activities and the solution acidity. Note that the experimental points in the diagram (shown in Fig. 9.8(a) only) correspond not to separate parageneses but to metasomatic columns of a given type. The lines in the diagrams separate the stability fields in columns that characterize a particular type or formation of metasomatites. In effect, these lines represent the main boundary transition reaction from one type of metasomatite to another. For example, the boundary between fields A and B corresponds to the reaction $1.5Kl + K^+ = Ser + H^+ + 1.5H_2O$; the boundary between B and G to the reaction $Ser + 3Q + K^+ = 1.5Ksp + H^+$; the line separating fields QS – B and QK – G marks the appearance of ankerite $Chl + Cc + CO_2 = Ank + Ser + H_2O$. The first two reactions take place in the rear zones, the third occurs in the middle zone. Comparison with the phase diagrams for the system Al_2O_3–SiO_2–H_2O–HCl–KCl shows that the positions for the reactions $KI = Ms$ and $Ms + Q = Ksp$ are somewhat displaced. In the right part of the diagram (increasing $m_{KCl}/m_{H_{quench}^+}$) they are driven into the higher-temperature region. This must have been due to the effect of carbon dioxide because the difference disappeared once the CO_2 molar fraction was decreased (Fig. 9.8(b)).

The experiments show that quartz–sericite metasomatites are very sensitive to an increase in the CO_2 pressure (Fig. 9.8(b, c, d)). With increasing X_{CO_2}, ankerite appears in paragenesis with quartz and sericite and the quartz–sericite metasomatites give way to beresites. This implies that the wide natural occurrence of ankerite-free quartz–sericite metasomatites (with or without calcite) has been due either to low CO_2 partial pressures (below 10–20 b) or to the effect of highly acidic (pH < 3) or high potassium ($m_K > 0.01$ M) fluids (Fig. 9.8(b)). Decreasing rock basicities causes the quartz–sericite stability field to expand into the higher CO_2 pressure region. Beresites and gumbeites form over a wide range with respect to solution composition but have a very sharp temperature limit of 350–370 °C, the likely X_{CO_2} values being not higher than 0.1–0.2.

Viewing the experimental diagrams, one realizes that the empirical geological classification of the wall-rock alteration types (metasomatic formations) is by no means accidental but rather rests on a sound physico-chemical basis. The way the fields are arranged in the diagrams suggests that gradual transitions and evolution relations are only possible for those metasomatic formations whose fields border on one another, otherwise the relations are superimposed.

3 Alkaline sodium metasomatism

In nature, alkaline metasomatism is often essentially sodium and involves the formation of Na-bearing minerals such as alkali amphiboles, alkali pyroxenes, albite, analcite, sodalite, nepheline, cancrinite. However the most widespread process of sodium metasomatism, albitization, proceeds not only in alkaline but in neutral and even acidic environments as well. Therefore, the border region between alkaline and acidic metasomatism is worthy of investigation.

Sodium metasomatism was studied experimentally for a wide range of acidity–alkalinity and temperature (Zaraisky, 1969, Zaraisky & Zyrianov, 1973, Zaraisky et al., 1984). Figs. 9.9 and 9.10 show the structure of main experimental columns derived from granodiorite (Table 9.1) under the action of sodium solutions with variable pH at 300 and 500 °C and constant pressure $P_{H_2O} = 1$ kb. The starting solution was 1.0 M NaCl to which NaOH or HCl had been added to yield the desired pH. The ratios m_{NaCl}/m_{NaOH} and m_{NaCl}/m_{HCl} were varied from 10^5 to 10 to cover the pH range 1.1–12.8.

The first mineral to react is K-feldspar which undergoes albitization as early as the front or middle zones of the column. Plagioclase also albitizes but its relict grains persist for a long time after. Quartz is unstable in the rear zones (no SiO_2 added to the solution). It dissolves most readily in alkaline media and at high temperatures is leached completely from all zones of the column to the bottom of the tube. The original actinolite amphibole and biotite are best preserved in near neutral solutions. With increasing acidity they chloritize and dissolve partially, and in alkaline solutions they are replaced by alkali amphibole and aegirine.

The composition of the newly formed phases in experimental columns depends

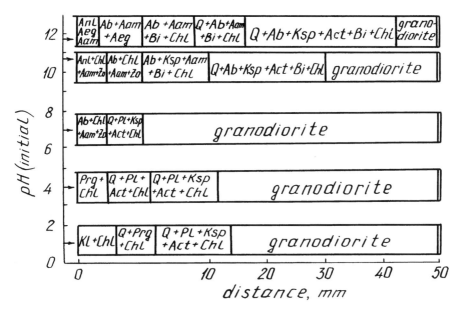

Fig. 9.9. Structure of experimental columns of medium–low temperature sodium metasomatism as a function of the acidity–alkalinity of the operating solution ($T = 300$ °C, $P_{H_2O} = 1$ kb, $m_{NaCl} = 1.0$ M, $t = 336$ h, solution is not saturated with SiO_2). For clarity the rear portion of the columns is to double scale.

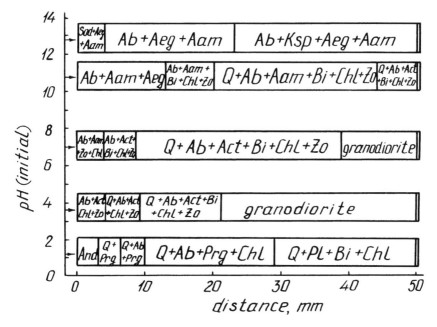

Fig. 9.10. Structure of experimental columns of high-temperature sodium metasomatism as a function of the acidity–alkalinity of the operating solution ($T = 500$ °C, $P_{H_2O} = 1$ kb, $m_{NaCl} = 1.0$ M, $t = 336$ h, solution is not saturated with SiO_2).

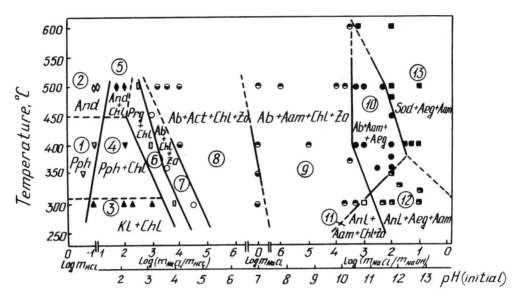

Fig. 9.11. Experimental diagram T–solution composition characterizing the formation conditions of sodium metasomatites. (1–13) – fields for sodium metasomatites of different types showing the mineral parageneses in the column rear zones at equilibrium with the solution ($P_{H_2O} = 1$ kb, $m_{NaCl} = 1.0$ M, $t = 336$ h, solution is not saturated with SiO_2).

strongly on the solution pH and temperature. Exposure to alkaline solutions at high temperatures gives rise to parageneses of alkali amphiboles and pyroxenes with albite and sodalite, while at medium and low temperatures parageneses contain albite and analcite. In neutral and weakly acidic sodium solutions feldspars undergo albitization, biotite is replaced by chlorite and the original actinolite amphibole remains stable. As the acidity continues to increase, chlorite replaces amphibole as well as biotite, and at length paragonite develops in place of the albite.

In the sodium solutions of maximum acidity, the rear zones in the columns lose all bases to the effect that the rock is eventually replaced by monomineralic kaolinite, pyrophyllite or x-andalusite, depending on the temperature of the run. The rear zones in such columns are identical to those in the columns of acidic metasomatism produced by the action of acidic potassium solutions, but the middle zones contain paragonite and albite rather than sericite and K-feldspar. Similarly, the mineral composition of the rear zones changes as the acidity of the solution decreases: kaolinite, pyrophyllite or x-andalusite are replaced by paragonite and then albite. This similarity indicates that the acidic solution is gradually neutralized as it diffuses deeper into the tube and that the zones in the column may attain local equilibrium.

The depth of granodiorite alteration in the tube is the smallest in a neutral NaCl solution and increases noticeably as both acidity and alkalinity of the solution increase.

Fig. 9.11 presents in terms of temperature and composition of the starting solution the experimental points and stability fields of the mineral parageneses in the rear zones of experimental columns. The abscissa is also calibrated in pH units of the starting solution without affecting the authenticity of the diagram. There are 13 fields altogether which differ in the parageneses of the rear zones. The general idea can be got of the column structure from comparing the diagram in Fig. 9.11 with Figs. 9.9 and 9.10 which give two isothermal sections of the diagram with respect to major types of metasomatic zoning. The field boundaries correspond to reactions by which stable mineral assemblages change one another in the rear zones. The diagram does not represent any particular system because the number of phases in the fields varies and so do the inert components in the system.

The middle part of the diagram covers a wide pH range of the starting solution, from 2.6–4.2 to 10.6–12.3 and characterizes the development of albite metasomatites. Several fields and, hence, types of the column can be recognized. It will be noted that the albite 'stability field' extends nearly as much into the acidic as into the alkaline region.

Amphibole is the most sensitive mineral indicator of the ambient acidity–alkalinity. In moderately acidic solutions with pH 4, amphibole disappears and the dark coloured mineral associated with albite is chlorite (field 7). With increasing alkalinity, the original actinolite amphibole in granodiorite (field 8) is replaced by an amphibole of the winchite or the edenite type (field 9). Field 10 corresponds to the most alkaline paragenesis of albite with alkali amphibole (glaucophane–crossite) and aegirine. As the alkalinity continues to increase, albite is replaced by feldspathoids: sodalite (fields 13), and analcite (fields 11, 12) at slightly lower temperatures.

The left part of the diagram depicts the region of acidic sodium metasomatism. This type is rarely observed in nature and we shall not discuss it here. It will only be noted that the position of the paragonite field is in good agreement with the experimental results of Hemley et al. (1961).

The chemical analysis showed that after the run the overall concentration of species leached out of the granodiorite into the solution was not higher than $5 \cdot 10^{-3}$ M–10^{-2} M. It was also found that owing to a large buffer volume of the solution the original concentration of the sodium and chlorine (1.0 M) changed only negligibly during the run.

Based on the starting system of the solution NaCl–HCl–NaOH–H_2O and assuming the H_2O concentration constant, the concentration and activity for the following seven species in the reaction solution were calculated at the P–T of the experiment: Na^+, H^+, Cl^-, OH^-, $NaCl^0$, HCl^0, $NaOH^0$.

Table 9.2 lists the dissociation constants used. The starting data for deriving the material balance in $\sum Na$ and $\sum Cl$ were taken from the coordinates of the boundary lines in the experimental diagram (Fig. 9.11) in 50 °C steps over the range 300–600 °C. Fig. 9.12 presents the recalculated experimental diagram in terms of T–$pH_{(P, T)}$. Comparison of these diagrams reveals considerable changes, primarily in the pH range which is narrower under the experimental P–T conditions than at room temperature and atmospheric pressure. The region of albite metasomatites is within the $pH_{(P, T)}$ range of 4.6–8.5. The line of alkaline metasomatic parageneses which separates fields 8

Table 9.2. *The values of dissociation constants for (log K_D) NaCl, HCl, NaOH and H₂O used in the computations (P = 1 kb; molal scale)*

$T\,°C$	NaCl	HCl	NaOH	H₂O
300	0.45	−0.8	−0.4	−10.50
350	−0.04	−1.4[a]	−0.8[a]	−10.54
400	−0.60	−2.0	−1.2	−10.77
450	−1.27	−2.7[a]	−1.7[a]	−11.19
500	−2.05	−3.5	−2.2	−11.81
550	−2.95	−4.3[a]	−2.8[b]	−12.59
600	−3.67	−5.1	−3.5[b]	−13.40
Reference	after Helgeson et al. (1981)	Montoya & Hemley (1975)	Montoya & Hemley (1975)	Marshall & Frank (1981)

Notes:
[a] interpolation
[b] extrapolation

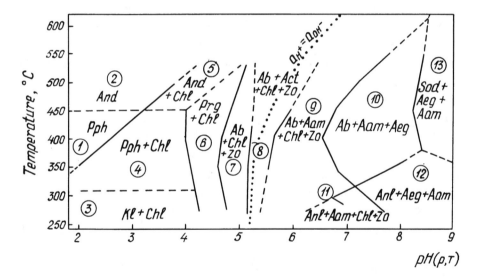

Fig. 9.12.   *T*–pH$_{(P,T)}$ diagram of sodium metasomatism calculated from the experimental diagram 11 (P_{H_2O} = 1 kb, m_{NaCl} = 1.0 м) The dash–dot line is for neutral pH ($a_{H^+} = a_{OH^-}$) at different temperatures and P_{H_2O} = 1 kb after Marshall & Frank (1981).

and 9 shifts slightly into the alkaline region and runs parallel to the neutral pH$_{(P, T)}$ line.

The results of the present study make it possible to estimate the maximum span the acidity–alkalinity of hydrothermal fluids covers during metasomatic processes. It amounts to 6 pH units: from pH = 2.5 for andalusite and pyrophyllite secondary quartzites to pH = 8.5 for solidate metasomatites. Changes in the NaCl concentration will not greatly affect this range because both limits shift in the same direction. Our

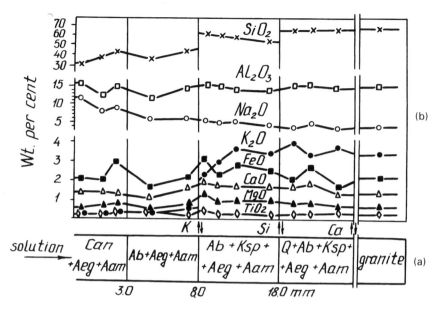

Fig. 9.13. (a) Schematic structure of experimental columns of high-temperature
sodium metasomatism of granite produced on exposure to alkaline
chloride–bicarbonate solution ($T = 500\ °C$, solution $0.1\ M\ NaCl + O$,
Na_2CO_3 not saturated with SiO_2, $t = 336\ h$). (b) Variation in the bulk
chemical composition throughout the column determined by the
microprobe scanning analysis. The position of symbols corresponds to
the scanning profiles positioned normal to the column axis. FeO here and
after refers to all iron recalculated to FeO.

experiments have confirmed the results by Ivanov (1970) and Wellman (1970) who
found that nepheline was unstable in $1.0\ M\ NaCl$ solution and replaced by sodalite over
the entire temperature range of the experiments (400–600 °C, $P = 1\ kb$). This implies
that metasomatic nephelinization can only take place in less concentrated chloride
solutions: below $0.3\ M\ NaCl$ at 400 °C and below $0.15\ M\ NaCl$ at 600 °C. For example, at
$T = 500\ °C$, on exposure to the $0.1\ M\ NaCl + 10^{-3}\ M\ NaOH$ solution, the following
column of granodiorite nephelinization was produced:

Solution→	Ne + Aeg + Aam	Ab + Aeg + Aam	Ab + Ksp + Aam + Aeg	granodiorite

In general, feldspathoids are highly sensitive to the anionic composition of the solution.
Addition of small amounts of Na_2CO_3 instead of NaOH to NaCl produced metasomatic columns with the cancrinite rear zone. Cancrinite forms instead of sodalite but
otherwise the structures are identical (Fig 9.13). The starting rock in this run was biotite
granite (Table 9.1). The plot of changes in the chemical composition derived from the
microprobe scanning reflects the main trends in component migration during alkaline
sodium metasomatism. It is seen that the process involves considerable gain in sodium

221

(4.3% Na_2O in the starting granite against 12% in the cancrinite zone) and noticeable loss in SiO_2 and K_2O. The loss is negligible in CaO, MgO, FeO (total), Al_2O_3 and TiO_2, slightly higher in Al_2O_3 and minimal in TiO_2. Titanium is present in all zones as a constituent in sphene not shown because of negligible amount, calcium enters into the composition of Sph, Can and is present in Ab, Aeg, Aam.

Although nearly all experimental columns of sodium metasomatism have their natural counterparts, the experimental results apply primarily to rocks of granite composition for which there is the best fit between the model and observed alteration patterns.

Fields 1–5 in the diagrams of Figs. 9.11 and 9.12 correspond to metasomatic formations of secondary quartzites, argillisites, greisens which were treated in more detail in relation to acidic metasomatism.

Paragonite metasomatites (field 6) do not occur as wall-rock alteration products because the hydrothermal solutions seldom are purely sodium, and equilibrium Ms–Prg in potassium–sodium solution is clearly shifted toward muscovite. The reaction constant $K_r = \log(Na^+/K^+) = 2.20 \cdot 10^{-5} T^2 - 1.486 \cdot 10^{-2} T + 4.179$ shows that for paragonite to form the sodium concentration should be 2.0–2.5 orders of magnitude higher than the potassium concentration (Zharikov et al., 1984). Moreover, the diagrams in Figs. 9.11 and 9.12 show that the paragonite field is reliably confined by albite as a result of increasing activity of sodium. The expansion of the paragonite field into the region of higher sodium activity is limited by the enhanced activity of silica commonly found in natural processes. The paragonite field is narrow even in experimental diagrams derived for quartz-free parageneses. Natural acidic metasomatites commonly contain parageneses of quartz with muscovite or sericite with small amounts of paragonite.

The mineral assemblages in fields 7 and 8 correspond most closely to albite propylites. According to the diagram in Fig. 9.12, propylitization is likely to occur in near neutral, weakly acidic and occasionally weakly alkaline conditions. Such solutions are not very reactive, the depth of alteration is not large and the alterations are similar to the greenschist metamorphism type.

Typical alkaline metasomatites are represented by fields 9–13. Albite parageneses with alkaline amphiboles and aegirine (fields 9–10) are rather specific to alkaline sodium metasomatism of granitoid rocks. A particular clear-cut example is uraniferous albites from the precambrian granite–gneiss complexes which usually display distinct lateral and vertical metasomatic zoning. The inner zones of the columns have the aegerine–riebekite–albite composition which corresponds to field 10 in the present diagrams. The middle and outer zones successively lose aegirine and alkali amphibole and instead contain epidote, chlorite and amphibole of the actinolite series. The experimental zoning of this type appears in granodiorites exposed to considerably alkaline solutions which under normal conditions have pH = 10.5–11.8 (Fig. 9.11). Under the experimental conditions, the calculated limiting values were found to be considerably lower: $pH_{(P, T)} = 6.6$–8.5. The temperature range favourable for aegirine riebekite albite formation is 350–450 °C because below and above these limits this field shrinks and finally closes.

Albitized, riebekitized granites with the tantalum–niobium and rare-earth mineralization can be assigned to fields 9 and 10. Some authors regard these rocks as alkaline granites but others give convincing evidence for their metasomatic origin. Alkaline amphibole is a more common alkaline dark-coloured mineral than aegirine to be found with them to suggest that these rocks had formed under slightly less alkaline conditions than uranium-bearing albites. They should be placed in field 9 rather than field 10 in the experimental diagrams.

Sodium metasomatites containing feldspathoids (sodalite, nepheline, cancrinite, analcite) have limited occurrence in nature. They only appear in specific geologic environments of maximum alkalinity and low silica activity typical of nepheline syenites and ultrabasic alkaline rocks. In the present experiments, sodalite and analcite developed in the granodiorite with alkaline solutions with the ratio log (m_{NaCl}/m_{NaOH}) $\leqslant 3.3$ and pH$_{(P, T)} > 7$–8.4 depending on the temperature.

Evidently, such a high level of alkalinity will not be achieved in postmagmatic processes related to granitoid magmatism (fields 12 and 13). However, with decreasing temperature, the field of analcite metasomatites expands quickly towards less alkaline solutions and at $T \leqslant 300$ °C approaches the neutral region (field 11). This is the reason for the common occurrence of metamorphic and metasomatic analcites in lavas, tuffs and volcanic sedimentary sequences.

4 Bimetasomatic skarn formation

Skarns are metasomatic rocks composed of calc–magnesian–iron silicates and aluminosilicates which develop within the high-temperature aureole of igneous rock masses through the interaction of carbonate rocks with magma, intrusive or other aluminosilicate rocks by the agency of magmatic solutions (Zharikov, 1968). Calcareous skarns with which iron, copper, tungsten, base metal and other ore deposits are associated are the most widespread in nature and have received the most study. Magnesian skarns form at contact with dolomites and contain magnetite, borate and phlogopite mineralizations. Our contemporary understanding of skarn genesis is largely due to Korzhinskii's theory on contact-reaction skarn formation (1945, 1955) in which the guiding principle is counter-diffusion: calcium (or calcium and magnesium) diffuses from limestones and dolomite into the aluminosilicate rock while silicon, aluminium and iron diffuse in the opposite direction – from the aluminosilicate into the carbonate rock. In the zone of counter-diffusion, both rocks at contact undergo metasomatic replacement to yield several reaction skarn zones (bimetasomatism).

The authors and co-workers have studied this process in detail (Zaraisky et al., 1972, 1978, 1986, Zaraisky and Stoyanovskaya, 1984). The experimentation covers a wide range of the physico-chemical conditions of mineral formation: $T = 400$–900 °C, $P = 1.0$–5.0 kb, $X_{CO_2} = 0$–0.5, the salt components in fluids NaCl, KCl, NaF, KHF$_2$, KF, NaOH, KOH, Na$_2$CO$_3$, Na$_2$SiO$_3$, CaCl$_2$, MgCl$_2$, FeCl$_3$, AlCl$_3$, NaAlO$_2$, HCl,

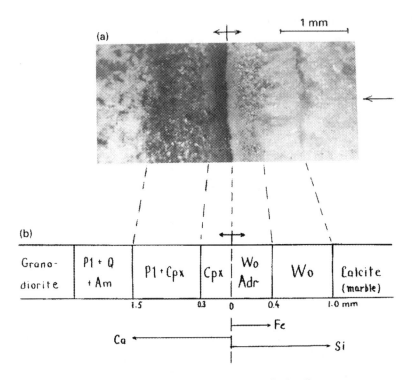

Fig. 9.14. Experimental skarn formation at the granodiorite–limestone contact ($T = 600\,°C$, $P_{H_2O+CO_2} = 1$ kb, $X_{CO_2} = 0.07$, solution 1.0 м NaCl, $t = 330$ h): (a) photograph of the experimental column; (b) schematic structure of the column.

concentration 0.05–5.0 м, pH range 1–13. The contact interaction of the following pairs of rocks has received most study: (1) granodiorite–limestone and (2) granodiorite–dolomite (see Table 9.1 for rock compositions). In most runs, granodiorite was placed near the open end of the tube.

In the presence of a salt solution, or even pure water, reaction zones always developed at the contact of the rock types. Solution species can enter into the composition of the minerals formed, shaping the facies types of skarn zoning, and yet the basic structural patterns of the experimental columns are determined not by the cation or anion composition of the solution but primarily by the composition of the rocks in contact and the solution pH.

Modelling calcareous skarns. The interaction between granodiorite and limestone produced columns which resemble calcareous skarns (Fig. 9.14). The photograph shows a series of newly formed zones with distinct boundaries. The arrows indicate the component counter migration and the distance from the contact where the components are fixed in mineral phases. As in natural skarns, garnet is essentially andraditic in the exocontact zones whereas in the endoskarn zones its iron content diminishes steadily with distance from the contact as the garnet approaches grossular composition (Fig.

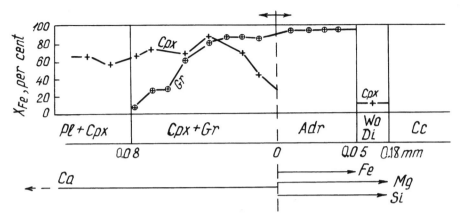

Fig. 9.15. Variation in the garnet and clinopyroxene composition in the endo- and exocontact zones of calcareous skarn experimental column as obtained by the data from local microprobe analysis ($T = 600$ °C, $P_{H_2O} = 1$ kb, solution 1.0 м NaCl, NNO oxygen buffer, $t = 671$ h). $X_{Fe}^{Cpx} = [Fe/(Fe + Mg)] \cdot 100$, $X_{Fe}^{Gr} = [Fe/(Fe + Al)] \cdot 100$.

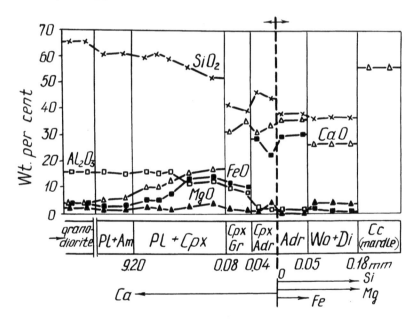

Fig. 9.16. Variation in the bulk chemical composition of calcareous skarns throughout the column obtained from microprobe scanning (run conditions as in Fig. 9.15).

9.15). Fig. 9.16 illustrates the migration of rock-forming species in the course of zone development in calcareous skarns as given by the microprobe analysis. Experimental skarn columns always display sharp zone boundaries and contain virtually no relict minerals.

The effect of solution composition was studied at 600 °C, $P_{H_2O} = 1$ kb (Table 9.3,

225

Table 9.3. *Structure of experimental bimetasomatic columns of calcareous skarns as a function of temperature and solution composition*

Column no.	T, °C	Composition concentration of solution mol/kg H_2O	Solution pH Before run	Solution pH After run	Structure of the column
1	600	0.1 M NaOH	12.95	7.0	Pl+Ksp+Am	1.7	Pl+Am	0	Wo	Cc — 3.3 mm
2	600	1.0 M NaCl	5.9	3.9	Q+Pl$_{25}$+Am$_{27}$	1.7	Pl$_{25-35}$ Cpx$_{50-70}$	0.05	Cpx$_{80}$	0	Wo Gr$_{95}$	0.3	Wo	Cc — 0.5 mm
3	600	0.1 M HCl	1.1	2.25	Q+Pl +Ksp+Bi	6.4	Pl+Q +Am	Pl Cpx$_{55}$	0.2	Cpx$_{58}$ Gr$_{60}$	0	Adr	0.1	Wo Adr	Cc — 1.0 mm
4	400	1.0 M NaCl	5.9	5.25	Pl+Q +Am+Bi	4.1	Pl+Am +Bi	1.5	Pl+Am	0	Wo	Cc — 0.4 mm
5	800	1.0 M NaCl	6.0	no data	glass	1.4	Pl$_{25-40}$ Ksp+Am	0.3	Pl$_{65-90}$ Cpx$_{40-50}$	0	Wo Cr$_{50-85}$	0.05	Wo	1.5	Mer	Cc — 1.7 mm
6	850	1.0 M NaCl	6.0	no data	glass	0.3	Pl+Cpx +Ksp	0	Wo+Mel$_{65}$ +Gr$_{10-70}$	3.2	Wo Mel$_{55}$	5.0	Mel$_{55}$	Cc — 6.1 mm

columns 1–3). Of course, in this series of experiments the solution composition at the place of contact was different from that in a free buffer volume because the place of contact was at half the height of the tube and the solution had earlier reacted with the granodiorite (Fig. 9.3(c)). Despite the reaction with the granodiorite skarn zoning depends strongly on the solution composition, mainly on its acidity–alkalinity. In alkaline solutions and pure water, silica displays intensive migration from granodiorite into limestone to produce a mono-mineral wollastonite zone (Table 9.3, column 1). Granodiorite loses quartz and, near the contact, K-feldspar which is replaced by plagioclase. Amphibole develops in biotite. Increasing alkalinity of the solution results in the desilicification of the granodiorite, appearance of nepheline, sodalite or cancrinite in the near-contact zones, depending on the anionic composition of the solution.

In neutral and weakly acidic solutions, the exocontact zones contain garnet and pyroxene in addition to wollastonite, owing to migration of iron, magnesium and, to some degree, aluminium, along with silicon. On the granodiorite side a pyroxene–plagioclase zone was formed, and the runs with $CaCl_2$ and $MgCl_2$ solutions yielded a typical skarn pyroxene–garnet assemblage. These columns reproduce most closely natural calcareous skarn (Table 9.3, columns 2–3).

In highly acidic media, the feldspars in granodiorite are replaced by basic plagioclase up to pure anorthite. In runs with $FeCl_2$ and $FeCl_3$ magnetite precipitates before replacement of calcite by garnet, i.e. an 'ore zone' develops.

Temperature dependence of the structure of the zoned column was studied in runs with 1.0 M NaCl solutions at $P_{H_2O} = 1$ kb (Table 9.3, columns 4–6). In this series, pyroxene first appeared at 500 °C and higher, garnet at 550 °C. With increasing temperature, pyroxene becomes depleted in iron, garnet enriches in alumina and plagioclase becomes more basic. Migration of alumina by diffusion becomes noticeable generally only from 800 °C and shows up in a higher alumina content of the garnet as well as in the formation of melilite with 55–65% gehlenite molecules in the exocontact zones. Granodiorite melts at these temperatures, however, owing to calcium diffusion from limestone, the granodiorite in the near-contact regions develops normally crystallized higher-melting zones which become narrower as the temperature rises.

Modelling magnesian skarns *Effect of solution composition.* The nature of the contact-reaction interaction between granodiorite and dolomite depends to a large extent on the acidity–alkalinity of the environment (Table 9.4, columns 1–3). Whatever the cation or anion composition of the solution might be, granodiorite readily loses silica under alkaline conditions. The near-contact zones lose quartz, K-feldspar and occasionally plagioclase; biotite becomes more magnesian and approaches phlogopite in composition. The dolomite is overprinted by a well-developed diopside zone which gives way to the calcite–forsterite one farther away from the contact. In fluoride solutions, fluorite, norbergite and cuspidine crystallize. The columns obtained under such conditions have much in common with natural magnesian skarns.

In near-neutral and acidic solutions, the granodiorite side develops zoning that is

Table 9.4. *Structure of experimental bimetasomatic columns of magnesian skarns as a function of temperature and solution composition*

Column no.	T, °C	Composition concentration of solution mol/kg H_2O	Solution pH Before run	Solution pH After run	Structure of the column
1	600	0.1 M NaOH	12.95	7.0	Pl+Ksp+Am+Bi	Pl+Phl+Am	Di+(Wo+Fo)	Cc / Fo	Cc / Brc (1.4 │ 0 │ 0 │ 3.8 │ 6.9 mm)
2	600	1.0 M KF	9.0	8.5	Ksp+Bi+Fl	Phl / Fl	Phl / Fl	Cc+Brc (0.5 │ 0 │ 0.6 mm)
3	600	0.1 M HCl	1.1	2.3	Q+Pl+Ksp+Bi+Am	Q+Pl+Am	$Pl_{100}+Cpx_{85}$	Fo / Mt	Fo	Cc / Fo	Cc+Brc	Dol (3.8 │ 1.0 │ 0 │ 0.2 │ 0.8 │ 1.8 │ 10.3 mm)
4	400	1.0 M NaCl	5.9	5.3	Pl+Q+Am+Bi	Pl+Am+Bi	Pl / Am	Tr+Srp	Cc+Srp	Cc / Brc (7.0 │ 0.3 │ 0 │ 1.5 │ 1.8 │ 5.0 mm)
5	800	1.0 M NaCl	6.2	no data	glass	Pl+Ksp+Am	Pl / Cpx	Mnt	Cc / Mnt	Cc+Per (0.15 │ 0.05 │ 0 │ 0.9 │ 1.5 mm)
6	850	1.0 M NaCl	6.0	no data	glass	$Pl_{35}+Ksp+Cpx_{50}$	$Pl_{100}+Cpx_{35}+Wo+(Cc)$	$Mel_{40}+Cc$	Mel_{75} / Mnt_3 Cc / Cc	Mnt_1 / Cc / +Per	Cc+Per +(Mnt_1) (2.1 │ 1.3 │ 0 │ 1.8 │ 2.6 │ 5.3 mm)

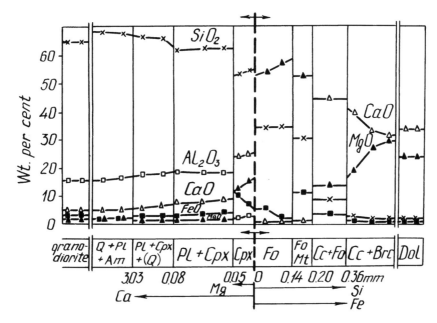

Fig. 9.17. Experimental skarn formation at the granodiorite–dolomite contact. Schematic structure of the magnesian skarn column and variation in the bulk chemical composition derived from the microprobe scanning data ($T = 600$ °C, $P_{H_2O} = 1$ kb, solution 1.0 M NaCl, NNO oxygen buffer, $t = 335$ h).

similar to the zoning at the contact of granodiorite and limestone, indicating that calcium diffusion from dolomite plays a dominant role, magnesium migration being clearly subordinate. Fig. 9.17 presents one such column which developed in a 1.0 M NaCl solution and was analyzed thoroughly on the microprobe. Note that a peculiar 'ore' magnetite zone overprinted dolomite as a result of the leaching and redeposition of iron from granodiorite.

The effect of temperature can be seen from consideration of the columns formed in a 1.0 M NaCl solution. At 400 °C, forsterite disappears and antigorite in assemblage with tremolite and calcite develops over dolomite. The amphibole–plagioclase zone grows on the granodiorite side. As the temperature increases to 500–600 °C, amphibole is replaced by a monoclinic pyroxene and serpentine gives way to forsterite. At 800 °C, the dolomite in the contact zone is replaced by monticellite and at 850 °C by melilite. Spinel, common in magnesian skarns, was not observed in these runs. It is believed that in nature the formation of spinels involves infiltration mass transfer which results in different component activity ratios (a higher alumina activity in the exoskarn zones).

Experimental modelling provides a better insight into the diversity of natural skarns (Zharikov, 1968). Table 9.5 is an attempt to systematize the main types (facies) of experimental skarn columns on the basis of the acidity–alkalinity of the reaction solutions. The observed patterns in the change of mineral parageneses in the exo- and

Table 9.5. *Mineral parageneses (facies) in calcareous and magnesian skarns as a function of the solution acidity–alkalinity (T = 600 °C; P = 1 kb; $X_{CO_2} \leqslant 0.005$)*

Solutions	Calcareous skarns			Magnesian skarns		
	Exoskarn zones	Endoskarn zones	Near-skarn zones	Endoskarn zones	Exoskarn zones	Near-skarn zones
Alkaline	Wo	—	Pl + Am	Di	—	Pi + Phl
				Cl + Fo		Pl + Phl + Am
				Phl		
Neutral or weakly acidic	Wo + Adr	Cpx + Gr	Pl + Cpx	Fo	Cpx	Pl + Cpx
	Wo + Di	Cpx		Fo + Mt		
				Cc + Fo		
Acidic	Adr	Cpx + Gr	Pl + Cpx	Fo	Cpx	Pl + Cpx
	Wo + Di	Cpx		Mt		Tlc + Pl

endoskarn zones make it possible to put brackets on the formation conditions of the relevant natural columns as well as to classify the skarn types (skarn zoning) on the facies acidity. Similar attempts can be made with regard to temperature, pressure (depth) facies, CO_2 activity and other variables.

A comparison of the results from the present study with the zoning observed in natural skarns suggests that the most widespread pyroxene–garnet calcareous skarns are preferentially formed in the 550–800 °C range under the action of near neutral or weakly acidic chloride solutions. The processes of magnesian skarn formation cover a larger temperature range, from 500 to 850 °C and up. On the whole, they occur in more alkaline environments, by the agency of alkaline and near-neutral solutions. Increasing the alkalinity favours the formation of diopside exoskarns while an increase in the acidity gives rise to calcite–forsterite and forsterite–magnetite skarns. Alumina-bearing bimetasomatic exoskarns only form under high-temperature conditions because during diffusion mass transfer alumina acquires much greater mobility at temperatures beyond 800 °C.

III General regularities of metasomatic zoning

1 Zoning as the more general feature of metasomatic formations

Experimental metasomatic studies have fully confirmed the basic concepts of Korzhinskii's theory of metasomatic zoning (1951, 1952, 1970). All structural features of the experimental columns are local chemical equilibrium attained within the zones and at their boundaries while on the whole the process of the rock metasomatic alteration is out of equilibrium. Whatever the process under study, the experimental columns are always zoned. All the zones appear contemporaneously and grow in time (their thicknesses increase). At the zone boundaries the mineral parageneses change and the rock chemical composition alters. The microprobe studies of the experimental columns have provided detailed information on the course of this process and its features which show up most clearly in bimetasomatic interaction, with the difference in the component contents in neighbouring zones attaining as much as tens of weight percent (Figs. 9.16, 9.17). Within the zones, the content either remains nearly constant or varies gradually with the mineral modes or changes in the minerals of variable composition. But these slight variations are much smaller in magnitude than the abrupt jumps in the component contents between the neighbouring zones. It can be inferred that the main interaction processes occur at the zone boundaries where metasomatic reactions take place, with the solid phase consuming some components from the solution and releasing other components into it.

It will be noted that in the present experiments the volume of the rock could change as a result of metasomatic replacement. Changes in volume were facilitated by the high porosity (about 40%) of the starting material and easy deformability of the thin tube walls. For example, in several bimetasomatic runs the tube walls developed circular bulging at the level where the contact-reaction zones were formed, i.e. their volumes were greater than that of the starting material. Unlike volume, the pressure in the system was held constant (isothermal–isobaric conditions).

2 The differential mobility of components

The diffusion interaction between a rock and a solution which is undersaturated with respect to the rock minerals, such as, for example, acidic leaching, involves differential passage of the rock components into the solution causing thereby a regular decrease in the number of minerals in the direction from forward zones invading the unaltered rock to those behind which are in contact with the reaction solution (Figs. 9.5, 9.6, 9.7, 9.11, 9.12). The rear zones in such columns are normally composed of only one or two minerals. According to the Korzhinskii phase rule (1936, 1959), the isothermal–isobaric system with perfectly mobile components will then be

$$r_{P,\,T,\,\mu\mathrm{m}} \leqslant k - k_{\mathrm{m}} \leqslant k_{\mathrm{in}} \tag{1}$$

where $r_{P,\,T,\,\mu\mathrm{m}}$ is the number of phases in the system at some definite temperature, pressure and the chemical potentials of the perfectly mobile components, k is the total number of the components in the system, k_{m} is the number of perfectly mobile components, k_{in} is the number of inert components. The maximum number of phases permitted under these conditions will be equal to the number of inert components. In other words, in the general case, each inert component will be related to a separate

Table 9.6. *Component mobility regime during experimental beresitization of quartz diorite* ($T = 300$ °C, $P = 1$ kb; solution: 0.1 M KCl + 0.02 M $\sum[S]$ + [quartz]$_{exc.}$; $X_{CO_2} = 0.1$)

Columns	Q+Ser +Py	Q+Ser+Chl +Py	Q+Ser+Ank +Chl	Q+Ser+Ank +Cc+Chl	Q+Pl+Am +Ser+Cc+Chl	Q+Pl+Am+Bi +Cc+Chl+Ser
Zone no.	6	5	4	3	2	1
Inert components	Si, Al, Fe	Si, Al, Fe, Mg	Si, Al, Fe, Mg	Si, Al, Fe Mg, Ca	Si, Al, Fe, Mg, Ca, Na	Si, Al, Fe, Mg, Ca, Na, K
Components passing in the perfectly mobile state		Mg↑		Ca↑	Na↑	K↑
Reactions at replacement fronts		Chl→Ser+Py+Q	Ank+Chl₁→Py+Chl₂	Cc→solution	Pl+Am →Ser+Chl+Ank+Q	Bi→Chl+Ser

phase (mineral). Therefore, for a multicomponent multiphase rock composition, a decrease in the number of phases in the rear zones down to 1 or 2 indicates that most components in the system are mobile.

The succession in which the mineral disappearance fronts occur throughout the column conforms to the sequence of the component differential mobility under the given factors of metasomatism in the experimentation. Which of the components passes into the perfectly mobile state at every replacement front is seen from the chemical reactions at the zone boundaries although the identification of the component regime may be difficult owing to the fact that normally minerals have multicomponent composition.

As an example, consider the experimental column produced during the beresitization of a quartz diorite (Table 9.6). Beresitization is a process of acidic leaching (or 'hydrogen metasomatism' after Hemley & Jones, 1964) and involves a gain in the hydrogen ion and loss of cations in equal mole quantities. The feature of beresitization which distinguishes it from other quartz–sericite ('phyllitic') alterations is the formation of calcium, magnesium and iron carbonates in the middle zones and pyrite in the rear zones of the metasomatic column. The forward zone (zone 1) differs from the original quartz diorite by the presence of some sericite and calcite. These changes are only due to the gain of H_2O and CO_2. Towards the rear zone of the column, the degree of rock alteration increases stepwise till the formation of a quartz–sericite–pyrite assemblage (zone 6).

The types of reactions at the zone boundaries (replacement fronts) are mainly of two types – hydrolysis and carbonation of the original minerals in quartz diorite. For brevity, the reactions in Table 9.6 are expressed by symbols with no stoichiometric coefficient or fluid components. For example, the reaction of biotite decomposition occurring at the zone boundary 1→2 can be written in its complete form as

$$6K(Mg, Fe)_3AlSi_3O_{10}(OH)_2 + 4H^+ + 6H_2O$$
$$\text{biotite}$$

$$\rightarrow 3(Mg, Fe)_6Si_4O_{10}(OH)_8 + 2KAl_3Si_3O_{10}(OH)_2 + 4K^+$$
$$\text{chlorite} \qquad\qquad\qquad \text{sericite}$$

Potassium passes from the inert to the perfectly mobile state at this replacement front and is partly lost into the solution while all the inert components in the biotite are inherited by the newly formed solid phases – chlorite and sericite.

All other reactions can be expressed in a similar way. The successive loss of Pl, Cc and Chl from the zone columns is related to the transition of, respectively, Na, Ca, Mg into the perfectly mobile state. The most inert elements are Al, Si and Fe with which the occurrence of quartz (Si), sericite (Al) and pyrite (Fe) in the rear zone is associated. Titanium also remains inert to the end of the process and its content is constant in all zones (about 0.4%). In the original quartz diorite titanium enters into the composition of biotite while in the middle and rear zones it occurs in the accessory sphene and rutile.

No decrease in the number of minerals takes place at the zone boundary 4→5. This is an example of 'additional zoning' which does not entail changes in the component mobility regime (Zharikov, 1966).

The sequence of the component mobility in the above beresitization can be represented as follows:

$$H_2O, CO_2, H^+, S \mid K, Na, Ca, Mg \parallel Fe, Si, Al, Ti$$

The components on the left of the single line are perfectly mobile in all zones; those on the right of the double line are inert throughout; the components in the middle pass successively from the inert to the perfectly mobile state all the way through the column.

The order of the transition of the components into the perfectly mobile state depends on the mineral solubility ratios. Korzhinskii (1951) showed that the crystallization pressure developed by a mineral during metasomatism is related to the degree of the solution saturation in the mineral species. For a one-component mineral

$$P_i = P_i^0 + \rho_m RT \ln \frac{C_i}{C_i^0} \tag{2}$$

where P_i is the crystallization pressure developed by the mineral growing in free solution under the saturation conditions, ρ_m is the mineral molal density, C_i is the component concentration in the reaction solution, C_i^0 is the concentration of saturation. The higher the C_i/C_i^0 ratio the more stable the mineral in the column which fixes the given component in its inert state. And conversely, the lower the degree of saturation, the higher the mobility of the respective component. Thus, for the ratios

$$\frac{C_a}{C_a^0} < \frac{C_b}{C_b^0} < \cdots < \frac{C_k}{C_k^0} \tag{3}$$

component a is the most and component k the least mobile in the series.

For multicomponent minerals, the solubility products rather than the saturation concentrations should be compared. The replacement front which is accompanied by the passage of the more mobile component into the solution moves fastest and is therefore closest to the fresh rock. The other replacement fronts follow one another in the sequence conforming to the mobility sequence.

The experimentation with rocks and solutions of different compositions has made it possible to evaluate the effect of various factors on the differential mobility of components. As a rule, Al and Ti are the most inert while Na and K the most mobile components of rocks. Si, Fe, Mg and Ca are intermediate. In this sense, the mobility sequence stated above for the beresitization column can be generalized. It represents the component mobility regime over a wide range of conditions – from strongly acidic to near neutral. In general, this series coincides with the sequence of component basicity decrease. However, the position of particular components in the series may change with the actual environmental conditions.

The experimental findings show, in general, that increasing both the content of the component in the rock and its concentration in the affecting solution contributes to the inertness of the component during the metasomatic process. These relationships are described conveniently by the approximate expression for the propagation rate of the complete dissolution front of a one-component mineral (Korzhinskii, 1970)

$$V = \frac{dx}{dt} = \sqrt{\left(D_i \frac{C_i^0 - C_i}{2t\rho_i}\right)},$$
(4)

where x is the distance along the column, t is the time, D_i is the diffusion coefficient for component i, ρ_i is the molar density (the content of component i in a rock unit volume). The rate at which the front moves is seen to decrease with increasing component content in the rock (ρ_i) and concentration in the solution (C_i).

It has also been found that the component mobility depends on the anionic composition of the solution. For example, the addition of sulphur makes iron more inert by binding it in pyrite. Fluorine fixes calcium in fluorite preventing thereby its loss from the rock. Exposure to carbon dioxide solutions gives rise to carbonates and as a result decreases the mobility of Ca, Mg, and Fe.

The prevailing process during alkaline sodium metasomatism is the introduction of sodium with the crystallization of Na-minerals: albite, Na-feldspathoids, Na-pyroxenes and amphiboles. However, even under such conditions of high alkalinity ($pH_{(P,T)} > 8$), the differential loss of components from the rock occurs, decreasing the number of minerals towards the rear zones of the column (Fig. 9.13). In the process, the component mobility sequence undergoes considerable changes relative to the acidic leaching series. One of the first minerals to disappear from the granite is quartz, which was present in all zones of acidic metasomatic columns. In contrast, iron–magnesian minerals aegirine and alkali amphibole retain their stability in all altered zones. Alumina remains inert and potassium is somewhat less mobile than in acidic alteration.

In Fig. 9.13 considerable differences in the mineral contents at the zone boundaries are seen to occur only for SiO_2 and K_2O, coinciding with the disappearance of quartz and potassium feldspar at the respective replacement fronts. The mobility sequence can be written as follows:

$$H_2O,\ Na\ \Big|\ Ca,\ Si,\ K\ \Big|\Big|\ Al,\ Mg,\ Fe,\ Ti$$

It differs from the acidic leaching series mainly by a high mobility of Si and Al, with Mg and Fe becoming clearly more inert. Due to the inert state of Ti, an independent accessory phase sphene occurs throughout the column. The relative position of calcium in this series is not quite clear. A greater mobility of the alumina relative to iron and magnesium is supported by the fact that, with a further increase in the alkalinity of the solution (to $pH_{(P,T)} = 9$), the rear zones lose completely all alumina-bearing minerals while retaining the iron–magnesian minerals. A metasomatic column developed under such conditions in the quartz–biotite schists (Fig. 9.4(e)) has the following structure:

Solution→	Aeg + Aam	Can + Aeg + Aam	Ab + Ksp + Aeg + Aam	quartz–biotite schist

Notice a decrease in the potassium mobility during alkaline sodium metasomatism which possibly could be due to an increase in the bulk activity coefficient of K_2O as well as other bases with increasing alkalinity of the solution (the acid–base interaction concept of Korzhinskii (1959)).

3 Local equilibrium

Metasomatic zoning as well as many other structural features of the experimental columns indicates that columns evolve when local equilibrium is either attained or clearly approached (Korzhinskii, 1950, 1959, 1970). The attainment of local equilibrium is evidenced by sharp zone boundaries, regular compositional changes in the minerals–solid solutions, relationship between the number of phases and the number of inert components uniquely expressed by the phase rule, the definite sequence of the metasomatic zones in a column depending on the component mobility regime, and the independence of the zone composition of run duration.

The attainment of equilibrium between the rock and pore solution along the column is also supported by the experimentally determined coincidence between the zone sequence in metasomatic columns and the succession in which the mineral parageneses in the rear zones change one another with decreasing acidic aggressiveness of the starting solution as the m_{KCl}/m_{HCl} or m_{NaCl}/m_{HCl} ratio increases (Figs. 9.5, 9.9, 9.10, 9.11).

The photomicrographs of the distribution of element concentrations across the zones in the experimental columns taken in the characteristic X-rays on the microprobe reveal remarkable sharpness of the replacement fronts between neighbouring zones, the transition regions being not wider than 0.005 mm (Fig. 9.18). Sharp boundaries between the zones can develop if the metasomatic reactions are confined to definite replacement fronts rather than occurring at every point of the rock volume.

All these structural regularities of the experimental metasomatic columns are best explained in terms of local equilibria which set up where the diffusion transport of components in the pore solution is the limiting stage of the process and the chemical reactions proceed faster than diffusion.

Because the diffusion profile in a pore solution is continuous the critical values of the component activities at which the solubility product for one or another mineral is achieved will correspond to a definite section on the profile moving in time in the direction of the diffusion flux. Such sections are specific for each mineral and furnish the metasomatic zone boundaries moving with different velocity along the columns as the run duration increases.

Within the metasomatic columns the composition of the minerals–solid solutions undergo regular changes. Notice that the changes in mineral composition correspond to the distribution of component concentrations in the diffusion flux. For example, in

236

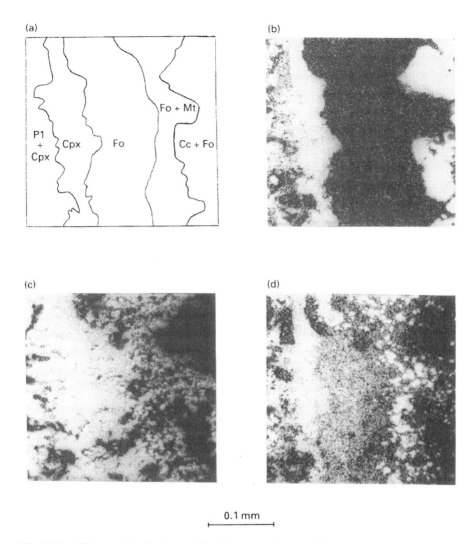

Fig. 9.18. Element distribution within the series of narrow bimetasomatic zones at the granodiorite–dolomite contact (Fig. 9.17). Microphotograph taken in characteristic X-rays on the microprobe; (a) scheme for section zoning; (b) calcium distribution (light-coloured); (c) silicon distribution in the same section (light-coloured); (d) iron distribution (light-coloured).

the calcareous skarn column the iron content of the clinopyroxene and the alumina content of the garnet decrease continuously with increasing distance from the granodiorite, the source of iron and alumina (Fig. 9.15). In the zones of bimetasomatic interaction between granodiorite and dolomite the iron content also decreases steadily while the magnesium increases in the clinopyroxene and forsterite in the direction from the granodiorite toward the dolomite (Fig. 9.19).

The mathematical model describing the formation of diffusion metasomatism zoning is rather complex. It involves a system of non-linear algebraic equilibrium

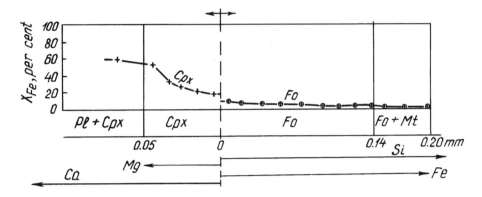

Fig. 9.19. Variation in the clinopyroxene and forsterite composition in the near-contact zones of the experimental magnesian skarn column as given by the local analysis on the microprobe (run conditions as in Fig. 9.17).

equations as well as differential transport equations. In the general case, such a description was first made by Korzhinskii (1952, 1970). Frantz & Mao (1979), using Korzhinskii's concept of local equilibrium, have obtained an ingenious computer-based numerical solution.

In recent years this problem has been treated by Balashov (1985), Lichtner (1985) and Zaraisky & Balashov (1987). See also Balashov's and Lebedeva's paper in this volume (Chapter 8).

4 The dynamics of the metasomatic zoning development

An important quantitative characteristic of metasomatism amenable to evaluation by experiment is the growth rate of zoned columns. The results from the special sets of experiments of various duration show that at the very beginning of metasomatic alteration as soon as the microscopic detection of secondary phases becomes feasible a metasomatic column contains all the zones. For example, the bimetasomatic interaction at the quartz–dolomite contact which furnishes a simple and suitable model for detailed investigation of the zone growth dynamics has produced the following columns in runs of only 1 h duration ($T = 600\,°C$, $P = 1$ kb, solution 1.0 м NaCl):

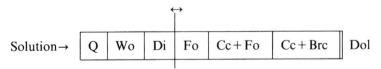

Although the total thickness of the contact-reaction zones formed over this period was as small as 0.1 mm, all the zones were distinct under the microscope, displayed sharp boundaries and contained no relict phases. Runs of various duration (up to 4 months) have shown that no new zones were formed or old ones lost in time. What actually takes place is the proportional growth of the zones which only slows down with the passage

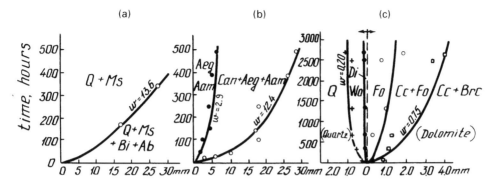

Fig. 9.20. Growth rates of experimental diffusion columns. (a) Acidic metasomatism of granite ($T = 500$ °C, $P_{H_2O} = 1$ kb, solution 0.3 M HCl). (b) Alkaline sodium metasomatism of quartz–biotite schists ($T = 500$ °C, $P_{H_2O} = 1$ kb, solution 0.2 M NaCl + 0.5 M Na$_2$CO$_3$). (c) Bimetasomatic interaction of quartz and dolomite ($T = 600$ °C, $P_{H_2O} = 1$ kb , solution 1.0 M NaCl). Experimental points indicate the distance of zone boundaries from the zero section in runs of various duration.

of time according to the diffusion mass transfer rules. Fig. 9.20 presents the dynamics of the development of bimetasomatic zoning as well as somewhat schematically studied growth patterns for columns of alkaline and acidic metasomatism.

The distance along the column from the zero cross section to every replacement front in the linear approximation of the diffusion flux is found from the expression (Korzhinskii, 1970)

$$x = \sqrt{\left(2D_i \frac{\Delta C_i}{\Delta \rho_i} t\right)},$$

(5)

where ΔC_i is the difference in the component concentration between the original solution and the solution at the replacement front ($\Delta C_i = C_i - C_i^0$), $\Delta \rho_i$ is the difference in the component content in the rock unit volume at either side of the section in question (in the neighbouring zones).

The experimentally established observation that the growth rates of columns of acidic and alkaline metasomatism are considerably higher than those of bimetasomatic columns is well accounted for by a greater ratio $\Delta C_i / \Delta \rho_i$ in the first two cases. Indeed, the ΔC_i values were far higher in runs modelling acidic and alkaline metasomatism than in bimetasomatic runs (owing to high component concentration in the original solution relative to the saturation concentration) whereas the differences in the component contents at the zone boundaries were much lower (Figs. 9.13, 9.17).

The expression $\sqrt{(2D_i \Delta C_i / \Delta \rho_i)}$ is a constant for the section in question and can be denoted by the symbol w (Korzhinskii, 1970). This constant characterizes the reduced rate of movement of the diffusion alteration front

$$w = \sqrt{(2D_i \frac{\Delta C_i}{\Delta \rho_i})} = \frac{x}{t}.$$

(6)

239

Table 9.7. *Growth rate for some diffusion metasomatic columns*

NN	Types of columns and zone boundaries	w, cm/y 1/2	Distance from the beginning of the column, m					
			Experimental results, years			Correction for porosity, years		
			100	1000	10000	100	1000	10000
1	Acidic metasomatism (Fig 9.20(a)), outer boundary for Q + Ms zone	13.6	1.4	4.3	13.6	0.14	0.43	1.4
2	Alkaline metasomatism (Fig. 9.20(b)), outer boundary for Can + Aeg + Aam zone	12.4	1.2	3.9	12.4	0.12	0.39	1.2
3	Bimetasomatism (Fig. 9.20(c)), the complete column of reaction zones	0.95	0.1	0.3	1.0	0.01	0.03	0.1

The experimental dependences of the movement of the metasomatic zone boundaries on time are satisfactorily described by the expression $x = w\sqrt{t}$.

Fig. 9.20 presents the w values for some of the most distinct zone boundaries in experimental columns. The temporal extrapolation using equation $x = w\sqrt{t}$ has made it possible to relate the thicknesses to the time of formation of diffusion metasomatic columns on the geological scale (Table 9.7). When applying the experimental results to natural phenomena, the thicknesses should be cut down at least to a tenth to allow for a lower porosity of rocks relative to the experimental powders.

It has been found that common near-vein and near-fissure aureoles of altered rocks occurring by diffusion may take some 1–10 th.y. to reach a thickness of 0.5–2.0 m. Narrow metasomatic rims of 1–5 cm often found around small ore veinlets may take several years to grow. Thick alteration aureoles to 10 m and more cannot occur by diffusion alone, infiltration must also be involved.

Fig. 9.21 compares the distances which will be covered by the fronts of infiltration and diffusion metasomatism over geologically practicable time. The change in the component content at the replacement front was set at $\Delta\rho_i = 10^{-2}$ mol/cm^3 (nearly 10 wt. % for Na, Mg, Al, Si), the concentration differential in the solution $\Delta C_i = 10^{-4}$ mol/cm^3, the fluid viscosity $\eta = 6.65 \cdot 10^{-2}$ cP (the H_2O viscosity at $T = 500$ °C, $P = 1$ kb – (Dudziak & Franck (1966)), pressure gradient, $-\Delta P = 2 \cdot 10^{-3}$ b/cm (given by the difference between the litho- and hydrostatic pressures). All these values appear sufficiently realistic for the metasomatic process. The movement of the diffusion front was calculated by equation (5), and the passage of the infiltration front was determined by Korzhinskii's equation (1951, 1970) and Darcy's law:

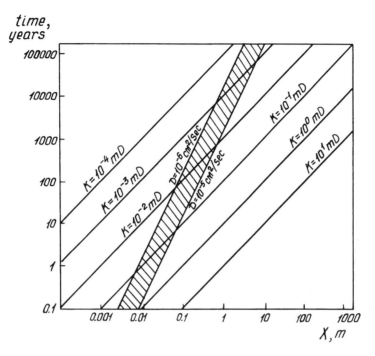

Fig. 9.21. Correlation of propagation rates of infiltration and diffusion metasomatism at different values of the rock permeability constant (K) and diffusion coefficients (D). Calculated by equations (5) and (7). See the text for explanation.

$$x = \frac{P \Delta C_i}{\eta \Delta \rho_i} Kt \qquad (7)$$

where x is the distance in cm, K is the permeability constant for the rock in D, t is the time in s.

At $T = 400$–600 °C, $P = 1$ kb, the coefficients for the diffusion of rock-forming components in chloride solution through granite membranes are 10^{-5}–10^{-6} cm^2/s (Balashov et $al.$, 1983). The calculated rates of the passage of the diffusion metasomatism front at these diffusion coefficients agree well with the experimental estimates for the growth rates of the acidic and alkaline diffusion columns (Fig. 9.21, Table 9.7), additional support for the correctness of the Korzhinskii local equilibrium model.

The relationships shown in Fig. 9.21 clearly demonstrate that infiltration metasomatism may extend a greater distance than diffusion alteration, provided however that the rock permeability is sufficiently high. Thus, for permeability 1 mD the infiltration metasomatism front can cover 100 m over 1000 y and will be overtaking the diffusion metasomatism front as early as the initial stages of the process (the first months). However, for rock permeability 10^{-3} mD and less, infiltration becomes negligible and diffusion metasomatism will predominate. It is shown that compact unaltered rocks enclosing ore deposits commonly have permeability in the range 10^{-4}–10^{-3} mD. The solution filtration through such rocks is extremely slow to the effect that infiltration metasomatism will be confined mainly to zones of deformations, jointing and cleavage.

Acknowledgements   The experimental evidence used in the present study has been obtained by the authors over a number of years in collaboration with Yu. Shapovalov, F. Stoyanovskaya, E. Ryadchikova, V. Balashov. It is a pleasure to express our thanks to our colleagues and to emphasize their participation.

The inspiration of the studies and the development of the new line of research – the experimental modelling of metasomatic process – are largely due to Korzhinskii's ideas.

The authors are grateful to Mrs Galina Lakoza for creative translation into English.

We are particularly indebted to Dr H.T. Haselton and to Prof. Raymond Joesten for their constructive review of the manuscript and many suggestions and criticisms which helped to improve the paper.

References

Ames, L.L., Jr. (1961). The metasomatic replacement of limestones by alkaline, fluoride-bearing solutions. *Econ. Geol.*, **56**, 730–9.

Balashov, V.N. (1985). On mathematical description of one model of metasomatic zoning with multicomponent minerals. *Doklady Akad. Nauk SSSR*, **280**, 746–50.

Balashov, V.N., Zaraisky, G.P., Tikhomirova, V.I. & Postnova, L.E. (1983). An experimental study of diffusion of rock forming components in pore solutions at $T = 250$ °C, $P = 100$ MPa. *Geokhimiya*, **1**, 30–42.

Creasey, S.C. (1959). Some phase relations in hydrothermally altered rocks of porphyry copper deposits. *Econ. Geol.*, **54**, 351–73.

Dudziak, K.H. & Franck, E.U. (1966). Messungen der Viskosität des Wassers bis 560 °C und 3500 bar. *Berichte der Bunses Gesellschaft*, **70**, 1120–8.

Ellis, A.J. & Mahon, W.A.J. (1964). Natural hydrothermal system and experimental hot-water/rock interaction. *Geochim. Cosmochim. Acta*, **28**, 1323–57.

Frantz, J.D. & Mao, H.K. (1979). Bimetasomatism resulting from intergranular diffusion: II. Prediction of multimineralic zone sequences. *Amer. J. Sci.*, **279**, 302–23.

Garrels, R.M. & Dreyer, R.M. (1962). Mechanism of limestone replacement at low temperatures and pressures. *Bull. Geol. Soc. America*, **63**, 375–80.

Gruner, J.M. (1944). The hydrothermal alteration of feldspar in acid solutions between 300 and 400 °C. *Econ. Geol.*, **34**, 578–89.

Helgeson, H.C., Kirkham, D.H. & Flowers, Y.C. (1981). Theoretical prediction of the thermodynamic behavior of aqueous electrolytes at high pressures and temperatures: IV. Calculation of activity coefficients, osmotic coefficients, and apparent molal and relative partial molal properties to 600 °C and 5 kb. *Amer. J. Sci.*, **281**, 1249–516.

Hemley, J.J. (1959). Some mineralogical equilibria in the system K_2O–Al_2O_3–SiO_2–H_2O. *Amer. J. Sci.*, **257**, 241–70.

Hemley, J.J. & Jones, W.R. (1964). Chemical aspects of hydrothermal alteration with emphasis on hydrogen metasomatism. *Econ. Geol.*, **59**, 238–69.

Hemley, J.J., Meyer, C. & Richter, D.H. (1961). Some alteration reactions in the system Na_2O–Al_2O_3–SiO_2–H_2O. U.S Geol. Surv. Profess. Paper **424-D**, pp. 338–40.

Hofmann, A. (1967). Alkali infiltration metasomatism in feldspar solutions. *Trans. Amer. Geophys. Union*, **48**, 230–1.

Ildefonse, J.P. & Gabis, V. (1976). Experimental study of silica diffusion during metasomatic reactions in the presence of water at 550 °C and 1000 bars. *Geochim. Cosmochim. Acta*, **40**, 291–303.

Ivanov, I.P. (1961). Experience on experimental modelling of mica schist albitization under hydrothermal conditions. *Bulletin Sci.-Techn. Information*, **2**, 64–6.

— (1962). Experiments on hydrothermal metamorphism of mica schists under dynamic conditions. In *Trudy VI Soveshchaniya po Eksperimentalnoy i Tekhnickeskoy Mineralogii i Petrographii*, pp. 53–60. Moscow: Izd. Akad. Nauk SSSR.

— (1970). *Problems of experimental investigations of mineral equilibria in metamorphic and metasomatic processes*. Chernogolovka, USSR.

Ivanov, I.P., Belyaevskaya, O.N. & Potekhin, V.Yu. (1974). A refined diagram of hydrolysis and hydration equilibria in the open multisystem KCl–HCl–Al_2O_3–SiO_2–H_2O at 1000 kg/cm². *Doklady Akad. Nauk SSSR*, **219**, 715–17.

Kazitsin. Yu.V., Chernoruk, S.G., Nechiporenko, G.O. & Dubik, O.Yu. (1967). A diffusion metasomatic argillization column. *Doklady Akad. Nauk SSSR*, **173**, 181–4.

Khitarov, N.I., Lebedev, E.B. & Lebedeva, P.V. (1962). Experimental data on the formation of wollastonite-bearing skarns. In *Experimental studies of deep-seated processes*, pp. 43–54. Moscow: Akad. Nauk SSSR.

Korzhinskii, D.S. (1936). Mobility and inertness of components during metasomatism. *Izvestiya Akad. Nauk SSSR, Ser. Math. Nat. Sci.*, no. 1, 35–60.

— (1945). Formation of contact deposits. *Izvestiya Akad. Nauk SSSR, Ser. Geol.*, **3**, 12–33.

— (1946). Metasomatic zoning during near-fissure metamorphism and veins. *Zapiski Vserossijskogo Mineral. Obshchestva*, **75**, 4, 321–32.

— (1950). Factors of equilibrium during metasomatism. *Izvestiya. Akad. Nauk SSSR, Ser. Geol.*, **3**, 21–49.

— (1951). Deriving the equation of infiltration metasomatic zoning. *Doklady Akad. Nauk SSSR*, **17**, 305–8.

— (1952). Deriving the equation of simple diffusion metasomatic zoning. *Doklady Akad. Nauk SSSR*, **14**, 761–4.

— (1955). The outline of metasomatic processes. In *Basic problems in the study of magmatogenic ore deposits*, 2nd edn., ed. A.G. Betekhtin, pp. 335–456. Moscow: Akad. Nauk SSSR.

— (1959). *Physicochemical basis of the analysis of the paragenesis of minerals*. New York: Consultants Bur.

— (1970). *Theory of metasomatic zoning*. Oxford: Clarendon Press.

Lapukhov, A.S. (1971). The kinetics of the hydrothermal process of magnetite zone formation during chloric iron diffusion in carbonate-bearing rocks. In *The physico-chemical dynamics of magmatism and ore formation processes*, pp. 145–57. Novosibirsk: Nauka.

Lichtner, P.C. (1985). Continuum model for simultaneous chemical reactions and mass transport in hydrothermal systems. *Geochim. Cosmochim. Acta.*, **49**, 779–800.

Lindgren, W. (1925). Metasomatism. *Geol. Soc. Amer. Bull.*, **36**, 247–61.

Marshall, W.L. & Frank, E.U. (1981). Ion product of water substance, −1000 °C, 1–1000 bars. New international formulation and its background. *J. Phys. and Chem. Ref., Data*, **10**, 295–304.

Meyer, C. & Hemley, J.J. (1967). Wall rock alteration. In *Geo-chemistry of hydrothermal ore deposits*, ed. H.L. Barnes. New York: Holt, Rinehart and Winston.

Montoya, J.W. & Hemley, J.J. (1975). Activity relations and stabilities in alkali feldspar and alteration reactions. *Econ. Geol.*, **70**, 577–99.

Morey, G.W. & Ingerson, E. (1937). The pneumatolytic and hydrothermal alteration and synthesis of silicates. *Econ. Geol.*, **32**, 607–76.

Shapovalov, Yu.B. & Zaraisky, G.P. (1974). An experimental study of diffusion metasomatic zoning during the acidic leaching of granites. In *Metasomatism and ore formation*, pp. 314–29. Moscow: Nauka.

Shapovalov, Yu.B. & Zaraisky, G.P. (1978). Experimental modelling of medium- and low-temperature metasomatites formed by the acidic leaching of granitoid rocks. In *Metasomatism and ore formation*, pp. 129–38. Moscow: Nauka.

Sharapov, V.N., Golubev, V.S. & Kalinin, D.S. (1970). On the dynamics of formation of bimetasomatic calcareous skarns. *Doklady Akad. Nauk SSSR*, **191**, 913–16.

Syromyatnikov, F.V. & Vorobiev, I.M. (1969). Experience in experimental modelling of processes of calcareous skarn formation. *Doklady Akad. Nauk SSSR*, **184**, 690–3.

Vidale, R. (1969). Metasomatism in a chemical gradient and the formation of calc-silicate bands. *Amer. J. Sci.*, **267**, 857–74.

Wellman, T.R. (1970). The stability of sodalite in a synthetic syenite plus aqueous chloride fluid system. *J. Petrol.*, **11**, 49–71.

Winkler, H.G.F. & Johannes, W. (1963). Experimentelle Metasomatose in einem Granitcontakt. *Naturwissenschaften*, **24**, 730–1.

Zaraisky, G.P. (1969). Experimental modelling of diffusion zoning during alkaline metasomatism. *Doklady Akad. Nauk SSSR*, **184**, 1409–12.

Zaraisky, G.P. (1979). On component differential mobility during experimental diffusion metasomatism. In *The problems of physico-chemical petrology*, vol. 2, pp. 118–45, Moscow: Nauka.

Zaraisky, G.P. & Balashov, V.N. (1987). Metasomatic zoning: theory, experiments and numerical solutions. In *Outlines of physico-chemical petrology*, vol. 14, pp. 136–82. Moscow: Nauka.

Zaraisky, G.P., Ryadchikova, E.V. & Shapovalov, Yu.B. (1984). Experimental modelling of sodium metasomatism of granodiorite. In *Outlines of physico-chemical petrology*, vol. 12, pp. 84–119. Moscow: Nauka.

Zaraisky, G.P., Shapovalov, Yu.B. & Belyaevskaya, O.N. (1981). On experimental study of acidic metasomatism. Moscow: Nauka.

Zaraisky, G.P. & Stoyanovskaya, F.M. (1984). An experimental study of zoning and physico-chemical conditions of bimetasomatic skarn formation. In *Metasomatism and ore formation*, pp. 283–309. Moscow: Nauka.

Zaraisky, G.P., Zharikov, V.A. & Stoyanovskaya, F.M. (1972). Experimental modelling of bimetasomatic skarn zoning. In *1st International Geochemical Congress*, vol. 3, Book I, *Metamorphism and metasomatism*, pp. 38–56, Moscow.

——— (1978). An experimental study of the effect of solution composition on skarn bimetasomatic zoning. In *Metasomatism and ore formation*, pp. 48–62. Moscow: Nauka.

Zaraisky, G.P., Zharikov, V.A., Stoyanovskaya, F.M. & Balashov, V.N. (1986). Experimental investigation of bimetasomatic skarn formation. Moscow: Nauka.

Zaraisky, G.P. & Zyrianov, V.N. (1973). An experimental study of alkaline metasomatism of granites. In *Phase equilibria and mineral formation processes*, pp. 119–56. Moscow: Nauka.

Zharikov, V.A. (1966). Some regularities of metasomatic processes. In *Metasomatic alterations in wall-rocks and their role in ore formation*, ed. N.I. Nakovnik, pp. 47–63. Moscow: Nauka.

— (1968). Skarn deposits. In *Genesis of endogenic ore deposits*, ed. V.I. Smirnov, pp. 220–320. Moscow: Nedra.

Zharikov, V.A., Ivanov, I.P. & Fonarev, V.I. (1972). *Mineral equilibria in the system K_2O–Al_2O_3–SiO_2–H_2O*. Moscow: Nauka.

Zharikov, V.A., Ivanov, I.P. & Zaraisky, G.P. (1984). Experimental studies of physico-chemical conditions of metasomatism. In *27th International Geological Congress. Moscow, August 4–14, 1984. Petrology Section 6.09. Doklady*, vol. 9, pp. 68–84. Moscow: Nauka.

Zharikov, V.A. & Zaraisky, G.P. (1973). Experimental studies of metasomatism: state, prospects. *Geol. of Ore Deposits*, **4**, 3–18.

10

The paragenesis of serendibite at Johnsburg, New York, USA: an example of boron enrichment in the granulite facies

EDWARD S. GREW, MARTIN G. YATES, GEORGE H. SWIHART, PAUL B. MOORE and NICHOLAS MARQUEZ

Introduction

In his studies of metasomatic deposits beginning in 1936, D.S Korzhinskiy proposed the concept of thermodynamic systems with 'perfectly mobile' components and the application of the phase rule to such systems. Much of Korzhinskiy's work concerned the theory of open systems, 'perfectly mobile' components, and rigorous application of the Gibbs method to understanding metasomatic deposits.

The principles developed by Korzhinskiy are ideally suited to study the paragenesis of serendibite, a Ca–Mg–Al borosilicate closely related to aenigmatite and sapphirine in crystal structure (Machin & Süsse, 1974). At all the known world localities (Table 10.1), including Johnsburg, New York, the subject of the present paper, serendibite formation was associated with metasomatism at high temperature over a range of pressures in the granulite facies, upper amphibolite facies (Melville Peninsula), or in contact aureoles (Riverside, California). For petrologists, the major questions concerning serendibite are the origin of the boron and the mechanism for its enrichment in the serendibite-bearing rocks. Boron-enriched rocks are rare in granulite-facies terrains, which are generally depleted in boron relative to amphibolite-facies terrains (e.g. Truscott et al., 1986). Thus the Johnsburg locality offers a unique opportunity to consider the question of boron enrichment under granulite-facies conditions.

The present study is based on field observations and samples obtained by G.H. Swihart and P.B. Moore in 1983 and 1984, and by E.S. Grew in 1985 and 1986. These materials were supplemented by Larsen and Schaller's original field notes and by samples collected by E. Larsen, W. Schaller, and E. Rowley, and purchased from commercial firms: Minerals Unlimited and Ward's.

In the present paper, we present chemical analyses of one rock type (provided by J. McLelland of Colgate University) and of the major minerals together with isotopic

Table 10.1. *World serendibite localities*

Locality	Associated minerals[a]	References
1 Gangapitiya, Sri Lanka	Di, Spl, Scp, Pl	Coomaraswamy (1902), Prior & Coomaraswamy (1903)
2 Johnsburg, NY USA	Di, Phl, Scp, Snh, Spl, Spr, Gdd, Tur, Prg, Cal, Kfs, Pl	Larsen & Schaller (1932)
3 Riverside, CA, USA	Di, Pl, Tur, Czo, Cal	Richmond (1939)
4 Tayozhnoye Deposit, Yakutia, USSR	Di, Prg, Phl, Cal, Spl, Tur, Pl, Mag, Ol, Chu, Zo	Shabynin (1956), Shabynin & Pertsev (1956), Pertsev & Nikitina (1959), Pertsev (1971), Pertsev & Boronikhin (1986)
5 Handeni District Tanzania	Snh, Spl, Tur, Tr, Fo, Cal	von Knorring (1967), Bowden *et al.* (1969)
6 Melville Penin. NWT, Canada	Di (Fst), Tur, Czo, Spl, Cal	Hutcheon *et al.* (1977)

Notes:
[a] Abbreviations given in Table 10.2. Also: Zo – zoisite, Czo – clinozoisite, Tr – tremolite, Fst – fassaite, Mag – magnetite, Ol – olivine, Chu – clinohumite.

analyses of carbon and oxygen in calcite (provided by J. Morrison and J.W. Valley of the University of Wisconsin, Madison) and of boron in tourmaline and serendibite. On the basis of these analyses and of thin section and field observations, we propose that the Johnsburg deposit is derived from the metamorphism of boron-enriched sediments. The critical factors leading to boron enrichment were a hypersaline environment during deposition and metasomatism during granulite-facies metamorphism.

Unless otherwise noted, the abbreviations for minerals given in Table 10.2 have been used throughout the text, tables, and figures.

Description of the Johnsburg locality

The area around the Johnsburg serendibite locality is underlain by quartzofeldspathic gneisses, marbles, and calc-silicate rocks, which are part of the 1.1 billion year old Proterozoic Grenville Province (e.g. Kreiger, 1937, McLelland & Isachsen, 1980, Wiener *et al.* 1984). These rocks have been deformed first into large recumbent folds and subsequently by three generations of upright folds. The serendibite locality is situated on the east flank of a dome cored by metamorphosed anorthosite (Fig. 10.1). According to the compilation by Bohlen *et al.* (1985), the rocks in the Johnsburg area were metamorphosed in the granulite facies at 6.5–8 kbar, 720–740 °C during the Grenville event.

Table 10.2. *Summary of the mineralogy of the Johnsburg serendibite rocks by zones*[a]

Zone	1	2	3	3'	4	5	5'
Apatite	T	T	T	—	T	—	—
Barite	—	—	—	—	T	—	—
Calcite (Cal)	X	X	X	X	X	X	X
Chalcopyrite[d]					T		
Diopside (Di)	X	X	X	T	X	T[b]	X
Dolomite (Dol)	—	—	—	—	—	X	—
Forsterite (serpentine) (Fo, Srp)	—	—	—	—	—	X	—
Grandidierite (Gdd)	—	—	X	—	—	—	—
Hyalophane (Hya)	T	—	—	—	—	—	—
Hydrogrossular	X	X	X	X	—	—	—
K-feldspar (Kfs)	X	T	—	—	—	—	T
Pargasite (Prg)	—	T	X	—	T	—	X[c]
Phlogopite (barian)	T	T	T	—	T	—	—
Phlogopite (and biotite) (Phl)	X	X	X	X	X	X	X
Plagioclase (Pl)	X	T	T	—	—	—	—
Pyrite[d]		T					
Pyrrhotite[d]					T		
Quartz (Qtz)	X	—	—	—	—	—	—
Rutile	—	—	T	—	—	—	—
Sapphirine (Spr)	—	—	—	X	—	—	—
Scapolite (Scp)	X	X	X	X	—	—	(?)[e]
Serendibite (Srd)	T	—	X	X	X	—	—
Sinhalite (Snh)	—	T	X	X	X	—	—
Spinel (Spl)	T	—	X	X	X	—	—
Tourmaline (Tur)	X	—	X	X	—	—	—
Zircon	T	—	T	—	T	—	—

Notes:

[a] The list of minerals for a given zone is not a mineral assemblage because not all the minerals will be found in a single thin section.

[b] In zone 5, diopside in nodules only.

[c] May be tremolite.

[d] Opaque minerals have not been identified in zones other than those indicated.

[e] No fresh material found.

X: present (at least in several sections).

T: in trace amounts or abundant in 1 or 2 sections only.

—: not found with any certainty.

Zones are defined by the dominant minerals. 1. K-feldspar. 2. Scapolite–diopside ± phlogopite. 3. Serendibite–diopside ± scapolite ± phlogopite. 3'. Serendibite–phlogopite. 4. Phlogopite ± calcite. 5. Calcite–dolomite (marble). 5'. Diopside. Primed zones – west of K-feldspar rock and sample no. SI–3. Unprimed – all other specimens.

Larsen & Schaller (1932 and unpublished field notes, 1922), who were the first to describe the Johnsburg locality, reported that the 'main serendibite outcrop' is found in a creek, while several small exposures are found some 45 m uphill to the north. E. Rowley (pers. comm.) also found serendibite on the hillside north of the main locality. Moore succeeded in relocating the main locality by the creek, but despite extensive search, none of us had any success in finding serendibite at the other reported localities uphill. The main outcrop is located less than 0.5 km west of Garnet Lake Road in the Thirteenth Lake, New York, 15 minute quadrangle.

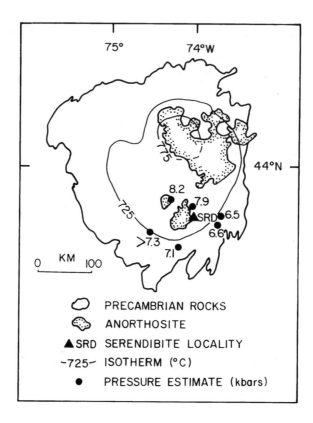

Fig. 10.1. Map of the Adirondack Mountains in northeastern New York State, showing location of the Johnsburg serendibite deposit relative to the isotherms and pressures estimated by Bohlen *et al.* (1985) for the peak metamorphic conditions.

Two or three meters northeast of the main serendibite outcrop (Fig. 10.2), layering in marble trends 330° (azimuth) and dips 30° NE, and about 50 m to the north, the trends are 335 to 360° and dips, 30 to 50°E. Biotitic quartzofeldspathic gneiss crops out west of the serendibite outcrop and structurally underlies the marbles and serendibite rocks; no contacts are exposed. Structurally overlying the serendibite rocks at the main locality is marble, in part dolomitic, and locally phlogopitic and graphitic. Centimeter-sized nodules of diopside in the marble are commonly surrounded by serpentine. The serendibite reported by Larsen & Schaller (1932; unpub. field notes, 1922) and by Rowley presumably originated around feldspathic lenses in the marble overlying the main serendibite locality. The marble in fresh exposures on Garnet Lake Road at BM 1388 northeast of (and structurally above) the serendibite locality contains minor tourmaline, graphite, quartz, and K-feldspar and is similar to other granulite-facies marbles in the Adirondacks (Valley & Essene, 1980). This marble is associated with garnet–clinopyroxene amphibolite, scapolite–clinopyroxene rock, and gneiss.

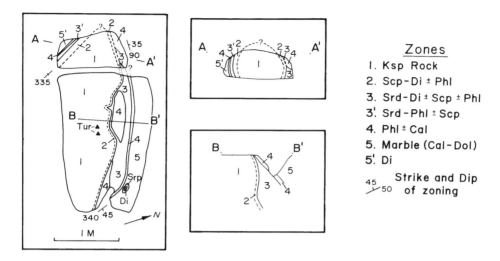

Fig. 10.2. Sketch map (left) and sections of the main serendibite locality at Johnsburg. Smaller features such as the Di pod have been enlarged in order to appear on the sketch. Ksp = K-feldspar.

The main serendibite outcrop is a mass roughly 2 × 1 m of gray, rather homogeneous K-feldspar rock containing sporadic dark-brown tourmaline grains up to 1 cm across (Fig. 10.2; cf. Larsen & Schaller, 1932, Figure 1). Serendibite is concentrated in a pod about 2 × 0.3 m of sky-blue massive serendibite rock. Within the pod, a zone 5 cm thick on the side towards the K-feldspar rock is indistinctly layered and discrete serendibite and diopside grains appear in a pale scapolite–diopside matrix. Phlogopite forms patches and discontinuous trains throughout the pod and a rind up to several cm thick between the pod and K-feldspar rock on one side, and the marble on the other.

Thin section and hand specimen study of the pod and of the other serendibite-bearing rocks made available to us indicates that three mineralogical zones intervene between the K-feldspar rock (zone 1) and marble (zone 5): (2) scapolite–diopside, (3) serendibite ± diopside, including the pod, (4) phlogopite (Fig. 10.2, Table 10.2). In places, the zoning appears to be repeated or includes subzones representing gradations between the above listed zones. The scapolite–diopside zone is gray or pale yellow and in the field is difficult to distinguish from the gray K-feldspar rock. The serendibite-± diopside zone is variable in thickness: it swells to 0.3 m in the pod, which pinches out such that to the west only scattered serendibite grains appear in the phlogopite zone close to the contact between the phlogopite and scapolite–diopside zones (Fig. 10.2). The contact between the serendibite ± diopside and phlogopite zones is irregular in detail and in places appears folded. Commonly the phlogopite flakes in the phlogopite zone (and less commonly, in other zones) are oriented perpendicular to the zoning trend; the colour of the phlogopite in zone 4 is typically yellow.

Zoning west of the K-feldspar rock is different because diopside rock instead of

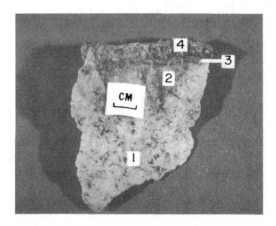

Fig. 10.3. Block of K-feldspar rock (1) from the main locality with scapolite–
diopside zone (2), serendibite zone (3), and phlogopite zone (4) as rind.
Collected by E.S. Grew.

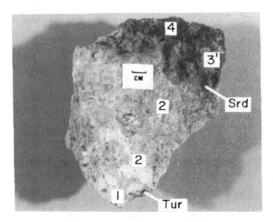

Fig. 10.4. Block of K-feldspar rock (1) from the main locality with scapolite–
diopside zone (2), serendibite in discrete crystals characteristic of zone 3′
and phlogopite zone (4) as rind. Tourmaline prisms 2 mm across (Tur)
are present in the K-feldspar rock. Sample SI–3 collected by G.H.
Swihart.

marble appears outside the phlogopite zone (Fig. 10.2) and serendibite forms discrete
cm-sized grains. This diopside rock may be equivalent to the diopside nodules observed
elsewhere in the marble.

The zoning (minus the marble, which has weathered away) is particularly well
developed as rinds on loose blocks of the K-feldspar rock (Figs. 10.3 & 10.4), which we
presume are broken off from the main *in situ* mass of K-feldspar rock. The zones are
generally only a few mm to a few cm thick. In a few cases, the scapolite–diopside zone
appears as a vein cutting the K-feldspar rock, while in one of Larsen and Schaller's
samples (no. 18946, Harvard Mineralogical Museum), the serendibite–diopside zone

forms a vein in the K-feldspar rock. Other blocks at the main locality include K-feldspar rock with irregular masses of scapolite, cm-sized green and brown tourmaline, quartz, and garnet in chalky masses, and a diopside pod engulfed in the phlogopite zone of a zoned rind.

We were also able to examine several specimens from Larsen & Schaller's (1932 and unpublished field notes, 1922) and E. Rowley's localities up the hill from the main outcrop. These are Harvard Mineralogical Museum Nos. 18959, 18962 and 18964 and N.Y. State Museum Nos. 2117A and B; 2232 and 2232A, 2148A and B, and 3992. The specimens purchased commercially apparently also originated from up the hill. Specimens 2148B and 18964 are zoned and the zoning fits the sequence deduced from specimens collected at the main outcrop. In contrast to specimens from the main outcrop, the K-feldspar and phlogopite in two specimens are barian and barite appears in zone 4 in a third, 2148B (Table 10.2). Serendibite and phlogopite in some of these rocks are darker in color than in the rocks from the main locality.

In summary, the zonal sequence around the K-feldspar rocks, both at the main locality and at the other localities, suggests that the K-feldspar rocks are enclosed in a phlogopite sheath, inside of which is a mantle of scapolite–diopside. This phlogopite sheath is commonly enriched in serendibite, generally leading to the formation of a discrete serendibite–bearing mantle between the scapolite–diopside mantle and phlogopite sheath.

Petrography

The present section is based on observations made on 140 thin sections of the serendibite and associated rocks from the Johnsburg locality. Only the salient features will be described here.

The *K-feldspar rock* (zone 1) at the main serendibite outcrop is a medium grained equigranular aggregate of microcline microperthite. Plagioclase mostly appears as selvages between the microcline grains or as small interstitial grains. In sample 18964 and a specimen purchased from Ward's (no. 5264), hyalophane and subordinate plagioclase constitute zone 1. Quartz is rarely a constituent of the K-feldspar matrix; more commonly it appears around larger tourmaline crystals. In contrast to the other zones, in zone 1 the ferromagnesian minerals, except diopside, are colored in thin section. Tourmaline grains (slender prisms 1 mm long to subequant ones 3 cm across) are, in places, euhedral and well zoned. Serendibite was found in one block as a few tiny (0.04–0.1 mm) grains sheathed in plagioclase or sericite enclosed in K-feldspar. These grains are highly pleochroic in pale yellow, deep blue and green-blue and show anomalous interference colors.

Scapolite and diopside form a symplectitic intergrowth (Fig. 10.5) in parts of the *scapolite–diopside zone* (zone 2). In places, scapolite (and rarely diopside) encloses rounded grains of K-feldspar, suggesting replacement of K-feldspar. Phlogopite is a

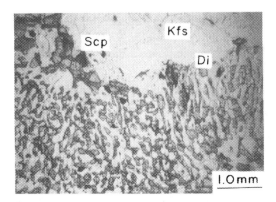

Fig. 10.5. Photomicrograph of symplectite of scapolite (low relief; one grain at edge of Kfs patch) and diopside (high relief) in zone 2 adjacent to K-feldspar of zone 1 (featureless patch). Plane light. National Museum of Natural History #94720–2.

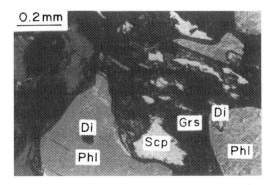

Fig. 10.6. Photomicrograph of hydrogrossular (Grs, black) replacing scapolite (white) in zone 2. Phlogopite contains small diopside grains (high relief). Crossed nicols. National Museum of Natural History #94720–2.

common constituent of zone 2 and in a few samples, e.g. National Museum of Natural History nos. 94720–2, there are distinct scapolite–diopside and scapolite–diopside–phlogopite zones. In the sample from Ward's (5264A), zone 2 is almost missing, and serendibite grains are nearly in contact with hyalophane. A texture characteristic of zones 2 and 3 in some samples is the partial replacement of scapolite by hydrogrossular (Fig. 10.6). Sinhalite occurs in zone 2 only in sample 18964, where it is found sparingly in contact with scapolite and diopside.

In the *serendibite–diopside zone* serendibite typically forms relatively coarse (up to at least one centimeter) anhedral, pale-blue grains displaying polysynthetic twinning. The spectacular blue–green–brown pleochroism and high dispersion characteristic of serendibite from other localities (and zone 1) are absent. Serendibite is commonly embayed, rimmed, and penetrated by tourmaline (Fig. 10.7), in places with spinel, phlogopite and calcite (Fig. 10.8), and less commonly, by pargasite and grandidierite

Fig. 10.7. Photomicrograph of serendibite (twinned) in a matrix of calcite and
mantled by tourmaline sieved with calcite and fine spinel (?). Crossed
nicols. Swihart #SI–3.

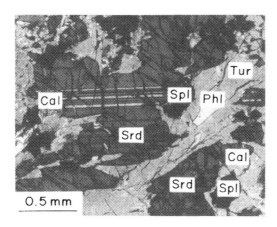

Fig. 10.8. Photomicrograph of serendibite (twinned) penetrated by tourmaline
(medium gray) with grains of spinel (black), calcite (small grains), and
phlogopite (white). Crossed nicols. Swihart #SI–3.

(Fig. 10.9). These minerals appear to be breakdown products of serendibite. It is also
partially replaced by a material in some of which polysynthetic twinning is still visible.

In places serendibite is intergrown with diopside in symplectites (Fig. 10.10), some of
which suggest replacement (Fig. 10.11), but convincing evidence of replacement is rare.
Diopside is commonly rimmed by pargasite (Fig. 10.12), a texture characteristic of
zone 3. Locally the tourmaline rims on serendibite and pargasite rims on diopside
touch, suggesting the breakdown of a serendibite–diopside assemblage to pargasite–
tourmaline. In zone 3′ west of the K-feldspar rock at the main locality, and in the
assemblage of sample SI–3, a block that we presume to be broken off from zone 3′,
diopside is rare.

Less abundant constituents of the serendibite–diopside zone are grandidierite and
sinhalite. Grandidierite is probably Larsen & Schaller's (1932) unknown mineral A, for

255

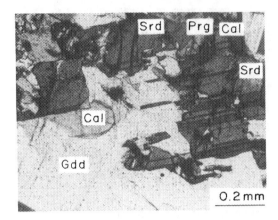

Fig. 10.9.   Photomicrograph of grandidierite (white, featureless) replacing serendibite (twinned); minor calcite (grain next to pargasite is black) and pargasite (paler patch in serendibite). Crossed nicols. Swihart #84–5.

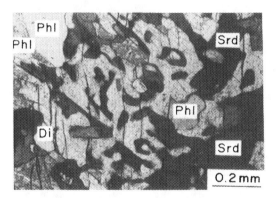

Fig. 10.10.   Photomicrograph of serendibite (dark)–diopside (light) symplectite. Phlogopite is minor. Crossed nicols. Harvard Mineralogical Museum #18961.

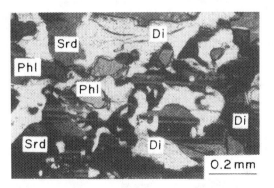

Fig. 10.11.   Photomicrograph of serendibite (twinned)–diopside intergrowth in which diopside appears to replace serendibite. Minor phlogopite. Crossed nicols. National Museum of Natural History #94720–7.

256

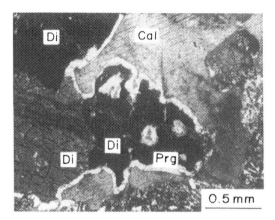

Fig. 10.12. Photomicrograph of pargasite rim (white) on diopside grains in calcite matrix of zone 3. Messy material along right side and bottom of photograph is altered serendibite. Crossed nicols. Harvard Mineralogical Museum #18955.

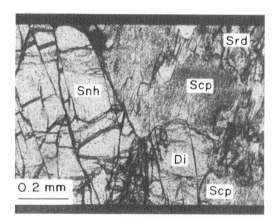

Fig. 10.13. Photomicrograph of sinhalite, diopside, and scapolite riddled with parallel acicular inclusions. Serendibite (penetrated by grandidierite or tourmaline) is nearby. Plane light. Grew #5223 collected and donated by E. Rowley.

the optical properties of grandidierite from other localities correspond closely to those reported by Larsen & Schaller (1932). The Johnsburg grandidierite is colorless, but anomalous dispersion is marked in grains viewed down the X optic axis. In places, it clearly embays serendibite (Fig. 10.9) or appears as patches inside serendibite, which we interpret to be from penetration of serendibite by grandidierite. With rare exceptions, grandidierite is closely associated with pargasite and calcite.

Sinhalite (Larsen & Schaller's (1932) unknown mineral B, identified as sinhalite by Schaller & Hildebrand, 1954) forms rounded grains up to 1 cm across. In some cases, sinhalite is enclosed in serendibite, in others it is enclosed in tourmaline. Nonetheless, in several sections sinhalite is found in contact with scapolite and diopside (Fig. 10.13),

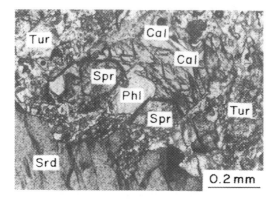

Fig. 10.14. Photomicrograph of sapphirine (subhedral, high relief), calcite and phlogopite in tourmaline (messy appearing) next to serendibite. Crossed nicols. Grew #5233.

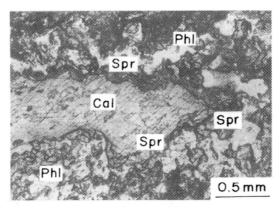

Fig. 10.15. Photomicrograph of sapphirine in margin around calcite and in phlogopite (white). Plane light. Swihart #SI–3.

suggesting the assemblage serendibite–sinhalite–diopside–scapolite–(calcite), the first reported occurrence of a sinhalite–scapolite association (cf. Pertsev, 1971).

Plagioclase occurs in patches in diopside and in one section with calcite and rutile between phlogopite and scapolite grains.

Sapphirine occurs as small, locally subhedral grains with phlogopite, tourmaline, calcite, and rare spinel in margins around serendibite in zone 3′ (Fig 10.14), and in one section, the grains form a ring around calcite near serendibite (Fig. 10.15). Sapphirine is rarely found near diopside. These textures suggest that sapphirine, together with tourmaline, calcite, phlogopite and spinel, is a breakdown product of serendibite.

In the *phlogopite zone* the proportion of calcite ranges from near zero to almost 50% of the rock, while diopside, spinel, sinhalite, and/or serendibite, are abundant in some cases, absent in others. Barite occurs sparingly in one section (2148B, New York State Museum) with calcite, pyrrhotite, barian phlogopite, and rare serendibite.

Chemistry

Methods

Serendibite and associated minerals were analyzed with energy dispersive (30 nanoamps sample current, 15 kV acceleration) and wavelength dispersive (50 nA, 15 kV) systems on an ARL-EMX electron microprobe by Swihart at the University of Chicago and with a wavelength dispersive MAC 400S microprobe (15 kV, 20–30 nA beam current) by Yates at the University of Maine. The probe data were corrected by Bence–Albee factors and minerals (diopside glass for Ca, Mg, Si at Chicago) were used as standards. In addition, serendibite, phlogopite, scapolite, and hyalophane were analyzed for Li, Be, B, F, Sr, Rb, and Ba with an ARL ion microprobe mass analyzer (IMMA) at the Aerospace Corporation, Los Angeles (method of Grew *et al.*, 1986, 1987).

Results

In general, the microprobe compositions of *serendibite* (Table 10.3) are close to the composition determined by classical chemical methods (Larsen & Schaller, 1932). The Johnsburg serendibites are the most magnesian reported and are distinctive in their sodium contents, equivalent to substitution of up to 15% of Ca by Na, possibly through the substitution $Na + Si = Ca + Al$. Compositional variations from sample to sample are attributed to $Fe^{2+} = Mg$ and $Mg + Si = 2$ Al (Tschermak) substitutions, while boron contents appear to be relatively constant.

Sapphirines are peraluminous relative to the 7:9:3 compositional ratio of MgO:Al_2O_3:SiO_2 and heterogeneous, showing marked variations from grain to grain and within a grain (Table 10.4). Compared to sapphirines from other localities, the Johnsburg sapphirines are unique in containing detectable Ca (up to 1.15% CaO) and boron contents of 1 to 2 wt % B_2O_3. Sapphirine composition varies on a micron scale and the variations among Al_2O_3, SiO_2, and CaO lie roughly along trends joining Ca-free sapphirine and serendibite. Possibly, these sapphirines originally contained serendibite in a solid solution and subsequently exsolved submicroscopic lamellae of serendibite.

Diopsides are remarkable for their relatively high and variable Na_2O and Al_2O_3 contents (Table 10.5). In the most sodic varieties, the calculated jadeite content is 10 mole %, and the calculated content of the Ca-Tschermak molecule reaches 8%.

Scapolites vary in Na/Ca ratio, Cl, and S (Table 10.6).

Phlogopites include Ba-poor and Ba-rich varieties (Table 10.7). The barian phlogopites vary widely in BaO (2.6–9.0%), TiO_2 (0.6–5.3%), and to a lesser extent, in Fe/Mg ratio from grain to grain in samples 18964 and 12148B (selected analyses given in the table). Incorporation of Ba is through the substitution $Ba + {}^{IV}Al = (K, Na) + Si$. The *hyalophane* (Table 10.8) and barian phlogopites are also sodic, e.g. three hyalophane

Table 10.3. *Serendibite*

Sample	SI–3		5231	18964	SI–4	84–5
Grain	SRD#16		SRD#11	SRD#6		
	Core	Rim				
Zone	3'	3'	3'	3	3	3
	Weight percent (electron microprobe)					
SiO_2	25.95	26.22	25.58	26.84	26.9	26.33
TiO_2	0.08	0.03	0.10	0.43	0.0	0.09
Al_2O_3	35.31	35.20	35.47	32.17	35.0	36.21
FeO	0.84	0.90	0.82	1.95	0.8	1.05
MnO	0.03	0.02	0.06	0.02	0.0	0.06
MgO	15.78	15.80	15.47	15.88	15.9	15.07
CaO	12.95	12.82	13.63	13.17	13.9	15.06
Na_2O	1.25	1.27	1.23	1.22	1.1	0.86
	Weight percent (ion microprobe)					
Li_2O	0.025	0.025[a]	0.019	0.017	n.d.	n.d.
BeO	0.013	0.013[a]	0.011	0.000	n.d.	n.d.
B_2O_3	7.64	7.64[a]	7.54	7.64	7.64[a]	7.64[a]
F	0.020	0.020[a]	0.007	0.000	n.d.	n.d.
Total	99.89	99.96	99.94	99.34	101.2	102.37
	Cations per 20 oxygens					
Si	2.989	3.016	2.954	3.128	3.024	2.976
Al	1.491	1.466	1.543	1.334	1.493	1.533
B	1.520	1.518	1.503	1.538	1.483	1.491
Total	6.000	6.000	6.000	6.000	6.000	6.000
Ti	0.007	0.003	0.009	0.038	0.000	0.008
Al	3.302	3.306	3.284	3.084	3.143	3.290
Fe	0.081	0.087	0.079	0.190	0.075	0.099
Mn	0.003	0.002	0.006	0.002	0.000	0.006
Mg	2.709	2.710	2.663	2.759	2.664	2.539
Li	0.0114	0.0114	0.0090	0.0078	n.d.	n.d.
Be	0.0035	0.0035	0.0032	0.0000	n.d.	n.d.
Total	6.117	6.123	6.053	6.081	5.882	5.942
Ca	1.598	1.580	1.686	1.644	1.915	1.824
Na	0.279	0.283	0.275	0.276	0.240	0.188
Total	1.877	1.863	1.961	1.920	2.155	2.012
Total Cations	13.994	13.986	14.014	14.001	14.037	13.954
	Anions per 20 oxygens					
F	0.0074	0.0074	0.0026	0.0000	n.d.	n.d.

Notes:
[a] Assumed values.
n.d.: Element not determined.
Totals corrected for F = O.

Table 10.4. *Sapphirine*

Sample	5231	SI–3	SI–3	SI–3
Grain	SPR#11	SPR#19–2	SPR#14	SPR#14
Zone	3′	3′	3′	3′
Weight percent (electron microprobe)				
SiO_2	11.96	11.36	13.07	11.87
TiO_2	0.10	0.06	0.05	0.04
Al_2O_3	64.84	66.42	62.16	64.20
FeO	0.65	0.61	0.60	0.61
MnO	0.05	0.03	0.08	0.07
MgO	18.63	18.82	18.79	18.76
CaO	0.10	0.03	1.15	0.39
Na_2O	0.06	0.05	0.09	0.05
Weight percent (ion microprobe)				
Li_2O	0.011	n.d.	0.012^a	0.011^a
BeO	0.091	n.d.	0.14^a	0.13^a
B_2O_3	1.09	n.d.	2.00^a	1.82^a
F	0.006	n.d.	0.034^a	0.031^a
Total	97.59	97.38	98.18	97.98
Cations per 20 oxygens				
Si	1.405	1.345	1.524	1.386
Al	4.374	4.655	4.073	4.248
B	0.221	n.d.	0.403	0.366
Total	6.000	6.000	6.000	6.000
Ti	0.009	0.005	0.004	0.004
Al	4.601	4.614	4.468	4.584
Fe	0.064	0.060	0.059	0.060
Mn	0.005	0.003	0.008	0.007
Mg	3.262	3.322	3.266	3.265
Li	0.0050	n.d.	0.0058	0.0053
Be	0.0256	n.d.	0.0398	0.0362
Total	7.972	8.004	7.851	7.962
Ca	0.013	0.004	0.144	0.049
Na	0.014	0.012	0.020	0.011
Total	0.027	0.016	0.164	0.060
Total Cations	13.999	14.020	14.015	14.022
Anions per 20 oxygens				
F	0.0021	n.d.	0.0126	0.0115

Notes:
[a] Same IMMA measurements applied to these analyses.
n.d.: Element not determined.
Totals corrected for F = O.

Table 10.5. *Diopside*

Sample	18964		84–5	SI–4	SI–3
Grain	DI#1	DI#3–2			
Zone	3	3	3	3	3′
	Weight percent (electron microprobe)				
SiO_2	53.14	51.20	55.67	54.14	53.81
TiO_2	0.32	0.90	0.05	0.00	0.00
Al_2O_3	4.93	7.34	2.52	4.34	1.54
FeO	1.07	1.18	0.65	0.55	0.42
MnO	0.03	0.04	0.00	0.02	0.13
MgO	15.76	14.80	16.61	15.51	17.43
CaO	22.54	22.09	25.20	24.60	25.22
Na_2O	1.39	1.46	0.57	0.80	0.51
Total	99.18	99.01	101.27	99.96	99.06
	Cations per 6 oxygens				
Si	1.928	1.863	1.980	1.949	1.965
Al	0.072	0.137	0.020	0.051	0.035
Total	2.000	2.000	2.000	2.000	2.000
Ti	0.009	0.025	0.001	0.000	0.000
Al	0.139	0.178	0.086	0.134	0.031
Fe	0.032	0.036	0.019	0.017	0.013
Mn	0.001	0.001	0.000	0.001	0.004
Mg	0.852	0.803	0.881	0.832	0.949
Total	1.033	1.043	0.987	0.984	0.997
Ca	0.876	0.861	0.960	0.949	0.987
Na	0.098	0.103	0.039	0.056	0.036
Total	0.974	0.964	0.999	1.005	1.023
Total Cations	4.007	4.007	3.986	3.989	4.020

grains in 18964 are Ab24–27 Or50–54 Cn22–25 and a phlogopite is 13 mole % Na-phlogopite, 61% K-phlogopite, and 25% Mn-free kinoshitalite, ideally $BaMg_3Al_2$ $Si_2O_{10}(OH)_2$ (Guggenheim, 1984). IMMA phlogopite Rb_2O contents range from 0.063 to 0.080 weight %. Only a trace of chlorine was found in scans of phlogopite in samples 18964 (#4) and 5231.

Analyses are also given for *pargasite, hydrogrossular, tourmaline, spinel, sinhalite* and *grandidierite* (Tables 10.8–10.10). *Calcite* contains up to 10 mole % $MgCO_3$ in solid solution. *Plagioclase* in 18964 (zone 2) is oligoclase (An 26).

IMMA boron contents of phlogopite, scapolite, and hyalophane do not exceed 0.09% B_2O_3. However, the barian phlogopite contains several times as much boron as the Ba-poor phlogopites in the Johnsburg rocks (0.04–0.09 vs 0.005% B_2O_3) and in kornerupine-bearing rocks from other localities (0.01% B_2O_3; Grew *et al.*, 1986). The hyalophane also contains more B_2O_3 than a Ba-poor K-feldspar found with kornerupine (Grew, unpub. data). This suggests that the tetrahedral sites in the vicinity of Ba ions are somewhat more able to accommodate B than sites next to K. Of the minerals

Table 10.6. *Scapolite*

Sample	SI–3	18964	SI–3	SI–3	SI–4	84–5
Grain	SCP#17	SCP#7–1	1	2		
Zone	2	3	2	3′	3	3

	Weight percent (electron microprobe)					
SiO_2	48.76	49.07	50.8	46.1	47.89	45.23
Al_2O_3	26.90	26.50	25.3	27.4	26.58	27.02
FeO	0.02	0.04	0.0	0.0	0.00	0.00
MgO	0.00	0.09	0.0	0.0	0.00	0.00
CaO	14.85	15.45	13.6	18.0	13.29	18.33
Na_2O	4.97	4.86	5.7	3.3	5.02	3.41
K_2O	0.82	0.18	1.0	0.3	0.72	0.20
Cl	1.21	0.76	1.6	0.6	1.11	0.50
SO_3	0.36	0.18	0.5	1.7	0.60	1.11

	Weight percent (ion microprobe)					
Li_2O	0.006	0.000	n.d	n.d.	n.d.	n.d
BeO	0.007	0.000	n.d.	n.d.	n.d.	n.d.
B_2O_3	0.018	0.045	n.d.	n.d.	n.d.	n.d.
BaO	0.036	0.344	n.d.	n.d.	n.d.	n.d.
F	0.016	0.000	n.d.	n.d.	n.d.[a]	n.d.
Total	97.97	97.52	98.5	97.4	95.21	95.69

	Cations normalized to $Al + Si + B = 12$					
Si	7.269	7.326	7.562	7.057	7.255	7.042
Al	4.726	4.662	4.438	4.943	4.745	4.958
B	0.0048	0.0116	n.d.	n.d.	n.d.	n.d.
Total	12.000	11.998	12.000	12.000	12.000	12.000
Fe	0.002	0.005	0.000	0.000	0.000	0.000
Mg	0.000	0.020	0.000	0.000	0.000	0.000
Li	0.0035	0.0000	n.d.	n.d.	n.d.	n.d.
Be	0.0025	0.0000	n.d.	n.d.	n.d.	n.d.
Total	0.008	0.025	0.000	0.000	0.000	0.000
Ca	2.372	2.471	2.169	2.952	2.157	3.058
Na	1.437	1.407	1.645	0.979	1.474	1.029
K	0.156	0.034	0.190	0.059	0.139	0.040
Ba	0.002	0.020	n.d.	n.d.	n.d.	n.d.
Total	3.967	3.932	4.004	3.990	3.770	4.127
Total Cations	15.975	15.957	16.004	15.990	15.770	16.127

	Anions normalized to $Al + Si + B = 12$					
Cl	0.3057	0.1923	0.4036	0.1557	0.2850	0.1319
S	0.0403	0.0202	0.0559	0.1953	0.0682	0.1297
F	0.0077	0.0000	n.d.	n.d.	0.0000	n.d.
Total	0.3537	0.2125	0.4595	0.3510	0.3532	0.2616

Notes:
[a] Electron microprobe yields no detectable F.
n.d.: Element not determined.
Totals corrected for F = O, Cl = O.

Table 10.7. *Phlogopite*

Sample	SI-3		18964		5231	2148B		84–5	SI–4
Grain	PHL#18	PHL#3	PHL#4	PHL#5	PHL#13	23	22		
Zone	2	3'	3	3	3'	4	4	3	3
								3	3
				Weight percent (electron microprobe)					
SiO_2	38.97	38.87	33.19	35.24	38.42	38.55	34.89	40.7	40.6
TiO_2	0.03	0.05	5.30	4.96	0.29	0.62	1.18	0.8	0.0
Al_2O_3	18.37	19.40	18.20	17.78	19.22	17.57	18.32	16.9	17.1
FeO	0.72	0.62	1.86	1.84	0.62	1.05	1.14	0.0	0.0
MnO	0.01	0.01	0.01	0.03	0.01	0.00	0.00	0.0	0.0
MgO	25.15	24.97	20.98	21.37	25.09	25.74	24.96	26.4	26.5
Na_2O	0.18	0.35	0.80	0.67	0.39	1.07	0.89	0.8	0.0
K_2O	10.61	10.36	6.77	7.13	10.45	8.14	6.25	10.5	10.5
BaO	n.d.	0.16	8.21	6.71	n.d.	2.64	8.45	n.d.	n.d.
Total	96.23	94.79	96.79	95.73	96.54	95.38	96.08	96.1	94.7
				Weight percent (ion microprobe)					
Li_2O	0.018	n.d.	0.011	n.d.	0.006				
B_2O_3	0.005	n.d.	0.087	n.d.	0.005				
BaO	0.450	n.d.	n.d.	n.d.	0.445				
F	1.72	n.d.	1.37	n.d.	1.59				
Total	96.23	94.79	96.79	95.73	96.54				

Cations per 22 oxygens

Si	5.505	5.451	4.936	5.147	5.403	5.450	5.108	5.621	5.666
Al	2.494	2.549	3.042	2.853	2.596	2.550	2.892	2.379	2.334
B	0.0013	n.d.	0.0223	n.d.	0.0013				
Total	8.000	8.000	8.000	8.000	8.000	8.000	8.000	8.000	8.000
Ti	0.003	0.005	0.593	0.545	0.031	0.066	0.130	0.083	0.000
Al	0.565	0.658	0.148	0.208	0.590	0.378	0.269	0.372	0.479
Fe	0.085	0.073	0.231	0.225	0.073	0.124	0.140	0.000	0.000
Mn	0.001	0.001	0.001	0.004	0.001	0.000	0.000	0.000	0.000
Mg	5.297	5.220	4.652	4.653	5.260	5.425	5.448	5.435	5.514
Li	0.0102	n.d.	0.0068	n.d.	0.0035				
Total	5.961	5.957	5.625	5.635	5.959	5.993	5.987	5.890	5.993
Na	0.049	0.095	0.231	0.190	0.106	0.293	0.253	0.214	0.000
K	1.912	1.854	1.285	1.329	1.875	1.468	1.167	1.850	1.870
Ba	0.025	0.009	0.479	0.384	0.025	0.146	0.485	n.d.	n.d.
Total	1.986	1.958	1.995	1.903	2.006	1.907	1.905	2.064	1.870
Total Cations	15.947	15.915	15.627	15.538	15.965	15.900	15.892	15.954	15.863

Anions per 22 oxygens

F	0.7701	n.d.	0.6449	n.d.	0.7094				

Notes:
n.d.: Element not determined.
Totals corrected for $F = O$.

Table 10.8. *Pargasite, hyalophane and hydrogrossular*

Mineral	Pargasite			Hyalophane	Hydrogrossular	
Sample	18964	SI–4	84–5	18964(2)	5231	SI–4
Zone	3	3	3	1	3′	3
	Weight percent (electron microprobe)					
SiO_2	43.03	42.7	44.38	55.66	36.13	36.3
TiO_2	2.18	0.0	0.18	n.d.	0.06	0.0
Al_2O_3	15.52	18.2	16.67	21.95	21.45	22.4
FeO	1.97	1.1	0.96	0.10	0.19	0.0
MnO	0.00	0.0	0.06	n.d.	0.30	0.0
MgO	17.90	18.8	18.22	n.d.	0.36	0.4
CaO	13.04	13.4	12.89	0.00	36.99	38.1
Na_2O	3.38	2.4	2.65	2.78	n.d.	n.d.
K_2O	0.59	2.2	1.70	7.96	n.d.	0.0
BaO	0.38	n.d.	n.d.	12.08	n.d.	n.d.
	Weight percent (ion microprobe)					
B_2O_3	n.d.	n.d.	n.d.	0.048	n.d.	n.d.
SrO	n.d.	n.d.	n.d.	0.162	n.d.	n.d.
Total	97.99	98.8	97.71	100.74	95.48	97.2
	Cations					
Oxygens	23	23	23	8	24	24
Si	6.076	5.966	6.225	2.733	5.748	5.671
Al	1.924	2.034	1.775	1.270	0.000	0.000
B	n.d.	n.d.	n.d.	0.0041	n.d.	n.d.
Total	8.000	8.000	8.000	4.007	5.748	5.671
Ti	0.231	0.000	0.019	n.d.	0.007	0.000
Al	0.659	0.963	0.981	0.000	4.022	4.125
Fe	0.233	0.129	0.113	0.004	0.025	0.000
Mn	0.000	0.000	0.007	n.d.	0.040	0.000
Mg	3.768	3.916	3.810	n.d.	0.085	0.093
Total	4.891	5.008	4.930	0.004	4.179	4.218
Ca	1.973	2.006	1.937	0.000	6.305	6.378
Na	0.925	0.650	0.721	0.265	n.d.	n.d.
K	0.106	0.392	0.304	0.499	n.d.	n.d.
Ba	0.021	0.000	0.000	0.232	n.d.	n.d.
Sr	n.d.	n.d.	n.d.	0.005	n.d.	n.d.
Total	3.025	3.048	2.962	1.001	6.305	6.378
Total Cations	15.916	16.056	15.892	5.012	16.232	16.267

Notes:
n.d.: Element not determined.

Table 10.9. *Tourmaline*

Sample	SI–3	18964	SI–4
Grain	1	2	
Zone	3′	3	3
	Weight percent (electron microprobe)		
SiO_2	36.82	36.98	36.9
TiO_2	0.05	0.43	0.0
Al_2O_3	33.57	31.24	31.4
FeO	0.43	1.06	0.0
MgO	11.06	11.54	13.2
CaO	2.87	2.87	4.4
Na_2O	1.58	1.57	0.6
K_2O	0.07	0.02	0.0
BaO	0.00	0.20	n.d.
Total	86.45	85.91	86.5
	Cations per 49 oxygens		
Si	11.673	11.873	11.719
Al	0.327	0.127	0.281
Total	12.000	12.000	12.000
Ti	0.012	0.104	0.000
Al	12.216	11.695	11.472
Fe	0.114	0.285	0.000
Mg	5.227	5.524	6.250
Total	17.569	17.608	17.722
Ca	0.975	0.987	1.497
Na	0.971	0.977	0.369
K	0.028	0.008	0.000
Ba	0.000	0.025	n.d.
Total	1.974	1.997	1.866
Total Cations	31.543	31.605	31.588

Notes:
n.d.: Element not determined.

analyzed with the IMMA, serendibite incorporates the most Li, which was reported to be present in the type material (Prior & Coomaraswamy, 1903). Serendibite in SI–3 and 5231 also contains a trace of Be, which was released on breakdown and incorporated in sapphirine (Tables 10.3 and 10.4).

In general, X_{Fe} (= atomic Fe/(Fe + Mg)) decreases as follows: Spl > Srd ≈ Prg > Tur > Spr > Phl ≈ Di > Gdd > Snh and X_{Na} (= atomic Na/(Ca + Na)) decreases as follows: Prg > Srd > Di, but Na–Ca relations among Tur, Scp, and Prg differ in each sample. Variations in Fe–Mg and Na–Ca fractionations from sample to sample, together with compositional variations from grain to grain in a given sample (notably diopside, phlogopite, and scapolite), imply that the fractionations observed in the Johnsburg rocks may not in all cases represent chemical equilibrium. Deviations from

Table 10.10. *Spinel, Sinhalite and Grandidierite*

Sample	5231	18964	84–5	84–5
Grain	Spinel	SNH	SNH	GDD
Zone	3′	2	3	3
	Weight percent (electron microprobe)			
SiO_2	0.11	0.02	0.00	21.53
TiO_2	0.02	0.00	n.d.	n.d.
Al_2O_3	70.31	41.11	42.06	54.57
FeO	3.79	1.43	0.66	0.41
MnO	0.00	0.04	n.d.	n.d.
MgO	26.17	30.52	30.76	13.01
CaO	n.d.	n.d.	0.00	0.12
	Weight percent (calculated)			
B_2O_3	n.d.	27.56	27.80	12.23
Total	100.40	100.68	101.28	101.87
	Cations			
Oxygens	4	4	4	9
Si	0.003	0.000	0.000	1.014
Ti	0.000	0.000	n.d.	n.d.
Al	1.988	1.016	1.029	3.030
Fe	0.076	0.025	0.011	0.016
Mn	0.000	0.001	n.d.	n.d.
Mg	0.936	0.954	0.952	0.914
Ca	n.d.	n.d.	0.000	0.006
B	n.d.	0.997	0.996	0.994
Total	3.003	2.993	2.988	5.974

Notes:
n.d.: Element not determined.

equilibrium fractionations observed elsewhere (e.g. for Fe–Mg, Grew (1986) reports Tur < Bi) are probably due to alteration and incomplete re-equilibration during later stages of metamorphism (see below), or in some cases, to other components, e.g. Ba in phlogopite or Na and Al in diopside could have affected Mg–Fe distribution in 18964.

Discussion

Metasomatic zoning and metamorphic history

The zonal distribution of scapolite, diopside, serendibite, and phlogopite around the K-feldspar rock is attributed to the metasomatic interaction between the K-feldspar rock and dolomite-bearing marble under granulite-facies conditions, e.g. Korzhinskiy's (1955) bimetasomatism. The zones have several of the features characteristic of metasomatic zoning (Korzhinskiy, 1955, Thompson, 1959); (1) simple mineralogy that contrasts with the overall chemical complexity and (2) embayments of the adjacent rocks: at Johnsburg the K-feldspar rock is penetrated by veins of zones 2 and 3.

Relict feldspars indicate that the scapolite–diopside zone developed by alteration of the K-feldspar rock. Rare zircons in zones 3 and 4 suggest that these zones may also be derived from alteration of the K-feldspar rock (endoskarn of Korzhinskiy, 1955), as zircon is rare in Adirondack marble (Valley & Essene, 1980). However, in one specimen, a diopside nodule similar to those found in the marble is enclosed by the phlogopite zone, implying that the phlogopite zone must have formed at the expense of marble (exoskarn), at least locally. Thus the original contact between the K-feldspar rock and marble may be located somewhere in zone 4.

Textures in the zones suggest at least three stages of metamorphism: (1) primary minerals formed at the highest temperature conditions in the granulite facies and during development of the metasomatic zoning, (2) secondary minerals formed by reactions among the primary minerals, under relatively high-temperature conditions of the upper amphibolite facies after metasomatic exchange had ceased, and (3) secondary minerals formed at low temperatures by breakdown of minerals formed at high temperatures.

In general, secondary minerals formed during stage 3, namely hydrogrossular, serpentine, and other phyllosilicates replacing serendibite, scapolite, and phlogopite, are relatively minor and have not destroyed textural relations developed during the first two stages. We will not further consider the reactions involved in stage 3. On the other hand, alteration during the second stage has been extensive and textural relations among primary minerals are commonly obscured; unambiguous interpretation of the primary mineral assemblages is difficult. This alteration also affected the composition of the minerals resulting in non-equilibrium fractionation and heterogeneity.

In zones 2–4, diopside, serendibite, scapolite, phlogopite (in part), sinhalite, apatite and barite are interpreted to be primary minerals, and grandidierite, tourmaline, sapphirine, pargasite, plagioclase, phlogopite (in part) and rutile to be high-temperature secondary minerals, largely on the basis of textures. Spinel appears to be largely a high-temperature secondary mineral in zone 3; primary spinel is found in zone 4, but only locally in zones 3 and 1. Calcite, which is present throughout, appears to be largely secondary (high and low temperatures) in zones 1–3, although the presence of primary calcite cannot be ruled out, especially in zone 3. In the K-feldspar rock, tourmaline appears to be primary, and in the marbles adjacent to K-feldspar rock, dolomite, phlogopite, and forsterite are primary. The serpentine margins around diopside nodules suggest that diopside could have reacted with dolomite to form forsterite (and calcite) that was largely altered to serpentine in the third stage; such forsterite margins would also be an example of metasomatic zoning developed during the granulite-facies metamorphism.

Primary assemblages

Because the primary mineral assemblages developed coevally with metasomatic zoning, we have adopted the diagram (Fig. 10.16) proposed by Korzhinskiy (1955,

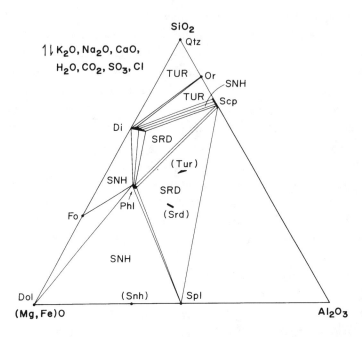

Fig. 10.16. Compositional projection of minerals in the primary assemblages of the Johnsburg rocks onto the plane SiO_2–Al_2O_3–(Mg, Fe)0 assuming K_2O, Na_2O, CaO, H_2O, CO_2, Cl, SO_3 are fully mobile components (adapted from Korzhinskiy (1957, Fig. 51), (1973, Fig. 56). The compositional ranges of the minerals (boron-bearing minerals in parentheses) are taken from the microprobe analyses (Tables 10.3–10.10); ideal compositions are assumed for Dol, Fo, orthoclase (Or), and Qtz. Tielines have been idealized to eliminate crossings. Capitals indicate two- and three-phase fields where a given boron-bearing mineral has been found.

Figure 10, 1957, Figure 51, 1973, Figure 56) for portraying mineral compatibilities in bimetasomatic magnesian skarns. Korzhinskiy deduced that K_2O, Na_2O, BaO, CaO and Fe, as well as the volatiles CO_2, SO_3, O_2, and H_2O were perfectly mobile components. In the Johnsburg rocks, the amount of Fe is in many cases too low to affect the mineral assemblages. The behavior of B_2O_3 has features characteristic of a perfectly mobile component and of an inert component.

Korzhinskiy's diagram is a fairly good first approximation to the boron-free mineralogy in the metasomatic zones at Johnsburg. The zonal succession can be represented as a marked decrease in SiO_2, a slight decrease in Al_2O_3 and an increase in MgO, starting from the assemblage K-feldspar–biotite or phlogopite–scapolite ± diopside ± quartz (zone 1), and passing through diopside–scapolite ± phlogopite (zones 2 and 3), phlogopite ± spinel (zone 4), to dolomite ± forsterite ± phlogopite. The phlogopite (biotite)–K-feldspar tieline in zone 1 violates the relations illustrated in Fig. 10.16, and could be due to inert behavior of K_2O within zone 1. Crossing of tielines between minerals in Fig. 10.16 (but not shown in order to simplify illustration) could be due to incomplete re-equilibration during later events as well to compositional variables not considered in the projection.

By analogy with Korzhinskiy's (1957, 1973) portrayal of Ti-bearing phases associated with the metasomatic rocks, a boron-bearing phase has been indicated for each 3-phase assemblage. Two boron-bearing phases could appear with 2-phase assemblages. In cases where the B-free phases have a variable composition (diopside, scapolite), only one particular composition of these phases would be stable with two boron-bearing phases under any given set of externally controlled conditions (T, P, μ_i) (see also Thompson (1972), Figure 4). This representation of the distribution of boron-bearing phases is also a fairly good first approximation of the observed mineral assemblages.

(together with diopside and rarely phlogopite or scapolite) e.g., in the marbles on Garnet Lake Road and elsewhere in the Adirondacks (Valley & Essene, 1980), as well as in the K-feldspar rock with scapolite and phlogopite or biotite, and less commonly quartz. Neither serendibite nor sinhalite has been found with K-feldspar or quartz, with the exception of the rare serendibite in a specimen of K-feldspar rock with tourmaline, spinel, biotite and scapolite. This assemblage could have been stabilized by the relatively high Fe/Mg ratio of the K-feldspar rock assemblage (deduced from the colors of the minerals) by comparison with the other Johnsburg rocks.

In zones 3 and 3', serendibite appears to be the only boron-bearing phase stable with the assemblage diopside–scapolite–phlogopite and scapolite–phlogopite, respectively. Sinhalite occurs with serendibite in diopside–scapolite assemblages (e.g. Fig. 10.13) and in scapolite–phlogopite assemblages, and the assemblage Srd–Snh–Di–Phl–Scp may have been a stable one. In zone 4, the assemblages sinhalite–serendibite (partially rimming sinhalite) – diopside–phlogopite and sinhalite–spinel–phlogopite were found. The sinhalite-bearing assemblages generally form isolated patches and could represent local boron concentrations; sinhalite is rarely an abundant constituent. Chemographic relations illustrated in Fig. 10.16 preclude the appearance of sinhalite in the two three-phase areas bounded by diopside, scapolite, phlogopite, and spinel, as well as the two-phase band phlogopite–scapolite. The sinhalite-bearing patches could represent volumes where variance decreased due to lower mobility of B_2O_3, thereby stabilizing an additional phase. The appearance of sinhalite instead of tourmaline in the relatively siliceous (although SiO_2-undersaturated) assemblage diopside–scapolite may be related to low H_2O or low CO_2 activities as suggested by Greenwood (1967) for mixed volatile reactions such as the following one (compositions of 18964, Table 10.11; some of the CO_2 for calcite probably comes from scapolite):

$$0.70 \text{ Scp} + 0.42 \text{ Di} + 2.53 \text{ Snh} + 0.24 \text{ B}_2\text{O}_3 + 1.63 \text{ CO}_2 +$$
$$2\text{H}_2\text{O} = \text{Tur} + 1.63 \text{ Cal} + 0.26 \text{ Na}_2\text{O}. \tag{1}$$

Boron is generally treated as a perfectly mobile component in boron-enriched metasomatic deposits (e.g. Marakushev, 1958, Pertsev, 1971). However, the compatibilities shown in Fig. 10.16 are consistent with the addition of B_2O_3 as a fourth inert

Table 10.11. *Compositions of minerals used in calculating reactions (simplified from Table 10.3–10.10, Fe combined with Mg, and K combined with Na).*

Mineral	Composition	Sample
Serendibite	$Ca_{1.6}Na_{0.3}Mg_{2.8}Al_{4.8}B_{1.5}Si_{3.0}O_{20}$	SI–3, 5231
	$Ca_{1.6}Na_{0.3}Mg_{3.0}Al_{4.4}B_{1.5}Si_{3.2}O_{20}$	18964
Tourmaline	$Ca_{0.5}Na_{0.5}Mg_{2.7}Al_{6.3}Si_{5.8}B_{3.0}O_{27}(OH)_4$	SI–3
	$Ca_{0.5}Na_{0.5}Mg_{2.9}Al_{5.9}Si_{5.9}B_{3.0}O_{27}(OH)_4$	18964[a]
Sapphirine	$Mg_{3.4}Al_{8.8}Si_{1.4}B_{0.4}O_{20}$	SI–3
Phlogopite	$K_{2.0}Mg_{5.3}Al_{3.2}Si_{5.4}O_{20}(OH)_4$	SI–3, 5231[a]
Pargasite	$Na_{1.0}Ca_{2.0}Mg_{4.0}Al_{2.6}Si_{6.1}O_{22}(OH)_2$	18964[a]
Diopside	$Na_{0.1}Ca_{0.9}Mg_{0.9}Al_{0.2}Si_{1.9}O_6$	18964[a]
Grandidierite	$Mg\ Al_3\ BSiO_9$	Ideal
Spinel	$Mg\ Al_2O_4$	Ideal
Calcite	$CaCO_3$	Ideal
Scapolite	$Na_{1.4}Ca_{2.5}Al_{4.7}Si_{7.3}(CO_2, Cl, SO_3)$	18964
Sinhalite	$MgAlBO_4$	Ideal

Notes:
[a] Charge balance in part made up by Ti.

component, analogous to TiO_2 in Korzhinskiy's (1957, Fig. 51, 1973, Fig. 56) projection. The configuration in Fig. 10.16 allows for the assemblage Phl–Scp–Srd–Di, a common assemblage in zone 3, to be stable, which is not possible if B_2O_3, as well as K_2O, is assumed to be fully mobile (compare Pertsev, 1971, Fig. 18b, where plagioclase instead of scapolite is plotted). Moreover, sinhalite, serendibite, and scapolite are nearly collinear in terms of MgO, Al_2O_3, and SiO_2, so that a sinhalite–scapolite–serendibite assemblage would not be expected in a system with perfectly mobile Na_2O, CaO, B_2O_3. Unfortunately, these arguments are not incontrovertible because the natural chemical system is far more complex than the model system used in Fig. 10.16; other factors, such as Na_2O (as an inert component), could stabilize assemblages such as Scp–Snh that appear to be forbidden. Another consideration is the buffering capacity of the serendibite- and sinhalite-bearing assemblages for B_2O_3. As B_2O_3 was introduced into the metasomatic zones, the boron-bearing assemblages controlled the activity of B_2O_3 in the solutions, even though B_2O_3 in these solutions was highly mobile. This interpretation is consistent with the relations deduced from Fig. 10.16.

Secondary assemblages

In contrast to the primary mineral assemblages, the high-temperature secondary assemblages appear to have formed after development of the metasomatic zoning. Nonetheless some of the components appear to have been relatively mobile, at least within a given zone. The reactions deduced from textures involve the breakdown of serendibite, diopside, and sinhalite, and are thus mostly in zones 3 and 3'. As a first approximation these reactions will be considered in terms of the model four-component system CaO–MgO–Al_2O_3–SiO_2. K_2O, Na_2O, B_2O_3, CO_2, H_2O, Cl, and SO_3 will be treated as perfectly mobile components, while FeO, BaO, and other

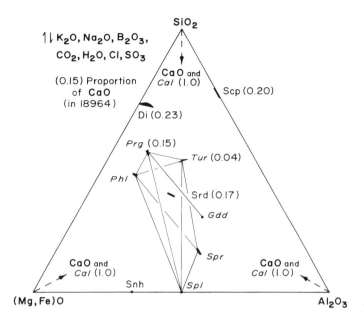

Fig. 10.17. Compositional projection of the minerals in the Johnsburg rocks into the tetrahedron SiO_2–Al_2O_3–(Mg, Fe)O–CaO, assuming Na_2O, K_2O, B_2O_3, CO_2, H_2O, Cl, and SO_3 are fully mobile components. The compositional ranges of the minerals are taken from the microprobe analysis (Tables 10.3–10.10). Proportion of CaO taken from data on sample 18964. Only compatibilities among the secondary minerals (indicated by italics) are shown and have been simplified to a single tieline.

components are generally present in too small quantities to have a significant effect on the reactions (Fig. 10.17).

A commonly observed reaction is the breakdown of serendibite to tourmaline, spinel, calcite and in some cases, phlogopite (Fig. 10.8). In its simplest form, this reaction can be written (based on the compositions of silicates in SI–3, Table 10.11):

$$Srd + 1.04\ H_2O + 1.42\ CO_2 + 0.15\ K_2O \rightarrow 0.37\ Tur + Spl +$$
$$0.15\ Phl + 1.42\ Cal + 0.06\ Na_2O + 0.19\ B_2O_3. \qquad (2)$$

In zone 3′, sapphirine is also a breakdown product of serendibite, and is associated with tourmaline, calcite, phlogopite (e.g. Fig. 10.14), and in rare patches, diopside. A simple reaction relating the diopside-free assemblage is (compositions of SI–3).

$$Srd + 0.94\ H_2O + 1.48\ CO_2 + 0.22\ K_2O \rightarrow 0.25\ Tur + 0.29\ Spr +$$
$$1.48\ Cal + 0.22\ Phl + 0.32\ B_2O_3 + 0.09\ Na_2O. \qquad (3)$$

A relatively MgO-rich phase is needed to balance the equation because serendibite lies on the MgO side of the Tur–Spl join, which is consistent with the appearance of phlogopite as a breakdown product of serendibite. MgO could also be accommodated in calcite, which in SI–3 contains up to 10 mole % $MgCO_3$. Serendibite probably did not incorporate this 'excess' MgO through the Tschermak substitution $Mg + Si = 2 Al$. No change in MgO content was found in a microprobe traverse across a serendibite–tourmaline contact in SI–3 (also in serendibite rim, Table 10.3), although SiO_2 and CaO seemed to decrease 5% (relative).

In other specimens, pargasite accompanies spinel as a secondary mineral, and the 'excess' MgO may be accommodated in pargasite in a reaction as follows (based on the compositions in 18964):

$$Srd + 0.06\ Na_2O + 0.76\ H_2O + 0.88\ CO_2 \rightarrow 0.23\ Tur +$$
$$1.13\ Spl + 0.30\ Prg + 0.88\ Cal + 0.40\ B_2O_3. \tag{4}$$

In some cases, diopside appears to have been a reactant, and thus an appropriate reaction probably was (compositions of 18964).

$$Srd + 3.41\ Di + 0.41\ Na_2O + 1.87\ H_2O + 1.9\ CO_2 \rightarrow 0.28\ Tur +$$
$$1.31\ Prg + 1.9\ Cal + 0.33\ B_2O_3. \tag{5}$$

In some specimens with diopside and pargasite, grandidierite is the more abundant secondary borosilicate, and a reaction consistent with textural evidence (Fig. 10.9) is (compositions of 18964):

$$Srd + 1.87\ Di + 0.25\ Na_2O + 0.99\ H_2O + 1.31\ CO_2 \rightarrow 0.74\ Gdd +$$
$$0.99\ Prg + 1.31\ Cal + 0.38\ B_2O_3. \tag{6}$$

Tourmaline also appears as rims on sinhalite in scapolite-bearing rocks, which suggests the following reaction (compositions of 18964; cf. reaction (1)):

$$2.53\ Snh + 0.7\ Scp + 0.42\ Di + 1.63\ CO_2 + 2\ H_2O + 0.24\ B_2O_3 \rightarrow$$
$$Tur + 1.63\ Cal + 0.26\ Na_2O. \tag{7}$$

Traces of interstitial sodic plagioclase are not uncommon in the serendibite zone; possibly this plagioclase forms from Na_2O released by reactions such as (2) and (7).

Physical and chemical conditions for formation of the secondary assemblages

The breakdown of serendibite to tourmaline and spinel is characteristic of other serendibite localities (Pertsev, 1971). On the other hand, the appearance of sapphirine and grandidierite has not been previously reported. By comparison with other serendibite-bearing rocks, the Johnsburg serendibite-bearing zones at the main locality are remarkably depleted in iron, such that serendibite contains only 0.6–0.8 wt % FeO (Table 10.3) compared to 3.5% and more at the other localities listed in Table 10.1 for which analyses are available. This depletion could favor the appearance of sapphirine instead of spinel because Mg is preferably incorporated in sapphirine relative to spinel (e.g. Higgins *et al.*, 1979). Another factor may be the rarity of diopside in zone 3′, where sapphirine appears, while in zone 3, diopside is relatively abundant, and the secondary assemblages include pargasite and spinel rather than sapphirine. In general, sapphirine would be expected in more aluminous bulk compositions than pargasite–spinel, as its composition is closer to the Al_2O_3 corner (Fig. 10.17). If diopside is rare or absent, the bulk composition of the reacting minerals would be more aluminous, that is, serendibite–scapolite lies closer to the Al_2O_3 corner than serendibite–diopside $\pm$ scapolite. In addition, the assemblage diopside–sapphirine may be restricted to a limited range of bulk compositions at the pressures and temperatures of metamorphism at Johnsburg. In general, the clinopyroxene–sapphirine assemblage is restricted to deep seated environments such as mantle xenoliths (15 kbar, Griffin & O'Reilly, 1986) and metamorphosed ultramafics (9 kbar, Sills et al., 1983).

In the case of grandidierite, the stability range of tourmaline is probably the critical factor. In order to determine how tourmaline stability affects grandidierite, we can write a reaction relating the assemblage Tur–Spl with the Gdd–Prg assemblage characteristic of Johnsburg (compositions of 18964):

$$Tur + 1.94\ Spl + 0.5\ Cal = 2.82\ Gdd + 0.5\ Prg +$$
$$1.5\ H_2O + 0.5\ CO_2 + 0.09\ B_2O_3. \tag{8}$$

Reaction 8 suggests that H_2O and CO_2 are probably the most important factors in controlling the distribution of these two assemblages. The more common Tur–Spl assemblage may be stable to relatively high temperatures except under conditions where $X(CO_2)$ or $X(H_2O)$ is very low, a characteristic feature of mixed-volatile reactions in which both volatile components are evolved (Greenwood, 1967). Thus grandidierite in the Johnsburg rocks probably developed where fluids happened to be unusually depleted in either H_2O or CO_2.

At first glance, the formation of sapphirine and grandidierite in relatively hydrous and carbonated assemblages appears incongruous with the paragenesis of these minerals elsewhere. Sapphirine is characteristic of upper amphibolite- to granulite-facies rocks in association with sillimanite, cordierite, gedrite, phlogopite, and orthopyroxene (e.g. review by Deer et al. 1978). Nonetheless, sapphirine parageneses involving relatively hydrous and carbonated minerals are known. For example, at Terakanambi in southern India, sapphirine in association with chlorite, biotite, anthophyllite, carbonate, and cordierite is interpreted to have formed by reaction of spinel and orthopyroxene with metamorphic fluids (Janardhan & Shadakshara Swamy, 1982, Grew, 1982). Grandidierite occurs in hornfels, granulite-facies rocks, and plutonic rocks (reviewed by van Bergen, 1980), crystallizing at temperatures above the breakdown of muscovite + quartz, that is 600–900 °C (Grew, 1983). However, this restriction to high temperatures may be due to the bulk compositions of the grandidierite-bearing rocks, most of which are Al-rich and Ca-poor, and thus may not apply to Fe-poor calcareous compositions such as the Johnsburg rock.

In summary, the secondary mineral assemblages Tur–Spl–Cal, Tur–Prg–Cal, Gdd–Prg–Cal, and Spr–Tur–Cal, formed by loss of B_2O_3 and gain of H_2O and CO_2 in the serendibite- and sinhalite-bearing assemblages. Pargasite, grandidierite, and sapphirine are characteristic of relatively high temperatures, probably upper amphibolite-facies. The appearance of these minerals suggests a discrete metamorphic event following the peak of metamorphism when serendibite formed. Textures also suggest a polymetamorphic origin for these secondary mineral assemblages. Tremolite and quartz rims, formed around diopside in a few Adirondack marbles (Valley & Essene, 1980), including those from Garnet Lake Road, could be a related development. However, Valley & Essene (1980, p. 757) attribute these rims to a local 'back-reaction of pyroxene', which presumably occurred during the isobaric cooling following peak conditions deduced by Bohlen *et al.* (1985). Possibly, the later metamorphic event has the greatest impact on the serendibite-bearing zones because serendibite and sinhalite were the most sensitive phases to reaction, particularly in the presence of fluids rich in H_2O and/or CO_2. The absence of an imprint in other zones and the limited extent of reaction in Adirondack marbles may be due to their mineralogy. For example, a scapolite–diopside assemblage would be stable under granulite- and upper amphibolite-facies conditions and thus would appear little affected.

Origin of the boron enrichment

The unusual concentration of boron at the Johnsburg locality is attributed to two stages. The first stage was during sedimentation and diagenesis and the second during metamorphism in the granulite-facies. Our proposed model for development of the Johnsburg deposit is based on the field relations, geochemistry, and isotope chemistry of the Johnsburg rocks.

The relations illustrated in Fig. 10.2 suggest that the K-feldspar rock is a lens in dolomitic marble. The low contents of rare earths and TiO_2, MgO, and total Fe (Table 10.11) are inconsistent with an igneous origin for the K-feldspar rock. A more likely precursor is sediment with a high clay content, most likely illite. Moreover, the boron originated in this illite-rich sediment. From a detailed study of clay fractions from a variety of sediments, Harder (1959, 1970) and Reynolds (1965) concluded that boron in carbonate rocks is largely incorporated in illite. In addition, Harder (1959) reported that the boron content of illite in limestone is about 200 ppm, in dolomite, 300 ppm, and in salt deposits, over 1500 ppm. This relation has led to the proposal that B/Al ratios in the clay fraction can be used as a paleosalinity indicator (B. Porthault in Moine *et al.*, 1981). In the case of the Adirondack marbles, the Johnsburg serendibite deposit and the disseminated tourmaline reported by Valley & Essene (1980) could have the same origin. The tourmaline is with rare exception associated with K-feldspar and thus the boron for the tourmaline could have originated with illite that subsequently recrystallized into K-feldspar. At Johnsburg, depositional conditions may have been hypersaline. The relatively high $\delta^{18}O$ value of 25.52 per mil for a dolomite collected 2m from the K-feldspar rock (Table 10.12) could be due to deposition in a hypersaline environment, where $\delta^{18}O$ values tend be higher than for other marine carbonates (e.g. Whelan *et al.*, 1984). As a result, the illitic sediment presumed to be the precursor to the K-feldspar rock could have been enriched in boron relative to the clay fraction in the precursors to the tourmaline-bearing calcite marbles.

Table 10.12. *Bulk compositions of the K-feldspar rock from the main serendibite locality, Johnsburg, N.Y.*

Major elements, weight %			Trace elements, ppm			
	1	2	3		2	
SiO_2	64.76	64.16	B	210	V	12
TiO_2	b.d.	0.04	La	1.3	Cr_2O_3	<15
Al_2O_3	18.39	18.82	Ce	2	Ni	<10
Fe_2O_3	0.14	0.03	Nd	6	Nb	6
FeO	—	0.18	Sm	0.3	Zr	60
MnO	—	0.1	Eu	<0.10	Y	6
MgO	b.d.	0.4	Gd	<1.0	Sr	549
CaO	0.59	0.96	Tb	<0.1	Rb	303
BaO	—	0.5722	Yb	0.19	Pb	16
Na_2O	2.0	1.71	Lu	<0.05	Th	<5
K_2O	12.03	12.11			U	<5
P_2O_5	0.081	0.04			Ce	8
L.O.I.	0.99	0.87			Ga	14
Total	98.98	99.99				

Notes:
b.d. below detection.
L.O.I. loss on ignition.
Columns 1. X-ray fluorescence by J.M. McLelland. Na_2O content is approximate.
      2. X-ray fluorescence by S.T. Ahmedali, McGill University, Montreal; FeO and Fe_2O_3 by titration.
      3. Neutron activation analysis, except for B and Gd, which were determined by Prompt Gamma Neutron Activation Analysis, both done at Nuclear Activation Service, Ann Arbor, Michigan.
Source: J.M. McLelland, pers. comm., 1987.

Our proposal that the boron in the Johnsburg deposit originally was incorporated in an illite-rich clay fraction is consistent with boron isotope data on tourmaline from the pod at the main locality and on serendibite from material collected by E. Rowley nearby. The $\delta^{11}B$ in the tourmaline and serendibite is -1.6 and $+1.9 \pm 2$ per mil, respectively ($\delta^{11}B$ expressed as per mil deviations from NBS SRM $951 = 4.0436$, Swihart, 1987). These $\delta^{11}B$ values are close to those for tourmaline, sinhalite, and warwickite concentrations in the Franklin Marble of New Jersey and New York, which is also part of the Grenville Province ($+0.5$ to $+4.1$ per mil, 4 samples – Swihart, 1987), but are somewhat lower than for tourmaline in marble near contacts with pelitic gneisses or in metapelite of the Grenville in the Adirondack Lowlands and in southeastern Ontario ($+7.5$ to $+13.4$ per mil, 4 samples – Swihart, 1987). The $\delta^{11}B$ values for the Johnsburg minerals are similar to those reported for non-desorbable B in marine sediments ($\delta^{11}B = -4.3$ to $+2.8$ per mil, Spivack *et al.*, 1987). In contrast, the boron in seawater from the Atlantic and Pacific is heavier ($\delta^{11}B = 39.5$ per mil –

Spivack & Edmond, 1987). Boron compositions in tourmalines from pegmatites range from -5.3 to -12.3 per mil (6 examples), and in tourmaline from two tourmalinites in Greenland of proposed submarine exhalative origin, -9.8 and -22.7 per mil (Swihart, 1987). Thus marine evaporite borates (for which $\delta^{11}B = 26.3 \pm 5$, 8 samples, Swihart *et al.*, 1986), pegmatitic fluids or submarine exhalations are unlikely sources for boron in the Johnsburg serendibite deposit.

This conclusion assumes that the boron isotope ratio changed little during metamorphism. As far as we know, fractionation of boron isotopes during metamorphism has not been studied. According to Spivack *et al.* (1987) and Spivack & Edmond (1987), who investigated the fractionation of boron isotopes in hydrothermal systems, boron has been leached under natural conditions from MORB with no resolvable fractionation, but ^{11}B is preferentially removed from sediments in Guayamas Basin (Gulf of California) with about a 15 per mil fractionation. However, behavior of boron isotopes under metamorphic conditions is undoubtedly quite different from either of these cases.

The second stage of boron enrichment occurred during the granulite-facies metamorphism. A portion of the boron remained in the K-feldspar rock; a boron determination on one sample gave 210 ppm (Table 10.12). The remainder was mobilized by fluids towards the periphery of the K-feldspar rock lens, where serendibite and sinhalite formed by reaction of the boron-bearing fluids with diopside, scapolite, and phlogopite in the metasomatic mantle. A plausible source for this fluid is breakdown of dolomite by reaction with the K-feldspar rock during development of the metasomatic zoning. The $\delta^{13}C$ and $\delta^{18}O$ values of calcite in the serendibite zone and K-feldspar rock (Table 10.13) indicate a sedimentary carbonate rock as a source for carbon and oxygen in the calcite. The $\delta^{13}C$ and $\delta^{18}O$ values of the calcites in the serendibite zone and K-feldspar rock are within the ranges for the three marbles at Johnsburg (Table 10.13) and for marbles elsewhere in the Adirondacks (Valley & O'Neil, 1984, Valley, 1986). The somewhat lower values for the $\delta^{18}O$ could be due to oxygen exchange with the silicate minerals in the K-feldspar rock and metasomatic zones, an explanation Valley & O'Neil (1984) proposed for variations in $\delta^{18}O$ values in Adirondack marbles. It could be argued that the calcites are largely secondary and thus provide little information on the source of CO_2 during formation of the primary minerals. However, no isotopic signature remains of a possible earlier source such as an igneous rock or meteoritic waters and we conclude that the marble was the main source of fluids during the granulite-facies metamorphism.

Origin of the barium enrichment

Several of the serendibite-bearing and associated rocks are enriched in barium. The Ba content of the main lens of K-feldspar rock (5100 ppm Ba, Table 10.12) is greater than that of shales, which range from 9 to 5000 ppm (average of 546 ppm Ba), and than that of metamorphic rocks, among which a maximum of 3800 ppm Ba was found in gneiss (reviewed by Puchelt (1972)). Even greater enrichment is indicated for some of the

Table 10.13. *Carbon and oxygen isotope analyses of calcite in rocks from the Johnsburg serendibite locality*

Grew sample no.	Rock type (zone)[a]	δ^{13} C per mil	δ^{18} O per mil
5203A	Dolomite-poor marble (5)	0.54	20.39
5208	Dolomite-rich marble (5)	1.96	25.52
5209A	Dolomite-poor marble (5)	0.29	21.04
5244	Diopside rock (5')	3.15	18.97
5200B	Serendibite pod (3)	2.11	19.73
5205	K-feldspar rock (1)	0.39	18.30

Notes:
[a] The analysed samples contained > 60% carbonate except 5205, which contained 0.26 wt %.
Source: J. Morrison and J.W. Valley, pers. comm. 1986.

smaller lenses at Johnsburg, in which hyalophane, barian phlogopite, and barite are found.

The simultaneous enrichment of boron and barium, as far as we are aware, has not been previously reported and unusual barium concentrations appear to be rare in evaporites and associated argillaceous rocks (Holser, 1979, Moine *et al.*, 1981). Korzhinskiy (1945) reported barian orthoclase, phlogopite and barite from the Slyudyanka phlogopite deposit near Lake Baikal. Reznitskiy & Vorobyov (1975) interpreted Ba-bearing calcite from veins at Slyudyanka to be endogenetic. We suggest that the barium in the Johnsburg rocks has a sedimentary origin; it could have been either incorporated in clay, like boron, or precipitated as barite during evaporation of sea water, subsequently redistributed into barite concentrations during diagenesis (Puchelt, 1972), and finally, redistributed again during metasomatism, as BaO is presumed to be a perfectly mobile component (Korzhinskiy, 1955). The unusual boron–barium enrichment in the Johnsburg serendibite rocks thus resulted from a combination of sedimentary and metasomatic processes.

Controls on serendibite stability

Fig. 10.16 indicates that an important control in serendibite formation is compositional. Serendibite appears to be restricted to bulk compositions within the area bounded by diopside, scapolite, phlogopite, and spinel. At other localities (Table 10.1) serendibite has not been reported from associations with quartz, dolomite, olivine and minerals of the humite groups (except for iron-rich ludwigite–magnetite–orthosilicate ore from the Tayozhnoye deposit–Pertsev & Boronikhin (1986)) corundum, Al_2SiO_5 polymorphs, orthopyroxene or primary sapphirine (Pertsev, 1971, Hutcheon *et al.*, 1977).

A second important control on serendibite stability is fluid composition. Breakdown reactions of serendibite to tourmaline, calcite, grandidierite and other minerals during the second stage at Johnsburg (reactions (2) to (6)) are all hydrations and carbonations. Consequently the appearance of serendibite instead of tourmaline (or grandidierite) may depend on relatively low H_2O *or* low CO_2 activities during metamorphism (Greenwood, 1967).

A third important control for serendibite is temperature; a minimum of 700 °C appears to be necessary. In addition to its occurrence in granulite-facies rocks (Sri Lanka, Johnsburg, Tayozhnoye), serendibite formed in the upper amphibolite-facies at 700–749 °C and 4.9–5.4 kbar in the Melville Peninsula, Arctic Canada (Henderson, 1983) and in contact aureoles of high-level intrusive rocks in Riverside, California. These intrusives are part of the Mesozoic batholithic complexes, which were probably emplaced at pressures not exceeding 3 kbar and at temperatures of 600–700 °C (Evernden & Kistler, 1970, Hammarstrom & Zen, 1986). Thus pressure seems to be less critical than temperature; possibly at lower pressures, temperatures need not be so high, as would be expected for devolatilization reactions.

Conclusion

In summary, our study suggests that a combination of a distinctive depositional environment and metamorphism under granulite-facies conditions, together with metasomatism, has resulted in boron mineralization dominated by tourmaline and the rare minerals serendibite and sinhalite. The precursors to the serendibite-bearing rocks are interpreted to be illite-rich sediments intercalated with dolomitic limestone deposited in a hypersaline environment. Enrichment of boron in the sedimentary precursor is attributed to this hypersalinity. During granulite–facies metamorphism, the illitic sediment recrystallized to a K-feldspar rock that interacted with the enclosing dolomitic marble to form metasomatic zones. Boron was mobilized during this metamorphism and was precipitated in the metasomatic zones as serendibite and sinhalite instead of tourmaline because of silica undersaturation and low CO_2 *or* low H_2O activities in the metamorphic fluids. During a later metamorphic event in the upper amphibolite facies, the serendibite rocks were partially recrystallized to tourmaline- and grandidierite-bearing assemblages as mixed CO_2–H_2O fluids were introduced. These fluids were probably depleted in boron, because boron was released during serendibite breakdown.

Clearly high-temperature fluids played an important role in the development of the serendibite deposit, as Korzhinskiy proposed for phlogopite deposits; and his students, e.g. A.A. Marakushev and N.N. Pertsev for the Tayozhnoye iron deposit, in which there is localized boron mineralization. However, in contrast to the deposits described by Korzhinskiy and his students, we have no evidence that the high-temperature fluids were associated with granitization or that boron was introduced from a source outside of the immediate vicinity of the K-feldspar rock. The available isotopic data suggest that the fluids and boron are derived from the metasedimentary rocks in the immediate vicinity of the serendibite deposit itself.

Moreover, we propose that the metasomatic zones and boron enrichment at

Johnsburg developed during the granulite-facies metamorphism. According to the Korzhinskiy model, many magnesian skarns are products of 'post-magmatic' metasomatic activity, that is, formed after the peak of metamorphism. In particular, serendibite and ludwigite in the Tayozhnoye deposit are interpreted to have formed after the peak of metamorphism when boron-rich fluids were introduced into magnesian skarns and rocks rich in olivine or humite-group minerals (Marakushev, 1958, Kulakovskiy & Pertsev, 1986).

An important consideration in the study of granulite-facies terrains is the depletion of certain elements, such as Rb, Cs, U and Th (e.g., Heier, 1973). Boron concentrations are rare in granulite-facies rocks and B may also be lost during granulite-facies metamorphism (Truscott *et al.*, 1986). High-temperature fluids percolating through the rocks could play an important role in transporting and mobilizing boron (Manning & Pichavant, 1983), thereby leading to loss of boron from the rocks. At Johnsburg, boron was mobilized by fluid activity, but significant loss from the boron-rich sedimentary precursor was probably averted by its concentration in the metasomatic mantle around the K-feldspar rock. The mantle provided a suitable environment for precipitation of boron in the form of serendibite and sinhalite.

We suggested that the serendibite-bearing assemblages could have buffered the activity of boron in the metasomatising fluid. Sufficient amounts of boron remained in the K-feldspar rock to form tourmaline. The overall system tourmaline $- B_2O_3$ (fluid)– serendibite $\pm$ sinhalite in the K-feldspar rock and its metasomatic mantle could have had an even stronger buffering capacity, and as a result, boron loss from the system was minimized. The retention of tourmaline in the K-feldspar rock, while serendibite formed in the metasomatic mantle, resulted from the original high bulk boron content of the K-feldspar rock. An analogous mechanism, albeit on a smaller scale, was proposed by Grew (1986) for retention of boron in the Waldheim kornerupine rock (Saxony, German Democratic Republic) by the buffering action of a tourmaline–kornerupine assemblage. The retention of boron under high-grade conditions appears to be facilitated by an original content of boron sufficiently great for more than one boron-rich mineral to be stabilized coevally for a period of time during the metamorphism.

Acknowledgements We thank the following institutions and people for samples: P.J. Dunn at the National Museum of Natural History, Smithsonian Institution, C.A. Francis at Harvard University, W.M. Kelly at the New York State Museum, and Elmer B. Rowley of Glens Falls, New York. We also thank Carl Francis at Harvard for providing Larsen and Schaller's unpublished field notes, and Elmer Rowley for permission to work on his samples in the New York State Museum and for his personal notes on his collecting at Johnsburg. We also thank J. McLelland for the whole rock analyses and J. Morrison and J.W. Valley for the oxygen and carbon isotope analyses. Many of the ideas in this paper were inspired through stimulating conversation with J.W. Valley and J. McLelland, including those on a field trip to the Adirondacks led by

J. McLelland in July, 1986, and supported by funds from the International Geological Correlations Program, Project 235, 'Metamorphism and Geodynamics'. ESG's and MGY's research was supported by a grant DPP84–14014 from the U.S. National Science Foundation (NSF) to the University of Maine at Orono. GHS's and PBM's research was supported by NSF grant EAR84–08164 to the University of Chicago and by a grant to GHS from the Division of Educational Programs of Argonne National Laboratory. Critical reviews of an earlier version of the manscript by N.N. Pertsev and J.W. Valley are much appreciated.

References

Bohlen, S.R., Valley, J.W. & Essene, E.J. (1985). Metamorphism in the Adirondacks. I. Petrology, pressure and temperature. *J. Petrol.*, **26**, 971–92.

Bowden, P., von Knorring, O. & Bartholemew, R.W. (1969). Sinhalite and serendibite from Tanzania. *Mineral. Mag.*, **37**, 145–6.

Coomaraswamy, A.K. (1902). The crystalline limestones of Ceylon. *Quart. J. Geol. Soc. London*, **58**, 399–422.

Deer, W.A., Howie, R.A. & Zussman, J. (1978). *Rock-forming minerals*, vol. 2A, 2nd ed., *Single-chain silicates*. London: Longman.

Evernden, J.F. & Kistler, R.W. (1970). Chronology of emplacement of Mesozoic batholithic complexes in California and western Nevada. *U.S. Geol. Survey Prof. Paper* 6923.

Greenwood, H.J. (1967). Mineral equilibria in the system $MgO-SiO_2-H_2O-CO_2$. In *Researches in geochemistry*, vol. 2, ed. P.H. Abelson, pp. 542–67. New York: Wiley.

Grew, E.S. (1982). Sapphirine, kornerupine, sillimanite, and orthopyroxene in the charnockitic region of South India. *J. Geol. Soc. India*, **23**, 469–505.

— (1983). A grandidierite–sapphirine association from India. *Mineral. Mag.*, **47**, 401–3.

Grew, E.S. (1986): Petrogenesis of kornerupine at Waldheim (Sachsen), German Democratic Republic. *Z. Geolog. Wissenschaften*, **14**, 525–58.

Grew, E.S., Hinthorne, J.R. & Marquez, N. (1986). Li, Be, B, and Sr in margarite and paragonite from Antarctica. *Amer. Mineral.*, **71**, 1129–34.

Grew, E.S., Herd, R.K. & Marquez, N. (1987). Boron-bearing kornerupine from Fiskenaesset, West Greenland: a re-examination of specimens from the type locality. *Mineral. Mag.* **51**, 695–708.

Griffin, W.L. & O'Reilly, S.Y. (1986). Mantle-derived sapphirine. *Mineral. Mag.*, **50**, 635–40.

Guggenheim, S. (1984). The brittle micas. In *Reviews in mineralogy*, vol. 13, *Micas*, ed. S.W. Bailey, pp. 61–104. Washington, D.C.: Mineral Soc. Amer.

Hammarstrom, J.M. & Zen, E-an (1986). Aluminum in hornblende: an empirical igneous geobarometer. *Amer. Mineral.*, **71**, 1297–313.

Harder, H. (1959). Beitrag zur Geochemie des Bors Teil II, Bor in Sedimenten. *Nachrichten Akad. Wissenschaften Göttingen, Math-phys. Klasse*, 1959, 123–83.

— (1970). Boron content of sediments as a tool in facies analysis. *Sedimentary Geology*, **4**, 153–75.

Heier, K.S. (1973). Geochemistry of granulite-facies rocks and problems of their origin. *Phil. Trans. Royal. Soc. London*, **A273**, 429–42.

Henderson, J.R. (1983). Structure and metamorphism of the Aphebian Penrhyn Group and its Archean basement complex in the Lyon Inlet area, Melville Peninsula, District of Franklin. *Geol. Survey Canada Bull.*, **324**, 50 pp.

Higgins, J.B., Ribbe, P.H. & Herd, R.K. (1979). Sapphirine I. Crystal chemical contributions. *Contrib. Mineral. Petrol.*, **68**, 349–56.

Holser, W.T. (1979). Mineralogy of evaporites. In *Reviews in Mineralogy*, vol. 6, *Marine minerals*, ed. R.G. Burns, pp. 211–94. Washington, D.C.: Mineral. Soc. Amer.

Hutcheon, I., Gunter, A.E. & Lecheminant, A.N. (1977). Serendibite from Penrhyn Group Marble, Melville Peninsula, District of Franklin. *Canad. Mineral.*, **15**, 108–12.

Janardhan, A.S. & Shadakshara Swamy, N. (1982). A preliminary report on the occurrence of sapphirine in an ultramafic enclave near Terankanambi, southern Karnataka. *Current Science*, **51**, 43–4.

Korzhinskiy, D.S. (1945). Zakonomernosti assotsiatsii mineralov v porodakh arkheya Vostochnoy Sibiri (Regularities in the mineral associations in Archean rocks of Eastern Siberia). *Trudy Instituta geol. nauk Akad. Nauk SSSR*, Vyp. 61.

— (1955). Ocherk metasomaticheskikh protsessov (Essay on metasomatic processes). In *Osnovniye problemy v uchenii o magmatogennykh rudnikh mestorozhdeniyakh*, 2nd ed. Moscow: Izdatelstvo Akad. Nauk SSSR.

— (1957). *Fiziko-khimicheskiye osnovy analiza paragenezisov mineralov*. Moscow: Nauka. (English translation: *Physicochemical basis of the analysis of the paragenesis of minerals*. New York: Consultants Bureau, 1959).

— (1973) *Teoreticheskiye osnovy analiza paragenezisov mineralov* (Theoretical basis for the analysis of mineral parageneses). Moscow: Nauka.

Kreiger, M.H. (1937). Geology of the Thirteenth Lake quadrangle, New York, *New York State Museum Bull.*, **308**, 124 pp.

Kulakovskiy, A.L. & Pertsev, N.N. (1986). Regeneratsiya ortosilikatov magniya i nekotoryye aspekty genezisa bogatykh zheleznykh rud Tayozhnogo mestorozhdeniya (Tsentralnyy Aldan) (Regeneration of magnesium orthosilicates and several aspects of the genesis of rich iron ores of the Tayozhnoye Deposit (Central Aldan)). *Geologiya Rudnykh Mestorozhdeniy*, 28, 36–46.

Larsen, E.S. & Schaller, W.T. (1932). Serendibite from Warren County, New York, and its paragenesis, *Amer. Mineral.*, **17**, 457–65.

Machin, M.P. & Süsse, P. (1974). Serendibite: a new member of the aenigmatite structure group. *N. Jahrb. Mineral. Monatsh.*, 1974, 435–41.

Manning, D.A.C. & Pichavant, M. (1983). The role of fluorine and boron in the generation of granitic melts. In *Migmatites, melting, and metamorphism*, ed. M.P. Atherton & C.D. Gribble, pp. 94–109. Cheshire: Shiva.

Marakushev, A.A. (1958). Petrologiya Tayozhnogo zhelezorudnogo mestorozhdeniya v Arkheye Aldanskogo shchita (Petrology of the Tayozhnoye iron-ore deposit in the Archean of the Aldan Shield). *Trudy Dalnevostochnogo filiala im. V.L. Komarova Akad. Nauk SSSR, Ser. Geol.*, vol. 5, Magadan, Far Eastern Krai, RSFSR.

McLelland, J.M. & Isachsen, Y.W. (1980). Structural synthesis of the southern and central Adirondacks: a model for the Adirondacks as a whole and plate tectonics interpretations. *Geol. Soc. Amer. Bull.*, **91** (part II), 208–92.

Moine, B., Sauvan, P. & Jarousse, J. (1981). Geochemistry of evaporite-bearing series: a tentative guide for the identification of metaevaporites. *Contrib. Mineral. Petrol.*, **76**. 401–12.

Pertsev, N.N. (1971). *Paragenezisy bornykh mineralov v magnezialnykh skarnakh* (Parageneses of boron minerals in magnesian skarns). Moscow: Nauka.

Pertsev, N.N. & Boronikhin, V.A. (1986). Singalit, turmalin, serendibit i lyudvigit v magnetitovoy rude Tayozhnogo mestorozhdeniya (Tsentralnyy Aldan) (Sinhalite, tourmaline, serendibite, and ludwigite in magnetite ore of the Tayozhnoye Deposit (Central Aldan)). In *Novyy dannyye o mineralakh, vyp. 33, Mineral. Muzey im. A. Ye. Fersmana Akad. Nauk SSSR*, pp. 143–7.

Pertsev. N.N. & Nikitina, I.V. (1959). Novyye dannyye o serendibite (New data on serendibite). *Zapiski Vsesoyuz. Mineral. Obshch.*, **88** (2), 169–72.

Prior, G.T. & Coomaraswamy, A.K. (1903). Serendibite, a new borosilicate from Ceylon. *Mineral. Mag.*, **13**, 224–7.

Puchelt, H. (1972). Barium, 56-K, abundance in common sediments and sedimentary rock types, and 56-M, abundance in common metamorphic rock types. In *Handbook of geochemistry*, vol. II/4, ed. K.M. Wedepohl, Heidelberg: Springer.

Reynolds, R.C. (1965). Geochemical behaviour of boron during the metamorphism of carbonate rocks. *Geochim. Cosmochim. Acta.*, **29**, 1101–14.

Reznitskiy, L.Z. & Vorobyov, Ye. I. (1975). Zakonomernyye mikrovklyucheniya barita v kaltsitakh flogopitovykh zhil Slyudyanki (Regular microinclusions of barite in calcites of phlogopitic veins at Slyudyanka). *Doklady Akad. Nauk SSSR*, **222**, 690–3.

Richmond, G.M. (1939). Serendibite and associated minerals from the New City Quarry, Riverside, California. *Amer. Mineral.*, **24**, 725–6.

Schaller, W.T. & Hildebrand, F.A. (1955). A second occurrence of the mineral sinhalite ($2MgO \cdot Al_2O_3 \cdot B_2O_3$). *Amer. Mineral.*, **40**, 453–7.

Shabynin, L.I. (1956). O nakhodke singalita ($2 MgO \cdot Al_2O_3 \cdot B_2O_3$) v SSSR (A find of sinhalite in the USSR). *Doklady Akad. Nauk SSSR*, **108** (2) 325–328.

Shabynin, L.I. & Pertsev, N.N. (1956). Varvikit i serendibit iz magnezialnykh skarnov yuzhnoy Yakutii (Warwickite and serendibite from magnesian skarns of southern Yakutia). *Zapiski Vsesoyuz. Mineral. Obshch.* **85**, 515–28.

Sills, J.D., Ackermand, D., Herd, R.K. & Windley, B.M. (1983). Bulk composition and mineral parageneses of sapphirine-bearing rocks along a gabbro–lherzolite contact at Finero, Ivrea Zone, N. Italy. *J. Metamorph. Geol.*, **1**, 337–51.

Spivack, A.J. & Edmond, J.M. (1987). Boron isotope exchange between seawater and oceanic crust. *Geochim. Cosmochim. Acta.*, **51**, 1033–43.

Spivack, A.J., Palmer, M.R. & Edmond, J.M. (1987). The geochemical cycle of boron isotopes. *Geochim. Cosmochim. Acta.*, **51**, 1939–49.

Swihart, G.H. (1987). Boron isotopic composition of boron minerals, and tracer applications. Ph.D. Dissertation, University of Chicago.

Swihart, G.M., Moore, P.B. & Callis, E.L. (1986). Boron isotopic composition of marine and nonmarine evaporite borates. *Geochim. Cosmochim. Acta*, **50**, 1297–301.

Thompson, J.B., Jr. (1959). Local equilibrium in metasomatic processes. In *Researches in geochemistry*, ed. P.H. Abelson, pp. 427–57. New York: Wiley.

— (1972). Oxides and sulfides in regional metamorphism of pelitic schists. *Reports, 24th International Geological Congress, Section 10*, pp. 27–35.

Truscott, M.G., Shaw, D.M. & Cramer, J.J. (1986). Boron abundance and localization in granulites and the lower continental crust. *Bull. Geol. Soc. Finland*, **58** (1), 169–77.

Valley, J.W. (1986). Stable isotope geochemistry of metamorphic rocks. In *Reviews in mineralogy*, vol. 16, *Stable Isotopes in high temperature geological processes*, J.W. Valley, H.P. Taylor, Jr. & J.R. O'Neil, pp. 445–89. Washington, D.C.: Mineral. Soc. Amer.

Valley, J.W. & Essene, E.J. (1980). Calc–silicate reactions in Adirondack marbles: the role of fluids and solid solutions. *Geol. Soc. Amer. Bull.*, **91** (part II), 720–815.

Valley, J.W. & O'Neil, J.R. (1984). Fluid heterogeneity during granulite facies metamorphism in the Adirondacks: stable isotopic evidence. *Contrib. Mineral. Petrol.*, **85**, 158–73.

van Bergen, M.J. (1980). Grandidierite from aluminous metasedimentary xenoliths within acid volcanics, a first record in Italy. *Mineral. Mag.*, **43**, 651–8.

von Knorring, O. (1967). A skarn occurrence of sinhalite from Tanzania. *Research Institute for African Geology, University of Leeds, 11th Annual Report* (1965–6), p. 40.

Whelan, J.R., Rye, R.O. & DeLorraine, W. (1984). The Balmat–Edwards zinc–lead deposits– synsedimentary ore from Mississippi Valley-type fluids. *Econ. Geol.*, **79**, 239–65.

Wiener, R.W., McLelland, J.M., Isachsen, Y.W. & Hall, L.M. (1984). Stratigraphy and structural geology of the Adirondack Mountains, New York: review and synthesis. In *The Grenville event in the Appalachians and related topics*, Geol. Soc. Amer. Spec. Paper 194, ed. M.J. Bartholomew, pp. 1–55.

11

The early history of the Adirondacks as an anorogenic magmatic complex

JAMES McLELLAND

Introduction

The Adirondack Mts (Fig. 11.1) consist of a NNE-trending, elongate dome of high-grade Proterozoic gneisses that connect with the Grenville Province of Canada via the Frontenac Arch across the St Lawrence River. The region is divided into the Adirondack Highlands and Lowlands with the boundary between the two defined by a belt of high strain referred to as the Carthage–Colton mylonite zone (CCMZ, Fig. 11.1). This zone is believed to be continuous with the Chibougamau–Gatineau line that separates the Central Metasedimentary Belt from the Central Granulite Terrane of the Grenville Province (Geraghty *et al.*, 1981). The Lowlands consist primarily of metasedimentary rocks, notably marbles, now at upper amphibolite grade; while the Highlands, which are dominated by orthogneisses with intervening synclinal keels, are predominantly at hornblende–granulite facies grade. Although metamorphic grade appears to be continuous across the CCMZ, continuity of lithic units across the zone remains uncertain and difficult to assess.

During recent years substantial progress has been made in recognizing and documenting the complex, polyphase deformation that characterizes both the Adirondack Highlands and Lowlands. These results are summarized by McLelland and Isachsen (1980, 1985, 1986) who emphasize the importance of large fold nappes in the structural framework of the region. Petrologic studies of high-grade metamorphism accompanying deformation have been summarized by Bohlen & Essene (1977) and Bohlen *et al.* (1985). Most Adirondack folding and metamorphism are believed to have occurred during the ~ 1100–1000 Ma Ottawan Orogeny of the Grenville Orogenic Cycle (Moore & Thompson, 1980), consistent with U–Pb zircon determinations (Silver, 1969; McLelland *et al.*, 1988a) as well as Rb–Sr age determinations that indicate high-temperature recrystallization and closure in the interval 1110–1000 Ma (Ashwal & Wooden, 1983).

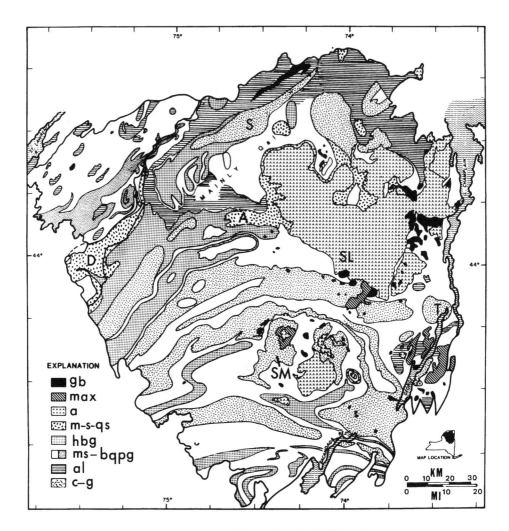

Fig. 11.1. Generalized geologic map of the Adirondack Mountains. c–g:
charnockitic and granitic gneisses. al: alaskitic and leucogranitic gneisses.
ms: undivided metasediments. bqpg: (Lowlands only) biotite–quartz–
plagioclase metapelites, migmatites and metavolcanis. hbg: hornblende
granitic gneiss. m–s–qs: mangeritic, syenitic, and quartz–syenitic gneiss
with occasional xenocrysts of blue-gray andesine. a: metanorthosite.
max: charnockitic, mangeritic, and jotunitic gneisses containing
abundant xenocrysts of blue-gray andesine plagioclase. gb: olivine
metagabbro. The Carthage–Colton mylonite zone (CCMZ) is indicated
by the heavy black line and by the Diana Complex [D] which occupies
the southern two-thirds of the zone. The CCMZ separates the
Adirondack Lowlands, on the west, from the Adirondack Highlands, on
the east. S: Stark complex, A: Arab Mountain anticline cored by acidic
AMCG suite rocks of the Tupper–Saranac complex, SM: Snowy
Mountain dome, T: Ticonderoga dome (From McLelland & Isachsen,
1986).

The purpose of this paper is to examine the nature and evolution of an important group of Highland orthogneisses that form magmatic complexes emplaced prior to the ~1100–1000 Ma dynamothermal metamorphism. These rocks consist of anorthosites, mangerites, charnockites, and granitic to alaskitic gneisses referred to by McLelland *et al.* (1988a) as the AMCG suite. Evidence is presented to show that the suite is bimodal with interaction between acid and mafic members limited to mixing in the vicinity of lithic contacts but that magmatic fractionation operated within each end-member series. Thermal modeling yields results consistent with the formation of the acidic rocks by deep crustal melting driven by gabbroic–picritic magmas parental to the anorthosites. Chemical trends are interpreted as indicating that the AMCG suite evolved under anorogenic conditions and may be related to a broad belt of mid-to-late Proterozoic anorogenic magmatism that traverses North America, Greenland, and the northern European shields (Emslie, 1978, 1985; Anderson, 1983).

Distribution and general characteristics of orthogneisses of the Adirondack Highlands

As shown in Fig. 11.1, the most widespread development of orthogneisses in the Adirondacks is in the central and northern Highlands where they comprise anorthosites, mangerites, charnockites, and hornblende granitic gneisses exposed in the Marcy massif as well as the Snowy Mt and Oregon domes (Fig. 11.1). Largely granitoid examples of the AMCG suite occur within the Diana, Stark, and Tupper–Saranac complexes (Fig. 11.1). Together with pink alaskitic facies of the hornblende granitic gneiss, this association forms a distinctive group of rocks referred to by McLelland *et al.* (1988a) as the AMCG suite, and is closely related to Anderson's (1983) 'anorogenic trinity', as well as to the 'rapakivi suite' as defined by Vorma (1971) and Emslie (1978, 1985). Barker *et al.* (1975) have discussed similar associations in the Pike's Peak batholith.

Although no single mode of field occurrence is always associated with the AMCG suite, its constituents are generally arranged in complexes which possess basic cores surrounded by acidic rocks in which silica content increases smoothly outward. As would be expected, faulting and complex multiple folding have caused interruptions and local reversals in this sequence which, nonetheless, may still be observed in major structures such as the Arab Mt and Stark anticlines, as well as the Snowy Mt and Oregon domes (Fig. 11.1). Within these anticlinal structures, anorthositic and gabbroic rocks are generally encountered in hinge regions, while alaskitic gneisses occur on the flanks, and intervening acidic types are arranged with silica content increasing outward.

The rocks of the AMCG suite are clearly of igneous origin. Igneous textures, feldspar phenocrysts, xenoliths, dikes, and crosscutting contacts (Fig. 11.2) can be recognized despite intense deformation and metamorphic recrystallization. The anorthositic and mangeritic–charnockitic rocks tend to alter to various shades of gray and brown on weathered surfaces, while on fresh surfaces both types exhibit the olive-green color that characterizes orthopyroxene-bearing granulites throughout the world. Texturally the anorthosites tend to be coarse-grained and exhibit ophitic to subophitic texture in more mafic portions. Local fine-grained facies may be the result of strain-related grain-size reduction or possibly of chilling near contacts. Mafic phases of the anorthosites generally include 20–30% pyroxene with orthopyroxene dominant. In his study of these rocks, Ashwal (1978, 1982) concluded that their late differentiates are rich in Fe, Ti, and P and include ferrodiorites, ferrogabbros, and oxide-rich pyroxenites. He also suggested that flow differentiation and fractional crystallization of a feldspathic parent magma could have given rise to the observed lithic variation. The feldspathic parent magma of the anorthosites may be characterized as a leucogabbro, itself derived through fractionation of tholeiitic magmas ponded at the crust–mantle interface (Emslie, 1978, 1985).

Mangeritic and charnockitic rocks are largely augen, or flaser, gneisses containing mesopherthitic feldspars 2.5–5.0 cm in length. Aside from the effects of high strain, these rocks are generally massive and homogeneous, although gradual compositional changes and occasional layering may be attributed to fractional crystallization (Buddington, 1939). Mafic clots and schlieren are believed to represent partially resorbed xenoliths. Visibly distinct from these rocks are pink hornblende ($\pm$ biotite) granitic gneisses that exhibit a variety of streaky zones and layers defined by variable concentrations of mafic minerals which give the rocks a migmatitic appearance. The mafics, which consist dominantly of Cl-bearing hornblende (ferrohastingite), generally account for 5–10% of the mode. Similar to these rocks in color, but more leucocratic and homogeneous in mineral content, are alaskitic gneisses which often contain only magnetite as a mafic phase and which commonly contain small quantities of fluorite. Except for the hornblende granites most of the granitoid members of the AMCG suite are anhydrous, and the more acidic varieties invariably contain hypersolvus feldspars ranging from microperthite in the mangerites to microcline perthite in the alaskites. Rapakivi feldspars are occasionally encountered, especially in Stark and Diana complexes (Buddington, 1939).

Most members of the AMCG suite can be readily distinguished from one another in the field, the exception being the mangeritic and charnockitic gneisses which are very similar in appearance. Chemically, a discontinuity appears to exist between the basic and acidic members of the suite, and this bimodality will be discussed below. According to evidence presented in this paper, differentiation, as evidenced by smooth chemical trends, appears to have taken place within, but not across, the acidic and basic members of the suite. Buddington (1972) and Ashwal (1978, 1982) have discussed the details of differentiation in the basic rocks, and trends towards low-silica/high-iron recognized by them are supported by the present study. The acidic rocks are discussed in further detail below, where it is known that a continuum exists between iron-rich, low-silica mangerites and charnockites (equivalent to orthopyroxene-bearing monzonites and granites) and evolved, high-silica granitic rocks. Olive-green mangerites and charnockites are represented by rocks that are relatively low in silica (55–65%) while pink, hornblende (ferrohastingite)-bearing granitic gneisses are comparatively silica-rich (65–76%). This correlation between modal orthopyroxene and silica content is reasonably consistent throughout the Adirondacks, and the transition between major bodies of charnockitic and granitic rocks occurs smoothly over relatively short distances of hundreds of meters. It appears that the 'green vs pink' variation of Adirondack AMCG-suite rocks is due primarily to chemical properties inherited from igneous precursors, and that the origin of charnockites within this suite is not, in general, the result of carbonic metamorphism of more hydrous parents such as proposed for the Indian examples discussed by Newton & Hansen (1983). On the other hand, McLelland et al. (1988b) have demonstrated that Adirondack charnockites may locally result from isochemical metamorphism in the presence of CO_2-rich fluids.

A minor, but petrologically significant, rock within the AMCG suite is the white-to-brown weathering fayalite granitic gneiss that occurs in two relatively small bodies, one just west of the Arab Mountain anticline and the other just beyond the northeastern margin of the Marcy massif (Fig. 11.1). These hypersolvus granites commonly contain small quantities of fluorite (Buddington, 1939; Bohlen & Essene, 1978a) reflecting the presence of halogens during their magmatic evolution.

Relative ages within the AMCG suite have been a matter of debate for years. Balk (1931) and deWaard & Romey (1969) argued that the mangeritic–charnockitic gneisses were comagmatic with associated anorthositic–gabbroic rocks and that both represent fractionates from dioritic–granodioritic parental magmas. The contemporaneity of the acidic and mafic rocks was believed to be reflected in the repeated close association of these types in complexes with mafic cores and acidic envelopes in which silica content increased smoothly outward. These investigators drew attention to the distinctive contacts between mangeritic and anorthositic members, wherein the acidic magmas clearly crosscut mafic rocks so as to produce intrusion breccias of anorthosite xenoliths within mangerite (Fig. 11.2). Furthermore, they pointed out that, in addition to xenoliths, single crystals of andesine plagioclase (Fig. 11.3) had been incorporated from the anorthosite and transported outward into the acidic magmas so that their abundance decreases away from the contact. It was argued that the widespread occurrence of these xenocrysts indicates that the anorthositic magmas were not wholly solidified when intruded by the acidic melts thus facilitating plucking off of single andesine grains from the crystal–liquid network. The xenocrysts, which consist of blue-gray andesine clouded with tiny specks of (Fe, Ti)-oxides, are typical of plagioclase in the anorthosite. They are out of equilibrium with mangeritic–charnockitic compositions, and complete resorption has been arrested by the development of oligoclase reaction rims (Fig. 11.3). Although most common near contacts, they occur within charnockites located 10–15 km from the nearest exposure of anorthosite. This observation implies that the charnockitic rocks were largely molten to these distances at the time when they incorporated the andesine xenocrysts.

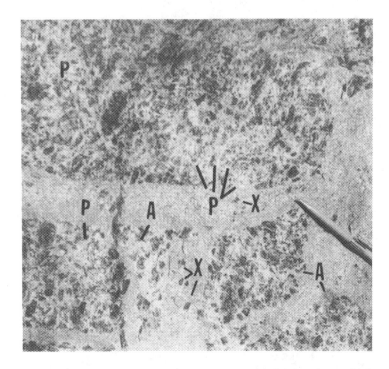

Fig. 11.2. Contact between mangeritic & jotunitic rocks (light) and anorthosite (mottled). Both xenoliths of anorthosite (A) and xenocrysts of blue-gray andesine (X) are visible. Although most contacts are sharply defined, mangeritic–jotunitic rock permeates (P) and anorthosite, filling interstices between plagioclase grains. Exposure is within a small quarry to the north of Rt. 30 four miles north of Tupper Lake Village. Pencil is 12 cm long.

Buddington (1939, 1972) presented a wide range of field and laboratory evidence against comagmatic relationships between the acidic and basic members of the AMCG suite. Based upon crosscutting relationships' he asserted that the anorthosites were emplaced first and were later intruded by the mangeritic and charnockitic rocks. He further postulated that many of the pink granitic gneisses represent a still younger suite of acidic magmas. The time intervals between these events were not specified, but the first two were believed to be temporally close and distinct from the third.

McLelland & Chiarenzelli (1990a) describe mutually crosscutting relationships between acidic and basic members of the AMCG suite and stress the magmatic contemporaneity demonstrated by these observations. One example of crosscutting relationships is provided by the dikes and anastomosing veins of mangerite that invade anorthosite and even permeate it so as to fill interstices between plagioclase grains (Fig. 11.2). The permeation indicates that the mangerites existed largely as liquids at the time of intrusion *and* that the anorthosites were not yet wholly solidified. The second set of crosscutting relationships includes sheets of gabbroic anorthosite that intrude manger-

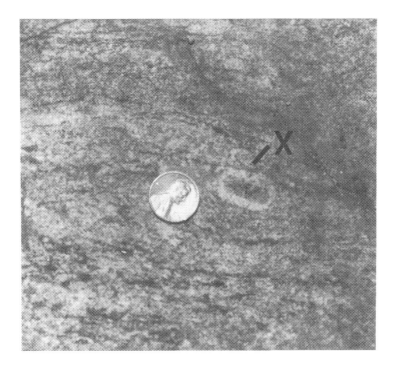

Fig. 11.3. Xenocryst (X) of blue-gray andesine, mantled by light-colored rim of oligoclase, within charnockitic gneiss. Exposure is in roadcut along Rt. 30 ten kilometers south of Tupper Lake Village and twelve kilometers from the nearest anorthosite exposure. Coin is 2 cm across.

ites and charnockites of the AMCG suite (McLelland & Chiarenzelli, 1990a). These examples of mutually crosscutting relationships demonstrate that acidic and basic members of the AMCG suite are coeval. In the following sections evidence is presented that, although coeval, these rocks are not comagmatic but constitute discrete members of bimodal magmatic complexes. Identical conclusions have been reached for a large number of similar anorthosite-bearing, mid-Proterozoic complexes (Barker *et al.*, 1975, Emslie, 1978, 1985, Morse, 1982, Hargraves, 1962, among others). This temporal and spatial association of AMCG-suite rocks, on a global scale, strongly suggests that a single, integrated sequence of events and processes is responsible for the evolution of all members of the suite.

Major-element chemistry and bimodality of AMCG-suite orthogneisses of the Adirondack Highlands

Table 11.1 summarizes major-element data for Adirondack Highland AMCG-suite rocks. Averages for various subgroups within the suite have been compiled from 78

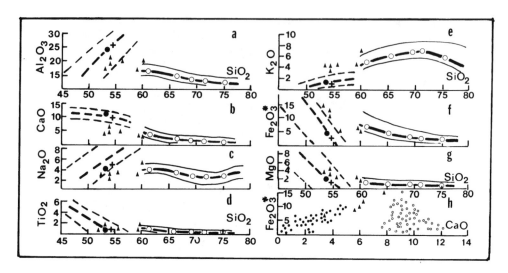

Fig. 11.4.   Harker variation diagrams (a)–(g) of basic (dashed curves) and acidic
(solid curves) members of the Adirondack Highland AMCG suite. The
heavy curves are drawn through the data averages and the narrower
curves indicate the spread of data for each suite. Large open circles (a)–
(g) refer to averages given in Table 11.1 and the filled large circle and
cross (a)–(g) refer to average values for the Whiteface and Marcy facies
of the anorthosite, respectively. The low silica rocks in the basic trend are
gabbros and norites. The filled triangles are jotunitic rocks collected from
anorthosite–mangerite contacts such as shown in Fig. 11.2. Fe_2O_3* is
total iron as ferric iron. In (h) the dark circles refer to the acidic
members, and the open circles to the basic members, of the AMCG suite.

new XRF-whole-rock analyses, as well as 47 analyses taken from the literature,
particularly Buddington (1939, 1953, 1972)*. These average values and ranges are
plotted in Harker variation diagrams shown in Fig. 11.4. As pointed out by Chayes
(1964), such diagrams cannot be used to unequivocally demonstrate liquid lines of
descent for igneous suites. In the present instance the ambiguity is greater because the
plotted samples have been collected from over a large region consisting of deformed,
granulite-facies rocks, and an undetermined portion of these could represent cumulates
rather than liquids. Nonetheless, the diagrams (Fig. 11.4) illustrate important features
of the AMCG suite. The patterns imply two distinct fractionation histories; one related
to gabbroic–anorthositic compositions and the other to mangeritic–granitic compositions. These two groups are indicated in Fig. 11.4 and their average trends plotted.
Within the gabbroic–anorthositic suite the average values for Whiteface and Marcy
facies of the anorthosite as defined by Buddington (1939), are indicated. In each
diagram, the high-silica members of the gabbroic–anorthositic trend are anorthositic
rocks, the majority of which are believed to represent plagioclase cumulates. The low-
silica members are gabbroic and noritic types.

* Printouts of all analyses may be obtained by writing to the author.

Within each variation diagram six closed triangles appear on the SiO_2-poor side of the mangeritic–granitic trend. These represent compositions of distinctive, orthopyroxene-bearing, iron-rich monzodioritic rocks found only near the contact of gabbroic–anorthositic and mangeritic–granitic rocks and referred to as jotunites (deWaard, 1969). The jotunites invariably crosscut anorthositic rocks and commonly occur as dikes within anorthosite (Fig. 11.3). They will be further discussed below.

For the purposes of the present investigation, the most important information conveyed by Fig. 11.4 is the *lack of continuity* between the gabbroic–anorthositic and monzonitic–granitic suites. This is clearly demonstrated by the K_2O and CaO vs SiO_2 plots which dramatically exhibit the discontinuity between the two suites. This discontinuity is similar to the 'Daly Gap' (Chayes, 1964) which refers to the relative paucity of andesitic rocks within many basalt–rhyolite terrains. Efforts have been made to account for this gap in a context consistent with continuous fractionation from gabbro to rhyolite (Clague, 1978); however, the mechanisms proposed do not offer convincing explanations for the widespread and common occurrence of the 'Daly Gap'. Within the AMCG suite of the Adirondack Highlands this gap is a reflection of the fact that rocks of intermediate SiO_2 content ($\sim 60\%$) are rich in K_2O and poor in CaO, i.e., dioritic rocks are missing and mangerites or charnockites occur instead. Moreover, in cases where it appears possible for the two suites to merge continuously, internal inconsistencies appear. For example, within the MgO–SiO_2 variation diagram, the two trends can be projected to overlap at about 53–55% SiO_2 and, neglecting the cumulate nature of the anorthositic rocks, a comagmatic history might be postulated. However, turning to the Fe_2O_3*– and Al_2O_3–SiO_2 variation diagrams, and applying the same reasoning, it is evident that overlap for these oxides, if it occurs, must take place within the low-SiO_2 portions of the gabbroic–anorthositic trends. Therefore the portion of the trend that matches for MgO does not match for Fe_2O_3* or Al_2O_3 and fractionation cannot lead from the basic to the acidic members of the AMCG suite.

Dioritic and tonalitic gneisses do occur within the southern Adirondacks but are far removed from anorthositic massifs and are not directly associated with other AMCG-suite rocks (McLelland & Isachsen, 1986). McLelland & Chiarenzelli (1990b) report that the tonalitic gneisses yield U–Pb zircon ages of ~ 1300 Ma that exceed the ages of AMCG-suite mangerites and charnockites by ~ 150 Ma (Table 11.3), and, therefore, the tonalitic rocks are not candidates for closure of the 'Daly Gap' within the AMCG suite. The existence of the gap is strongly supported by the large quantity of data summarized in Fig. 11.4. Further evidence of bimodality within the AMCG suite is provided by the plot of SiO_2 vs $FeO/(FeO + MgO)$ (Fig. 11.5) which shows anorthositic differentiates moving toward high-iron, low-silica values while acidic rocks move toward high-iron, high-silica values. The divergent trends shown in Fig. 11.5 were initially discerned by Buddington (1972) and Ashwal (1978). The present analysis, which includes six of Ashwal's analyses of mafic differentiates, further substantiates the divergence in, as well as the bimodality of, the AMCG suite.

Table 11.1.

	Average composition of acidic AMCG-suite members in weight %[a]						Compositions of basic AMCG-suite members[a]					Granitoid averages[b]		
	1 Jotunite	2 Mangerite	3 Quartz Mangerite	4 Charnockite	5 Granite	6 Hi-SiO_2 alaskite	Average Marcy anorthosite	Average Whiteface anorthosite	Norite	Gabbro	Ultramafic	I-type	S-type	A-type
Range of SiO_2	<60	60–63	63–68	68–70	70–73	73–76	53–55	53–54	49–52	45–50	—	53–78	63–74	70–74
No. of Samples	6	11	20	15	14	7	4	7	27	15	1	532	316	31
SiO_2	54.63	61.52	66.23	68.93	72.37	75.04	54.54	53.34	50.72	49.34	39.00	67.98	69.08	73.60
TiO_2	1.35	0.81	0.71	0.51	0.41	0.15	0.67	0.72	1.03	2.17	8.20	0.45	0.55	0.33
Al_2O_3	16.69	16.04	15.17	13.89	13.73	12.57	25.61	22.50	20.80	16.15	5.92	14.49	14.30	12.69
Fe_2O_3	3.56	1.53	1.31	0.99	1.18	0.59	1.00	1.26	2.41	1.53	6.58	1.27	0.73	0.99
FeO	7.30	5.15	3.23	3.16	2.29	1.67	1.26	4.14	7.82	12.21	18.13	2.57	3.23	1.72
MnO	0.14	0.11	0.08	0.05	0.07	0.03	0.02	0.07	0.12	0.16	0.34	0.08	0.06	0.06
MgO	1.77	0.96	0.68	0.49	0.41	0.35	1.03	2.21	3.01	4.7	10.65	1.75	1.82	0.33
CaO	5.42	3.45	2.37	1.60	0.98	0.89	9.92	10.12	8.32	9.43	9.90	3.78	2.49	1.09
Na_2O	3.75	4.01	3.98	2.84	2.91	3.22	4.53	3.70	3.02	2.48	0.80	2.95	2.20	3.34
K_2O	3.79	4.89	5.76	6.14	6.24	5.30	1.01	1.19	1.37	0.87	0.15	3.05	3.63	4.51
P_2O_5	0.50	0.31	0.15	0.15	0.13	0.11	0.09	0.13	0.27	0.32	0.07	0.11	0.13	0.09
H_2O	0.31	0.35	0.38	0.47	0.35	0.58	0.55	0.12	0.33	0.29	0.27	—	—	—
Total	99.21	99.13	100.05	99.22	101.07	100.50	100.23	99.5	99.22	99.65	100.01	98.48	98.22	98.75

Notes:
[a] Data from Buddington (1939, 1953) and this work.
[b] From White & Chappell (1983).

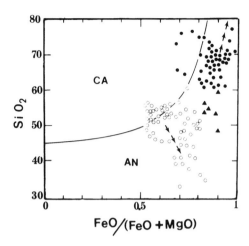

Fig. 11.5.   FeO/(FeO + MgO) vs weight percent SiO₂ for AMCG-suite rocks from the Adirondack Highlands. The curve separates fields for calcalkaline (CA) and anorogenic (AN) rocks. Filled triangles are jotunites. Arrows indicate differentiation trends discussed in text. (After Ashwal, 1978, and Anderson, 1983.)

Jotunitic rocks (Figs. 11.4, 11.5) were included by Buddington (1939, 1953) in his mafic pyroxene syenite (quartz ≤ 5%, mafics ≥ 25%). These rocks form a 1–2 km thick zone that directly overlies the Marcy anorthosite and grades upward into mangeritic and charnockitic rocks (quartz ≥ 12%, mafics ≤ 20%). Buddington (1953, pp. 78–80) reports that this belt of jotunite can be traced almost all the way around the Marcy massif and is present even where it has not yet been mapped along the remote southwestern margin of the Marcy massif from the Arab Mt anticline to Sanford Lake (A to SL, Fig. 11.1). Jotunites in this belt invariably crosscut the anorthosite resulting in xenoliths as well as xenocrysts of blue-gray andesine (Figs. 11.2, 11.3) which decrease in quantity away from the contact as discussed earlier. Buddington (1953) interpreted this belt of jotunitic rocks as due to contamination of quartz–syenite magma by incorporation of amphibolite, metagabbro, and skarn followed by fractional crystallization that yielded iron-enriched magmas and mafic cumulates near the base of the proposed chamber. It is suggested here that the jotunitic rocks mark a relatively narrow zone of variable mixing between the acidic and basic members of the bimodal AMCG suite. A similar proposal was made by Davis (1969). Previously cited evidence for intrusion and permeation suggests that mangeritic magma incorporated both solid andesine and interstitial melt still present within the anorthosites. This mixing resulted in a decrease in silica content of the jotunites (Fig. 11.4) and the hybrid rocks occupy the 'Daly Gap' referred to earlier. Mixing of acidic and mafic magmas can help to explain the anomalously high, mantle-type K/Rb ratios of jotunitic (1000–750) and mangeritic (800–600) rocks in the Adirondacks. This suggestion was made by Reynolds *et al.* (1969) who pointed out that the variation of K/Rb with percent K for AMCG-suite rocks is very similar to mixing curves derived for contamination of mangerite by

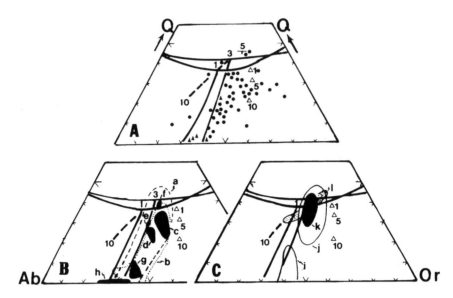

Fig. 11.6. CIPW-normative plots in the system albite–orthoclase–quartz (Ab–Or–Q). Open triangles are ternary minima for $P_{H_2O} = 0$. Dashed curve gives ternary minima for $P_{H_2O} = P_{total}$. Ternary minima at 3 and 5 are percent An for $P_{H_2O} = P_{total} = 1$ kb. P in kilobars. (A) acidic members of the Adirondack Highland AMCG suite shown as filled circles. (B): (a) east Australian anorogenic volcanics, (b) Puklen complex, (c) Pike's Peak potassic granites, (d) Pike's Peak fayalite granites, (e) Cripple Creek granite, (f) Pike's Peak peralkaline granites, (g) Pike's Peak quartz syenite, (h) Pike's Peak gabbro and syenite. (C): (i) Flowers Bay syenites, (j) Flowers Bay peralkaline granites, (k) Flowers Bay hypersolvus olivine–pyroxene granites, (l) Flowers Bay subsolvus granites. Data sources: ternary minima–Tuttle & Bowen (1958), James & Hamilton (1969), Luth (1969); (a) Ewart (1981); (b) Parsons (1972); (c)–(h) Barker et al. (1975); (i)–(l) Collerson (1982). Q = Qtz

anorthositic material (K/Rb ~ 1500). Quartz-rich charnockitic and granitic rocks fall off of these curves, suggesting that mixing is limited to magmas in proximity to anorthosite.

Evidence presented to this point has been used to establish the bimodal nature of the AMCG suite. In addition, and despite the arguments of Chayes (1964), the continuity evident in the binary variation curves (Fig. 11.4) also argues strongly for the operation of within-group fractional crystallization during the evolution of the AMCG suite. This is especially true of the acidic rocks which appear to be less dominated by cumulates than the anorthositic–gabbroic rocks. The strong positive correlation between FeO and CaO is a relationship characteristic of differentiated suites (Fig. 11.4). Fractional crystallization is also strongly suggested by the plot of Q–Or–Ab CIPW compositions shown (Fig. 11.6(A)). The acidic members of the AMCG suite plot close to the minimum trough of the Q–Or–Ab system that approximates the trend of the two-feldspar cotectic defined by James & Hamilton (1969). The continuum of acidic compositions from jotunites to granites and alaskites, and their coincidence with the liquidus trend, suggest that these rocks are related by fractional crystallization. These trends are compared with several anorogenic suites (Figs. 11.6(B), (C)) and the similarity to the Adirondack data is striking. The coincidence of the most evolved AMCG compositions with the piercing point in the projection of the system [AbOr-Q]$_{97}$An$_3$ at 1 kb $P_{H_2O} = P_{Total}$ suggests that the suite was ultimately emplaced at shallow crustal levels. Shallow emplacement has also been suggested by oxygen isotope analyses obtained by Valley & O'Neill (1982) as well as by Morrison & Valley (1987) and is consistent with the anhydrous, fluorine-bearing nature of the suite.

Depth of origin of the AMCG suite

Although the bimodality and late petrogenetic evolution of the AMCG suite are reasonably well documented, the processes resulting in the formation of these rocks are less well understood. Both neodymium (Ashwal & Wooden, 1983, Basu & Pettingill, 1983) and strontium (Heath & Fairbairn, 1969, Hills & Gast, 1964, Hills & Isachsen, 1975) isotope ratios (Table 11.3) are consistent with a depleted, upper-mantle source for the anorthositic rocks and a deep crustal source for the acidic types. While the details of anorthosite genesis are not agreed upon, many investigators believe that these rocks evolve from deep-seated, anhydrous picritic to gabbroic magmas ponded at the crust–mantle interface. These magmas differentiate into feldspathic gabbros that eventually ascend into the crust and precipitate intermediate plagioclase to form anorthositic cumulates (Emslie, 1978, 1975, Morse, 1982). This generalized model is adopted here.

The general agreement concerning crustal sources for the acidic rocks does not extend to mechanisms responsible for their generation or to the depths at which they formed. One model, proposed by Barket *et al.* (1975) and Emslie (1978, 1985), as well as others, suggests that heat from ponded mafic magmas at the crust–mantle interface partially melts overlying lower crustal rocks thus producing acidic magma. McLelland (1986) suggested that this process may yield syenitic to quartz–syenitic melts likely to form at high pressure in the anhydrous Q–Ab–Or system or in tonalitic–granodioritic source rocks in which K-feldspar is a near-solidus phase (Luth, 1969, Green, 1969, Robertson & Wyllie, 1971, Huang & Wyllie, 1975). Thermally induced crustal weakening, together with decreasing magma density, results in the eventual ascent of these magmas either as discrete plutons, or as bimodal diapirs, which evolved into the AMCG suite.

The sequence of events described above stands in contrast to a back-intrusion model proposed by Hargraves (1962) and applied to Adirondack rocks by Isachsen *et al.* (1975). These investigators assert that the mangeritic and charnockitic gneisses *pre-date* the anorthosites, which intrude them at depth, causing melting of the already hot acidic country rocks. As the anorthosite crystallizes, the still molten mangeritic and charnockitic magmas intrude back into it. This overall scenario is referred to as the *in situ* melting model, because the present location of the acidic rocks relative to the surrounding Adirondack terrain is essentially the same as existed when they first came in contact with the anorthosites. In his original proposal made for the Allard Lake area, Hargraves (1962) stressed the need for elevated P_{H_2O} in the back-intrusion model. Within the Adirondacks the evidence is that the AMCG suite has always been essentially anhydrous (Bohlen & Essene, 1978b). Many mangeritic and charnockitic rocks are essentially quartz–mesoperthite gneisses of hypersolvus nature and must have crystallized from dry, high-temperature magmas.

Both the *in situ* melting and the bimodal-diapir models may be tested against observations and quantitative constraints. As previously discussed, acidic rocks of the AMCG suite display a consistent association with anorthosites as well as a configuration about them in which enveloping jotunites grade out smoothly into mangerites, charnockites, and granitic gneiss. These recurrent relationships, both in the Adirondacks and elsewhere, can be produced by the *in situ* model only if (a) by repeated coincidence anorthosites consistently intrude into the centers of pre-existing, compositionally zoned mangeritic–charnockitic complexes, or (b) the country rocks invaded by the anorthosites were melted almost totally so as to become homogenized and then underwent fractionation into the observed configuration. The first choice is clearly *ad hoc* and is dismissed. On a qualitative basis the second alternative appears to be consistent with a wide range of data, including the occurrence of andesine xenocrysts in charnockitic rocks many kilometers away from contacts with anorthosite. In fact, one of the few differences between the *in situ* and the bimodal-diapir models is that the former relies on a specific quantity of anorthosite as the heat source for melting while the latter utilizes an unspecified quantity of gabbroic–picritic magma. As discussed below, this crucial difference may be analyzed quantitatively and the results used to choose between models.

The thermal effect of the anorthosite on the country rocks depends upon the size, initial temperature, and degree of solidification of the intrusive body, as well as on the initial temperature of the country rocks. The heat capacities, as well as the densities and heats of crystallization of AMCG-suite rocks and magmas, are sufficiently alike that treating each of these parameters as identical will not introduce significant errors into the thermal calculations undertaken below.

The anorthosite is approximated as a 4 km thick slab measuring 100×50 km, consistent with the gravity results of Simmons (1964). Models are computed for anorthosites with initial temperatures of 1300 °C and 1200 °C. For the 1300 °C case the calculations are made for a totally molten intrusion (heat of crystallization, $L = 400$ kJ kg^{-1}) as well as one initially 50% crystalline ($L = 200$ kJ kg^{-1}). For the 1200 °C initial temperature only the 50% crystalline ($L = 200$ kJ kg^{-1}) case is computed. The top of the slab is emplaced into a crustal temperature gradient of 30 °Ckm^{-1} and is initially set at four different depths from 30 to 20 km corresponding to initial crustal temperatures, T_O, of 900°, 800°, 700°, and 600 °C, respectively. The country rocks are approximated as anhydrous acidic igneous rocks similar to the hypersolvus granitoids observed in the field. In the absence of H_2O the solidi of the acidic rocks all lie close to 1000 °C over the pressure range of 5–10 kb (Luth, 1969, Green, 1969, Robertson & Wyllie, 1971, Huang & Wyllie, 1975).

Given these data, the solidification temperature at the contact of the slab and the maximum temperatures in the surrounding country rock may be computed by solving the heat-flow equations given by Turcotte & Schubert (1982, pp. 172–7) for a sill in a semi-infinite half-space. The results are summarized in Table 11.2 for the first 5 km above the top of the anorthosite slab. No results are shown beyond 5 km, because calculated temperatures in this region are below the country-rock solidus of $T \sim 1000$ °C. Rocks heated to $T_m > 1000$ °C can undergo anatexis if all, or part of, the temperature in excess of 1000 °C, ΔT_{XS}, can be transformed into latent heat of melting, ΔH_M (~ 400 kJ kg^{-1}). Quantitatively the transformation is given by $\Delta H_M = C \cdot \Delta T_{XS}$, where C (~ 1 kJ kg^{-1} °C^{-1}) is the specific heat. For every temperature exceeding 1000 °C in Table 11.2, two values of percentage melting, M, are calculated. The largest of these assumes that 100% of ΔT_{XS} is partitioned into country-rock melting, while the smaller value assumes that 50% of ΔT_{XS} is partitioned to melting and 50% to raising temperature. The latter partitioning is consistent with melting experiments (Wright & Okamura, 1977), but both percentages are given for comparison.

The results given in Table 11.2 do not favor the *in situ* melting hypothesis. Even in the most extreme favorable case (i.e., $T_i = 1300$ °C, $L = 400$ kJ kg^{-1}, $T_0 = 900$ °C), melting is restricted to the inner 3 km. At best, only 54% of the country rock melts, and this is at the contact, with far lesser percentages of melt occurring at greater distances. The severity of this flaw becomes increasingly evident in less extreme models. In addition, there is no evidence that metasedimentary rocks of the Adirondacks were ever at regional temperatures of 900 °C or pressures of 10 kb, both of which would have resulted in mineral assemblages and melting that are not observed. It is far more likely that the peak regional $P-T$ conditions were close to 7.5 kb and 750 °C, as observed. Under these conditions melting is restricted to the inner 2 km, and the maximum amount of melt produced is 35% with a more likely maximum value being 10–15%. These results are inconsistent with the presence of andesine xenocrysts in clearly igneous mangeritic and charnockitic rocks 10–15 km from the nearest anorthosite. They are also inconsistent with the absence of leucosomes, restites, and migmatites in the mangeritic and charnockitic rocks. Finally, the *in situ* model fails to account for the large quantity of almost *total* melting that would be required to produce a magmatic complex capable of fractionating into the mangeritic and charnockitic bodies of the scale and configuration observed in the Adirondacks.

The principle reasons that the *in situ* melting model fails to explain the observations, and hence the evolution of the AMCG suite, is the explicitly restricted heat source represented by the Marcy anorthosite slab, or any other anorthosite slab present in the Adirondacks. The shortcoming of a limited thermal budget is not shared by the preferred model of deep crustal melting and bimodal diapirism. In this instance large quantities, and repeated intrusions, of mafic magma at the crust–mantle interface can

Table 11.2. *Maximum country rock temperatures, T_m, and percent anatexis, M, at various initial conditions and depths of emplacement*

Initial melt conditions				Distance, Z_-, above top of slab (km) for contact					
T_i(°C)	L(kJ kg^{-1})	T_0(°C)		0	1	2	3	4	5
1300	400	900	$T_{maximum}$, $T_m =$ $f(T_i, T_0, L, Z^{-1})$	1220	1160	1100	1040	980	940
		800		1160	1080	1020	970	920	870
		700		1120	1020	960	900	850	800
		600		1020	920	860	800	750	710
			Partitioning	1.0 0.5	1.0 0.5	1.0 0.5	1.0 0.5	1.0 0.5	1.0 0.5
		900	Percent melt, $M = f(T_m - 10^3)/10^2$	54 27	40 20	24 12	12 6	No crustal melting	
		800		40 20	20 10	5 3			
		700		30 15	5 3				
		600		5 3					
1300	200	900	$T_{maximum}$, $T_m =$ $f(T_i, T_0, L, Z^{-1})$	1160	1100	1050	990	940	900
		800		1110	1050	990	940	890	840
		700		1070	990	930	870	820	780
		600		1000	900	840	780	730	700
			Partitioning	1.0 0.5	1.0 0.5	1.0 0.5	1.0 0.5	1.0 0.5	1.0 0.5
		900	Percent melt, $M = f(T_m - 10^3)/10^2$	40 20	25 12.5	12 6	No crustal melting		
		800		28 14	12 6				
		700		18 9					
1200	200	900	$T_{maximum}$, $T_m =$ $f(T_i, T_0, L, Z^{-1})$	1110	1060	1010	970	930	890
		800		1060	1000	950	900	860	820
		700		1020	940	880	840	800	770
		600		930	850	800	760	702	690
			Partitioning	1.0 0.5	1.0 0.5	1.0 0.5	1.0 0.5	1.0 0.5	1.0 0.5
		900	Percent melt, $M = f(T_m - 10^3)/10^2$	28 14	16 8	2 1	No crustal melting		
		800		16 8	2 1				
		700		5 3					

Note: T_0: initial country rock temperature.

evolve essentially unlimited heat as mafic rocks cool and fractionate into feldspathic leucogabbro. Prior to diapirism, this mechanism is able to partially melt large quantities of lower crust, with melts segregating into liquid batches while leaving restites behind. The assembled batches of acidic melt can either rise into the crust as discrete plutons or form the acidic envelopes of bimodal diapirs. The process will be enhanced by halogens, especially fluorine, present in lower crustal biotite and hornblende. The presence of halogens, and the absence of H_2O, result in diapirs able to ascend to shallow crustal levels (Valley & O'Neil, 1982, McLelland & Husain, 1986) and to develop the extreme high-SiO_2 fractionates of the AMCG suite as well as the Fe-, Ti- and P-rich late differentiates of the anorthositic rocks. In this sequence of events rock compositions, configurations, and crosscutting relationships are easily understood as part of a simple, quantitatively consistent process of deep-seated crustal melting and bimodal diapirism in response to ponded mafic magmas. As pointed out by Emslie (1978, 1985) ponding of this magma is unlikely in an environment of rapid rifting or horizontal compression. On the other hand, abortive rifting, or anorogenic conditions, will enhance ponding and the development of bimodal complexes similar to those described here.

Tectonic setting during emplacement of the Highland AMCG suite

The AMCG suite of rocks examined in this paper have a Peacock Index of 51–52 which places them at the lower end of the alkalic–calcic group (Peacock, 1931) suggestive of evolution in a non-compressional, or extensional, environment. This suggestion is examined below and found to be consistent with the evidence provided by tectonic discrimination diagrams (Pearce & Cann, 1973, Pearce *et al.*, 1984, Anderson, 1983, White & Chappell, 1983). The AFM distribution of the AMCG suite is presented in Fig. 11.7 with gabbroic–anorthositic members of the suite plotted in Fig. 11.7(A) and monzonitic–granitic members plotted in Fig. 11.7(B). Note in Fig. 11.7(C) that the gabbroic–anorthositic rocks exhibit a trend that is principally calcalkaline. However, this is deceiving since these rocks are dominantly plagioclase cumulates and cannot, therefore, represent a liquid line of descent. The gabbroic, and least differentiated, members of the suite plot near, or above, the tholeiitic–calcalkaline divide and are more representative of magmatic trends. A suite of 36 olivine metagabbros is plotted in Fig. 11.7(D), together with eight plagioclase-rich differentiates that are demonstrably descended from the metagabbros. Note that six of these cumulates fall along a calcalkaline trend, although the suite is strongly tholeiitic. We therefore conclude that the AFM trend of the gabbroic–anorthositic suite corresponds to a tholeiitic rather than a calcalkaline configuration.

Fig. 11.7(B) shows the AFM trend for the mangeritic to granitic members of the AMCG suite. The distribution of points is clearly not calcalkaline but lies close to a

J. McLelland

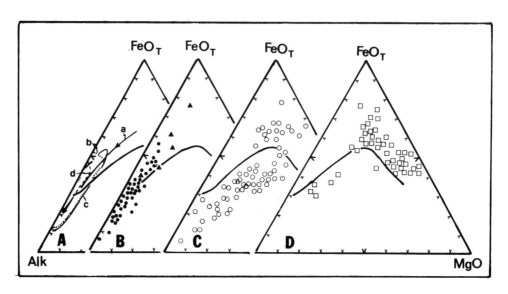

Fig. 11.7. AFM diagrams for AMC-suite and other anorogenic rocks. (A): (a) line
of liquid descent of Klokken complex, (b) dotted curve encloses
anorogenic syenites and granites from Labrador, (c) solid curve endorses
hypersolvus granites from Colorado, New England, Nigeria, and
Scandinavia, (d) dashed curve encloses Puklen complex. (B): Filled circles
are compositions of acidic members of AMCG-suite rocks from the
Adirondack Highlands; filled triangles are jotunites from anorthosite–
mangerite contact. (C): Open circles are compositions of basic members
of AMCG-suite rocks from the Adirondack Highlands. (D): Open
squares are compositions of olivine metagabbros and associated rocks
from the Adirondack Highlands. Data sources in A: (a) Parsons (1979);
(b) Collerson (1982); (c) Barker et al. (1975); (d) Parsons (1972).

highly evolved, tholeiitic trend. As shown in Fig. 11.7(A) the trends of mangeritic to
granitic rocks resemble those of classic anorogenic complexes (Barker et al., 1975) as
well as the granitic rocks of the St François Mts. (Bickford et al., 1981).

Fig. 11.8 shows total alkalies vs silica for AMCG-suite rocks. The trend from
moderately alkaline to subalkaline with increasing SiO_2 is characteristic of anorogenic
suites (Upton, 1974; Anderson, 1983). Fig. 11.9 shows A/CNK (molecular Al_2O_3/
$(Na_2O + K_2O + CaO)$) versus silica for which the mangeritic to granitic rocks vary
from peralkaline through metaluminous to mildly peraluminous. The dashed curves in
Figs. 11.8 and 11.9 enclose trends for well established proterozoic anorogenic, bimodal
complexes compiled by Anderson (1983), and both figures reflect the discontinuity
between acidic and basic members of the AMCG suite. It has been suggested that the
peraluminous aspect of some high-silica granitic rocks may be the result of relatively
high fluorine contents with the magmas (Christiansen et al., 1983). According to this
model fluorine complexes aluminum and transports it upward into the top portion of
magma chambers. Increased concentrations of fluorine will also lower the viscosity and
solidus temperature of the magma thus facilitating further differentiation evident

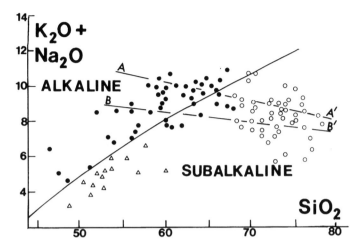

Fig. 11.8.   Total alkalies vs weight percent SiO₂ for AMCG-suite rocks in the
            Adirondack Highlands. Open triangles – anorthositic rocks; closed circles
            – mangeritic and charnockitic rocks; open circles – granitic and alaskitic
            rocks. The lines AA′ and BB′ define the trends for anorogenic complexes
            in the Ragunda massif, the St François Mountains, and the Wolf River
            batholith (Anderson, 1983). (After McLelland, 1986).

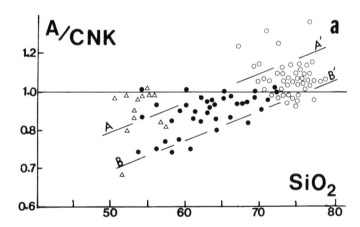

Fig. 11.9.   Molecular Al₂O₃(A) divided by the sum of molecular CaO(C), Na₂O(N),
            and K₂O(K) vs SiO₂ content in wt %. Symbols and lines AA′, BB′ as in
            Fig. 11.8. (After McLelland, 1986).

within the high-silica AMCG granitic gneisses. The presence of small quantities of fluorite within high-silica granites and alaskites lends support to this model (Buddington, 1939, Postel, 1952, Bohlen & Essene, 1978a). The development of such peraluminous granitic rocks appears to be characteristic of anorogenic magmatism.

Anderson (1983) presented plots of FeO/(FeO + MgO) vs SiO₂ for several series of rocks and, on the basis of these plots, was able to subdivide the data into calcalkaline and tholeiitic fields, with anorogenic rocks plotting in the latter. Fig. 11.5 shows

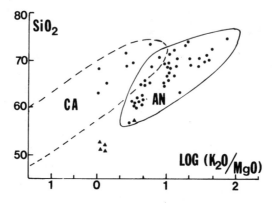

Fig. 11.10. Plot of log (K_2O/MgO) ratio vs SiO_2 content in weight percent for AMCG-suite rocks of the Adirondack Highlands. Filled triangles refer to jotunites. The field of A-type granites is indicated by AN and calcalkaline granites by CA (Rogers & Greenberg, 1981).

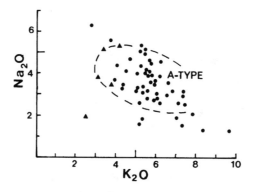

Fig. 11.11. Na_2O vs K_2O in weight percent for AMCG-suite rocks from the Adirondack Highlands. Filled triangles refer to jotunites. The field of A-type granites is outlined by a dashed curve (White & Chappell, 1983).

Anderson's subdivision together with data for the Adirondack AMCG suite. Clearly the rocks show a strong preference for the anorogenic field and this is true for gabbroic–anorthositic types as well as mangeritic–granitic varieties. The discontinuity between the acidic and basic trends has already been discussed. The strong iron enrichment indicated in Fig. 11.5 is typically developed in water-poor, halogen-bearing, bimodal, anorogenic magmatism.

Fig. 11.10 shows the variation of log (K_2O/MgO) vs SiO_2 for AMCG-suite monzonitic–granitic rocks (Rogers & Greenberg, 1981). This plot is subdivided into anorogenic (alkalic) and orogenic (calcalkalic) fields with the AMCG-suite rocks falling within the anorogenic region of the diagram.

The granitic members of the AMCG suite closely resemble A-type granitoids (White & Chappell, 1983; Loiselle & Wones, 1979). They are high in silica and alkalies; marginally corundum-normative; low in lime and magnesia; and often contain fluorite.

Fig. 11.11 represents a plot of Na_2O vs K_2O for AMCG-suite mangeritic–granitic rocks. They almost all fall within the field of A-type granitoids (White & Chappell, 1983). Table 11.1 compares the whole-rock compositions of I, A, and S granites with AMCG-suite averages. The similarity between A-granitoids and high-silica members of the AMCG suite is evident, and the data available is consistent with an anorogenic I-type source as was also concluded by Basu & Pettingill (1983) on the basis of Nd-isotopic ratios.

This section has presented evidence based upon major-element whole-rock analysis that the AMCG-suite of the Adirondack Highlands bears chemical signatures characteristic of anorogenic, or aborted-rift, environments described by Barker *et al.* (1975), Emslie (1978), Anderson (1983), and White & Chappell (1983). Whitney (1983, 1986) has reached similar conclusions for granitoids throughout the Adirondacks. This evidence is consistent with McLelland's (1986) suggestion that the Adirondacks experienced an early history of bimodal magmatism characterized by gabbroic–anorthositic magmatic cores enveloped by syenitic to granitic melts. As shown in the preceding section the acidic melts are probably generated in deep crustal rocks heated by ponded mafic magmas. An anorogenic setting is consistent with this model, because it maximizes long term residence of the ponded magmas thus enhancing crustal melting and mafic differentiation. In contrast, active rifting tends to provide magmatic access to the surface, and horizontal compression will tend to mix or displace magmas. In these instances, crustal melting is inhibited and bimodal complexes are disrupted.

Age and regional correlation of Adirondack AMCG-suite magmatism

Until recently the emplacement age of the Adirondack AMCG suite has been controversial due to differences in absolute ages (Table 11.3) and their interpretation. While it is generally agreed the Rb/Sr whole-rock isochron ages of 1110–1000 Ma represent metamorphic resetting, there is no consensus on the interpretation of Sm–Nd isochron ages. Recent acquisition of over 30 U–Pb zircon ages by McLelland *et al.* (1988a) has helped to resolve this uncertainty.

Silver (1969) and McLelland *et al.* (1988a) report U–Pb zircon ages of ~ 1070–1000 Ma for anorthositic rocks and associated (Fe, Ti)-rich gabbros (Table 11.3). On the basis of rock fabric and zircon morphology, these small, sparse zircons have been interpreted as metamorphic. In contrast, the large, elongated, and doubly terminated zircons of the mangerite–charnockite gneisses yield ages of 1160–1130 Ma (McLelland *et al.*, 1988a; Grant *et al.*, 1986) and 1113 ± 10 Ma (Silver, 1969) which are interpreted to be the time of emplacement of these rocks (Table 11.3). By reason of mutually crosscutting relationships, the anorthositic rocks were emplaced at the same time (McLelland & Chiarenzelli, 1990a).

Shown in Table 11.3 are Sm–Nd whole-rock and mineral isochron ages recently

Table 11.3. *Summary of geochronology for the Adirondack Highlands*

Method	Rock type	Locality	Age (Ma)	Comment/interpretation	Reference
(1) Rb–Sr whole-rock	Mangerite	Snowy Mt Dome	1174 ± 14	$I_{Sr} = 0.7045$	Hills & Isachsen (1975)
(2) Rb–Sr whole-rock	Hornblende granitic gneiss	Lake George Pluton	1075 ± 72	$I_{Sr} = 0.7098$	Hills & Gast (1964)
(4) U–Pb zircon	Charnockite	Ticonderoga Dome	113 ± 10	Emplacement	Silver (1969)
(5) U–Pb zircon	Syenite dike	Marcy Massif	1084 ± 15	Metamorphic	Silver (1969)
(6) U–Pb zircon	Anorthosite pegmatite	Marcy Massif	1064 ± 15	Metamorphic	Silver (1969)
(7) U–Pb zircon	Metanorite	Snowy Mt Dome	1064 ± 15	Metamorphic	Silver (1969)
(8) U–Pb zircon	Ti-magnetite mixed ore	Sanford Lake	1005 ± 10	Metamorphic	Silver (1969)
(9) U–Pb zircon	Fayalite granite	Ausable	1089 ± 26	Emplacement	McLelland et al. (1988a)
(10) U–Pb zircon	Mangerite–charnockite	Highlands	1160–1125	Emplacement (8 sites)	McLelland et al. (1988a)
(11) U–Pb zircon	Mangerite	Diana Complex	1154 ± 4	Emplacement	Grant et al. (1986)
(12) U–Pb zircon	Anorthosite	Marcy Massif	1050 ± 20	Metamorphic	McLelland et al. (1988a)
(13) U–Pb zircon	Oxide-rich gabbro	Marcy Massif	1000 ± 20	Metamorphic	McLelland et al. (1988a)
(14) U–Pb zircon	Metatonalite	Eastern Adirondacks	1327 ± 37	Emplacement	McLelland & Chiarenzelli (1989b)
(15) Sm–Nd	Anorthosite	Marcy Massif	1190 ± 130	Emplacement at $\geq 1288 \pm 36$	Ashwal & Wooden (1983)
(16) Sm–Nd	Anorthosite–charnockite	Snowy Mt Dome	1095 ± 7	Emplacement-metamorphic	Basu & Pettingill (1983) Ashwal & Wooden (1983)

obtained by Basu & Pettingill (1983) and Ashwal & Wooden (1983). Basu & Pettingill (1983) interpret their ages of 1095 ± 7 Ma as the time of anorthosite emplacement, but as pointed out by Ashwal & Wooden (1983) the slope of their isochron is determined by garnet and, therefore, gives a metamorphic age. Ashwal & Wooden (1983) obtained a number of whole-rock and mineral Sm–Nd isochrons which yielded an emplacement window of 1190 ± 130 Ma for the Marcy massif anorthosite. They interpreted their oldest, and most pristine, four-point isochron age of 1288 ± 36 Ma as the minimum age of emplacement of the anorthosite. These authors argue that all younger U–Pb zircon ages in the Adirondacks have been reset by the granulite-facies metamorphism. However, as discussed by Silver (1969) and McLelland et al. (1988a), resetting of these zircons is extremely unlikely, and it seems more probable that the Sm–Nd system has been disturbed by the 1100–1000 Ma granulite-facies event.

McLelland & Chiarenzelli (1990a) document the emplacement age of the Marcy massif within the range 1135–1113 Ma. The entire AMCG suite is believed to have been emplaced within the interval 1160–1130 Ma. This interval is similar to ages of 1160–1120 Ma reported by Emslie & Hunt (1989) for the Morin and Lac St Jean anorthosite massifs in Quebec but is younger than the 1450 and 1300 Ma ages reported for the Harp Lake and Nain Complexes in eastern Labrador (Krogh & Davis, 1968, 1973, Simmons et al., 1986). The Adirondack ages are similar to those from the Flowers River Complex, Labrador (Collerson, 1982), the Gardar complex, Greenland (Blaxland et al., 1978, Upton, 1974), and the South Rogaland complex, Norway (Demaiffe & Michot, 1985). As pointed out by Herz (1969) and Bridgewater & Windley (1973) these complexes, together with those of the Grenville Province, form an apparently continuous belt across a Laurasia-like, mid-Proterozoic supercontinent. This belt appears to continue to the west through the ~ 1485 Ma Wolf River batholith (Anderson & Cullers, 1978), the ~ 1460 Ma St François Mts (Bickford et al., 1981), and into the 1480–1340 Ma granite–rhyolite belt of the mid-continent (Silver et al., 1977). Other well known exposures within this belt occur in the ~ 1050 Ma Llano Uplift, Texas (Garrison et al., 1979, Mosher, 1988), the 1030 ± 13 Ma Pike's Peak complex, Colorado (Hedge, 1970), and the ~ 1400 Ma Laramie complex, Wyoming (Peterman & Hedge, 1968).

The 1000–1500 Ma belt outlined above is characterized by anorogenic magmatic activity (Van Schmus et al., 1987, Silver et al., 1977, Emslie, 1978, 1985, Anderson, 1983) closely resembling that described in the Adirondack Highlands and other Grenville anorthosite–mangerite–charnockite complexes. Causes underlying the origin of the belt remain obscure, but its existence is a matter of observation. The best analogue for AMCG-suite magmatism is the 1600 km $\times$ 200 km belt of anorogenic magmatism in Niger and Nigeria (Windley, 1989). In the northern section there exist ring complexes of syenitic and granitic rocks which envelop anorthositic plutons coring caldera complexes up to 60 km across (Bowden & Turner, 1974, Husch & Moreau, 1982). Magmatic activity in this belt lasted from ~ 490 Ma to ~ 140 Ma (Husch & Moreau, 1982), a span of ~ 350 Ma, similar to the 300–400 Ma magmatic interval that apparently existed over an anorogenic belt extending from Nain, Labrador, to the

Adirondack Mts during mid-Proterozoic time. The development of the mid-Protero-zoic belt was diachronous (Bickford & Van Schmus, 1986) so that several magmatic provinces of the Niger–Nigeria type may have contributed to the continental-scale magmatic feature seen today.

Summary

Whole-rock chemical data, together with field observations, lead to the following conclusions for the AMCG suite of the Adirondack Highlands:

1. The suite is bimodal with complexes consisting of basic rocks in the core surrounded by an envelope of acidic rocks;
2. The acidic envelope appears to have undergone differentiation whose products range from mafic mangerites to high-iron, high-silica granites that occasionally contain rapakivi-textured feldspars;
3. Basic rocks evolve toward anorthosite cumulates and ultramafic differentiates rich in Fe, Ti, P and poor in silica;
4. Intrusion of anorthosite by mangerite and mangerite by late ultramafics indicates the members of the AMCG suite are coeval, as is also indicated by andesine xenocrysts in acidic rocks and by permeation of anorthosite by mangerite;
5. Acidic members of the AMCG suite cannot be produced by *in situ* anatexis but are probably the product of deep crustal melting caused by mafic magmas ponded at the crust–mantle interface; and
6. Geochemical signatures of AMCG-suite rocks indicate emplacement in an anorogenic tectonic environment similar to that associated with a mid-Proterozoic magmatic belt extending from western North America through Labrador and Greenland, and into northern Europe.

Acknowledgements   Much of the research described in this paper has been supported by NSF research grant EAR-8515739, by the author's tenure on the United States Geological Survey Glens Falls 2° sheet CUSMAP project from 1983 to 1985, and by the Colgate University Research Council. These sources are gratefully acknowledged.

Appendix: Reactions in metagabbros of the Adirondacks

Adirondacks olivine metagabbros exhibit extremely well developed coronal assemblages. The most common and spectacular of these coronas form between olivine and plagioclase and are shown in Figs. 11.12–11.17. Mass-balanced reactions accounting for coronite assemblages were developed by Whitney & McLelland (1973) and McLelland & Whitney (1980a, b). Johnson & Essene (1982) provided thermodynamic

Fig. 11.12.   Olivine metagabbro with preserved subophitic texture. Spinel clouding
of plagioclase laths visually enhances zoning and twinning. Clear areas
consist of plagioclase with An < 32. Dark, ilmenite-clouded
clinopyroxenes fill wedge shaped interstices at upper left, lower right,
and top center. An opaque grain of ilmenite occurs at right. Centers of
clear interstices consist of sodic plagioclase, clinopyroxene, and apatite.
Plane light, bar scale = 0.25 mm.

phase equilibria showing that garnet-bearing coronite assemblages formed at $P–T$
conditions of 7.5 ± 1 kb and 750 ± 50 °C, consistent with granulite-facies metamorphism. Despite this high-grade metamorphism and regional isoclinal folding, the
interiors of metagabbro bodies preserve delicate igneous textures such as those shown
in Fig 11.12. The preservation is due to the anhydrous nature of these rocks as well as
their great strength and rigidity. The development of coronal assemblages within
metagabbros is, in part, the result of mobile and immobile species and is, in this respect,
reminiscent of Korzhinskii's (1959) treatment of such components.

As is commonly the case with high-grade coronites, plagioclase tends to be clouded
with minute rods of hercynitic spinel whose concentration outlines twinning and
zoning (Fig. 11.12). As noted by McLelland & Whitney (1980b), spinel clouding is
absent from plagioclase with anorthite content less than An_{32}. Within the clouded
regions of plagioclase, anorthite content varies between An_{50} and An_{30}, although
normative plagioclase has an anorthite content of close to An_{65}. These results are
interpreted by McLelland & Whitney (1980b) as the result of chemical interchanges
that take place during metamorphic reactions that produce the corona assemblages.
The partial reaction responsible for the clouding is as follows (Whitney, 1972):

Fig. 11.13.   Olivine metagabbro with two-pyroxene–spinel corona developed at right. The upper two-thirds of the olivine core is black due to oxidation; the lower third is bright with high relief. The olivine is surrounded by a gray shell of orthopyroxene some of which (right hand side) is igneous in origin. Dark, symplectites of clinopyroxene and spinel embay into plagioclase lightly dusted with spinel rods. At the bottom center a hornblende forms a corona around ilmenite. Subophitic pyroxene appears to right. Plane light, bar scale = 0.25 mm.

$$2Na^{+1} + 2(Fe, Mg)^{+2} + 3\ CaAl_2Si_2O_8 =$$

Albite   Spinel

$$2NaAlSi_3O_8 + 2(Mg, Fe)Al_2O_4 + 3\ Ca^{+2}. \tag{1}$$

This reaction is written on the basis of aluminum immobility together with mobility of the other metallic species (arbitrarily chosen as ionic). Partial reaction (1) balances and closes a series of several partial reactions developed by Whitney & McLelland (1973) and McLelland & Whitney (1980a, b) to account for coronas in terms of a system that is chemically closed over the scale of several centimeters. Thus spinel clouding of anorthite-depleted plagioclase is considered to be a necessary consequence of corona formation in olivine-bearing rocks during high-grade metamorphism. Evidently plagioclases with anorthite contents less than An_{32} are too sodic to participate in reaction (1) and remain as clear, primary oligoclases and albite generally situated in outermost zones and late interstices (Fig. 11.12).

Fig. 11.13 shows the development of a spinel–clinopyroxene–orthopyroxene corona

Fig. 11.14. Close-up of clinopyroxene–spinel symplectite embaying into dark,
spinel-clouded plagioclase. Note the radial orientation of small domains
within the symplectite. Primary clinopyroxene is in sharp contrast with
the symplectite, the more curved contacts are with orthopyroxene which
gives way to olivine near the top of the picture. Plane light, bar
scale = 0.125 mm.

around a partially oxidized olivine core and its orthopyroxene shell. A close-up of a similar symplectite is shown in Fig. 11.14. The partial development of the symplectite in Fig. 11.13 is consistent with the fact that some of the orthopyroxene shell is igneous in origin. Igneous and metamorphic orthopyroxene in coronas can be distinguished texturally, with the latter exhibiting a radial configuration about olivine cores (Fig. 11.14). The production of metamorphic orthopyroxene arises from the breakdown of olivine which releases excess $(Mg, Fe)^{+2}$ into neighboring plagioclase where it reacts to form symplectites of clinopyroxene ($\pm$ orthopyroxene) and spinel. Still unreacted $(Mg, Fe)^{+2}$, together with Na^{+1} from reacting plagioclase, migrates outward into surrounding plagioclase and reacts with the anorthite molecules to form albite and spinel as in reaction (1). Ca^{+2} released in reaction (1) migrates to evolving symplectites where it enters into clinopyroxene. Iron oxides (magnetite, ilmenite) often participates in this reaction which may be summarized as

$$\text{Anorthite} \qquad \text{Olivine}$$
$$CaAl_2Si_2O_8 + 2k(Mg, Fe)_2 SiO_4 + 2(1-k)\text{Fe-oxide}$$

$$\text{Spinel} \qquad \text{Clinopyroxene} \qquad \text{Orthopyroxene}$$
$$= (Mg, Fe)Al_2O_4 + Ca(Mg, Fe)Si_2O_6 + 2k(Mg, Fe)SiO_3. \qquad (2)$$

313

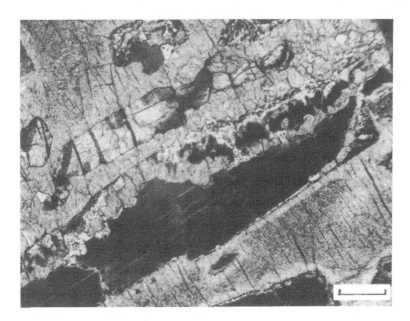

Fig. 11.15. Large, primary ilmenite-clouded clinopyroxene containing several olivine grams including one that is elongate from lower left to upper right. Below the olivine is a spinel-clouded plagioclase embayed by garnet. Remnants of dark, still unreacted clinopyroxene–spinel symplectite may be seen in the garnet thus implying that the symplectite-forming reaction preceded garnet formation. The absence of any symplectite from this sample is consistent with garnet formation directly from olivine and plagioclase. Plane light, bar scale = 0.25 mm.

The quantity k varies between 0 and 1 depending on the relative amounts of olivine and Fe-oxide. Detailed probe analyses of phases indicates that reaction (2) is consistent with mineral compositions in spinel–two-pyroxene coronas of the Adirondacks.

Upon entering the stability field of garnet, two-pyroxene–spinel symplectites become unstable and, together with olivine, plagioclase, and Fe-oxides, react to form the coronal assemblage garnet–clinopyroxene. Reaction (1) remains an integral part of this system (McLelland & Whitney, 1980a,b) and results in further spinel clouding of plagioclase. It is for this reason that garnet-bearing coronas (Figs. 11.14–11.17) contain more heavily clouded plagioclase than the two-pyroxene–spinel coronites (Fig. 11.13). It is important to realize that garnet coronas can form directly from plagioclase and olivine without going through an initial two-pyroxene–spinel phase. The net reactions turn out to be the same regardless of path (McLelland & Whitney, 1980a).

By treating aluminum as an immobile component, Whitney & McLelland (1980a, b) were able to generate a generalized garnet-producing reaction which holds for all plagioclase compositions (y = number of moles of anorthite per two moles albite), all olivine to Fe-oxide ratios (k), and all ranges of silicon mobility, w (perfectly mobile, $w = 0$, immobile, $w = 2$):

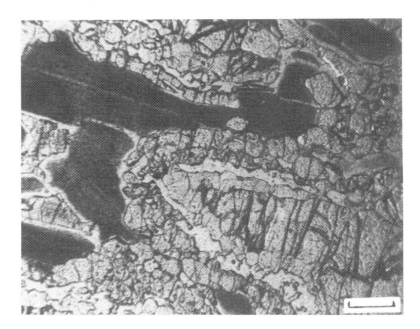

Fig. 11.16.   In the lower right a shell of garnet embays into spinel-clouded
plagioclase and is separated by a moat of clear, sodic plagioclase from
an inner core of olivine (high relief) and pyroxene (gray). A second
corona is partially seen at center top and a third barely enters the field
of view at bottom left. Corona pyroxene consists of inner
orthopyroxene and outer clinopyroxene. Plane light, bar scale = 0.25
mm.

$$\text{Anorthite} \qquad\qquad \text{Olivine}$$
$$4(y+1+w)\text{CaAl}_2\text{Si}_2\text{O}_8 + 4k(y+1+2w)(\text{Mg, Fe})_2\text{SiO}_4 + 4(1-k)(y+1+2w)\text{Fe-oxide}$$
$$\text{Orthopyroxene}$$
$$+ (8(y+1) - 4k(y+1+2w))(\text{Mg, Fe})\text{SiO}_3$$
$$\text{Garnet} \qquad\qquad \text{Clinopyroxene}$$
$$= 2(y+1)\text{Ca}(\text{Mg, Fe})_5\text{Al}_4\text{Si}_6\text{O}_{24} + 2(y+1+2w)\text{Ca}(\text{Mg, Fe})\text{Si}_2\text{O}_6$$
$$\text{Spinel}$$
$$+ 4w(\text{Mg, Fe})\text{Al}_2\text{O}_4. \tag{3}$$

The most significant and common form of reaction (3) is that for which $k = 1$, $y = 1$, and $w > 0$. In this case, we have

$$(2+w) \text{ Anorthite} + (2+2w) \text{ Olivine} + (2-2w) \text{ Orthopyroxene} = \text{Garnet} + (w+1)$$
$$\text{Clinopyroxene} + w \text{ Spinel}. \tag{4}$$

In the event that silicon is completely immobile, $w = 2$, and reaction (4) reduces to

$$4 \text{ Anorthite} + 6 \text{ Olivine} = 2 \text{ Orthopyroxene} + 3 \text{ Clinopyroxene} +$$
$$\text{Garnet} + 2 \text{ Spinel}. \tag{5}$$

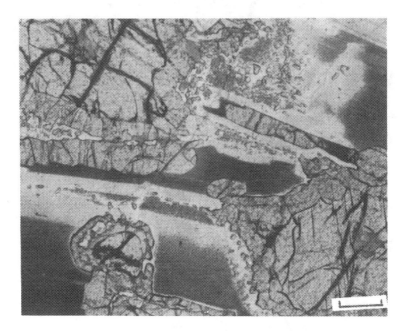

Fig. 11.17. Partial coronas of garnet (high relief, gray) around cores of olivine (high relief) and pyroxene (gray). The garnet forms only in plagioclase with anorthite content greater than An_{30} and is therefore restricted almost entirely to spinel-clouded plagioclase. Small, droplet-like grains are late clinopyroxene and apatite in sodic plagioclase. Corona pyroxene consists of inner orthopyroxene and outer clinopyroxene. Plane light, bar scale = 0.25 mm.

Reaction (5) probably accounts for most garnet coronas in olivine metagabbros, since silicon mobility is necessarily limited within olivine- and spinel-bearing rocks. Probe analyses of coronitic phases indicate that reaction (5) is consistent with assemblages in olivine metagabbros of the Adirondacks.

As pointed out by McLelland & Whitney (1980b), garnet within Adirondack coronites is not observed to grow in plagioclase with anorthite contents less than approximately An_{30}. This is believed to be the explanation for moats of clear plagioclase developed *between* garnet and pyroxene in many coronas (Figs. 11.16, 11.17). The clear plagioclase is believed to be primary and is optically continuous with plagioclase outside of the garnet shells (McLelland & Whitney, 1980b).

Several other corona types develop within olivine metagabbros. The most frequently encountered of these involves kaersutitic hornblende enclosing ilmenite (Fig. 11.13). These amphibole-bearing coronas appear to interact with anhydrous assemblages and are discussed by Whitney & McLelland (1983).

All of the foregoing reactions are characteristic of granulite-facies conditions and are believed to have formed during prograde metamorphism (Johnson & Essene, 1982).

References

Anderson, J. (1983). Proterozoic anorogenic granitic plutonism of North America. In Medaris (1983), pp. 133–54.

Anderson, J. & Cullers, R. (1978). Geochemistry and evolution of the Wolf River batholith, a late Precambrian rapakivi massif in North Wisconsin, U.S.A. *Precambrian Res.*, 7, 287–324.

Ashwal, L. (1978). Petrogenesis of massif-type anorthosites: crystallization history and liquid line of descent in the Adirondack and Morin Complexes. Ph.D. thesis, Princeton Univ.

— (1982). Mineralogy of mafic and Fe–Ti oxide-rich differentiates of the Marcy anorthosite massif, Adirondacks, N.Y. *Am. Mineral.*, 67, 14–27.

Ashwal, L. & Wooden, J. (1983). Sr and Nd isotope geochronology, geologic history, and origin of Adirondack anorthosite. *Geochim. Cosmochim. Acta.*, 47, 1875–87.

Balk, R. (1931). Structural geology of the Adirondack anorthosite. *Mineral. Petrol. Mitt.*, 41, 308–434.

Barker, F., Wones, D., Sharp, W. & Desborough, G. (1975). The Pikes Peak batholith, Colorado Front range, and a model for the origin of the gabbro–anorthosite–syenite–potassic granite suite. *Precambrian Res.*, 2, 97–160.

Basu, A.R. & Pettingill, H.S. (1983). Origin and age of Adirondack anorthosites re-evaluated with Nd isotopes. *Geol.*, 11, 514–18.

Bickford, M., Sides, J. & Cullers, R. (1981). Chemical evolution of magmas in the Proterozoic of the St. François Mts., S.E. Missouri, 1. Field, petrographic, and major element data. *J. Geophys. Res.*, 86, 10365–6.

Bickford, M. & Van Schmus, R. (1986). Resetting of whole rock and mineral Rb–Sr ages by subsequent Proterozoic orogenies. *Geol. Soc. Am. Abstracts with Programs*, 17, 523.

Blaxland, A., van Breemen, O., Emeléus, H. & Anderson, J. (1978). Age and origin of the major syenite centers in the Gardar province of south Greenland: Rb–Sr studies. *Geol. Soc. Am. Bull.*, 89, 231–44.

Bohlen, S. & Essene, E. (1977). Feldspar and oxide thermometry in the Adirondack Highlands. *Contrib. Mineral. Petrol.*, 62, 153–69.

—— (1978a). The significance of metamorphic fluorite in the Adirondacks. *Geochim. Cosmochim. Acta*, 42, 1669–78.

—— (1978b). Igneous pyroxenes from metamorphosed anorthosite massifs. *Contrib. Mineral. Petrol.*, 65, 433–42.

Bohlen, S., Valley, J. & Essene, E. (1985). Metamorphism in the Adirondacks I. Petrology, pressure, and temperature. *J. Petrol.*, 26, 971–92.

Bowden, P. & Turner, D. (1974). Peralkaline and associated ring-complexes in the Nigeria–Niger Province, West Africa. In Sorenson (1974), pp. 330–51.

Bridgewater, D. & Windley, B. (1973). Anorthosites, post-orogenic granites, acid volcanic rocks, and crustal developments in the North Atlantic Shield during the mid-Proterozoic. In *Symposium on granites and related rocks*, ed. L. Lister, pp. 307–18. Geol. Soc. S. Africa Spec. Publ. 3.

Buddington, A.F. (1939). *Adirondack igneous rocks and their metamorphism*. Geol. Soc. Am. Mem. 7.

— (1953). *Geology of the Saranac Lake 15' Quadrangle*. New York State Museum Bull. 346.

— (1969). Adirondack anorthositic series. In Isachsen (1969), pp. 215–32.

— (1972). Differentiation trends and parental magmas for anorthosite and quartz mangerite series. In *Studies in Earth and space sciences*, ed. R. Shagam *et al.*, p. 477–88. Geol. Soc. Am. Mem. 132.

Carl, J. & Van Diver, B. (1975). Precambrian Grenville alaskite bodies as ash-flow tuffs, northwest Adirondacks, N.Y. *Geol. Soc. Am. Bull.*, **86**, 1691–707.

Chappell, B. & White, A. (1974). Two contrasting granite types. *Pacific Geol.*, **8**, 173–4.

Chayes, F. (1964). A petrographic distinction between Cenozoic volcanics in and around the open ocean. *J. Geophys. Res.*, **69**, 1573–88.

Christiansen, E., Burt, D., Sheridan, N. & Wilson, R. (1983). The petrogenesis of topaz rhyolites from the western United States. *Contrib. Mineral. Petrol.*, **81**, 126–47.

Clague, D. (1978). The ocean basalt-trachyte association: an explanation of the Daly Gap. J. *Geol.*, **86**, 739–43.

Collerson, K. (1982). Geochemistry and Rb–Sr geochronology of associated proterozoic peralkaline and sub-alkaline gronites from Labrador. *Contrib. Mineral. Petrol.*, **81**, 126–47.

Davis, B. (1969). Anorthosite and quartz syenitic series of the St. Regis quadrangle, New York. In Isachsen (1969) pp. 281–8.

Demaiffe, D. & Michot, J. (1985). Isotope geochronology of the Proterozoic crustal segment of southern Norway: a review. In Tobi & Touret (1985), pp. 411–35.

deWaard, D.. (1969). The anorthosite problem: the problem of the anorthosite–charnockite suite of rocks. In Isachsen (1969), pp. 71–91.

deWaard, D. & Romey, W. (1969). Petrogenetic relationships in the anorthosite–charnockite series of the Snowy Mt. Dome, south-central Adirondacks. In Isachsen (1969), pp. 307–15.

Emslie, R. (1978). Anorthosite massifs, rapakivi granites, and late Precambrian rifting of North America. *Precambrian Res.*, **7**, 61–98.

Emslie, R. (1985). Proterozoic anorthosite massifs. In Tobi & Touret (1985), pp. 39–61.

Emslie, R. & Hunt, P. (1989). The Grenvillian event: magmatism and high grade metamorphism. *Current Res. Part C., Geol. Surv. Can. Pap.*, **89–1C**, 11–17.

Ewart, A. (1981). The mineralogy and chemistry of the anorogenic Tertiary silicic volcanics of S.E. Queensland and N.E. New South Wales, Australia. *J. Geophys. Res.*, **86**, 10242–56.

Garrison, J., Long, L. & Richmond, D. (1979). Rb–Sr and K–Ar geochronologic and isotopic studies, Llano uplift, central Texas. *Contrib. Mineral. Petrol.*, **69**, 361–74.

Geraghty, E., Isachsen, Y. & Wright, S. (1981). Extent and character of the Carthage–Colton mylonite zone, Northwest Adirondacks, N.Y. Nuc. Reg. Comm. Final Rept., NUREG/CR-1865.

Grant, N., LePak, R., Maher, T., Hudson, M. & Carl, J. (1986). Geochronological framework of the Grenville rocks of the Adirondack Mountains. *Geol. Soc. Am. Abstracts with Programs*, **18**, 620.

Green, T. (1969). Experimental fractional crystallization of quartz diorite and its applications to the problem of anorthosite origin. In Isachsen (1969), pp. 23–9.

Hargraves, R. (1962). Petrology of the Allard Lake anorthosite suite, Quebec. In *Petrologic studies; a volume in honor of A.F. Buddington*, Geol. Soc. Am. Buddington Volume, ed. A. Engle, H. James & B. Leonard, pp. 163–91.

Heath, S. & Fairbairn, H. (1969). Sr^{87}/Sr^{86} ratios in anorthosites and some associated rocks. In Isachsen, (1969), pp. 99–110.

Hedge, C. (1970). Whole rock Rb–Sr age of the Pikes Peak batholith, Colorado. *U.S. Geol. Surv. Prof. Pap.*, **700-B**, 86–9.

Herz, N. (1969). Anorthosite belts, continental drift, and the anorthosite event. *Science*, **164**, 894–947.

Hills, A. & Gast, P. (1964). Age of pyroxene–hornblende granite gneiss of the eastern Adirondacks by rubidium–strontium whole-rock method. *Geol. Soc. Am. Abstracts with Programs*, **7**, 73.

Hills, A. & Isachsen, Y. (1975). Rb–Sr isochron data for mangeritic rocks from the Snowy Mt. massif, Adirondack Highlands. *Geol. Soc. Am. Abstracts with Programs*, 7, 73.

Huang, W. & Wyllie, P. (1975). Melting reactions in the system $NaAlSi_3O_8$–$KAlSi_3O_8$, dry and with excess water. *J. Geol.*, 83, 737–48.

Husch, J. & Moreau, C. (1982). Geology and major element geochemistry of anorthosite rocks associated with Paleozoic hypabyssal ring complexes. *J. Volcan. Geotherm. Res.*, 14, 47–66.

Isachsen, Y. (ed.) (1969). *Origin of anorthosite and related rocks*. New York State Museum Me. 18.

Isachsen, Y., McLelland, J. & Whitney, P. (1975). Anorthosite contact relationships in the Adirondacks and their implications. *Geol. Soc. Am. Abstracts with Programs*, 7, 78–9.

James, R. & Hamilton, D. (1969). Phase relations in the system Ab–Or–An–Q at 1 kilobar water vapor pressure. *Contrib. Mineral. Petrol.*, 21, 111–41.

Johnson, C. & Essene, E. (1982). The formation of garnet in olivine-bearing metagabbros from the Adirondacks. *Contrib. Mineral. Petrol.*, 81, 240–51.

Korzhinskii, D. (1959). *Physicochemical basis of the analysis of the paragenesis of minerals*. Consultants Bureau.

Krogh, T. & Davis, G. (1968). Geochronology of the Grenville Province. In *Carnegie Institution, Washington, Yearbook 67*, pp. 224–30.

—— (1973). The significance of inherited zircons and the origin of igneous rocks; an investigation of the age of Labrador adamellites. In *Carnegie Institution, Washington, Yearbook 72*, pp. 610–613.

Loiselle, M. & Wones, D. (1979). Characteristics and origin of anorogenic granites. *Geol. Soc. Am. Abstracts with Programs*, 11, 468.

Luth, W. (1969). The systems $NaAlSi_3O_8$–SiO_2 and $KAlSi_3O_8$–SiO_2 to 20 kb and the relationship between H_2O content, P_{H_2O} and P_{Total} in granitic magmas. *Am. J. Sci.*, 267-A, 325–41.

McLelland, J. (1986). Pre-Grenvillian history of the Adirondacks as an anorogenic, bimodal caldera complex of mid-proterozoic age. *Geol.*, 14, 229–33.

McLelland, J. & Chiarenzelli, J. (1990a). Isotopic constraints on the emplacement age of the Marcy anorthosite massif, Adirondack Mts., NY. *J. Geol.* 98, 19–111.

—— (1990b). Geochronological studies in the Adirondack Mts. and the implications of a Middle Proterozoic Tonalite suite. In *Mid-Proterozoic geology of the southern margin of Proto-Laurentia-Baltica*, ed. C. Gower, T. Rivers & B. Ryan. Geol. Assoc. Can. Spec. Pap. (in press).

McLelland, J., Chiarenzelli, J., Whitney, P. & Isachsen, Y. (1988a). U–Pb zircon geochronology of the Adirondack Mountains and implications for their geologic evolution. *Geol.*, 16, 920–4.

McLelland, J., Hunt, W. & Hansen, E. (1988b). The relationship between charnockite and marble near Speculator, Central Adirondack Mountains, N.Y. *J. Geol.*, 96, 455–67.

McLelland, J. & Isachsen, Y. (1980). Structural framework of the southern and central Adirondacks: a model for the Adirondacks as a whole and plate tectonics implications. *Geol. Soc. Am. Bull.*, 91, part I, 68–72, part II, 208–92.

—— (1985). Geological evolution of the Adirondack Mountains: a review. In Tobi & Touret (1989), 175–217.

—— (1986). Geological synthesis of the Adirondack Mts., and their tectonic setting within the Grenville Province of Canada. In *The Grenville Province*, ed. J. Moore, A. Baer & A. Davidson, pp. 75–95. Geol. Assoc. Can. Spec. Pap. 31.

McLelland, J. & Whitney, P. (1980a). A generalized garnet-forming reaction for meta-igneous rocks on the Adirondacks. *Contrib. Mineral. Petrol.*, 73, 111–22.

—— (1980b). Compositional controls on spinel clouding and garnet formation in plagioclase of olivine metagabbros. *Contrib. Mineral. Petrol.*, 73, 243–51.

Medaris, L. (ed.) (1983). *Proterozoic geology; selected papers from an international symposium.* Geol. Soc. Am. Mem. 161.

Moore, J. & Thompson, P. (1980). The Flinton Group: a late Precambrian metasedimentary succession in the Grenville Province of eastern Ontario. *Can. J. Earth Sci.*, **17**, 1685–707.

Morrison, J. & Valley, J. (1987). Contamination of the Marcy anorthosite massif, Adirondack Mts., New York; petrologic and isotopic evidence. *Contrib. Mineral. Petrol.* (in press).

Morse, S. (1982). A partisan review of Proterozoic anorthosites. *Am. Mineral.*, **67**, 1087–100.

Mosher, S. (1988). The Grenville (Llano) Orogeny – Texas style. *Geol. Soc. Am. Abstracts with Programs*, **20**, 57.

Newton, R. & Hansen, E. (1983). The origin of Proterozoic and Archean charnockites – evidence from field relations and experimental petrology. In Medaris (1983), pp. 167–79.

Parsons, I. (1972). Petrology of the Puklen syenite, alkali granite complex, Nunarssuit, south Greenland. *Medd. Gron.*, **195**, 1–73.

— (1979). The Klokken gabbro–syenite complex, south Greenland. *J. Petrol*, **20**, 653–94.

Peacock, I. (1931). Classification of igneous rock series. *J. Geol.*, **39**, 54–67.

Pearce, J. & Cann, J. (1973). Tectonic setting of basic volcanic rocks investigated using trace element analyses. *Earth Planet. Sci. Lett.*, **19**, 290–300.

Pearce, J., Harris, N. & Tindle, A. (1984). Trace element discrimination diagrams for the tectonic interpretation of granitic rocks. *J. Petrol.*, **25**, 956–83.

Peterman, Z. & Hedge, C. (1968). Chronology of Precambrian events in the Front Range, Colorado. *Can. J. Earth Sci.*, **5**, 749–56.

Postel, W. (1952). The geology of the Clinton County Magnetite District, N.Y. *U.S. Geol. Surv. Prof. Pap.*, **237**, 1–88.

Reynolds, R., Whitney, P. & Isachsen, Y. (1969). K/Rb ratios in anorthositic and associated charnockitic rocks of the Adirondacks. In Isachsen (1969), pp. 267–80.

Robertson, J. & Wyllie, P. (1971). Experimental studies on rocks from the Deboullie Stock northern Maine, including melting relations to a water-deficient environment. *J. Geol.*, **79**, 549–71.

Rogers, J. & Greenberg, J. (1981). Trace elements in continental margin magmatism. Part III. Alkali granites and their relationship to cratonization; summary. *Geol. Soc. Am. Bull.*, Part I, **92**, 6–9.

Silver, L. (1969). A geochronologic investigation of the anorthosite complex, Adirondack Mts., N.Y. In Isachsen (1969), pp. 233–52.

Silver, L., Bickford, M., Van Schmus, W., Anderson, J., Anderson, T. & Medaris, L. (1977). The 1.4–1.5 b.y. transcontinental anorogenic plutonic perforation of North America. *Geol. Soc. Am. Abstracts with Programs*, **9**, 1176–7.

Simmons, G. (1964). Gravity survey and geological interpretation, northern New York. *Geol. Soc. Am. Bull*, **75**, 81–98.

Simmons, K., Wiebe, R., Snyder, G. & Simmons, G. (1986). U–Pb zircon age for the Newark island layered intrusion, Nain anorthosite complex, Labrador. *Geol. Soc. Am. Abstracts with Programs*, **18**, 751.

Sorenson, H. (ed.) (1974). *The alkaline rocks*, Wiley.

Tobi, A. & Touret, J. (ed.) (1985). *The deep Proterozoic crust of the North Atlantic Provinces.* Reidel.

Turcotte, D. & Schubert, G. (1982). *Geodynamics: applications of continuum physics to geological problems.* Wiley.

Tuttle, O. & Bowen, N. (1958). *Origin of granite in the light of experimental studies in the system* $NaAlSi_3O_8$–$KAlSi_3O_8$–SiO_2–H_2O. Geol. Soc. Am. Mem. 74.

Upton, B. (1974). The alkaline province of southwest Greenland. In Sorenson (1974), pp.221–38.

Valley, J. & O'Neil, J. (1982). Oxygen isotope evidence for shallow emplacement of Adirondack anorthosite. *Nature*, 300, 497–500.

Van Schmus, R., Bickford, M. & Zeitz, I. (1987). Early and Middle Proterozoic Provinces in the central United States. *Am. Geophys. Union, Geodynamics Series*, **17**, 43–68.

Vorma, A. (1971). Alkali feldspars of the Wiborg rapakivi massif in southeastern Finland. *Bull. Comm. Geol. Fin.* **246**, 1–72.

White, A. & Chappell, B. (1983). Granitoid types and their distribution in the Lachlan fold belt, southeastern Australia. In *Circum-Pacific plutonic terranes,* ed. J. Roddick, pp. 21–34. Geol. Soc. Am. Mem. 159.

Whitney, P. (1972). Spinel inclusions in plagioclase of metagabbros from the Adirondack Highlands. *Am. Mineral.*, **57**, 1392–436.

—— (1983). A three-stage model for the tectonic history of the Adirondack region, New York. *Northeastern Geol.*, **5**, 61–72.

— (1986). Geochemistry of proterozoic granitoids from the western Adirondacks, New York State. *Geol. Soc. Am. Abstracts with Programs*, **18**, 788.

Whitney, P. & McLelland, J. (1973). Origin of coronas in metagabbros of the Adirondack Mts., N.Y. *Contrib. Mineral. Petrol.*, **39**, 81–98.

—— (1983). Origin of biotite–hornblende–garnet coronas between oxides and plagioclase in olivine metagabbros. *Contrib. Mineral. Petrol.*, **82**, 34–41.

Windley, B. (1989). Anorogenic magmatism and the Grenville Orogeny. *Can. J. Earth Sci.*, **26**, 479–89.

Wright, T. & Okamura, W. (1977). Cooling and crystallization of tholeiitic basalt, 1965. Makaopuhi lava lake, Hawaii. *U.S. Geol. Surv. Prof. Pap.*, **1004**.

12

An essay on metamorphic path studies or Cassandra in P–T–τ space

RALPH A. HAUGERUD and E-AN ZEN

Introduction: rocks as flight recorders

Metamorphic rocks can be thought of as flight recorders, black boxes recovered from the wreckage of an orogen that tell us something of that orogen's history – primarily the pressure or depth (P), temperatures (T), and times (τ) of the journey. P, T and τ are linked by the physics of heat transfer and the thermodynamics and kinetics of possible transformations in the rocks mass. For a simplified one-dimensional rock mass, this relation can be visualized as a surface in depth($= P$)–T–τ space (Fig. 12.1(A)). P–T–τ paths of rocks are lines on this surface. Tectonic histories (P–τ paths) studied by the structural geologist, P–T loops determined by the metamorphic petrologist, and the T–τ histories examined by the geochronologist are all projections of the P–T–τ path onto the appropriate surface (Fig. 12.1(B)).

Forward modelling of the thermal response to tectonism (for example, England & Thompson, 1984, Haugerud, 1986) illustrates the kinds of P–T–τ paths rocks must follow for certain tectonic scenarios and heat-transfer mechanisms. Much of metamorphic petrology addresses the inverse problem: determining P and T from rocks and from these data inferring tectonic history. When combined with radiometrically or kinetically determined τ, the resulting data set is directly comparable to a P–T–τ path obtained by forward modelling. Combining the forward and inverse approaches to metamorphic path studies promises to extend our understanding of orogenic history. However, at present our ability to model the 'flight' of a given rock, or read the black box to see what path it followed, is limited.

In this essay we review the interrelation of tectonism and P–T conditions; describe ways in which forward modelling of P–T–τ paths can be improved; discuss the limits of some available geochronologic tools, and, because of its considerable promise for extracting P–T information from zoned minerals, examine in detail the Gibbs method recently developed by Spear *et al.* (1982, Spear & Selverstone, 1983, Spear & Rumble,

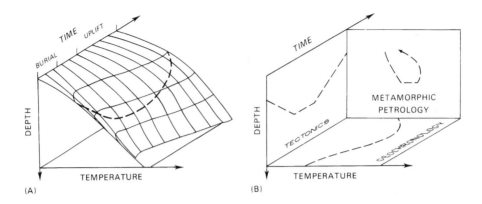

Fig. 12.1. (A) $P–T–\tau$ surface in a one-dimensional Earth, illustrating the effect of a simple tectonic cycle of subsidence balanced by deposition, then stasis, then uplift balanced by erosion. Solid curves extending from upper left to lower right are instantaneous geotherms. Solid curves extending from lower left to upper right are temperatures, throughout time, at constant depth. Both sets of curves lie in the $P–T–\tau$ surface. The heavy dashed line is $P–T–\tau$ path followed by a single rock. (B) Projection of this $P–T–\tau$ path onto the bounding surfaces produces the $P–T$ path, thermal history, and tectonic history studied by, respectively, the petrologist, geochronologist, and tectonicist.

1986). Finally, we outline a few of the tectonic and petrologic problems to be addressed by $P–T–\tau$ path studies.

Geotherms as dynamic entities

Rocks are poor conductors of heat, thus deformation at reasonable rates can lead to significant tectonic transport of heat and change of the geotherm with time. The effectiveness of heat-transfer by tectonism can be estimated by considering the Peclet number, Pe:

$$Pe = Ul/\kappa \tag{1}$$

where κ = thermal diffusivity (in units of length^2/time), U = velocity, here the decompression rate (units of length/time), and l is a characteristic length for thermal conduction, typically the depth (z) of the point of interest. Peclet numbers greater than one indicate that steady-state tectonic transport of heat is more effective than conductive thermal relaxation (Oxburgh & Turcotte, 1974). The assumption that the geotherm is insensitive to tectonism is reasonable only when the Peclet number is much less than one. The relation between decompression rate, depth, and the Peclet number is illustrated in Fig. 12.2.

Fluid circulation, anatexis and melt migration, metamorphic devolatilization, and

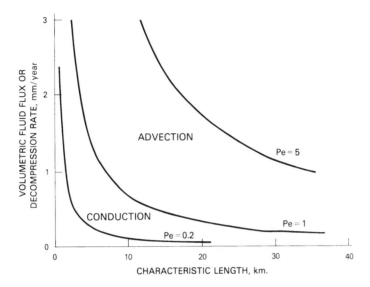

Fig. 12.2.   Isopleths of Peclet number plotted against decompression rate (or volumetric fluid flux) and characteristic length for diffusive relaxation (typically the depth of the point of interest), calculated with thermal diffusivity $(\kappa) = 9 \times 10^{-3} \text{ m}^2 \text{ s}^{-1}$. Conductive heat transfer dominates for Pe < 1, advective transfer dominates for Pe > 1.

strain heating will produce transient disturbances in the geotherm. Change in crustal thickness, also a likely consequence of tectonism, will change total crustal heat generation. Migration of melts and (or) aqueous fluids may redistribute radiogenic heat sources within the crust. Both processes will modify the steady-state geotherm. The magnitudes and timing of all these effects are not generally known.

Numerical studies of the relation between tectonism and *P-T-τ* paths

Simple one-dimensional conductive–advective models

Heat transfer in a one-dimensional, fluid-free Earth can be described by the equation

$$\frac{\partial T}{\partial \tau} = \kappa \frac{\partial^2 T}{\partial z^2} - U \frac{\partial T}{\partial z} + \frac{A}{\rho c} \tag{2}$$

that is,

change of	conductive	advective	internal
temperature =	heat	+ heat	+ heat
with time	transfer	transfer	generation

where z = depth, κ = thermal diffusivity, U = velocity in the z direction (negative for upwards motion), ρ = density, c = heat capacity, and A = rate of internal heat production (for example, radioactive self-heating, heating or cooling by chemical reaction). For many situations of geologic interest, equation (2) must be solved numerically, usually by finite-difference methods.

Such forward models (for example, Albarede, 1976, England & Richardson, 1977, Draper & Bone, 1981, Parrish, 1982, England & Thompson, 1984, Zeitler, 1985, Chamberlain & England, 1985, Day, 1985) are useful to prescribe the general forms that $P–T–\tau$ paths must take; however, their application to specific situations is often limited. Hands-on experience with forward modeling (available using the program of Haugerud, 1986) is an excellent reminder of the uncertainty produced by our ignorance of many of the properties (thermal diffusivity, radiogenic heat production, etc.) that control the thermal structure of the crust. The modeling studies cited also do not address a number of potentially important controls on thermal structure; these are outlined below.

Towards more comprehensive numerical models

Anatexis, magma migration, and plutonism If rocks are heated, they eventually melt. The heat of fusion temporarily retards the local increase in temperature. If the melting reaction is discontinuous, the geotherm in the zone of reaction will be buffered along the univariant reaction curve until the reaction has gone to completion (Fig. 12.3(A)). Continuous reactions will be less effective at local buffering of the geotherm because temperatures can increase as melting proceeds (Fig. 12.3(B)). Melt will in most cases be less dense than the surrounding rock and tend to migrate upwards, advecting heat towards the Earth's surface. The net effect will be a tendency to evolve geotherms having compound curvatures (Fig. 12.3(C)). The complete thermal history of this process has not been analyzed quantitatively, although Hanson & Barton (1986), Lux *et al.* (1986), and Zen (1988 c) have investigated parts of it.

To analyze the effects of anatexis, magma migration, and plutonism, we need to simplify the thermodynamics of the melting process and to model magma mobilization and transport. Thermal effects of melting and crystallization can be incorporated into equation (2) via the heat source term or incorporated via the increased (or decreased) heat capacity of the reacting material and the resulting change in thermal diffusivity (J. DeYoreo, manuscript, R.B. Hanson & M. Barton, manuscript). With the second treatment, diffusivity is no longer a constant, and the conductive heat transfer term in equation (2) must be replaced by

$$\frac{\partial}{\partial z}\left(\kappa \frac{\partial T}{\partial z}\right)$$

(2a)

(Carslaw & Jaeger, 1959, p. 10). At middle and lower crustal conditions ($P \geq 5$ kb) the temperature of melting can be assumed, to a first approximation, to be pressure-independent, and we need only to prescribe the melt fraction at any given T and the heat capacities of solid and melt.

Magma probably will not migrate from the source rock until most source-rock grain

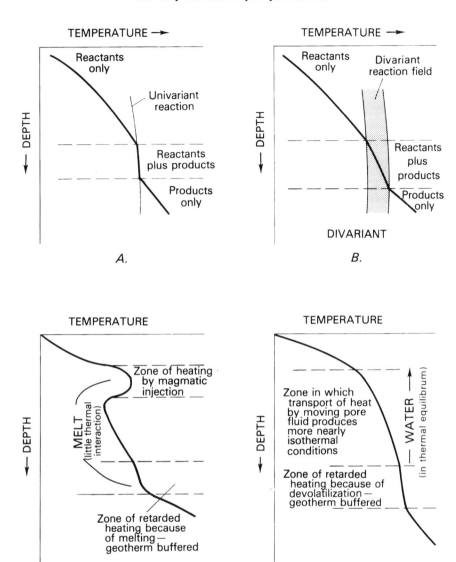

Fig. 12.3. (A) Buffering of geotherm by a univariant reaction. In region where reaction is in progress, geotherm must coincide with univariant reaction curve. (B) Buffering of geotherm by a divariant reaction. (C) Transient geotherm having compound curvature, produced by melting at depth and emplacement of evolved magma at a higher level. Magma is presumed to not interact with intervening rock column. Not to scale. (D) Transient geotherm produced by metamorphic devolatilization at depth and upward percolation of evolved fluid. Fluid is presumed to be in thermal equilibrium with wallrock. Not to scale.

boundaries are wetted by melt. The theory of polyphase aggregates (Smith, 1964) and experiments by Waff & Bulau (1982), Jurewicz & Watson (1985) and others indicate that the amount of melt required to achieve this interconnectedness depends on the composition of the system. Because many magmas contain a fraction of entrained crystals, the separation of melt from restite is probably rarely perfect.

Flow of magma is strongly focused; the spacing of igneous centers in magmatic arcs is on the order of tens of kilometers. Because magmas obviously did not congeal on their way to the final sites of consolidation, cooling in transit must have been limited. From these observations, we infer that, in a one-dimensional Earth, magma migration may be adequately modeled as an adiabatic process. However, more detailed consideration of the physics of intrusion (Marsh, 1984) leads to the conclusion that thermal interaction with the wallrock must be significant and reminds us that thermal histories in the vicinity of the intrusive conduit cannot be adequately simulated with a one-dimensional model.

Lux *et al.* (1986) and Hanson & Barton (1986) have numerically modelled the thermal effects of pluton intrusion. Both investigations demonstrate the possibility of producing regional andalusite–sillimanite facies-series prograde metamorphism by contact-metamorphism, either around regionally extensive thick sills (Lux *et al.*, 1986) or in the overlapping aureoles of plug-like plutons (Hanson & Barton, 1986).

Moving aqueous fluid   The heat capacity (per volume) of water is similar to that of rock, thus a given vertical aqueous-fluid volume has effects similar to that of decompression, or rock volume flux, at the same rate (Peacock, 1987). We can calculate the Peclet number of this fluid flux in the same fashion as for a rock volume flux (equation 1), replacing U with u_f ($=$ fluid flux, in units of volume area^{-1} time^{-1}). Peclet numbers > 1 (Fig. 12.2) are required for hydrothermal effects to dominate the steady-state geotherm.

For quantitative calculations, the transport of heat by a migrating fluid can be easily incorporated into equation (2) by adding a second advective term

$$ -U_f \frac{\partial T}{\partial z} \frac{\rho_f c_f}{\rho_r c_r} \tag{2b} $$

where U_f is the mass flux of the fluid, in this case relative to the rock matrix; c_r and c_f are the heat capacities (per mass) of matrix + fluid and fluid; and ρ_r and ρ_f are densities of matrix + fluid and fluid.

Predicting the mass flux of aqueous fluid as a consequence of tectonic evolution is difficult. Tectonism will directly influence fluid motion by (1) compacting and dewatering sedimentary rocks, (2) creating a topographic hydraulic head that drives regional ground-water flow (Garven & Freeze, 1984a,b), and, indirectly, by (3) producing hot spots (plutons, basement uplifts) or permeability anomalies that may localize convection systems (for example, Norton & Knight, 1977). With the exception of compaction of a sedimentary basin with laterally uniform stratigraphy, such effects cannot be realistically simulated with a one-dimensional model. Volume increase because of thermal expansion and differences in fluid density resulting from variable fluid temperature and dissolved solid content will also affect fluid fluxes. These problems have occupied ground-water modellers for several decades (for example, Norton & Knight, 1977, Fyfe *et al.*, 1978, Garven & Freeze, 1984a, b, and Bethke, 1985).

Metamorphic devolatilization and fluid migration Metamorphic devolatilization will tend to produce geotherms having compound curvatures, for the same reasons anatexis does. The thermal effects of devolatilization reactions do not, in principle, differ from melting reactions, and their thermodynamics may be simplified in the same way. Added complications are (1) reactions cannot be assumed to be insensitive to pressure and (2) fluid pressure and chemical activities may be strongly affected by metamorphic history and thus will not be simple linear functions of depth.

We expect metamorphic fluid migration will not be as episodic or as strongly focussed as magma flow (see, however, Chamberlain & Rumble, 1986, Paterson, 1986). We may consider the flow of fluid to be uniform, pervasive, and in thermal equilibrium with its matrix, a situation that can be adequately simulated in one dimension. The thermal consequences of such flow can be simulated by adding a second advective term to equation (1). The challenge in such calculations lies in adequately prescribing fluid velocity U_f as a response to the kinematic, thermal, and chemical evolution of the numerical model. Interesting and significant results probably can be obtained by simplification and omission of some of the processes involved.

Migration of radionuclides If the regular distribution of heat production with depth (Birch *et al.*, 1968, Lachenbruch, 1968) is real, then some sort of redistribution of radioactive elements (K, U, Th) during tectonism is required. Redistribution will change the equilibrium thermal structure of the crust, and the effects will be more pronounced at greater depth. If we are concerned with tectonic cycles that last long enough for deeper parts of the crust to approach thermal equilibrium, we must consider this temporal variation of internal heat production. Investigation of these effects by numerical modeling is desirable, but there are no well-understood mechanisms for radionuclide redistribution (see Albarede, 1975).

Strain heating Shortening of a 100-km-wide orogen at 1 cm/year requires average principal strain rates of $\sim 3 \times 10^{-15}\,\mathrm{s}^{-1}$, or an average shear strain rate of $\sim 6 \times 10^{-15}$ s^{-1}. Deformation at this rate and an average shear stress of 100 bars transforms work into heat at a rate of $6 \times 10^{-8}\,\mathrm{W\,m}^{-3}$, 1/30 the typical rate of radioactive self-heating for upper-crustal rocks. Such deformation throughout 30 km of crust results in a total power output of $\sim 2\,\mathrm{mW\,m}^{-2}$. Typical surface heat fluxes are $\sim 60\,\mathrm{mW\,m}^{-2}$, thus at these deformation rates, strain heating will not significantly affect temperatures at depth.

Strain heating in conjunction with tectonic transport of heat (advection) may be significant. Two-dimensional numerical modelling by Haugerud (1985) suggests that, at shear stresses greater than ~ 100 bars and slip rates of 1 to 2 cm/year, strain heating in an oblique-slip high-angle fault zone is sufficient to counteract the cooling effects of uplift and significantly modify the P–T–τ paths of rocks in and near the fault zone.

Two-dimensional numerical models   Many tectonothermal regimes cannot be adequately simplified to one dimension, for example crustal-scale tilt blocks, strike-slip faults, non-horizontal layers of rocks with differing thermal properties, and steep-sided plutons. Significant lateral advective heat transport may occur in large thrust and normal-fault complexes that develop over periods of millions of years; such regimes also are not adequately simulated in one dimension. The primary difficulty with two-dimensional models is a geometric increase in the number of computations required, although Harrison & Clarke (1979), Haugerud (1985), and Hanson & Barton (1986) have developed such numerical models.

P–T–τ determinations from rocks

Time and temperature

Temperature-dependent loss, or reequilibration, of a daughter product of radioactive decay (for example, K–Ar and fission-track systems) can be exploited as a thermochronometer, a system that measures the time since an already-formed mineral cooled below a closure temperature (Dodson, 1973).

K–Ar   K–Ar geochronology can measure the time since a mineral cooled below closure temperatures that range between $\sim 500\,°C$ (closure temperature of hornblende, Harrison, 1981) and $\sim 150\,°C$ (closure temperature of potassium feldspar, Harrison & McDougall, 1982). In some cases crystallization, or recrystallization, of all or part of a mineral population at temperatures below its closure temperature may turn a K–Ar system into a reaction-progress clock (defined below). Muscovite and some metamorphic amphiboles are the only commonly dated minerals that are likely to have crystallized below their K–Ar closure temperatures.

Application of the K–Ar system to P–T–τ path studies is limited by poorly determined closure temperatures for most samples (see, for example, a review by Cliff, 1985) and evidence, from ^{40}Ar–^{39}Ar step-heating studies, that the behavior of many K–Ar systems is not simple (for example, Zeitler & FitzGerald, 1986, Harrison & FitzGerald, 1986).

Closure temperatures are functions of the intrinsic diffusion parameters, the effective diffusion radius, and the cooling history (Dodson, 1973). Closure temperatures for minerals *that remain stable during heating in a vacuum* can be estimated by analyzing the step-heating data themselves as a diffusion experiment (Berger & York, 1981). The only routinely dated mineral for which this can be done is potassium feldspar. Amphiboles and micas are not stable at high temperatures in a vacuum and closure temperatures for these minerals must be estimated by other means. Some data exist on the effects on closure temperature of Fe–Mg substitution in amphiboles and biotite (Harrison, 1981,

Harrison *et al.*, 1985), but the effects of Tschermak's substitution, Ti content, redox state, Fe–Mg substitution in muscovite, etc., have not been studied. Some evidence (Harrison & FitzGerald, 1986) indicates that exsolution lamellae in amphiboles may significantly affect closure temperatures. Hydrothermal experiments that examine compositional and structural effects on argon diffusivity would be useful; careful ^{40}Ar–^{39}Ar step-heating analyses and complete chemical and structural characterization of a range of mineral compositions from a small region having a well-understood thermal history might be even more productive.

'Excess argon' and partial argon loss pose another set of problems. Unexpectedly old conventional K–Ar ages and some types of discordant ^{40}Ar–^{39}Ar age spectra are commonly ascribed to 'excess ^{40}Ar'. The problem can be especially severe for the metamorphic and deep-seated igneous rocks of interest in many cooling-history studies (see Zeitler & FitzGerald 1986, for examples). Haugerud (1987) suggested that 'excess argon' reflects incorporation in the sample of a ^{40}Ar-enriched intergranular argon at the time the mineral became closed to argon diffusion and noted that step-heating experiments can routinely provide the data necessary to identify the isotopic ratio of this initial argon and remove its effects when calculating sample ages.

Unexpectedly young K–Ar ages are often ascribed to partial argon loss during later reheating. Diffusion theory predicts that the composition of the first infinitesimal amount of gas released in the step-heating experiments will reflect the age of the later heating event. Following the example of Turner (1968), Harrison (1983, Harrison & McDougall, 1980) has used step-heating data to recognize partial argon loss in a later heating event and to calculate the age of later heating. The analysis assumes that the effective diffusion volume (usually taken to be a mineral grain or subgrain) is surrounded by an infinite, homogeneous argon reservoir. Some rocks may satisfy this assumption; others, including many of the samples of interest in cooling-history studies, apparently do not (Haugerud, unpublished data). Minerals, as they degas, modify the composition and concentration of the argon reservoir that surrounds them to an unknown degree. The step-heating technique can be used to recognize partial argon loss, but cannot always identify the time of argon loss.

Monotonically increasing age spectra from some K-feldspars have been interpreted as reflecting slow cooling. Numerical modeling (Haugerud, unpublished) suggests that in general this interpretation is not unique: such spectra can also be created by partial argon loss in a later heating event, with an appropriate distribution of effective grain radii.

Geospeedometry If thermally-activated diffusion is the rate-limiting process in a metamorphic reaction, the extent of diffusive relaxation of a chemical potential gradient can be used to calculate an integrated time–temperature factor (Lasaga, 1983). To do so, both the relevant diffusion coefficients and the original potential gradient must be determined. If the chemical potential gradient is the result of changes in ion-exchange equilibrium upon cooling (for example, Fe–Mg partitioning between garnet and biotite), and if temperature is some simple function of time, a cooling rate can be calculated. If the chemical potential gradient is the result of a pre-metamorphic inhomogeneity, such as a serpentinite block in a mudrock, the width of the resulting reaction zone can be used to estimate the total time–temperature factor for a metamorphic event.

This type of kinetic analysis has been used (for example, Spear & Chamberlain, 1986) to set bounds on the rate of post-metamorphic cooling based on zoning of garnets adjacent to an Fe–Mg exchange reservoir. At present, geospeedometric analysis seems most useful in analyzing assumptions regarding the attainment and preservation of local chemical equilibrium (Korzhinskiy, 1959, Lasaga, 1983).

Time and reaction history

Some isotopic systems (for example, U–Pb in zircon, K–Ar in muscovite crystallized below its closure temperature) are not prone to diffusive loss of daughter products. These systems record the time since crystallization or recrystallization of a rock; they are not thermochronometers. Correlation of ages from these systems with other aspects of a rock's history presumes an understanding of the controls on crystallization and recrystallization.

U–Pb zircon   Preservation of older U–Pb systems in xenocrystic zircons in melts (for example, Pidgeon & Aftalion, 1978), persistence of U–Pb systems in zircons that have undergone upper amphibolite- and granulite-facies metamorphism, and complex zoning of Pb in zircon from high-grade environments (Williams *et al*, 1984) indicate that the diffusivity of Pb in zircon is very low at crustal temperatures ($\leq \sim 800$ °C). Most discordant U–Pb zircon ages appear to be the result of low-temperature Pb loss related to alpha-particle damage to the zircon structure, analysis of mixtures of older and younger zircon, or both. This low diffusivity of Pb makes the U–Pb zircon system uniquely able to record the ages of igneous crystallization and high-temperature metamorphic recrystallization.

Limits to the use of U–Pb zircon dating include the following. (1) Ability to physically separate zircon that crystallized at one time. Abrasion experiments and ion-probe analyses demonstrate that some zircon grains contain regions having different ages. (2) Uncertainty as to the mechanism and effects of low-temperature Pb loss. (3) Limited understanding of the controls on zircon stability. Temperature cannot be the sole controlling factor, as zircon crystals survive anatexis and also are known to have formed during amphibolite-facies metamorphism (R. Parrish, oral communication, 1986). Likewise, effects of the bulk composition of a rock on zircon stability (see also Watson & Harrison, 1983) are probably not simple. (4) Difficulty in recognizing discordance in Cretaceous and younger zircons. Because this portion of the concordia curve is relatively straight, Pb loss or presence of a slightly older inherited component moves analyses along trends sub-parallel to concordia. Unless the amount of Pb loss or inherited component varies among several analyzed fractions, even very precise analyses may be inadequate to recognize discordant behavior.

Ages by other means

Consistent relations between K–Ar and Rb–Sr ages in the Alps (Armstrong *et al.*, 1966, Jäger, 1970, Purdy & Jäger, 1976) have led to the use of Rb–Sr mineral systems as thermochronometers. This appears unwise: Sr isotope homogenization at sub-green-schist facies conditions has been demonstrated on the scale of an outcrop (Gebauer & Grünenfelder, 1974), while lack of homogenization during amphibolite- and granulite-facies metamorphism has been observed on the scale of a hand specimen (for example, Hofmann & Grauert, 1973). Thermally activated diffusion cannot be the sole factor governing Sr-isotope homogenization in most rocks. Cliff *et al.* (1985) suggest that reaction-induced fluid fluxes are the most important factor.

Where diffusion can be demonstrated to be the mechanism of Sr-isotope homogenization, an additional complication to Rb–Sr mineral dating is the requirement for other phase(s) in the rock to act as exchange reservoirs. For a rock that contains biotite and one additional Sr-bearing phase, Sr-isotope equilibration between biotite and the rock will cease when either biotite or the exchange reservoir closes to diffusion. For example, the 'age' calculated from biotite and whole-rock analyses of a biotite gneiss may well reflect the closure temperature for Sr diffusion in feldspar. If there are several Sr reservoirs in a slowly cooled rock, a meaningful age can be calculated only from analyses of the last two phases to close to Sr diffusion, or the assumption of closed-system behavior and analyses of the whole rock and the first phase to close. Giletti (1986) has discussed this problem with regard to oxygen-isotope exchange.

Mattinson (1978, 1986) has suggested that U–Pb systems in sphene and apatite may have potential as thermochronometers. Besides uranium, these minerals commonly incorporate a substantial amount of common lead at the time of crystallization, so that extracting meaningful ages requires an isotope-correlation analysis analogous to that used for interpreting Rb–Sr data (Mattinson, 1986, Cliff, 1985). Sphene and apatite U–Pb systems are subject to the same limitations as Rb–Sr systems: apparent ages depend on the characteristics of the Pb-exchange reservoir, and it has yet to be demonstrated that temperature-activated volume diffusion is the dominant Pb-transfer mechanism. Different 'blocking temperatures' reported for U–Pb sphene systems (Mattinson, 1978, Cliff & Cohen, 1980) may reflect these problems.

An isotopic system that allows direct dating of individual growth zones in garnets would be an extremely powerful tool for P–T–τ path studies. If such a system is defined, it must be shown whether it is a thermochronometer or a reaction-progress clock. In either case, intelligent use of such a system will require knowledge of other reservoirs for the relevant isotopes, diffusivities in all phases involved, and the potential role of inclusions as isotopic hosts.

Pressure and temperature, with special reference to the Gibbs method

Mineral assemblages and mineral compositions can be used: (1) to assign rocks to mineral facies (for example, amphibolite), zones (for example, staurolite) and batho-

R.A. Haugerud and E-an Zen

zones (Carmichael, 1978); (2) in conjunction with experimental phase equilibrium data, to more tightly circumscribe the P and T of formation (the petrogenetic grid approach used by Marakushev, 1965, Winkler, 1974, Glebovitskiy, 1977, and others); and (3) in conjunction with thermochemical data and thermochemical modeling of mineral solutions, to define specific P–T conditions of final equilibration of the mineral assemblage. The latter approach, which we shall call 'classical thermobarometry', depends on the assumption of equilibrium throughout the domain of interest ('local equilibrium' of Korzhinskiy, 1959); therefore, no more than a single P–T condition can be evaluated per domain.

Pseudomorph relations or other reaction textures indicate changes in P–T conditions. Different generations of fluid inclusions can indicate changes in P, T and activities of fluid species (for example, Hollister & Crawford, 1981). If different samples, or domains within samples (such as solid inclusions in low-diffusivity host minerals), cease to equilibrate at different times, the locus of P–T conditions defined by different domains can be used to define the P–T history of a rock mass (for example, Rosenfeld, 1970, Hodges & Royden, 1984, Perchuk *et al.*, 1985, St-Onge & King, 1985).

The Gibbs method Building on the work of Korzhinskiy (1959, 1973) and Thompson (1959, 1967), Rumble, Spear, and others (Rumble, 1974, 1976, Spear *et al.*, 1982, Spear & Selverstone, 1983, Selverstone *et al.*, 1984, Spear, 1986a) developed the 'Gibbs method', which allows the inference of a continuous time sequence of P–T points from compositional zoning of low-diffusivity minerals in certain low-variance assemblages.

As summarized by Spear *et al.* (1982), the Gibbs method relates compositional zoning in minerals to equilibrium changes in state (P, T, and chemical potentials) on the assumption that zoning resulted from continuous reactions among the minerals that now constitute the assemblage. If the pressure and temperature of formation of the rim of a zoned mineral can be determined, the Gibbs method can be applied to 'read' the zoning history inward and thereby to derive a continuous P–T history for the assemblage.

The method involves repeated numerical solution of a set of simultaneous linear equations (Table 12.1) that describe thermochemical equilibrium in the assemblage (Spear *et al.*, 1982). Known quantities in the equations are entropies and volumes of the various phases, mole fractions of different components in each phase, and stoichiometric coefficients for species involved in heterogeneous reactions within the assemblage. The unknown quantities are the differentials of temperature, pressure, and chemical potentials in the assemblage. The variance of this set of equations is equal to the variance of the assemblage. Additional equations that arise from solid-solution phases relate the differentials of compositions of these minerals (and curvatures of their Gibbs functions) to these other differentials; they do not change the variance of the system of equations, yet relate changes in mineral composition (for example, zoning) to changes in the intensive parameters of metamorphism. Solution of the set of equations relates changes in a small number of measurable compositional variables (for example,

334

Table 12.1. *Equations, knowns, and unknowns of the Gibbs method*

Type of equation	Number of equations	Knowns	Unknowns
(a) Gibbs–Duhem	NP	$S_k \ V_k \ X_i$	$dT \ dP \ d\mu_i$
(b) heterogeneous equilibrium	$(\sum_k NPC) - NC$	p_j	$d\mu_j$

(variance of system of equations (a) + (b) is equal to phase-rule variance of assemblage)

Type of equation	Number of equations	Knowns	Unknowns
(c) composition–chemical potential	$\sum_{ss}(NPC - 1)$	$S_i \ V_i$ $\partial^2 G / \partial X^2$	$d\mu_i \ dT \ dP \ dX_i$

(number of equations of type (c) is equal to number of new unknowns dX_i)

Type of equation	Number of equations	Knowns	Unknowns
(d) mass-balance	NC	$M_k \ X_i \ n_{hk}$	$dX_i \ dm_h \ dM_k{}^a$

(See note 1. If changes in system composition are known, that is, dm_h are known, the number of equations of type (d) minus the number of new unknowns dM_k is equal to $NC - NP$. In this case, addition of these equations to the system (a) + (b) + (c) reduce variance of system of equations to two. If dm are unknown, the number of new unknowns is NC.)

Notation:
NP number of phases
NPC number of phase components in phase (= degrees of compositional freedom in phase)
NC number of system components
S entropy
V volume
M modal fraction of phase in rock
X mole fraction of component in phase
G Gibbs energy
μ chemical potential
n stoichiometric coefficient of system component in phase
m mole fraction of system component in rock
p_j stoichiometric coefficient of phase component j in reaction
h system components
i phase components in phase
j phase components in reaction
k phases
ss solid-solution phases (= phases for which $i > 1$)
[a] M, dM, and dm are not independent. If any combination of $2 \times NC$ of these variables is specified, the remainder are determined.

X_{alm} in garnet) to changes in the intensive parameters of metamorphism. For a divariant system, knowledge of the changes of two independent compositional variables yields a history of the change of P, T, and activities of all species. A trivariant system requires three independent compositional variables for application of the Gibbs method, and so forth.

Use of the Gibbs method requires: (1) correct identification of the reacting assemblage and its variance; (2) knowledge that the present rim compositions of all phases reflect equilibrium at some final P–T condition; (3) determination of final P–T conditions, commonly by classical thermobarometry; and (4) the assumption that a zoned mineral represents a unidirectional temporal progression. A complete Gibbs-method analysis also must include an estimate of the sensitivity of the results to assumptions and input data.

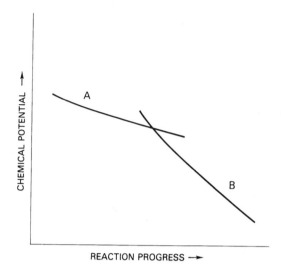

Fig. 12.4.   Chemical potential of a phase component as a function of reaction
progress, showing cusp in potential at change in assemblage. Diagram
could, for example, represent chemical potential of almandine in garnet
as assemblage changes from garnet–biotite–chlorite–quartz–muscovite–
H_2O (curve A) to garnet–biotite–kyanite–quartz–muscovite–H_2O (curve
B).

What is the assemblage?   The assemblage during equilibration of mineral rims
presumably includes all minerals present within the domain of interest (excluding
inclusions), plus fluid(s). For interior portions of zoned minerals, the assemblage is not
so easily determined. Abundant and varied minerals as solid inclusions within a given
zone of the host crystal may reflect the assemblage in equilibrium with that zone. If such
inclusions have not undergone subsequent cation exchange with the host crystal, their
compositions may also allow calculation of $P–T$ values independent of the Gibbs-
method extrapolation from rim conditions. Zoning of other minerals in the assemblage
can in some cases be correlated with that of the zoned crystal in question (Spear &
Rumble, 1986). Congruence of predicted and observed compositional trends in other
zoned phases suggests maintenance of equilibrium during evolution of the mineral
assemblage.

Compositional trends of zoned minerals themselves can provide clues to changes in
assemblage. During the progress of a continuous reaction, the chemical potentials of
relevant components are continuous functions of P, T and mole fractions. At the point
where a discontinuous reaction occurs, chemical potentials are identical in both
products and reactants, but these potentials are most certainly not smooth functions of
reaction progress. A cusp – a discontinuity in the first derivative – in the compositions
of zoned minerals is a likely indicator of the change of mineral assemblage (Fig. 12.4). If
the assemblage changes because progress of a continuous reaction has led to loss of a
phase, the reaction and crystal growth will stop until growth by another reaction

commences. At the point of renewed growth, the stable composition is not likely to be that at which growth stopped, and a step-like discontinuity in composition is probable. Recognizing such a cusp or discontinuity in mineral composition depends on both the spacing and precision of the microprobe analyses and the lack of subsequent diffusive smoothing.

A discontinuity in compositional zoning does not necessarily indicate a change in assemblage. A garnet rim, with its composition buffered by a constant divariant assemblage, may grow, be resorbed, and then grow again as a result of changing P–T conditions. The stable composition of garnet at the point of renewed growth may not be the same as that exposed by resorption, yet the assemblage has not changed. Gibbs-method analysis of such a grain would correctly identify the net P–T change associated with the discontinuity but would not disclose the intervening P–T history.

Determination of variance   The Gibbs method provides only as much information as the variance allows. Three independent compositional variables can be monitored with an ordinary four-component garnet, allowing us to examine trivariant systems. Monitoring systems with higher variance requires (1) that there is a zoned monitor phase with well-understood solution behavior and a larger number of significant components, (2) that P–T behavior of the system is insensitive to variation in one of the monitor components (for example, Spear & Selverstone (1983) assumed changes in $X_{anorthite}$ of plagioclase have no effect) or (3) that a given compositional zone in the interior of one mineral can be shown to be coeval with a certain compositional zone in another mineral and that zoning in neither has been affected by later diffusion.

The equations of the Gibbs method are written in terms of the intensive parameters of metamorphism (P, T, the chemical potential μ or the mole-fractions X). They contain no assumptions, and make no predictions, about changes in relative mass (metasomatism) or mode within a system. Spear (1986b and oral communication, 1987) pointed out that mass-conservation equations can be added to the Gibbs method (Table 12.1); by using these equations and a modal analysis, we can monitor metasomatism. If the composition of the reacting system is fixed, Duhem's theorem reduces the variance of the sytem to two, because only two intensive parameters must be specified to uniquely determine the state of the system if its bulk composition is known (Korzhinskiy, 1959).

The composition of the reacting system generally does not correspond to that of the whole rock. The reacting chemical system analyzed by the Gibbs method consists of the intergranular medium and those portions of the solid phases immediately in contact with this medium. Only if diffusion inside solid phases is rapid with respect to the rate of change of chemical potentials (Thompson, 1959, Korzhinskiy, 1959) in the system are the interiors of grains part of the reacting system. Dissolution or crystallization of grains in which diffusion cannot keep pace with changing grain-boundary concentrations is, in effect, open-system behavior. The usefulness of the Gibbs method for P–T path studies depends on the existence of zoned grains, thus open-system behavior must be the rule in such studies.

Assumptions of equilibrium   Two assumptions of equilibrium are as critical to the Gibbs method as they are to classical thermobarometry. The first is that heterogeneous equilibrium was achieved and preserved. This assumption can be tested by comparing mineral rim compositions from different places in a volume of rock: if equilibrium was achieved and preserved within this volume, the compositions should be the same.

The second assumption is that the reactions critical to determination of P and T (for example Fe–Mg exchange between biotite and garnet and transfer of Ca and Al from grain to grain for growth of grossular at the expense of plagioclase) ceased equilibrating at the same time. If movement of reacting species inside a rock is limited by thermally activated diffusion, this assumption is almost certainly false, because diffusion rates are not all identical; only by extreme coincidence should multi-reaction thermobarometry give meaningful results (see Dodson, 1973, Thompson & England, 1984). However, classical thermobarometry using reactions with more than one diffusing species does seem to work, so we are forced to conclude that mass transfer is aided by some other factor, probably an intergranular fluid film, and that its loss stops reactions and preserves 'final' equilibrium (Zharikov & Zaraisky, Chapter 9 in this volume). To what extent can we assume that this facilitating factor was present throughout the growth of a zoned mineral? Clearly we need a better understanding of reaction mechanisms and kinetics in rocks. In the interim, any reaction history derived from Gibbs-method analysis that does not imply at least minor progressive devolatilization should be regarded with some skepticism.

Difficulties in estimating final P and T   In some circumstances, P and T of final equilibration can be closely estimated from the mineral assemblage by reference to a petrogenetic grid. More commonly, P and T must be determined by mineral composition data and classical thermobarometry. The Gibbs method builds upon classical thermobarometry and thus is sensitive to all its limitations.

While straightforward in principle, classical thermobarometry is imprecise in application (Essene, 1982). The question of equilibrium is addressed above. Thermochemical data on pure phases can be extracted from phase equilibrium experiments because calculated thermochemical properties are not very sensitive to the P and T of equilibrium; conversely, estimated P and T are very sensitive to errors in thermochemical values (Zen, 1977). Best-fit sets of thermochemical data estimated by consideration of large amounts of experimental data (Haas, 1974, Helgeson *et al*, 1978, Berman *et al*, 1985) overcome some of these problems, but data for several minerals of interest (for example, many Fe-bearing minerals) are not yet available. There is little agreement on the solid-solution models necessary to extend the data beyond end-member phases, even for such a common and important mineral as garnet (see Ganguly & Kennedy, 1974, Newton & Haselton, 1981, Hodges & Spear, 1982, Bohlen *et al.*, 1983, Ganguly & Saxena, 1984). Workers often fail to discuss, specify, and calculate the cumulative uncertainties in the quantitative analysis, or to ensure that the proposed equations for thermobarometry are mutually consistent. Some of these difficulties are discussed by Powell (1985).

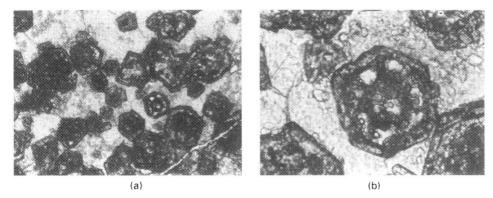

Fig. 12.5.   Photomicrographs of skeletal garnet from a fine-grained Upper
Cretaceous pelite that has been contact-metamorphosed by intrusion of a
granodiorite of the Pioneer batholith, Montana. Sample 776–1 (Zen,
1988 a). The spoke-and-ring garnet is about 0.5 mm across. (A) Various
stages of annular growth and infilling of euhedral garnet; (B) close-up of
the spoke-and-ring garnet showing the fine-grained matrix occluded in
the spaces between the spokes.

Time's arrow in zoned crystals   Interpretation of mineral zoning as a time sequence is
based on an assumption of the crystal's growth direction. However, crystals can grow
in fashions other than radially outwards! Fig. 12.5 shows garnets in a calcareous pelite
within 2 km of a plutonic contact in the Pioneer Mountains, Montana (Zen, 1988 a).
The idioblastic garnet crystals have atoll, skeletal, or ring-and-spoke shapes. Matrix
material is still present in the interstices of some skeletons; the shapes are interpreted to
reflect skeletal growth. Why are the garnets skeletal? Perhaps this occurrence is peculiar
to contact-metamorphic rocks and would never be seen in regional metamorphism;
perhaps it reflects some unusual composition or rapid heating and cooling. The
possibility of skeletal growth, now obliterated by infilling, in other rocks cannot be
dismissed. If so, we must question the infallibility of unique time direction given by
shells of zoned garnet. Because of possible diffusional modification of the compositions
of the various zones, the former presence of skeletal zones of the same composition
(because they were grown at the same time) might not be detected by examining
microprobe profiles of the garnet for matching compositions among annular rings.

For the same reason, if infilling of skeletal crystals is suspected, then the time
significance of solid inclusions also must be subject to question. Inclusions could be
partly the result of recrystallization of the enclosed matrix, with or without exchange
with the matrix outside.

Error analysis   The extensive computations of the Gibbs method make error analysis
more difficult than usual. Sources of error can be divided into two groups: quantifiable
errors in input variables (compositions, formula calculations, thermochemical data,
solution model parameters, initial P and T) and qualitative 'geologic errors' (faulty
assumptions about equilibrium, variance, and crystal growth direction).

Effects of errors in input variables can be computed by standard techniques if the probable distributions of errors are known, but to our knowledge this computation has only been done for analytical errors in a single Gibbs-method analysis of a zoned garnet: Spear & Rumble (1986) performed a Monte Carlo analysis of the effect of a standard error in calculated mole fractions of 0.005 for major elements and 0.001 for minor elements. Their analysis showed that, whereas the qualitative shape of the inferred P–T path was relatively insensitive to analytical errors, the quantitative uncertainty in the pressure at which interior zones of the garnet grew was large, $\pm \sim 40$ °C and $\pm \sim 1$ kb (not including the uncertainty in rim P–T conditions).

The effects of qualitative geologic errors cannot be so easily assessed. Tests of consistency will be most useful. For example, do analyses of zoning in other garnets from the same assemblage, garnets in slightly different assemblages, and amphiboles from different rocks in the same outcrop all give similar P–T paths?

Recovering P–T–τ regimes of metamorphic terranes

The isolated T–τ, τ-reaction history, and P–T measurements described above are not complete P–T–τ regimes. To obtain such complete information, P–T data must be linked to time measurements, and then extrapolated to a complete P–T–τ regime for the metamorphic terrane of interest.

Correlated pressure–temperature–time   In certain circumstances P–T data and isotopic age measurements can be directly linked. Occasionally we can relate a P–T fix to an age determined by U–Pb zircon geochronology; for example, if we can relate P–T conditions in a metamorphic aureole to zircon crystallization in a pluton. Some greenschist and low-amphibolite facies rocks never exceeded the closure temperatures for amphibole (and perhaps white mica) K–Ar systems. For these rocks, the K–Ar systems can be treated as reaction-progress clocks, and direct dating of different portions of zoned amphiboles by laser-probe ^{40}Ar–^{39}Ar methods (Sutter & Hartung, 1984) might allow correlation of τ with P–T conditions. If we have P–T data for the sub-500 °C cooling history of a hornblende-bearing rock, we can in principle directly relate these data to T–τ data determined by K–Ar methods. However, mineral assemblages commonly do not extensively reequilibrate over this portion of the metamorphic path, thus the required P–T data are usually unobtainable. Moreover, we need better knowledge of reaction kinetics before we can interpret P–T conditions from partial retrograde alteration.

Beyond these relatively rare situations, relating P, T, and τ requires careful geologic reasoning, working from available information on the ages of structural and intrusive events, and correlating these events with P–T conditions. Potentially, a great deal could be learned if, for example, we knew enough about the phase relations of zircon to relate the ion-microprobe U–Pb ages of different zones of a zircon grain to events in the reaction history of a rock.

We do not have tools to decipher complete P–T paths, or to date most of the thermal history of an outcrop. Large parts of the prograde metamorphic path probably will remain hidden from us because the records in our petrologic 'flight recorders' have been erased by further metamorphism.

The extrapolation problem   Given limited information on the P–T and T–τ histories of a rock and some assumptions regarding the significance of possible heat-transfer mechanisms, a plausible P–T–τ regime can be developed by trial-and-error methods using forward numerical techniques. There is, however, no guarantee that such a regime is unique.

Construction of a complete P–T–τ regime over the volume and time of interest is possible if several P–T–τ determinations can be made. The quality of such extrapolations would be improved if they included knowledge of the physical mechanisms of heat transfer and some reasonable assumptions about boundary conditions; we now lack the theoretical tools to do this. The inversion technique developed by Royden & Hodges (1984, McNutt & Royden, 1987) for situations in which P–τ relations are accurately known (that is, constant decompression rate) exemplifies this approach. Further studies of this type would be useful.

Metamorphic path answers to tectonic and petrologic problems

Mechanisms of crustal thickening   Two models of crustal thickening during orogeny include the Coulomb-wedge model of Davis *et al.* (1983) and the viscous thickening model of England & McKenzie (1982, 1983). Both of these mechanisms imply diachronous thickening across an orogen. However, Schweickert (1981, p. 127, 129) hypothesized that the Nevadan orogen of eastern California thickened uniformly during a single late Jurassic folding event. Each hypothesis also implies certain burial histories. If we can read the rock 'flight recorders' with sufficient accuracy, we may be able to distinguish which, if any, of these mechanisms operated in a given orogen.

Penetrative ductile deformation as a burial and denudation mechanism   Pervasive large strains in many metamorphic terranes show that the rigid-block models of orogeny used for much thermal modeling are often poor approximations. If metamorphic recrystallization is synkinematic, thin-section textures may relate mineral growth to the strain history of the rock. If these same minerals record a P–T path, then the P–T path can be correlated with the strain history. Such strains, integrated across an orogen, can reflect substantial displacements. How much burial of rocks is associated with the steep regional-metamorphic foliations observed in some orogens? To what extent is the decompression of some metamorphic terrains the result of vertical thinning and associated development of regional sub-horizontal foliation?

Heat transfer by moving fluids   Combined numerical modeling studies of conductive heat transfer and P–T–τ studies of rocks have the potential to identify and set limits to the extent of heat transfer by moving fluids. If predicted and observed paths do not coincide, then either our observations are in error or the assumption of heat transfer solely by conduction is wrong.

Causes of orogenic magmatism   Arc magmatism is commonly assumed to result from the process of subduction. Consequently, the Cretaceous magmatic arc in the northern North American Cordillera is often ascribed to subduction of the floor of the paleo-Pacific Ocean. However, petrologic and regional-tectonic considerations suggest that the crust in this region was thicker during the Cretaceous (Zen & Hammarstrom, 1984, Zen, 1988 b) and that at least some magmatism probably resulted from anatexis following crustal thickening. Is this process of magma generation energetically feasible? Are the temporal relations of deformation, metamorphism, and plutonism appropriate for anatexis? These questions might be answered by numerical modelling of crustal P–T–τ regimes and by P–T–τ studies of the plutons and associated metamorphic rocks.

Summary

Tectonics and P–T conditions within the Earth's crust are related. This relation can be explored by numerical modelling of P–T–τ regimes and by integrated geochronologic, petrologic, and structural studies. Comparison of the two approaches can expand our knowledge of both petrogenesis and tectonics.

Problems to be addressed by further numerical modeling include the thermal effects of moving aqueous fluids, magmatism, metamorphism, strain heating, and radionuclide redistribution, though these effects may for the most part be minor. Extensional tectonic regimes should be examined more closely. Structurally correct two-dimensional models of low-angle thrust and normal-fault complexes should be constructed.

P–T–τ studies should include error analyses. Many of the searched for P–T–τ effects are probably at the limit of our analytic resolution; moreover, in many cases we do not know the precision of the analysis, thus we cannot always distinguish signal from noise. Full-fledged error analyses of the Gibbs method are needed.

Geochronologic studies should carefully distinguish between thermochronometers (fission-track, most K–Ar analyses) and reaction-progress clocks (U–Pb in zircon, perhaps other isotopic systems). More work directed at characterizing closure temperatures for K–Ar systems will be necessary if the system is to be fully exploited. The rocks of interest in cooling-path studies are especially susceptible to 'excess Ar' problems; these can be resolved, but to do so will require ^{40}Ar–^{39}Ar step-heating studies with low, well-characterized analytical blanks.

Intra- and inter-granular mass transport mechanisms and kinetics during metamorphism are not understood. A useful first step would be to routinely characterize the spatial extent of equilibration during peak metamorphism and later cooling. To do this, microprobe analyses should include sample maps and x–y locations for the analyzed points. Averaged analyses are almost useless.

Inconsistencies and incomplete knowledge of classical thermobarometry do not yet allow us to use it in a routine way. The results are strongly dependent on the thermochemical data and solution models used, as well as the too-often unquestioned assumption of whole-phase thermodynamic equilibrium. Published $P-T$ estimates should be accompanied by discussions of the effects of these uncertainties. Use of the Gibbs method, as formulated by Spear *et al.* (1982), requires a large number of geologic assumptions. These assumptions are best tested by repeating the analysis on other crystals in the same sample and on crystals in other samples having the same history.

Much of our ability to interpret mineral parageneses depends on the postulate of finite domains of local equilibrium (see discussion of Korzhinskiy's work by Thompson, 1959). When we attempt to extract a continuous record of $P-T$ conditions from a zoned garnet, we implicitly assume that diffusion has been incapable of keeping the interior of the garnet in equilibrium with the exterior; that is, equilibrium domains have become infinitesimal. With modern analytical techniques (electron microprobe, laser probe, etc.) the hypothesis of local equilibrium, first proposed by Korzhinskiy, may be empirically tested.

Our ability to read the 'black boxes' – metamorphic rocks – needs to be improved by greater theoretical sophistication, analytical and experimental advances, and, especially, common-sense application of geologic, petrologic, and geochronologic techniques. Rocks do not lie, but our ignorance may condemn us to be suspicious of their truths for a while longer.

Acknowledgements Zen recalls with pleasure his many stimulating geologic and non-geologic discussions with D.S. Korzhinskiy. Conversations with Page Chamberlain, Brooks Hanson, Peter Kelemen, Josh Lieberman, Randy Parrish, Doug Rumble, Frank Spear, John Sutter, and others have helped our education. We thank L.Ya. Aranovich, Brian Patrick, L.L. Perchuk, Richard Sanford, John Sutter, Priestley Toulmin III, and especially Sorena Sorenson for their helpful reviews of the manuscript. Some of the experience on which this paper is based was acquired while Haugerud was supported by a National Research Council – U.S. Geological Survey Research Associateship.

References

Albarede, F. (1975). The heatflow/heat generation relationship; an interaction model of fluids with cooling intrusions. *Earth and Planetary Science Letters*, **27**, 73–8.

— (1976). Thermal models of post-tectonic decompression as exemplified by the Haut-Allier granulites (Massif Central, France). *Bulletin de la Société Géologique Française*, **18**, 1023–32.

Armstrong, R.L., Jäger, E. & Eberhardt, P. (1966). A comparison of K–Ar and Rb–Sr ages on Alpine biotites. *Earth and Planetary Science Letters*, **1**, 13–19.

Berger, G.W. & York, D. (1981). Geothermometry from $^{40}Ar/^{39}Ar$ dating experiments. *Geochimica et Cosmochimica Acta*, **45**, 795–811.

Berman, R.G., Brown, T.H. & Greenwood, H.J. (1985). An internally-consistent thermodynamic data base for minerals in the system $Na_2O–K_2O–CaO–MgO–FeO–Fe_2O_3–Al_2O_3–SiO_2–TiO_2–H_2O–CO_2$. Atomic Energy Authority of Canada Technical Report TR-377.

Bethke, C.M. (1985). A numerical model of compaction-driven groundwater flow and heat transfer and its application to the paleohydrology of intracratonic sedimentary basins. *Journal of Geophysical Research*, **90**, 6817–28.

Birch, F., Roy, R.F. & Decker, E.R. (1968). Heat flow and thermal history in New England and New York. In *Studies in Appalachian geology, northern and Maritime*, ed. E. Zen, W.S. White, J.B. Hadley & J.B. Thompson, Jr, pp. 437–51. New York; Wiley-Interscience.

Bohlen, S.R., Wall, V.J. & Boettcher, A.L. (1983). Experimental investigation and application of garnet granulite equilibria: *Contributions to Mineralogy and Petrology*, **83**, 52–61.

Carmichael, D.M. (1978). Metamorphic bathozones and bathograds: a measure of the depth of post-metamorphic uplift and erosion on the regional scale. *American Journal of Science*, **278**, 769–97.

Carslaw, H.S. & Jaeger, J.C. (1959). *Conduction of heat in solids*, 2nd ed. New York; Oxford University Press.

Chamberlain, C.P. & England, P.C. (1985). The Acadian thermal history of the Merrimack synclinorium in New Hampshire. *Journal of Geology*, **93**, 593–602.

Chamberlain, C.P. & Rumble, D., III (1986). Thermal anomalies in a regional metamorphic terrane: evidence for convective heat transport during metamorphism. *Geological Society of America, Abstracts with Programs*, **18**, 561.

Cliff, R.A. (1985). Isotopic dating in metamorphic belts. *Journal of the Geological Society of London*, **142**, 97–110.

Cliff, R.A. & Cohen, A. (1980). Uranium–lead isotope systematics in a regionally metamorphosed tonalite from the Eastern Alps. *Earth and Planetary Science Letters*, **50**, 211–18.

Cliff, R.A., Jones, G., Choi, W.C. & Lee, T.J. (1985). Strontium isotopic equilibration during metamorphism of tillites from the Ogcheon Belt, South Korea. *Contributions to Mineralogy and Petrology*, **90**, 346–52.

Davis, D., Suppe, J. & Dahlen, F.A. (1983). Mechanics of fold-and-thrust belts and accretionary wedges. *Journal of Geophysical Research*, **88**, 1153–72.

Day, H.W. (1985). Distinguishing among major controls on the apparent thermal and barometric structure of metamorphic belts. *Geological Society of America, Abstracts with Programs*, **17**, 559.

Dodson, M.H. (1973). Closure temperature in cooling geochronological and petrological systems. *Contributions to Mineralogy and Petrology*, **40**, 259–74.

Draper, G. & Bone, R. (1981). Denudation rates, thermal evolution, and preservation of blueschist terrains. *Journal of Geology*, **89**, 601–13.

England, P. & McKenzie, D. (1982). A thin viscous sheet model for continental deformation. *Geophysical Journal of the Royal Astronomical Society*, **70**, 295–321.

—— (1983). Correction to: a thin viscous sheet model for continental deformation. *Geophysical Journal of the Royal Astronomical Society*, **73**, 523–32.

England, P.C. & Richardson, S.W. (1977). The influence of erosion upon the mineral facies of rocks from different metamorphic environments. *Journal of the Geological Society of London*, **134**, 201–13.

England, P. & Thompson, A.B. (1984). Pressure–temperature–time paths of regional metamorphism, Pt. I. Heat transfer during the evolution of regions of thickened crust. *Journal of Petrology*, **25**, 894–928.

Essene, E.J. (1982). Geologic thermometry and barometry. In Ferry (1982), pp. 153–206.

Ferry J.M. (ed.) (1982). *Characterization of metamorphism through mineral equilibria*. Mineralogical Society of America, Reviews in Mineralogy 10.

Fyfe, W.S., Price, N.J. & Thompson, A.B. (1978). *Fluids in the Earth's crust*. Amsterdam: Elsevier.

Ganguly, J. & Kennedy, G.C. (1974). The energetics of natural garnet solid solution, pt. 1, mixing of the aluminosilicate end-members. *Contributions to Mineralogy and Petrology*, **48**, 137–48.

Ganguly, J. & Saxena, S.K. (1984). Mixing properties of aluminosilicate garnets: constraints from natural and experimental data, and applications to geothermo-barometry. *American Mineralogist*, **69**, 88–97.

Garven, G. & Freeze, R.A. (1984a). Theoretical analysis of the role of groundwater flow in the genesis of stratabound ore deposits, pt 1, mathematical and numerical model. *American Journal of Science*, **284**, 1085–124.

—— (1984b). Theoretical analysis of the role of groundwater flow in the genesis of stratabound ore deposits, pt 2, quantitative results. *American Journal of Science*, **284**, 1125–74.

Gebauer, D. & Grünenfelder, M. (1974). Rb–Sr whole-rock dating of late diagenetic to anchimetamorphic, Paleozoic sediments in southern France (Montagne Noire). *Contributions to Mineralogy and Petrology*, **47**, 113–30.

Giletti, B.J. (1986). Diffusion effects on oxygen isotope temperatures of slowly cooled igneous and metamorphic rocks. *Earth and Planetary Science Letters*, **77**, 218–28.

Glebovitskiy, V.A. (1977). Mineral facies as criteria of estimates of parameters of metamorphism. In *Thermo- and barometry of metamorphic rocks*, ed. V.A. Glebovitskiy, pp. 5–39. Leningrad: Nauka.

Haas, J.L., Jr (1974). Phas20, a program for simultaneous multiple regression of a mathematical model to thermochemical data. U.S. Department of Commerce, National Technical Information Service, AD-780-301.

Hanson, R.B. & Barton, M.D. (1986). A model for regionally extensive low-*P* metamorphism in plutonic belts with applications in the Cordillera. *Geological Society of America, Abstracts with Programs*, **18**, 627–8.

Harrison, T.M. (1981). Diffusion of ^{40}Ar in hornblende. *Contributions to Mineralogy and Petrology*, **78**, 324–41.

Harrison, T.M. (1983). Some observations on the interpretation of ^{40}Ar/^{39}Ar age spectra. *Isotope Geoscience*, **1**, 319–38.

Harrison, T.M. & Clarke, G.K.C. (1979). A model of the thermal effects of igneous intrusion and uplift as applied to Quottoon pluton, British Columbia. *Canadian Journal of Earth Sciences*, **16**, 411–20.

Harrison, T.M. & FitzGerald, J.D. (1986). Exsolution in hornblende and its consequences for ^{40}Ar/^{39}Ar age spectra and closure temperature. *Geochimica et Cosmochimica Acta*, **50**, 247–53.

Harrison, T.M. & McDougall, I. (1980). Investigations of an intrusive contact, northwest Nelson, New Zealand, pt II. Diffusion of radiogenic and excess ^{40}Ar in hornblende revealed by ^{40}Ar/^{39}Ar age spectrum analysis. *Geochimica et Cosmochimica Acta*, **44**, 2005–20.

—— (1982). The thermal significance of potassium feldspar K–Ar ages inferred from ^{40}Ar/^{39}Ar age spectrum results. *Geochimica et Cosmochimica Acta*, **46**, 1811–20.

Harrison, T.M., Duncan, I. & McDougall, I. (1985). Diffusion of ^{40}Ar in biotite: temperature, pressure and compositional effects. *Geochimica et Cosmochimica Acta*, **49**, 2461–8.

Haugerud, R.A. (1985). Geology of the Hozameen Group and the Ross Lake shear zone, Maselpanik area, North Cascades, southwest British Columbia. Unpublished Ph.D. dissertation, University of Washington.

—— (1986). 1DT – interactive screen-oriented micro-computer program for simulation of 1-dimensional geothermal histories. U.S. Geological Survey, Open-file Report 86-511, 19 pp. and 5¼" diskette.

—— (1987). The initial-argon correction and interpretation of ^{40}Ar/^{39}Ar step-heating data [abstract]. *EOS*, **68**, 431.

Helgeson, H.C., Delany, J.M., Nesbitt, H.W. & Bird, D.K. (1978). Summary and critique of the thermodynamic properties of rock-forming minerals. *American Journal of Science*, **278A**, 1–299.

Hodges, K.V. & Royden, L. (1984). Geologic thermobarometry of retrograded metamorphic rocks: an indication of the uplift trajectory of a portion of the northern Scandinavian Caledonides. *Journal of Geophysical Research*, **89**, 7077–90.

Hodges, K.V. & Spear, F.S. (1982). Geothermometry, geobarometry and the Al_2SiO_5 triple point at Mt. Moosilauke, New Hampshire. *American Mineralogist*, **67**, 1118–34.

Hofmann, A.W. & Grauert, B. (1973). Effect of regional metamorphism on whole-rock Rb–Sr systems in sediments: Carnegie Institution of Washington, Yearbook 72, pp. 299–302.

Hollister, L.S. & Crawford, M.L. (ed.) (1981). *Short course in fluid inclusions: applications to petrology*. Mineralogical Association of Canada, Short Court Handbook 6.

Jäger, E. (1970). Rb–Sr systems in different degrees of metamorphism: *Eclogae Geologicae Helvetiae*, **63**, 163–72.

Jurewicz, S.R. & Watson, E.B. (1985). The distribution of partial melt in a granitic system: application of liquid phase sintering theory. *Geochimica et Cosmochimica Acta*, **49**, 1109–21.

Korzhinskiy, D.S. (1959). Physicochemical basis of the analysis of the paragenesis of minerals. New York: Consultants Bureau.

— (1973). Theoretical basis for the analysis of minerals. Moscow: Nauka.

Lachenbruch, A.H. (1968). Preliminary geothermal model of the Sierra Nevada. *Journal of Geophysical Research*, **73**, 6977–89.

Lasaga, A.C. (1983). Geospeedometry: an extension of geothermometry. In *Advances in physical geochemistry*, vol. 3, *Kinetics and equilibrium in mineral reactions*, ed. S.K. Saxena, pp. 81–114. New York: Springer–Verlag.

Lux, D.R., DeYoreo, J.J., Guidotti, C.V. & Decker, E.R. (1986). Role of plutonism in low-pressure metamorphic belt formation. *Nature*, **323**, 794–7.

Marakushev, A.A. (1965). *Problems in mineral facies of metamorphic and metasomatic rocks*. Moscow: Nauka.

Marsh, B.D. (1984). Mechanics and energetics of magma formation and ascension, in *Explosive volcanism: inception, evolution, and hazards*, pp. 67–83. National Academy of Sciences Press, Studies in Geophysics.

Mattinson, J.M. (1978). Age, origin, and thermal histories of some plutonic rocks from the Salinian block of California. *Contributions to Mineralogy and Petrology*, **67**, 233–45.

— (1986). Geochronology of high-pressure–low-temperature Franciscan metabasites: a new approach using the U–Pb system. In *Blueschists and related eclogites*, ed. E.H. Brown & B.W. Evans, pp. 95–105. Geological Society of America Memoir 164.

McNutt, M. & Royden, L. (1987). Extremal bounds on geotherms in eroding mountain belts from metamorphic pressure–temperature conditions. *Geophysical Journal of the Royal Astronomical Society*, **88**, 81–95.

Newton, R.C. & Haselton, H.T. (1981). Thermodynamics of the garnet–plagioclase–Al_2SiO_5-quartz geobarometer. In *Thermodynamics of minerals and melts*, ed. R.C. Newton, A. Navrotsky & B.J. Wood, pp. 131–47.

Norton, D. & Knight, J. (1977). Transport phenomena in hydrothermal systems: cooling plutons. *American Journal of Science*, **277**, 937–81.

Oxburgh, E.R. & Turcotte, D. (1974). Thermal gradients and regional metamorphism in overthrust terrains with special reference to the eastern Alps. *Schweizerische Mineralogische und Petrographische Mitteilungen*, **54**, 641–62.

Parrish, R. (1982). Cenozoic thermal and tectonic history of the Coast Mountains of British Columbia as revealed by fission track geological data and quantitative thermal models [Ph.D. dissertation]. Vancouver: University of British Columbia.

Paterson, C.J. (1986). Controls on gold and tungsten mineralization in metamorphic-hydrothermal systems, Otago, New Zealand. In *Turbidite-hosted gold deposits*, ed. J.D. Keppie, R.W. Boyle & S.J. Haynes. Geological Association of Canada, Special Paper 32.

Peacock, S.M. (1987). Advective heat transfer by metamorphic fluid flow [abstract]. *EOS*, **68**, 466.

Perchuk, L.L., Aranovich, L.Ya., Podlesskii, K.K., Lavrant'eva, I.V., Gerasimov, V.Yu., Fed'kin, V.V., Kitsul, V.I., Karsakov, L.P. & Berdnikov, N.V. (1985). Precambrian granulites of the Aldan shield, eastern Siberia, USSR. *Journal of Metamorphic Geology*, **3**, 265–310.

Pidgeon, R.T. & Aftalion, M. (1978). Cogenetic and inherited zircon U–Pb systems in granites: Palaeozoic granites of Scotland and Britain. In *Crustal evolution in northwestern Britain and adjacent regions*, pp. 183–220. Geological Journal, Special Paper 10.

Powell, R. (1985). Geothermometry and geobarometry: a discussion. *Journal of the Geological Society of London*, **142**, 29–38.

Purdy, J.W. & Jäger, E. (1976). K–Ar ages on rock-forming minerals from the Central Alps. *Memorie degli Istituti di Geologia e Mineralogia dell'Universita di Padova*, **30**.

Rosenfield, J.L. (1970). *Rotated garnets in metamorphic rocks*. Geological Society of America, Special Paper 129.

Royden, L. & Hodges, K.V. (1984). A technique for analyzing the thermal histories of eroding orogenic belts: a Scandinavian example. *Journal of Geophysical Research*, **89**, 7091–106.

Rumble, D., III (1974). Gibbs phase rule and its application in geochemistry. *Journal of the Washington Academy of Sciences*, **64**, 199–208.

— (1976). The use of mineral solid solutions to measure chemical potential gradients in rocks. *American Mineralogist*, **61**, 1167–74.

Schweickert, R.A. (1981). Tectonic evolution of the Sierra Nevada Range. In *The geotectonic development of California, Rubey volume I*, ed. W.G. Ernst, pp. 87–131. Englewood Cliffs, New Jersey: Prentice-Hall.

Selverstone, J., Spear, F.S., Franz, G. & Morteani, G. (1984). High pressure metamorphism in the SW Tauern window, Austria: *P–T* paths from hornblende–kyanite–staurolite schists. *Journal of Petrology*, **25**, 501–31.

Smith, C.S. (1964). Some elementary principles of polycrystalline microstructure. *Metallurgical Reviews*, **9**, 1–48.

Spear, F.S. (1986a). PTPATH: a FORTRAN program to calculate pressure–temperature paths from zoned metamorphic garnets. *Computers and Geosciences*, **12**, 247–66.

— (1986b). The Gibbs method, Duhems Theorem, and $P–T–X$(Fe–Mg–Mn) relations in pelites [abstract]. *EOS*, **67**, 407.

Spear, F.S. & Chamberlain, C.P. (1986). Metamorphic and tectonic evolution of the Fall Mountain nappe complex and adjacent Merrmack synclinorium. In *Regional metamorphism and metamorphic phase relations in northwestern and central New England*, ed. P. Robinson, pp. 121–43. Department of Geology and Geography, University of Massachusetts, Amherst. Contribution 59.

Spear, F.S., Ferry, J.M. & Rumble, D., III (1982). Analytical formulation of phase equilibria: the Gibbs method. In Ferry (1982), pp. 105–52.

Spear, F.S. & Rumble, D., III (1986). Pressure, temperature, and structural evolution of the Orfordville Belt, west-central New Hampshire. *Journal of Petrology*, **27**, 1071–93.

Spear, F.S. & Selverstone, J. (1983). Quantitative *P–T* paths from zoned minerals: theory and tectonic applications. *Contributions to Mineralogy and Petrology*, **83**, 348–57.

St-Onge, M.R. & King, J.E. (1985). Zoned poikiloblastic garnets: documentation of *P–T* paths of syn-metamorphic uplift, Wopmay Orogen, N.W.T. *Geological Association of Canada, Programs and Abstracts*, **10**, A60.

Sutter, J.F. & Hartung, J.B. (1984). Laser microprobe $^{40}Ar/^{39}Ar$ dating of mineral grains *in situ*. *Scanning Electron Microscopy*, **4**, 1525–9.

Thompson, J.B. (1959). Local equilibrium in metasomatic processes. In *Researches in geochemistry*, ed. P.H. Abelson, pp. 427–57. New York: Wiley.

— (1967). Thermodynamic properties of simple solutions. In *Researches in Geochemistry*, ed. P.H. Abelson, pp. 340–61. New York: Wiley.

Thompson, A.B. & England, P.C. (1984). Pressure–temperature–time paths of regional metamorphism II. Their inference and interpretation using mineral assemblages in metamorphic rocks. *Journal of Petrology*, **25**, 929–55.

Turner, G. (1968). The distribution of potassium and argon in chondrites. In *Origin and distribution of the elements*, ed. L.H. Ahrens, pp. 387–98. London: Pergamon.

Waff, H.S. & Bulau, J.R. (1982). Experimental determination of near-equilibrium textures in partially-molten silicates at high pressures. In *High pressure research in geophysics*, ed. S. Akimoto & Manghnani, pp. 229–36. Center for Academic Publications.

Watson, E.B. & Harrison, T.M. (1983). Zircon saturation revisited: temperature and composition effects in a variety of crustal magma types. *Earth and Plantary Science Letters*, **64**, 295–304.

Williams, I.S, Compston, W., Black, L.P., Ireland, T.R. & Foster, J.J. (1984). Unsupported radiogenic Pb in zircon: a cause of anomalously high Pb–Pb, U–Pb and Th–Pb ages. *Contributions to Mineralogy and Petrology*, **88**, 322—7.

Winkler, H.G.F. (1974). *Petrogenesis of metamorphic rocks*. New York: Springer–Verlag.

Zeitler, P. (1985). Cooling history of the NW Himalaya, Pakistan. *Tectonics*, **4**, 127–51.

Zeitler, P. & FitzGerald, J.D. (1986). Saddle-shaped $^{40}Ar/^{39}Ar$ age spectra from young, microstructurally complex potassium feldspars. *Geochimica et Cosmochimica Acta*, **50**, 1185–99.

Zen, E-an (1977). The phase-equilibrium calorimeter, the petrogenetic grid, and a tyranny of numbers. *American Mineralogist*, **62**, 189–204.

Zen, E-an (1988 a). Bedrock geology of the Vipond Park 15 minute, Stine Mountain 7½ minute, and Maurice Mountain 7½ minute quadrangles, Pioneer Mountains, Beaverhead County, Montana. U.S. Geological Survey, Bulletin 1625.

— (1988 b). Tectonic significance of high-pressure plutonic rocks in the western cordillera of North America. In *Metamorphism and crustal evolution of the western United States, VII Rubey Symposium*, ed. W.G. Ernst. Englewood Cliffs, New Jersey: Prentice-Hall, pp. 41–67.

— (1988 c). Thermal modelling of stepwise anatexis in a thrust-thickened sialic crust. *Royal Society of Edinburgh, Trans., Earth Sciences*, **79**, 223–35.

Zen, E-an & Hammarstrom, J.M. (1984). Magmatic epidote and its petrologic significance. *Geology*, **12**, 515–18.

PART III

The mantle and magmatic processes

13

Complications in the melting of silicate minerals from atmospheric to high pressures*

ART MONTANA, R.W. LUTH, B.S. WHITE, S.L. BOETTCHER,
K.S. McBRIDE and J.F. RICE

Introduction

It is a pleasure to contribute to a volume honoring the late Academician Korzhinskii. He has made major advances to our knowledge of volatile components, not only in systems undergoing metasomatic changes, but in igneous systems as well. Although our understanding of melting and the role of fluids has increased noticeably in recent years, significant gaps remain in our understanding of the melting behaviors of the feldspars and most of the other rock-forming minerals, not only at high pressures, but at atmospheric pressure as well. Largely missing is a knowledge of the temperatures of the solidi and the liquidi. Also unknown in any detail are the structures, thermal properties, and physical properties of the liquids *in equilibrium with crystals*. Methods such as calorimetry and spectroscopy have not been used to study liquids *in situ* at elevated pressures. In this paper, we will consider our use of phase equilibria in chemically simple systems to shed light on the structural and thermal properties of magmas.

Experimental methods

One sample of anorthite (sample B) was synthesized from gel, prepared using the method of W.C. Luth & Ingamells (1965) and crystallized hydrothermally at 5 kbar, 1050 °C for at least 48 hours. Another was crystallized from a glass synthesized from oxide reagents, the analysis of which is in Table 13.1 (sample Sc). Synthetic sanidine (sample Ss) was similarly prepared from glass and crystallized at 2 kbar, 700 °C for 30 days (see Table 13.1). Three different samples of albite were used. One was a natural

* Institute of Geophysics and Planetary Physics. Contribution No. 3030.

Table 13.1. *Compositions of starting materials*

	Albite CL	Sanidine Ss	Anorthite Sc
SiO_2	68.75	64.75	43.12
TiO_2	0.00	0.00	0.0
Al_2O_3	19.51	18.03	35.76
Fe_2O_3	0.01	0.02	0.05
MgO	0.00	0.00	0.0
CaO	0.03	0.00	20.27
Na_2O	11.76	0.07	0.16
K_2O	0.04	16.52	0.0
H_2O^-	0.00	0.19	—
H_2O^+	—	0.59	—
Total	100.10	100.17	99.36

low albite (see Table 13.1, sample CL) (see R.W. Luth & Boettcher, 1986); another was converted to high albite by treating this low albite at 20 kbar, 1175 °C for 26 hours in an assembly buffered by hematite (sample Ch). The third sample was prepared from gel and crystallized hydrothermally at 2 kbar, 700 °C for at least 30 days (sample S). The quartz was prepared from natural Brazilian material (see Boettcher, 1984). All starting materials were dried for 24 hours at 800 °C prior to using. Diopside was crystallized from gel of $CaMgSi_2O_6$ composition.

For the high-pressure experiments, ~ 5 mg of the starting material were loaded into an open, 2.0-mm Pt capsule and dried at 900 °C for 16 hours, and the capsule was welded shut. In some experiments, $Ag_2C_2O_4$ added to the sample capsule served as the source of CO_2. This capsule, together with ~ 350 mg of Fe_2O_3, was sealed into a 3.5-mm Pt capsule, inhibiting the ingress of H_2 into the sample. For the H_2O–H_2 experiments, iron–wustite, hematite–magnetite, nickel–nickel oxide, wustite–magnetite, or pure H_2O were used to buffer f_{H_2}. These experiments were performed in piston–cylinder apparatus, with 2.54-cm furnace assemblies composed of NaCl, pyrex, graphite, BN, and MgO (Boettcher *et al.*, 1981). The capsules were run horizontally, with temperature monitored by WRe_3/WRe_{25} or $Pt/Pt_{90}Rh_{10}$ thermocouples, encased in 99.8% pure alumina tubing and in contact with the top of the capsule. We used the hot-piston-in technique to bring the experiments to run conditions, increasing the pressure to $\sim 90\%$ of the final temperature, then increasing the pressure and finally the temperature to the desired values. The pressure was controlled to ± 0.05 kbar, and the temperature to ± 3 °C. After the outer capsule was opened, the buffer was examined microscopically to ensure that $< 50\%$ hematite had converted to magnetite. To demonstrate the attainment of equilibrium, we reversed the solidi at several pressures, using two-stage experiments. First, we placed the sample under conditions known from previous experiments to melt or partly melt the sample. Then the temperature was lowered to sub-solidus conditions, again determined from previous experiments; successful reversals contained no quenched liquid.

For experiments at atmospheric pressure, ~ 5 mg of the starting material were loaded into Pt or Ag–Pd tubing welded at one end and then crimped, but not completely sealed, at the other. For several experiments with albite at 1064 °C, the sample was dried in the capsule at 900 °C for 24 hours and then the capsule was welded shut. This ensured that the melting observed at similar temperatures in other experiments was not the result of volatilization of Na from the albite. The capsules, together with a Pt/Pt_{90}–Rh_{10} thermocouple, were partly immersed in Al_2O_3 powder contained in a large alumina crucible. This assembly was run in custom-built furnaces with $MoSi_2$ elements and with temperature controlled to ± 1 °C. The temperature-measuring system was calibrated before and after every experiment using a standard thermocouple that was periodically calibrated against the melting of Au.

Using a petrographic microscope, the appearance of glass that was interpreted to be quenched liquid indicated that the sample was above the temperature of the beginning of melting. The structural state of the albite in the run products was determined using the method of Kroll & Ribbe (1980).

Results

Albite at atmospheric pressure

The results for albite are in Table 13.2. Surprisingly, the temperature of the solidi for the natural and synthetic high albites is 955 ± 5 °C, and that for the low albite is ~ 950 °C. This is in marked contrast to all previously reported values, which range from 1100 to 1120 °C (Bowen, 1913, Greig & Barth, 1938, Schairer & Bowen, 1956, Boettcher et al., 1982, Navrotsky et al., 1982). To ensure that something was not dreadfully wrong with our temperature-measuring system, Prof. Edward Stolper at Caltech corroborated our results in his laboratory using our natural and synthetic high albites (see runs 3272, 3273, 3298, and 3299). Because of slow reaction rates near the beginning of melting, none of our near-solidus experiments resulted in complete melting; i.e., crystals coexist with liquid stably or metastably above the melting temperature. To investigate whether this observation may be the result of stable, incongruent melting of albite, the durations of a series of experiments were varied to see if longer runs produced more melting, as would be the case if only kinetic factors were involved. These results are inconclusive; e.g., run #3299 (1064 °C/12 days) produced $\sim 30\%$ glass, compared to the $\sim 3\%$ glass in #3252 (1055 °C/7 days) but run #3369 (1056 °C/30 days) also produced $\sim 30\%$ glass. It is possible that albite melts incongruently over a wide temperature range, and experiments of much longer duration are in progress to establish this. Microprobe analyses of the quenched liquids suggest that this partial melting produces a liquid richer in SiO_2 than is albite, but the results are not yet precise enough to require incongruent melting. Previously, Pugin & Soldatov (1973, 1976) proposed on the basis of high-pressure experiments that albite and sodic plagioclase melt incongruently to mullite + Na-rich liquid, with a solidus of ~ 1050 °C at

Table 13.2. *Definitive experiments for the melting of feldspar-bearing assemblages at atmospheric pressure.*

Run no.	Material	Temp °C	Duration	Results
Albite				
3034	S	1100	6 days	~100% glass
3035	Ch	1100	6	~50% glass
3496	Ch	1081	120	30–40% glass
3497	CL	1081	120	~50% glass
3498	S	1081	120	30–40% glass
3495	LL	1081	120	some glass
3272[a]	S	1064	12	~30% glass
3298[a]	Ch	1064	12	>30% glass
3299	S	1064	12	>30% glass
3273	Ch	1064	12	~30% glass
3369	S	1056	30	~30% glass
3252	S	1055	7	~3% glass
3380	CL	961	62	>5% glass
3381	S	961	62	5–10% glass
3382	Ch	961	62	~1% glass
3357	S	951	32	no glass
3358	CL	951	32	trace glass
3359	Ch	951	32	no glass
Albite + quartz				
3148	S	940	51 days	~25% glass
3519	Ch + Qz	930	75	~1% glass
3518	CL + Qz	930	75	~2% glass
3520	S + Qz	930	75	~2% glass
3405	Ch + Qz	901	76	no glass
3406	CL + Qz	901	76	trace glass
3407	S + Qz	901	76	trace glass
Sanidine				
3364	Ss	1060	30 days	~3% glass
3402	Ss	1055	42	~2% glass
2708	Ss	1050	17	no glass
3409	Ss	1040	50	no glass
Sanidine + quartz				
3258	Ss + Qz	980	50 days	~30% glass
3383	Ss + Qz	965	56	~30% glass
3343	Ss + Qz	962	53	no glass
2954	Ss + Qz	950	50	no glass
Anorthite				
3436	Sc	1555	30 min	100% glass
3420	B	1555	30	100% glass
3433	Sc	1550	60	no glass
3434	B	1550	60	100% glass
3432	B	1540	90	no glass
Anorthite + quartz				
3410	$B_{40} + Qz_{60}$	1120	39 days	100% glass
3515	B + Qz	1116	138	~2% glass
3415	B + Qz	1110	34	no glass
3404	B + Qz	1101	40	no glass

Table 13.2. (*cont.*)

Notes:
Abbreviations: See also text and Table 13.1. S = synthetic albite. CL = Coleman's law albite. Ch = Coleman's albite converted to high albite. Qz = natural quartz. Ss = synthetic sanidine. B = synthetic anorthite crystallized from gels. Sc = synthetic anorthite crystallized from glass.
[a] Run in the laboratory of Ed Stolper at the California Institute of Technology.

atmospheric pressure. However, it is possible that the high-temperature treatment of their glass starting materials resulted in the volatilization of Na, with subsequent crystallization of mullite during experimentation.

In response to some of our previously published results on the melting of albite, Navrotsky *et al.* (1982) pointed out that our low melting temperatures conceivably could be the result of quartz contaminant in the natural albite and excess SiO_2 in the synthetic albite. This is *conceptually* possible, but it is unlikely for the following reasons. As shown in Table 13.1, chemical analyses of the natural and the synthetic albites reveal no excess SiO_2. Very small proportions of excess SiO_2 that are not detected by such chemical analyses could not be expected to lower the temperature of the solidus by ~ 150 °C. In addition, the temperature of the albite–quartz eutectic of Schairer & Bowen (1956) is considerably higher than that of our revised solidus for albite (their eutectic for albite–tridymite is at ~ 1062 °C–the albite–quartz eutectic would be at a slightly lower temperature). Finally, as discussed below, our revised value for the temperature of the albite–quartz solidus is also considerably lower than previously published values. This cannot be ascribed to excess SiO_2 because the presence of quartz in the reactants and the products buffers the $aSiO_2$ for any given set of pressure and temperature.

Albite + quartz at atmospheric pressure

The results of our determination of the beginning of melting of high albite + quartz are in Table 13.2. Our tentative results yield a value for the eutectic of about 910 °C, compared to our value of 955° for the albite solidus. Schairer & Bowen (1956) obtained 1062 °C for the albite–critobalite eutectic (illustrated in their Fig. 6 as albite–tridymite – see their page 162); assuming ideal behavior, we calculate that this corresponds to 1040 °C for the albite–quartz eutectic. Because of the low temperature, melting at the solidus is very sluggish, and runs of at least two months are required to produce significant proportions of liquid. Somewhat higher temperatures significantly increase the reaction rates; e.g. one run (#3148) about 30 °C above the solidus produced about 25% liquid in 51 days.

Although it is the albite–tridymite assemblage that is stable at the one-atmosphere eutectic, we are interested in the metastable albite–quartz eutectic, because this is the assemblage that is stable at high pressures. Fortunately, this is experimentally realizable because quartz in the reactants persists throughout the longest runs and shows no tendency to invert to tridymite.

Sanidine at atmospheric pressure

At pressure below ~ 15 kbar, sanidine melts incongruently to leucite + SiO_2-rich liquid; all liquids on the join leucite–silica are extremely viscous, and Schairer & Bowen (1955) were unable to crystallize any glasses on this join, even in experiments of as long as five years. We find that experiments of at least one month duration are required to produce $\sim 2\%$ liquid within 10 °C of the solidus. Although we have yet been unable to reverse the reaction, our results yield a value of 1053 ± 5 °C for the solidus of sanidine.

Although the value of 1150 ± 20 °C shown by Schairer and Bowen is the accepted temperature for the solidus of sanidine at atmospheric pressure, it appears that they did not crystallize any $KAlSi_3O_8$ glass below 1250 °C, and they did not attempt to melt *crystalline* $KAlSi_3O_8$. Morey & Bowen (1922), the first to recognize the nature of the incongruent melting of sanidine, attempted to locate the solidus using a hydrothermally crystallized sanidine and three natural potash feldspars in runs up to eight-days duration. They tentatively placed the solidus at 1170 °C, even though they did note traces of glass in the run products at lower temperatures.

Sanidine + quartz at atmospheric pressure

Our experiments to determine the solidus of sanidine + quartz yield a value of 965 ± 3 °C. The stable solid assemblage at the eutectic is sanidine + tridymite, but it is the metastable quartz-bearing assemblage that is of interest in this study because it is stable at higher pressures. Fortunately, as in the case of albite + quartz, the quartz in this system persisted metastably for the longest run durations with no evidence of transforming to tridymite.

Our solidus for sanidine–quartz at 965 ± 3 °C compares with the accepted value of Schairer & Bowen (1955) of 990 ± 20 °C for sanidine–tridymite (actually determined for sanidine–cristobalite) which we calculate to be equivalent to 965 ± 20 °C for the quartz-bearing eutectic assemblage, assuming ideal mixing. However, this agreement is fortuitous, as they obtained this number by extrapolating the liquidi of sanidine and cristobalite from 1250 °C and 1218 °C, respectively, with no liquidus data at lower temperatures.

Anorthite at atmospheric pressure

Our experiments on anorthite yield a value for the solidus of 1553 ± 3 °C. This agrees with the accepted value of 1553 ± 2 °C (on the Geophysical Laboratory scale – comparable to 1558 ± 2 °C on the 1968 International Temperature Scale), generally ascribed to Schairer & Bowen (1947), but which appears to have been determined by Bowen (1913) and by Osborn (1942).

The general agreement on the melting temperature of anorthite is mostly attributable to the ease with which equilibrium is attained at these high temperatures. Experiments just several °C above the solidus for 30 minutes will produce complete melting. Also, the absence of alkalis in anorthite greatly reduces any change in composition resulting from incongruent volatilization during firing and drying of glasses or gels.

Anorthite + quartz at atmospheric pressure

Our results for the determination of the anorthite–quartz eutectic are the only data for the melting of this assemblage, and they indicate a temperature of 1113 ± 10 °C. Reaction rates are moderate, and a run at 1120 °C for 39 days produced complete melting. As is true for the other compositions described earlier, we are interested in the quartz-bearing assemblages, although they are metastable relative to those containing cristobalite. Even at temperatures in excess of 1300 °C, the quartz in our run products showed no inversion to other polymorphs in runs up to five days.

Schairer & Bowen (1947) obtained a value of 1368 ± 2 °C for the anorthite–cristobalite eutectic, which agrees with the later results of Longhi & Hays (1979). Using this value for the anorthite–cristobalite eutectic and assuming ideal mixing in the liquid, we calculate that the eutectic for anorthite–quartz is 110 °C lower, or ~ 1255 °C. In a different approach, using only calorimetric data (Richet & Bottinga, 1986) and the melting temperatures of the end-members, and assuming ideal mixing in the liquid, we calculate a temperature of the anorthite–quartz eutectic of 1337 °C (which compares with our experimental value of 1113 °C). It appears that the mixing of $CaAl_2Si_2O_8$ and SiO_2 components in the liquid is *very* nonideal at the low temperatures of our eutectic, and calculating equilibria in this system without experimental data to pin it on would not be particularly useful.

Why are our revised melting temperatures so much lower than all previous determinations? At least with albite, the melting may be incongruent, with solidus temperatures much lower than those of the liquidus. Most importantly, equilibrium was not attained in the earlier experiments. Few of the solidi determined previously were obtained using crystalline starting materials, and none of the experiments was reversed. In addition, we know now that runs of hours or even several days are insufficient to produce detectable proportions of liquid in these feldspar and feldspar–quartz systems at temperatures near the solidi. For example, using two-day experiments to melt albite, which is longer than those used by previous investigators, quenched liquid was not detected in experiments 100 °C above our revised solidus, whereas $\sim 20\%$ glass appeared in the products of 20-day experiments at the same temperatures. The degree to which these crystalline assemblages superheat for long periods is even greater than we imagined. Also, the high viscosities of glasses and liquids in systems such as $NaAlSi_3O_8$–SiO_2 at temperatures below ~ 900 °C and the extreme sluggishness of the melting reactions bring up the possibility that the stable solidi in these systems may not be determinable during the duration of laboratory experiments; geologic time may be required to produce significant amounts of liquid at temperatures near the beginning of melting.

Silicate–H₂O–H₂ systems at high pressures

As described in detail by Luth & Boettcher (1986), we determined the solidi of albite, diopside, and quartz in the presence of H_2O–H_2 vapors over a range of f_{H_2} from 5 to 30 kbar (see Figs. 13.1, 13.2, and 13.3). These experiments were designed to assess the

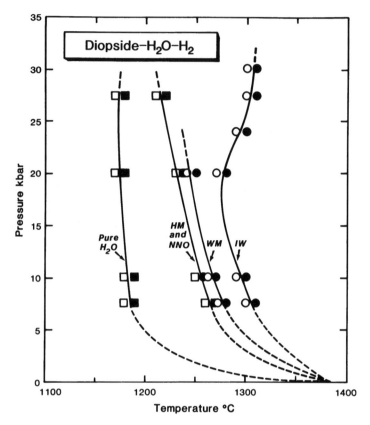

Fig. 13.1. Pressure–temperature projection of the vapor-saturated solidi of diopside in the system $CaMgSi_2O_6$–H_2O–H_2 for various values of the activity of hydrogen in the vapor (after Luth & Boettcher, 1986). Closed symbols represent hypersolidus conditions. Buffers: hematite-magnetite (HPl); Ni–Nio(NNO), mestite magnetite (WM) and iron-wustite (IW).

solubility and solubility mechanisms of H_2 relative to those of H_2O and CO_2. In all three of these systems, increasing f_{H_2} increases the temperatures of the solidi much more than is calculated assuming ideal dilution of the H_2O. This is consistent with the observations of Persikov & Epel'baum (1985) at pressures below 2 kbar that the solubility of H_2 in albite liquid is much less than is that of H_2O.

For diopside, increasing f_{H_2} from that of pure H_2O, to HM (and N–NO), to WM, to IW progressively increases the temperature of the solidus, the change being $\sim 110\,°C$ at 10 kbar. For the IW buffer, there is a pronounced temperature minimum above 15 kbar, and the increase is $\sim 130\,°C$ at higher pressures.

We examined albite–H_2O–H_2 to only ~ 17 kbar because albite reacts to jadeite + quartz at higher pressures (Fig. 13.2). The temperature difference between the solidus for pure H_2O and that for IW conditions is $\sim 50\,°C$ over most of the pressure range.

For quartz–H_2O–H_2, increasing f_{H_2} also increases the temperature of the solidus, with measurable differences between the solidi for pure H_2O and HM and between

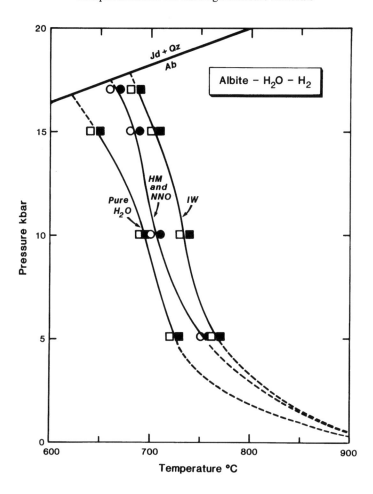

Fig. 13.2. Pressure–temperature projection of the vapor-saturated solidi of albite in
the system $NaAlSi_3O_8$–H_2O–H_2 (after Luth & Boettcher, 1986). Closed
symbols represent hypersolidus conditions. Jd is jadeite.

those for HM and IW (Fig. 13.3). All of the solidi terminate in critical end-points or
critical points, but the most unusual features are the inflections in these curves, which
have been tightly constrained by the reversal experiments of Luth & Boettcher (1986).
They attributed these inflections to pressure-induced structural changes in the liquids
that increase the solubility of H_2O and/or H_2. Goldsmith & Jenkins (1985) observed a
metastable inflection in the solidus of low albite–H_2O at ~ 6 kbar, and similar
phenomena may be detected in other systems when they are examined in detail.

Raman spectroscopic studies of quenched $NaAlSi_3O_8$–H_2 liquids (glasses) and
others in the system Na_2O–Al_2O_3–SiO_2–H_2 (Luth et al., 1987) suggest that hydrogen
dissolves primarily as molecular H_2, thus differing from CO_2 and H_2O (e.g., Fine &
Stolper, 1985, Mysen & Virgo, 1980). However, E.S. Persikov (written communication,
1987) believes that H_2 in the spectra of the quenched albitic glasses probably results
from submicroscopic bubbles.

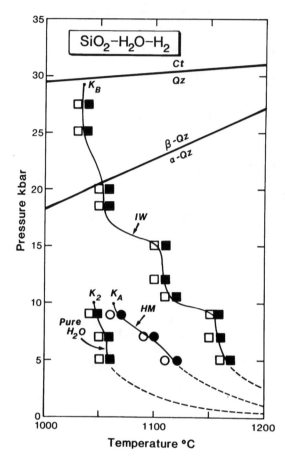

Fig. 13.3. Pressure–temperature projection of the vapor-saturated solidi of quartz in the system SiO_2–H_2O–H_2 (after Luth & Boettcher, 1986). Closed symbols represent hypersolidus conditions. K_2 is a critical end-point in the system SiO_2–H_2O, and K_A and K_B are critical points in the system SiO_2–H_2O–H_2. Ct is coesite.

Silicate–CO_2 systems at high pressures

Because of the differing melting behaviors of albite, anorthite, and sanidine, Boettcher *et al.* (1987a) investigated melting relationships in the systems $NaAlSi_3O_8$–CO_2, $CaAl_2Si_2O_8$–CO_2, and $KAlSi_3O_8$–CO_2 to shed light on the solubility mechanisms of carbon in aluminosilicate liquids. The results appear in Figs. 13.4, 13.5, and 13.6. The solidi are based on the first appearance of glass (interpreted to be quenched liquid) in the run products. For albite–CO_2, experiments within $\sim 10\,°C$ of the solidus contained $\sim 5\%$ glass; experiments at higher temperatures produced greater percentages. These results may reflect the metastable persistence of albite at temperatures above those of the equilibrium liquidus. For example, at 15 kbar for 24-hour runs, there was $\sim 5\%$ glass at $1220\,°C$, but $\sim 40\%$ glass at $1230\,°C$. In several experiments, we demonstrated

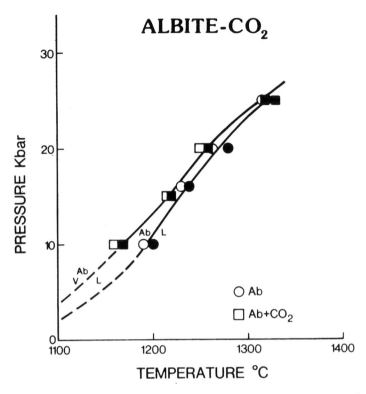

Fig. 13.4.   Pressure–temperature projection of the vapor-saturated solidus of albite
in the system $NaAlSi_3O_8$–CO_2 (after Boettcher *et al.*, 1987a), compared
to the vapor-absent solidus of albite (Boettcher *et al.*, 1982). Closed
symbols represent hypersolidus conditions. V is vapor and L is liquid.

that longer durations increased the percentage of liquid, which suggests that at least
part of the observed melting interval is the result of slow reaction rates.

For sanidine–CO_2, only 3–5% glass appeared in experiments quenched from as
much as 20 °C above the solidus, with larger percentages appearing at higher
temperatures. For example, runs at 15 kbar for 6 hours at 1220°, 1230°, 1250°, and 1270
°C contained 3, > 5, > 10, and > 30% glass, respectively. The observed melting interval
probably results at least in part from slow reaction rates.

For anorthite–CO_2, the higher melting temperatures produced much more rapid
melting, and the anorthite completely disappeared within 10 °C of the solidus in most
sets of runs.

The solidus of albite–CO_2 coincides with that of albite (Boettcher *et al.*, 1982) at 25
kbar within the combined uncertainty of 20 °C. However, at 10 kbar, CO_2 produces a
freezing-point depression of 30 °C, which is greater than experimental uncertainty.
Above a pressure of ~30 kbar, albite melts to jadeite + SiO_2-rich liquid (Bell &
Roseboom, 1969). Boettcher *et al.* (1984) proposed that the activity of SiO_2 component
in the liquid ($a_{SiO_2}^l$) changes relatively rapidly with pressure because the liquid

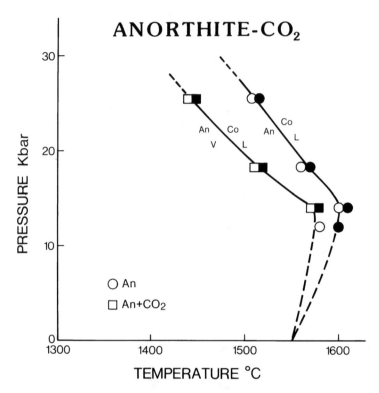

Fig. 13.5. Pressure–temperature projection of the vapor-saturated solidus of anorthite in the system $CaAl_2Si_2O_8$–CO_2, compared to the vapor-absent solidus of anorthite (after Boettcher *et al.*, 1987a). Closed symbols represent hypersolidus conditions. Co is corundum.

transforms over a pressure interval from one with an albite-like structure to one more nearly like quartz and jadeite, or some other form with at least some of the Al coordinated with more than four oxygen ions. Throughout this paper, $a^l_{SiO_2}$ is defined relative to a hypothetical SiO_2 liquid at the pressure and temperature of interest.

In contrast to that of albite, the freezing-point depression of anorthite by CO_2 increases with increasing pressure, from ~ 30 °C at 14 kbar to ~ 70 °C at 25.5. kbar. Based on our new data for the melting of anorthite in the absence of a vapor at 12 and 14 kbar, we have revised slightly the previous results of Boettcher *et al.* (1984). Anorthite melts congruently up to ~ 10 kbar, above which it melts incongruently to corundum + a SiO_2-rich liquid (Boettcher, 1970, Goldsmith, 1980). The $a^l_{SiO_2}$ probably increases quite rapidly with increasing pressure up to ~ 10 kbar as the liquid becomes more siliceous and transforms to a structure with at least some of the Al in greater than four-fold coordination (see Boettcher *et al.*, 1984).

The melting behavior of sanidine differs from that of albite and anorthite in several key ways. Between atmospheric pressure and the singular point (S) at ~ 15 kbar, sanidine melts incongruently to leucite + silica-rich liquid. As pressure increases toward S, the liquid approaches $KAlSi_3O_8$ composition, and the $a^l_{SiO_2}$ decreases.

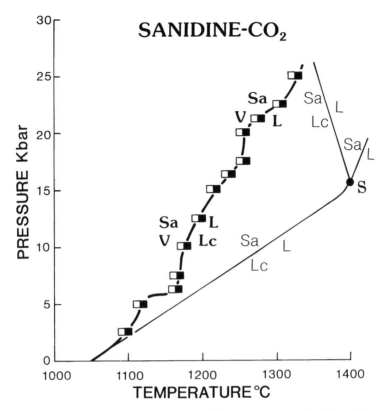

Fig. 13.6. Pressure–temperature projections of the vapor-saturated solidus of
sanidine in the system $KAlSi_3O_8$–CO_2 (after Boettcher *et al.*, 1984); the
vapor-absent reactions are after Lindsley (1966) and Boettcher *et al.*,
(1984). S is a singular point where the vapor-absent curves meet
tangentially. Closed symbols represent hypersolidus conditions.

Above the pressure of **S**, where sanidine melts congruently, the $a_{SiO_2}^l$ remains nearly
constant (Boettcher *et al.*, 1984). In addition, sanidine remains stable to pressures of
~ 120 kbar or more (Ringwood *et al.*, 1967), whereas albite and anorthite transform
below 35 kbar at liquidus temperatures to phases such as jadeite, corundum, kyanite,
and grossularite that have Al in six-fold coordination (Al^{VI}). As shown in Fig. 13.6,
CO_2 is far more effective in depressing the freezing point of sanidine than it is for albite
or anorthite. This depression increases to a maximum of about 200 °C at 15 kbar,
remaining fairly constant at > 130 °C at higher pressures. Curious features of the CO_2-
saturated solidus are the inflections between 5 and 10 kbar and between 20 and 25 kbar.
Reversal experiments closely spaced in pressure confirm that the inflections are real;
they are similar to those described above in the system SiO_2–H_2O–H_2.

All of the previous studies of the solubility of CO_2 in silicate liquids were conducted
at temperatures at least 100 °C, and typically > 200 °C, above the liquidus. Using
infrared spectroscopy, Stolper *et al.* (1987) determined that the solubility of CO_2 in
$NaAlSi_3O_8$ liquid increases with increasing pressure at constant temperature and

decreases with increasing temperature at constant pressure in the range of 15–30 kbar and 1450–1625 °C. Because of this difference between the temperature at which the solubility of CO_2 was measured and those of the solidus at a given pressure, it may be unwarranted to use these data to evaluate our hypothesis that the solubility of CO_2 decreases above 10 kbar along the $NaAlSi_3O_8$–CO_2 solidus. Boettcher *et al.* (1987a) extrapolated the data of Stolper *et al.* (1987) down to the temperatures of the solidus, obtaining a value for the solubility of CO_2 in the liquid in equilibrium with albite and vapor of 0.91 wt % (5.2 mole %) at 15 kbar, but 0.87 wt % (5.0 mole %) at 20 kbar. This apparent negative pressure derivative may be more pronounced from 15 to 10 kbar than from 20 to 15 kbar. Thus, Boettcher *et al.* (1987a) concluded that coupling the increase in the isothermal solubility of CO_2 with increasing pressure with a decrease in the isobaric solubility with increasing temperature might explain the decrease in the apparent solubility of CO_2 *along the solidus* with increasing pressure. Nevertheless, this is probably an unwarranted approach, and extrapolation of the data of Stolper *et al.* (1987) does not indicate a decrease in the solubility of total CO_2 with increasing pressure above 20 kbar, which conflicts with this proposal. In addition, Boettcher *et al.* (1987b) have shown that the solubility of as much as 1.2 wt % argon in albitic liquid produces a negligible freezing-point depression. It is conceivable that the solubility of molecular CO_2 increases with pressure, as proposed by Stolper *et al.*, but that the effect on the temperature of the solidus decreases, as we propose.

Whereas anorthite and albite are unstable above ~30 kbar, sanidine is stable to at least 120 kbar, where it transforms to the hollandite structure with Al^{VI} (Ringwood *et al.*, 1967). Rather than *directly* reflecting the size difference between Na^+ and K^+ or the relative abilities of these ions to charge-balance Al^{IV}, these differences in the stability relationships of sanidine and albite result from the absence of a dense, potassic phase with Al^{VI} at pressures below those where the hollandite structure is stable, suppressing the appearance of Al^{VI} in $KAlSi_3O_8$ until well above 100 kbar.

The factors that possibly control the solubility of CO_2 and CO_3^{2-} include the number of sites sufficiently large to accommodate the large CO_2 molecule, the availability of mono- and divalent cations, and the degree of polymerization, which can be represented by the ratio of non-bridging oxygen ions per tetrahedral cation (NBO/T). These latter two factors are related, especially for fully polymerized aluminosilicate liquids, where the cations that could complex with CO_3^{2-} are those that charge-balance the network-forming Al ions.

The low solubility of carbon in all silicate liquids relative to that of H_2 (R.W. Luth & Boettcher, 1986) and H_2O (Mysen, 1976) must reflect a low concentration of sites in the liquid appropriate for carbon in any form. It is known that decreasing polymerization of the liquids enhances the solubility of carbon (Mysen *et al.*, 1976, Spera & Bergman, 1980) and that increasing the concentration of SiO_2 (i.e. the $a_{SiO_2}^l$) increases the ratio of CO_2/CO_3^{2-} (Fine & Stolper, 1985).

The influence of the availability of mono- and divalent cations is evident in the large depression of the solidus of anorthite by CO_2 versus the relatively small depression of the solidus of albite by CO_2, reflecting the lower stability of the Na_2CO_3 complexes relative to $CaCO_3$ ones.

The complexing of CO_3^{2-} with mono- and divalent cations also plays a role in the dissolution of CO_2 in polymerized liquids. Previous studies (Mysen et $al.$, 1982, Fine & Stolper, 1985) indicate that CO_2 dissolves in fully polymerized liquids, such as $NaAlSi_3O_8$, in part by forming carbonate complexes with cations that previously charge-balanced Al^{IV}. As a result, Al is forced out of the four-fold, network-forming sites into a network-modifying role.

On the composition join $NaAlSi_3O_8$–SiO_2–CO_2, the CO_2/CO_3^{2-} ratio decreases as the bulk composition becomes less silicic, i.e. as the $a_{SiO_2}^l$ decreases (Fine & Stolper, 1985). A decrease in this ratio may result from an absolute increase in the concentration of CO_3^{2-} as $a_{SiO_2}^l$ decreases and the activity of the charge-balancing cations is concomitantly increased. If the $a_{SiO_2}^l$ increases with increasing pressure in $NaAlSi_3O_8$ liquid as proposed by Boettcher et $al.$ (1982), this may result in an increased $CO_2/$ CO_3^{2-}, because Mysen (1976) and Fine & Stolper (1985) determined experimentally that this ratio is dependent on the SiO_2 content of the liquid. If the solubility of molecular CO_2 decreases with increasing temperature as proposed by Fine & Stolper (1985), then the positive dP/dT of the albite solidus, coupled with our inferred increase in CO_2/CO_3^{2-} with increasing pressure, may account for our inferred decline in the solubility of carbon at high pressures. Although the $a_{SiO_2}^l$ also increases with pressure in molten anorthite, the negative dP/dT of the anorthite–CO_2 solidus coupled with the ready availability of Ca^{2+} to complex with CO_3^{2-} may account for our inferred increase in the solubility of carbon (CO_2 and CO_3^{2-}) with increasing pressure.

For sanidine, two factors at high pressures lower the CO_2/CO_3^{2-} ratio. First, above ~ 15 kbar, sanidine melts congruently, and the a_{SiO2}^l is less than it is at lower pressures where sanidine melts to leucite plus a SiO_2-rich liquid. We assume that a decrease in a_{SiO2}^l lowers the proportion of molecular CO_2 in sanidine liquid, as discussed above for albite. In addition, the large, positive dP/dT of the sanidine–CO_2 solidus likely has the same effect. We propose that the large freezing-point depression of the solidus of sanidine at high pressures results from the dissolution of carbon in the liquid largely as K_2CO_3 complexes. As a concomitant, when K^+ associates with the CO_3^{2-}, it becomes unavailable to charge-balance Al^{IV}, and depolymerization of the aluminosilicate network must follow. This is consistent with the concepts established by Mysen & Virgo (1980) for the solubility of carbon in other polymerized, aluminosilicate liquids.

The contrast in the behaviour of the two alkali feldspars in the presence of CO_2 can thus be explained by the differences in the distribution of the alkali cations in the liquid between association with Al–O–Si complexes and association with CO_3^{2-} complexes. In the system $NaAlSi_3O_8$–CO_2, Na^+ prefers the former role. In the system $KAlSi_3O_8$– CO_2, the liquid depolymerizes with increasing pressure, because K^+ progressively associates with CO_3^{2-}. In the case of $CaAl_2Si_2O_8$–CO_2, at high pressures where corundum is a liquidus phase, Ca^{2+} is sufficiently abundant in the liquid to play both roles. In general, the solubility of carbon as CO_3^{2-} is enhanced if monovalent and divalent cations are available in the liquid to complex with CO_3^{2-}.

Conclusions

Although melting is a process that is fundamental to many processes in the origin and evolution of the Earth, our knowledge of the mechanisms of melting and the nature of the silicate liquids is primitive, even for compositionally simple systems at atmospheric pressure.

Much remains to be learned about the role of volatile components in melting processes. Hydrogen fugacity is an important parameter in the melting of compositionally simple, hydrous silicate systems, and this variable must be evaluated for a wide range of compositions and conditions. Much has been learned recently about the mechanisms of solution of H_2O and CO_2 in silicate liquids, but pressure, temperature, and compositional derivatives are poorly known. Nevertheless, the prospects are good for a much better understanding of the nature of liquids and glasses through continued experimentation and the application of new techniques.

Acknowledgements We appreciate the invitation from Prof. L.L. Perchuk to contribute to this Korzhinskii Volume. We also appreciate the very helpful reviews of the manuscript by Dr. E.S. Persikov (USSR Institute of Experimental Mineralogy, Chernogolovka) and Dr Mark Brearley (UCLA). Prof. Bob Coleman of Stanford University donated albite sample CL; Dr Dave Stewart of the USGS provided sanidine glass sample Ss; Prof. Chris Scarfe of the University of Alberta provided anorthite glass sample Sc. This research was supported by NSF Grants EAR83–06410 and EAR87–05870 to ALB.

References

Bell, P.M. & Roseboom, E.H., Jr. (1969). Melting relationships of jadeite and albite to 45 kilobars with comments on melting diagrams of binary systems at high pressures. *Mineral. Soc. Am. Spec. Paper*, **2**, 151–61.

Boettcher, A.L. (1970). The system $CaO–Al_2O_3–SiO_2–H_2O$ at high pressures and temperatures, *J. Petrol.*, **11**, 337–79.

— (1984). The system $SiO_2–H_2O–CO_2$: melting, solubility mechanisms of carbon, and liquid structure to high pressures. *Am. Mineral.*, **69**, 823–33.

Boettcher, A.L., Burnham, C., Wayne, Windom, K.E. & Bohlen, S.R. (1982). Liquids, glasses, and the melting of silicates to high pressures. *J. Geol.*, **90**, 127–38.

Boettcher, A.L., Guo, A., Bohlen, S. & Hanson, B. (1984). Melting in feldspar-bearing systems to high pressures and the structure of aluminosilicate liquids. *Geol.*, **12**, 202–4.

Boettcher, A.L., Luth, R.W. & White, B.S. (1987a). Carbon in silicate liquids: the systems

$NaAlSi_3O_8$–CO_2, $CaAl_2Si_2O_8$–CO_2, and $KAlSi_3O_8$–CO_2. *Contrib. Mineral. Petrol.*, **97**, 297–304.

Boettcher, Art, White, B.S. & Brearley, M. (1987b). Nitrogen, argon, and other volatiles in silicate systems at high pressures. *EOS*, **68**, 1452–3.

Boettcher, A.L., Windom, K.E., Bohlen, S.R. & Luth, R.W. (1981). Low friction, anhydrous, low- to high-temperature furnace sample assembly for piston-cylinder apparatus. *Rev. Sci. Instruments*, **52**, 1903–4.

Bowen, N.L. (1913). The melting phenomena of the plagioclase feldspars. *Am. J. Sci.*, **35**, 577–99.

Fine, G. & Stolper, E. (1985). The speciation of carbon dioxide in sodium aluminosilicate glasses. *Contrib. Mineral. Petrol.*, **91**, 105–21.

Goldsmith, J.R. (1980). The melting and breakdown reactions of anorthite at high pressures and temperatures. *Am. Mineral.*, **65**, 272–84.

Goldsmith, J.R. & Jenkins, D.M. (1985). The hydrothermal melting of low and high albite. *Am. Mineral.*, **70**, 924–33.

Greig, J.W. & Barth, T.F.W. (1938). The system Na_2O–Al_2O_3–$2SiO_2$ (nephelite, carnegieite) – Na_2O–Al_2O_3 (albite). *Am. J. Sci.*, **35-A**, 93–112.

Kroll, H. & Ribbe, P.H. (1980). Determinative diagrams for Al, Si order in plagioclase. *Am. Mineral.*, **65**, 449–57.

Lindsley, D.H. (1966). Melting relations of $KAlSi_3O_8$: effect of pressures up to 40 kilobars. *Am. Mineral.*, **51**, 1793–9.

Longhi, J. & Hays, J.F. (1979). Phase equilibria and solid solution along the join $CaAl_2Si_2O_8$–SiO_2. *Am. J. Sci.*, **279**, 876–90.

Luth, R.W. & Boettcher, A.L. (1986). Hydrogen and the melting of silicates. *Am. Mineral.*, **71**, 264–76.

Luth, W.C. & Ingamells, C.O. (1965). Gel preparation of starting materials for hydrothermal experimentation. *Am. Mineral.*, **50**, 255–8.

Luth, R.W., Mysen, B.O. & Virgo, D. (1987). A Raman spectroscopic study of the solubility behavior of H_2 in the system Na_2O–Al_2O_3–SiO_2–H_2. *Am. Mineral.*, **72**, 481–6.

Morey, G.W. & Bowen, N.L. (1922). The melting of potash feldspar. *Am. J. Sci.*, **4**, 1–21.

Mysen, B.O. (1976). The role of volatiles in silicate melts: solubility of carbon dioxide and water in feldspar, pyroxene, and feldspathoid melts to 30 kb and 1625 °C. *Am. J. Sci.*, **276**, 969–96.

Mysen, B.O., Eggler, D.H., Seitz, M.G. & Holloway, J.R. (1976). Carbon dioxide in silicate melts and crystals, Part I. Solubility measurements. *Am. J. Sci.*, **276**, 455–76.

Mysen, B.O. & Virgo, D. (1980). Solubility mechanisms of carbon dioxide in silicate melts: a Raman spectroscopic study. *Am. Mineral.*, **65**, 885–99.

Mysen, B.O., Virgo, D. & Seifert, F.A. (1982). The structure of silicate melts: implications for chemical and physical properties of natural magmas. *Rev. Geophys. Space Phys.*, **20**, 353–82.

Navrotsky, A., Capobianco, C. & Stebbins, J. (1982). Some thermodynamic and experimental constraints on the melting of albite at atmospheric and high pressure. *J. Geol.*, **90**, 679–98.

Osborn, E.F. (1942). The system $CaSiO_3$–Diopside–Anorthite. *J. Sci.*, **240**, 751–88.

Persikov, E.S. & Epel'baum, M.B. (1985). Melt differentiation mechanisms in experiments under hydrogen pressure. *Geochem. Intl.*, **22**, 26–32.

Pugin, V.A. & Soldatov, I.A. (1973). Melting and crystallization of albite at pressures up to ten thousand atmospheres. *Geochem. Intl.*, **10**, 974–8.

—— (1976). Incongruent melting of acid plagioclase at elevated pressures. *Geochem. Intl.*, **13**, 122–6.

Richet, P. & Bottinga, Y. (1986). Thermochemical properties of silicate glasses and liquids: a review. *Rev. Geophys.* **24**, 1–25.

Ringwood, A.E., Reid, A.F. & Wadsley, A.D. (1967). High pressure $KAlSi_3O_8$, an aluminosilicate with 6-fold coordination. *Acta Crystal.*, **23**, 1093–5.

Schairer, J.F. & Bowen, N.L. (1947). The system anorthite–leucite–silica. *Bull. Comm. Géol. Fin.*, **20**, 67–87.

—— (1955). The system $K_2O–Al_2O_3–SiO_2$. *Am. J. Sci.*, **253**. 681–746.

—— (1956). The system $Na_2O–Al_2O_3–SiO_2$. *Am. J. Sci.*, **254**, 129–95.

Spera, F.J. & Bergman, S.C. (1980). Carbon dioxide igneous petrogenesis: I. Aspects of the dissolution of CO_2 in silicate liquids. *Contrib. Mineral. Petrol.*, **74**, 55–66.

Stolper, E., Fine, G., Johnson, T. & Newman, S. (1987). Solubility of carbon dioxide in albitic melt. *Am. Mineral.*, **72**, 1071–85.

14

Evolution of the lithosphere, and inferred increasing size of mantle convection cells over geologic time*

W.G. ERNST

Planetary evolution: the early Earth as an example

For the inner planets of the solar system, surface features which reflect tectonism and the interplay of lithosphere with atmosphere (and hydrosphere, if present), versus the preservation of ancient impact features, is very much a function of planetary mass and distance from the Sun (Head & Soloman, 1981). Larger bodies such as Venus and the Earth have experienced geologically recognizable physiographic modification, whereas surficial activity declined or ceased on Mars, Mercury and the Moon several Ga ago. Smaller masses lose heat more rapidly, and are less able to retain gassy constituents as completely as larger bodies. Volatiles, particularly H_2O, lower fusion temperatures because of solubilities in silicate melts, thus enhancing the planetary capacity for crystal–liquid–vapor fractionation. Partial melting, in turn, is probably related to thermally-created density instabilities in the mantle. Such convective flow, an important driving force for lithospheric plates, is a manifestation of the escape of internal heat within a gravitational field.

Isotopic data from meteorites and the Moon indicate that the solar system is approximately 4.5–4.6 Ga old (Patterson, 1956, Papanastassiou & Wasserburg 1971). Gravitational self-attraction of a locally dense cloud of interstellar gas resulted in condensation of the solar nebula which, to preserve angular momentum, initially formed a rotating disc (Safronov, 1972, Cameron, 1978). Accretion of planetesimals in progressively more peripheral regions of the disc gave rise to planetary bodies orbiting the newly-formed star. Although initially these condensates may have been rather cold, burial of heat during successive impacts probably resulted in increasingly hot accretion (Wetherill, 1976, Kaula, 1979). Growth is assumed to have been homogeneous, but

* Institute of Geophysics and Planetary Physics Publication No. 3005.

need not have been so. This primordial phase of formation of the solar system probably was completed by about 4.5 Ga, at which time Earth differentiation began.

The relative abundance of long-lived radioactive isotopes of elements (Lee, 1970) such as uranium, potassium and thorium, and the initial presence of short-lived radioisotopes, would have resulted in additional planetary self-heating during and directly following the accretionary stage. Because iron melts at lower temperatures than do silicates under high confining pressure, and because silicates and Fe(+ Ni) liquids are immiscible, a dense ferrous metal-rich melt must have formed at depth within the Earth. Due to gravitational instability, iron-rich liquid would have migrated downwards, displacing silicates upwards, producing a molten metallic core overlain by a largely solid silicate mantle (Elsasser, 1963). Infall of the core evidently accompanied later stages of planetesimal accretion (Stevenson, 1983). The conversion of potential energy to heat during this process would have liberated substantial amounts of additional thermal energy, contributing to the fusion and more complete separation of metal core and silicate mantle (Birch, 1965, Flaser & Birch, 1973), and, probably, partial melting of near-surface portions of the latter (Ringwood, 1979). Slight cooling since this earliest stage of planetary differentiation has resulted in solidification of the inner core. Judging by analogy with the Moon, the separation of Fe(+ Ni) core and encompassing, largely solid, silicate mantle, and conjectured generation of a partly molten upper mantle – the magma ocean – were completed by about 4.4 Ga at the latest. The molten silicate layer, if present, would have been located near the Earth's surface because the adiabatic gradient appropriate for convection apparently possesses a larger dP/dT value than the mantle solidus. Experiments at high pressures suggest that under deep mantle conditions, some melts actually may be denser than the associated, more refractory crystalline assemblages, but the evidence is still equivocal (e.g., see Stolper et al., 1981, Herzberg, 1987).

No terrestrial rocks have survived from this primeval stage of planetary evolution; the most ancient samples now known, from the Isua supracrustal belt of western Greenland, are no older than about 3.8 Ga (Moorbath et al., 1972, Moorbath, 1975). Froude et al. (1983), Myers & Williams (1985), and Compston & Pidgeon (1986) have reported 4.1–4.2 Ga old zircons extracted from Precambrian metasediments in western Australia, so now obliterated continental crust had begun to form by this time. Evidence of this initial stage of development is preserved on the Moon however: the ferroan anorthositic impact breccias of the lunar highlands contain rock fragments at least as old as 4.4 Ga, and the highlands are strongly pulverized, attesting to continued, intense meteorite bombardment there, and presumably on Mercury, Venus, the Earth and Mars to about 3.9 Ga (Wetherill, 1972, Burnett, 1975, Shoemaker, 1984). Such meteoritic accretion probably was most intense during the earliest stages of planetary formation and declined as meteoritic debris in the solar system was swept up by the growing planets. The less-heavily cratered lunar maria are floored by Early Archean basalts; thus, the thermal budget of the Moon allowed local production of mafic magma within the lunar mantle as late as 3.8 to 3.3 Ga ago. The more ancient highlands

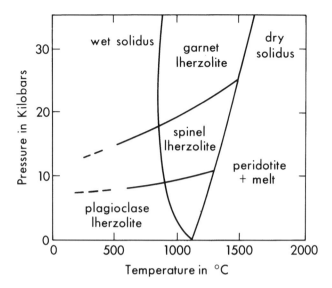

Fig. 14.1.   Phase equilibria for mantle compositions (Wyllie, 1970). The anhydrous
solidus is after Kushiro *et al.* (1968), the H$_2$O-saturated solidus from
Mysen & Boettcher (1975). Subsolidus assemblages are generalized after
Green & Ringwood (1967a, b), Ito & Kennedy (1967), O'Hara (1967,
1968) and MacGregor (1968).

appear to have formed from the accumulation of calcic plagioclase which rose
buoyantly towards the surface of the hypothesized lunar magma ocean (e.g., Longhi,
1978, Smith, 1979).

Because of the far greater mass of the Earth, higher pressure equilibria would have
dominated crystallization other than in the most surficial parts of a postulated
terrestrial magma ocean. Phase equilibria for a chondritic (mantle-like) bulk composition are illustrated in Fig. 14.1. On the Earth, early crystallization of garnet rather
than calcic plagioclase would have resulted in the settling of dense aluminous
crystalline phases rather than flotation. In addition, the thin, transitory, basaltic crust
that would have solidified at the upper cooling surface, lying upon less-dense, partly
molten material, would have foundered during continued meteoritic bombardment of
the Earth. Therefore, dense, refractory ferromagnesian constituents should have been
stabilized in relatively deeper portions of the gradually thickening, largely solid mantle
of the primitive Earth, with more fusible LIL, incompatible, and volatile elements
concentrated toward the surface in a relatively silicic, globe-encompassing rind and
dense primitive atmosphere (e.g., see Hargraves, 1976, Shaw, 1976, Fyfe, 1978).
Probably during this early, very hot stage, ephemeral, relatively sialic material would
continue to have been reincorporated in the rapidly convecting, rehomogenizing
mantle (Jacobsen & Wasserburg, 1981); eventually a more silicic, gravitatively stable,
molten but congealing scum would have been produced (Morse, 1986). Outgassing
must have been intense, with a considerable portion of the volatiles transferred from
the mantle to the hydrosphere + atmosphere. By about 4.2 Ga at the latest, the

W.G. Ernst

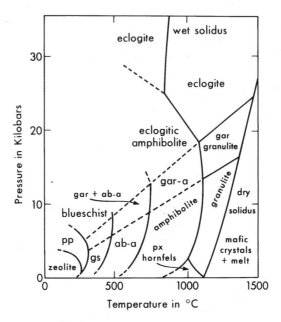

Fig. 14.2. Phase equilibria for basaltic compositions based on available
experimental and oxygen isotopic data (e.g., see Ernst, 1976).
Metamorphic zone boundaries are shown as univariant lines for
simplicity, but are actually multivariant *P–T* zones; their locations and
widths are sensitive functions of numerous chemical variables. Facies
abbreviations are a = amphibolite, gs = greenschist, pp = prehnite-
pumpellyite. Where aqueous fluid pressure is equal to P_{total}, amphibole-
dominant assemblages are stable over most reasonable crustal- and
uppermost-mantle-type conditions.

postulated magma ocean had solidified and a lithospheric skin was in existence, as
required to explain the occurrence of zircon-bearing granitoids in western Australia. It
is clear from Figs. 14.2 and 14.3 that the presence of continental crust requires oceanic
crust and mantle lithosphere also to have been in existence at 4.2 Ga, because the solidi
of granitic materials lie at temperatures well below those of basaltic and peridotitic rock
types.

At 3.8 Ga and probably earlier, water-laid sediments containing silicic, fusible, LIL-
rich continental detritus were being deposited on the Earth. The surface temperature,
therefore, was less than the boiling point of H_2O. Thus, no matter how the initial
chemical differentiation of the Earth was accomplished, an outer rind of crust and
overlying watery and gassy envelopes must have existed at least locally by 3.8 Ga
(Jacobsen, 1984). The fusion of mantle peridotite requires higher temperatures than are
necessary for formation of continental crust, so at least a thin layer of solid mantle
basement must also have been present by this time (Fig. 14.2). Thin lithospheric
platelets clearly were extant by Early Archean time.

Calculated geothermal gradients depend critically on the assumptions employed
(e.g., see Lambert, 1980), but the early Earth unquestionably had a higher average

372

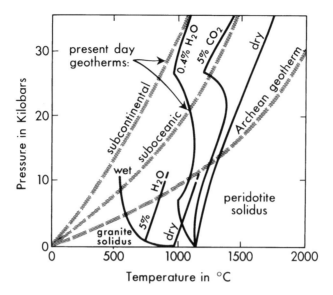

Fig. 14.3. Phase relations for the initiation of melting of undepleted peridotite, anhydrous and in the presence of minor amounts of H_2O and of CO_2, and of granite, anhydrous and in the presence of minor and excess H_2O (Luth *et al.*, 1964, Ito & Kennedy, 1967, Green & Ringwood, 1967b, O'Hara, 1968, Lambert & Wyllie, 1972, Wyllie 1979, 1981). Also shown are computed present-day lithospheric intraplate geothermal gradients (Clark & Ringwood, 1964) and an estimated 3.6 Ga geotherm assuming approximately three times the current radiogenic heat production and an isotopically stratified Archean Earth (Lambert, 1976).

temperature/depth curve than currently exists, as required by infall of the core and both intense radioactive and impact heating. Continental crust and mantle would have begun to melt at shallower depths than at present because of the high heat generation and consequent elevated thermal structure. Therefore, Archean continents (Bridgwater & Fyfe, 1974) and especially their solid mantle underpinnings, on the average, must have been relatively thin. On the other hand, metamorphic mineral assemblages suggest the existence of some Precambrian continental crustal thicknesses approaching those of Phanerozoic orogenic belts (Newton, 1978, 1983, Grew, 1980). Evidence for ancient, especially thick mantle lithosphere remains equivocal (Green, 1981, Boyd *et al.*, 1985, Moore and Gurney, 1985).

Compositions of near-solidus melts derived from fertile, chondrite-like peridotite source regions are complex functions of the depth of derivation, extent of partial fusion, absence or presence of H_2O, CO_2 and O_2, and degree of later, shallow-level crystal–liquid fractionation (Boettcher, 1973, Ringwood, 1975, Eggler, 1978, Wyllie, 1979, 1981, Takahashi, 1986). In general, more profound depths of fusion, very low percentages of partial melting and relatively high CO_2 activities promote the formation of alkalic and picritic magmas, whereas shallow levels of fusion, moderate degrees of partial melting, and high H_2O and/or O_2 activities favor more tholeiitic liquids.

Extensive partial melting of a mantle protolith favors production of ultramafic liquids. Subsequent crystal fractionation during ascent, and assimilation of wall rocks introduce chemical modifications of these primary melts.

Heat budget and mantle flow

The chemical/density differentiation of a planet is largely a consequence of thermal energy transfer operating in a gravitational field. The high heat production provided by primordial abundances of radioactive elements, meteorite impacts, and rapid infall of the Fe(+Ni) core would have resulted in elevated temperatures throughout the Earth shortly after accretion. Wanke et al. (1984) estimated the relative energy contributions of these processes since the time of origin as: accretion of planetesimals > core formation > decay of K, Th and U.

The Earth has lost heat by several transfer mechanisms, chiefly through radiation, conduction and convection. The last appears to be the most efficient process for transporting thermal energy from the deep interior towards the surface, and may have operated since the time of core formation (McKenzie & Weiss, 1975). Although not understood in detail, the motions of lithospheric plates are regarded as a consequence of such bodily flow. Layered convective cells seem to best fit the geophysical data (Anderson, 1987).

Today most of the terrestrial heat loss occurs in the ocean basins in the vicinity of the ridge systems. The latter mark the rising limbs of mantle convection cells. In contrast, the more radioactive continents are characterized by lower thermal fluxes, requiring their siting over rather depleted mantle material. Descending lithospheric slabs are thermal sinks, so heat flow near trenches is low. Elevated flux on the landward side of convergent plate junctions coincides with the calc-alkaline arc, and reflects advective heat transport. The early Earth probably was not more homogeneous in terms of heat generation; in fact, the much higher overall energy fluxes might have resulted in more marked differences between regimes of high thermal dissipation compared to low heat flow zones. Ultramafic lavas typical of Archean tholeiitic greenstone belts required temperatures approaching 1650 °C for their production (Green, 1975, Nisbet et al., 1977), and reflect moderate to large degrees of melting of upwelling mantle. Such komatiitic liquids could only have been generated in oceanic regions of very high heat flow. Thick Archean sialic crustal sections of high-grade granulitic gneiss, on the other hand, contain phase assemblages comparable to some Phanerozoic metamorphic geothermal gradients (Lambert, 1976, Boyd et al., 1985), hence certain portions of the continents were not subjected to especially high thermal flux.

Early in Earth history, small, rapidly convecting cells apparently would have allowed the efficient transfer of large quantities of heat from the upper mantle towards the surface. Lowered rigidity in a hotter Archean mantle could well have resulted in smaller-scale convection and in cells with lower aspect ratios (e.g., see Turcotte & Schubert, 1982, p. 330). However, horizontal dimensions of such cells probably would have been governed more by lithospheric plate dimensions than by fluid dynamic properties; hence as shown in Fig. 14.4, the thinner, smaller lithospheric plates of the early Earth would have promoted the occurrence of diminutive, more nearly circular, convection cells. In contrast to the current situation involving lithospheric plates of great thickness, aspect ratio, and lateral extent, moving sedately in response to large-scale laminar convection, the primitive Earth evidently was characterized by oceanic areas underlain by small, very rapidly flowing, perhaps initially poorly organized upper mantle convection cells, with continental debris accumulating in cooler eddies above regions of asthenospheric stagnation or downwelling (e.g., see Baer, 1981a, b, Boak & Dymek, 1982, Wanke *et al.*, 1984, Richter, 1984). This situation could explain observed geologic relationships: the discontinuous fabric of Archean granite–greenstone micro-belts and tonalitic gneiss terranes (Goodwin, 1981a, b) has given way to remarkably long paired metamorphic zones produced during Phanerozoic time (Miyashiro, 1961, 1972).

Generation of continental crust over geologic time

Modern ocean basins contain crust as old as about 0.2 Ga, but most is much younger. Basaltic material forms at spreading centers by decompression partial fusion accompanying the buoyant rise of somewhat depleted mantle. The oceanic crust and its further-depleted substrate are recrystallized, hydrated in part, and continuously returned to the mantle today; probably such ophiolitic assemblages were similarly recycled during the early stages of terrestrial evolution. The fate of oceanic crust + uppermost depleted peridotite on descent into the present mantle is still a matter of debate, but evidently devolatilization and partial melting have produced calc-alkaline magmas characteristic of active continental margins and island arcs (Green & Ringwood, 1967a, Martin, 1986). Such sialic material is less refractory and less dense than subjacent portions of the lithosphere, hence it tends to decouple and remain near or at the surface due to buoyancy. It has accumulated over time, along with obducted ophiolitic scraps, and – although reprocessed – constitutes the main source of information concerning crustal evolution.

However, during the pre-Archean, mantle circulation apparently was so vigorous that virtually all of the sialic material was returned to the mantle because of viscous drag (Condie, 1980, Hargraves, 1981). Reworking as a consequence of meteorite impacts was also significant until about 3.9 Ga. Isotopic evidence indicates that complete upper mantle rehomogenization continued until about 3.6–3.8 Ga, that the average age of the continental crust is approximately 1.8–2.8 Ga, and that much of the Earth's sialic material – perhaps upwards of 70 percent or more – had been generated by Early Proterozoic time (Veizer, 1976, Lambert, 1976, Jacobsen & Wasserburg, 1981, DePaolo, 1981, McLennan & Taylor, 1982, Moorbath, 1984, Nelson & DePaolo, 1985, Turcotte & Kellogg, 1986). Thus, although generated, net preservation of sial was nil until the Early Archean, due to reincorporation in the mantle. By approximately 3.6–

3.8 Ga ago, the rate of return to the mantle appears to have fallen below the rate of formation of continental crust, and increasing amounts of sial were preserved due to high, but decelerating, rates of mantle convection. This process gradually cleansed the mantle of LIL and incompatible elements, and, as energy input to the thermally-driven gravitative instability waned, the vigorous production of both oceanic and continental crust from progressively more depleted mantle declined toward the present low and very low generation rates, respectively. The net addition of sialic material seems to have risen to a maximum in the Late Archean and Early Proterozoic and has gradually lessened ever since (O'Nions *et al.*, 1980, Lambert, 1981, Nelson & DePaolo, 1984, Reymer & Schubert, 1986; however, for a different view, see Armstrong, 1968).

Mantle-derived igneous rocks constitute the primary additions to the Earth's crust. Properly sampled, such materials provide an accurate measure of the nature of this growth. Sedimentary processes rework the uppermost crust and produce chemically disparate units whose aggregate bulk composition, including the associated volatile species, reflects the chemistry of the source. The relative abundance of volcanogenic units evidently has declined steadily over the course of geologic time; immature clastic sediments are volumetrically important in Archean units, whereas younger deposits are typified by a gradually increasing proportion of multicycle, chemically more differentiated strata (Ronov, 1964, Ronov & Migdisov, 1971, Condie, 1982, Knoll, 1984). Cratonization apparently was a unidirectional process which resulted in extensive continental assemblies and concomitant sedimentary differentiation by the onset of Proterozoic time.

Evolution of the Earth's lithosphere

Petrologic associations most readily related to modern-type plate tectonic regimes appear to be uncommon from all but the latest Precambrian and more recent rock record (Ernst, 1983). The preservation of heavily cratered surfaces on the Moon, Mercury, and Mars demonstrates that these bodies were not subjected to major crustal reworking through erosion, sea-floor spreading and plate subduction, at least subsequent to the marked decline in meteoritic bombardment approximately 3.9 Ga ago. Of course, all of these planets have considerably smaller masses than the Earth and thus could not long have sustained core-formation-, impact- and radioactivity-induced internal temperatures as elevated as on the Earth; neither could they have been subjected to as strong a thermal gradient, presumably required for the onset of plate tectonic processes, for as extended a period of time. On the Earth, retention of the hydrosphere allows the continuous reintroduction of small amounts of H_2O into the mantle via the subduction process, thus lowering the mantle solidus temperature and perpetuating the existence of asthenosphere beneath the capping lithosphere. Thus, although the mantle has been progressively devolatilized over geologic time, a small amount of H_2O is nevertheless present in its upper portions.

Preliminary radar imagery from Venus suggests that its topography is more or less unimodal, hence the Earth's sister planet evidently is not characterized by continents and ocean basins; although plate tectonics may have attended early stages of development, such processes apparently ceased there long ago, presumably because of its very thick lithospheric shell (Phillips *et al.*, 1981, McGill, 1983). The possible existence of oceanic spreading centers on Venus has recently been suggested by Head & Crumpler (1987); even if true, modern activity has not been documented, however. Thus, although volcanic features demonstrate that primordial melting occurred on the other inner planets, the Moon, Mercury, and Mars evidently cooled to the stage at which their outer portions become too rigid to allow differential plate motions prior to the effective termination of planetesimal sweep-up. The larger mases of the Earth and Venus allowed maintenance of high temperatures and strong thermal gradients, and consequently circulation within the mantle – expressed on the Earth, at least, as plate tectonics. H_2O probably is absent from the Venusian lithosphere and atmosphere due to proximity to the Sun; thus, volatilization of CO_2 (coupled with dissociation of H_2O and escape of H_2 into space), resulted in a runaway greenhouse effect (Phillips *et al.*, 1981). The lack of strongly bimodal topography on Venus could be accounted for by the inability of an anhydrous planet to maintain asthenospheric flow and in the attendant subduction process, to manufacture sialic material (Campbell & Taylor, 1983).

Because advective heat transfer has almost certainly existed since formation of the Earth's core 4.4–4.5 Ga ago, a peripheral solid rind and small, thin, soft platelets must have bounded the surface beginning quite early in Earth history, a supposition consistent with the preserved Archean rock record and with experimentally determined phase equilibrium relations (Figs. 14.1–14.3). Extremely rapid, and initially, perhaps, poorly organized mantle overturn probably would have driven such platelets against and beneath one another, but because of elevated near-surface temperatures and thinness of the plates, together with the small magnitude of the lithospheric/asthenospheric density inversion and the consequent very minor gravitative instability, the lithosphere would not have been subducted to appreciable depths before rising temperature caused its transformation to asthenosphere. This accounts for the lack of terrestrial Archean and Early Proterozoic high-pressure metamorphic rocks such as blueschists and eclogites, and the absence of alkalic igneous suites, as well as for occurrence in greenstone belts of highly refractory komatiites and related magnesian lavas. Models for the generation of andesite and tonalitic continental crust by partial melting of subducted eclogite (Green & Ringwood, 1967a, Ringwood, 1975) therefore seem to be unlikely for the Archean. Tonalitic compositions could have been derived at relatively shallow depths by fractional crystallization of mafic magma (Bowen, 1928), through the partial fusion of basaltic amphibolite (Holloway & Burnham, 1972, Boettcher, 1973), or by incipient melting of hydrous upper mantle material (e.g., see Mysen & Boettcher, 1975, Arculus, 1981).

Due to the elevated thermal structure of the primitive Earth, volcanic activity would have been much more intense than at present, resulting in the concomitant rapid recycling of oceanic crust and consequent development of sialic crust and the hydrosphere/atmosphere. Although Archean gray gneisses were undoubtedly remobilized during accretion and thickening of the cratons, mantle-like trace elements and isotopic geochemistries could reflect their partial fusion derivation from fertile peridotitic, basaltic, and amphibolitic precursors in a hydrous environment. Perhaps they represent the amalgamation of many granite + greenstone belts, thoroughly recrystallized and devolatilized at deep crustal levels and broadly invaded by anatectic melts. However formed, accumulation, annealing, and preservation as relatively thick (20–30 km) Archean continental crust must have taken place above nearly anhydrous, cool portions of the mantle lithosphere and asthenosphere, for in the presence of abundant H_2O or high temperatures, subsequent extensive partial melting would have ensued (see Figs. 14.1–14.3).

By Early Proterozoic time, small sialic masses and immature island arcs, produced chiefly during earlier mantle overturn and chemical segregation, evidently had been assembled into thick, buoyant continental cratons. As freeboard became increasingly important (Schubert & Reymer, 1985), the processes of mechanical erosion and sedimentation began to play major roles in the production of chemically-differentiated sedimentary facies. The Archean to Early Proterozoic transition seems to have been characterized by a gradual change from small, thin, hot Archean platelets, driven about by – and controlling the dimensions of – numerous rapidly convecting asthenospheric cells, to relatively thicker, cooler, laterally more extensive, coherent plates, the motions of which may have been a function partly of asthenospheric flow on a grander scale.

Four transitional stages in the plate tectonic evolution of the Earth seem to be recognizable (Smith, 1981, Kröner, 1981, Goodwin, 1981a, b, Condie, 1982): (a) a pre-Archean tenuous lithosphere stage, characterized by a partly molten planet that was intensely bombarded by meteorites and that very early, during accretion of planetesimals, underwent profound gravitative separation to form a metallic Fe(+ Ni) core, a ferromagnesian silicate mantle (± magma ocean) and a continuously reworked ephemeral crustal scum; (b) an Archean microplate stage, in which the Earth's surface was dominated by hot, soft, relatively thin, nearly unsubductable platelets, and that aggregated island arcs and microcontinental terranes to form protocontinents; (c) a Proterozoic supercratonal stage, characterized by the emergence of broad continental shields, development of freeboard, intracratonal orogeny, and by the drift of supercontinents; and (d) a latest Precambrian–Phanerozoic Wilson cycle of plate tectonics involving rifting, the dispersal and suturing of continental fragments (exotic terranes), and the generation of long, linear, paired mobile belts at convergent plate margins. Stage (a) was probably typified by ill-organized – perhaps chaotic – but extremely rapid mantle flow, stage (b) by small, rapidly overturning, better organized convective cells, stage (c) by intermediate-sized, laminar-flowing convection, and stage (d) by enormous, more slowly overturning cells. Transitions from stage to stage probably were gradual and continuous. Lithospheric thickening and increasing lateral dimensions accompanying cooling, and a concomitant increase in the negative buoyancy of oceanic-crust-capped lithospheric plates attended and guided the evolution towards larger convective cells (characterized by larger aspect ratios) in the upper mantle.

Convection models and the thermal history of the Earth are matters of considerable debate (Schubert, 1979). At issue are the extent of dependence of mantle viscosity on temperature and H_2O content (e.g., see Jackson & Pollack, 1987), the rate of heat loss relative to energy production, and the scale of overturn. If the convection cells belong to two separate systems – upper mantle + transition zone versus lower mantle – as seemingly required by isotopic and some seismic evidence (Anderson, 1987), then the response time for thermal dissipation is lengthened considerably relative to whole-mantle circulation (McKenzie & Richter, 1981, Spohn & Schubert, 1982). Even if mantle heat loss shows only a weak dependence on temperature (Christensen, 1985), Archean heat transfer, mantle temperatures and plate velocities probably exceeded those of today, perhaps by substantial amounts. Moreover, because H_2O, which lowers viscosity, would have been more abundant in mantle material prior to its progressive devolatilization, more rapid convection should have typified the early Earth. As previously emphasized, a higher geothermal gradient and H_2O content in the mantle would have resulted in diminutive Archean lithosphere plates, which in turn would have had an important influence on the scale of upper mantle circulation.

With these provisos in mind, and employing geologic relationships previously described (e.g., Ernst, 1983), schematic illustrations of the hypothesized geometry of mass and energy flow in the upper mantle as a function of time are shown in Fig. 14.4. The proposed scenario for mantle circulation involves several undirectional trends: (1) primordial, chaotic, poorly organized, extremely vigorous flow was replaced by more orderly, laminar circulation early in Earth history; (2) the velocities of convective cells diminished with time; (3) the thicknesses, areal dimensions, and differential buoyancies of the predominantly continental- and oceanic-crust-capped lithospheric plates increased with time; (4) the vertical dimensions of the upper layer of cells increased modestly as the declining heat budget (= declining geothermal gradient) caused the depth of the transition zone to become greater with time; and (5), the lateral dimensions of the near-surface convecting cells and aspect ratios increased significantly as the thermal flux and H_2O content diminished towards the present. This model seems to be compatible with observed geologic relationships and the inferred thermal evolution of the planet. Gradual dessication of the mantle with the passage of time would also tend to increase mantle stiffness, and therefore the size of the circulating cells.

While consistent with geologic data currently available, the scenario presented in Fig. 14.4 is obviously speculative. Several conclusions seem reasonable, however. As today, two distinct crustal regimes were present in the Archean Earth, but the heat flow, and lithospheric areal extent + thickness contrasts may have been more pronounced then that now: crusts in the oceanic and circumoceanic regions were relatively thin and were characterized by very high T/P gradients; the protocratonal regions carried thick crust and were typified by moderate but not low T/P gradients. The

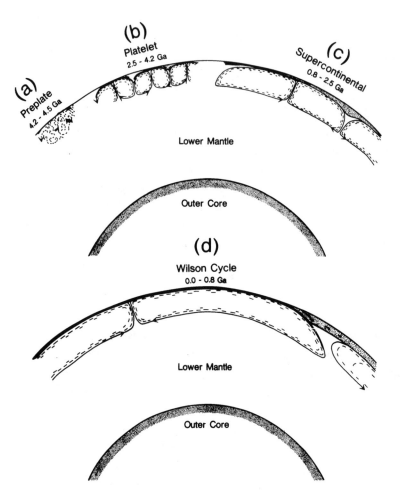

Fig. 14.4. Schematic, gradational, four-stage history of the lithosphere (oceanic-crust-capped plate shown in black, continental-crust-capped plate indicated by stippled pattern). (a) Vigorous, poorly organized (chaotic?) convection, *pre-continental stage*. 4.5–4.4 Ga: rapid accretionary and radioactive heating, nearly simultaneous infall of Fe(+Ni) core, probable formation of magma ocean in the upper mantle, massive devolatilization. 4.4–4.2 Ga: crystallization of hypothetical magma ocean and extremely rapid convectional reworking of protocrust into high-T, H_2O bearing, low-viscosity mantle, continued extensive outgassing. (b) Very rapid, small-scale, low aspect ratio convection, *microcontinental stage*. 4.2–3.8 Ga: stabilization of soft, globe-encircling lithospheric platelets surmounted by thin simatic and sialic crust, covered by hydrosphere; deposition of chemically-primitive sediments; platelets and crust ultimately returned to mantle; ongoing exuberant volcanism, continued loss of mantle volatiles; rapid decrease in rate of planetesimal sweep-up; reworking by meteorite impact becoming negligible. 3.8–2.7 Ga: increasing crustal preservation of products of platelet tectonics, involving small, thin, ductile, hot lithospheric slabs capped by komatiitic and tholeiitic lavas; accretion of sialic veneer marginal to vigorously convecting, small suboceanic asthenospheric cells to form primitive calc-

progression of lithologic assemblages over the course of geologic time has been a response chiefly to ongoing chemical differentiation of the entire Earth in the context of a gradually declining thermal budget + concomitant expulsion of H_2O from the deep interior, and consequent increase in the viscosity, aspect ratio and scale of mantle convection. The plate tectonic evolution of the Earth seems to reflect a gradual but substantial increase in the lateral dimensions of upper mantle convection cells, which is reflected in the changing nature of lithotectonic and petrologic assemblages.

Acknowledgements This synthesis is based on an earlier review (Ernst, 1983) of Precambrian terranes investigated by many workers, and has been partially supported through NSF grant EAR83-12702. I thank F.H. Busse for discussions, A.P.S. Reymer, Eugene Artyushkov, and Gerald Schubert for reviewing draft versions of the manuscript. The speculative ideas presented here regarding mantle convection are my own, evoked by the puzzling contrasts among Archean, Proterozoic, and Phanerozoic crustal sections.

Fig. 14.4. (*cont.*)

alkaline arcs and growing, largely submerged protocontinents, diminishing proportions of the sialic crust being returned to the mantle; deposition of immature, first-cycle sedimentary aprons around volcanic arcs. Greenstone belts form in oceanic regimes characterized by mantle upwelling, and accrete in circumoceanic subduction zones along with superjacent metasediments and subjacent anatectic granitic intrusions – all subjected to low volatile activities but high H_2O/CO_2; thick, high-grade granulitic gray gneiss terranes develop at depth from annealed, transformed granite + greenstone collages in protocratonal environments typified by underlying stagnant or downwelling depleted mantle and high CO_2/H_2O. 2.7–2.5 Ga: lessening production of new sial from mantle, but concomitant rapid decline in return to mantle, resulting in substantial net continental accretion; emplacement of large-scale layered mafic intrusions, dikes and sills reflecting cooling, thickening and embrittlement of sialic crust; gradual attainment of continental freeboard, leading to formation of epeiric seas and production of chemically-mature sediments. (c) Enlarging, laminar flowing, slowing convection, *megacontinental stage*. 2.5–0.8. Ga: decelerating continental growth; deposition of increasingly abundant multicycle sediments; aggregation of supercontinents embedded in thickening and areally more extensive lithospheric plates; orogenic belts forming along intracratonal weakness zones due to incipient convergence and lithospheric delamination; drifting of emergent supercontinents, promoting formation of glacial deposits in polar regions. (d) Gigantic, laminar flowing convection cells, continental rifting and reconnection stage, *Wilson cycle*. 0.8 Ga–present: continued lateral enlargement of cooling lithospheric plates, resulting in onset of present mode of plate tectonics characterized by enormous, relatively cool, stiff, slowly overturning mantle convective cells of high aspect ratio, continental fragmentation and migration of exotic terranes, formation of paired metamorphic belts and linear orogenic zones near convergent plate margins.

References

Anderson, D.L. (1987). Thermally induced phase changes, lateral heterogeneity of the mantle, continental roots, and deep slab anomalies. *J. Geophys. Research*, **90**, 13,965–80.

Arculus, R.J. (1981). Island arc magmatism in relation to the evolution of the crust and mantle. *Tectonophys.*, **75**, 113–33.

Armstrong, R.L. (1968). A model for the evolution of strontium and lead isotopes in a dynamic earth. *Rev. Geophys.*, **6**, 175–99.

Baer, A.J. (1981a). Geotherms, evolution of the lithosphere and plate tectonics. *Tectonophys.*, **72**, 203–27.

— (1981b). A Grenvillian model of Proterozoic plate tectonics. In *Precambrian plate tectonics*, ed. A. Kröner, pp. 353–85. Amsterdam: Elsevier.

Birch, F. (1965). Speculations on the Earth's thermal history. *Bull. Geol. Soc. Amer.*, **76**, 133–54.

Boak, J.L. & Dymek, R.F. (1982). Metamorphism of the ca. 3800 Ma supracrustal rocks at Isua, west Greenland: implications for early Archean crustal evolution. *Earth Planet. Sci. Lett.*, **59**, 155–76.

Boettcher, A.L. (1973). Volcanism and orogenic belts – the origin of andesites. *Tectonophys.*, **17**, 223–40.

Bowen, N.L. (1928). *The evolution of the igneous rocks*. New York: Dover Pubs.

Boyd, F.R., Gurney, J.J. & Richardson, S.H. (1985). Evidence for a 150–200-km thick Archaean lithosphere from diamond inclusion thermobarometry. *Nature*, **315**, 387–9.

Bridgwater, D. & Fyfe, W.S. (1974). The pre-3 b.y. crust: fact–fiction–fantasy. *Geosci. Canada*, **1**, 7–11.

Burnett, D.S. (1975). Lunar Science: the Apollo legacy. *Rev. Geophys. Space Phys.*, **13**, 13–34.

Cameron, A.G.W. (1978). Physics of the primitive solar disk. *Moon and Planets*, **18**, 5–40.

Campbell, I.H. & Taylor, S.R. (1983). No water, no granites – no oceans, no continents. *Geophys. Res. Lett.*, **10**, 1061–4.

Christensen, V.R. (1985). Thermal evolution models for the Earth. *Jour. Geophys. Res.*, **90**, 2995–3007.

Clark, S.P. & Ringwood, A.E. (1964). Density distribution and constitution of the mantle. *Rev. Geophys.*, **2**, 35–88.

Compston, W. & Pidgeon, R.T. (1986). Jack Hills, evidence of more very old detrital zircons in western Australia. *Nature*, **321**, 766–9.

Condie, K.C. (1980). Origin and early development of the Earth's crust. *Precambrian Res.*, **11**, 183–97.

— (1982). Plate tectonics and crustal evolution. New York: Pergamon.

DePaolo, D.J. (1981). Nd isotopic studies: some new perspectives on Earth structure and evolution. *Trans. Amer. Geophys. Union*, **62**, 137–40.

Eggler, D.H. (1978). The effect of CO_2 upon partial melting of peridotite in the system Na_2O–CaO–Al_2O_3–MgO–SiO_2–CO_2 to 35 kb, with an analysis of melting in a peridotite–H_2O–CO_2 system. *Amer. Jour. Sci.*, **278**, 305–43.

Elsasser, W.M. (1963). Early history of the Earth. In *Earth science and meteorites*, ed. J. Geiss & E. Goldbert, pp. 1–30. Amsterdam: North-Holland Pub. Co.

Ernst, W.G. (1976). *Petrologic phase equilibria*. San Francisco: W.H. Freeman.

— (1983). The early Earth and Archean rock record. In *The Earth's earliest biosphere: its origin and evolution*, ed. J.W. Schopf, pp. 41–52. Princeton, N.J.: Princeton Univ. Press.

Flaser, F.M. & Birch, F. (1973). Energetics of core formation: a correction. *Jour. Geophys. Res.*, **78**, 6101–3.

Froude, D.O., Ireland, T.R., Kinny, P.D., Williams, I.S., Compston, W., Williams, I.R. & Myers, J.S. (1983). Ion microprobe identification of 4100 to 4200 Ma-old terrestrial zircons. *Nature*, **304**, 616–18.

Fyfe, W.S. (1978). The evolution of the earth's crust: modern plate tectonics to ancient hot spot tectonics? *Chem. Geol.*, **23**, 89–114.

Goodwin, A.M. (1981a). Archean plates and greenstone belts. In *Precambrian plate tectonics*, ed. A. Kröner, pp. 105–35. Amsterdam: Elsevier.

— (1981b). Precambrian perspectives. *Sci.*, **213**, 55–61.

Green, D.H. (1975). Genesis of Archean peridotitic magmas and constraints on Archean geothermal gradients and tectonics. *Geol.* **3**, 15–18.

— (1981). Petrogenesis of Archean ultramafic magmas and implications for Archean tectonics. In *Precambrian plate tectonics*, ed. A. Kröner, pp. 469–89. Amsterdam: Elsevier.

Green, D.H. & Ringwood, A.W. (1967a). An experimental investigation of the gabbro to eclogite transformation and its petrological applications. *Geochim. Cosmochim. Acta*, **31**, 767–833.

—— (1967b). The stability fields of aluminous pyroxene peridotite and garnet peridotite and their relevance in upper mantle structure. *Earth Planet. Sci. Lett.*, **3**, 151–60.

Grew, E.S. (1980). Sapphirine + quartz association from Archean rocks in Enderby Land, Antarctica. *Amer. Mineral.*, **65**, 821–36.

Hargraves, R.B. (1976). Precambrian geologic history. *Sci.*, **193**, 363–71.

— (1981). Precambrian tectonic style: a liberal uniformatarian interpretation. In *Precambrian plate tectonics*, ed. A. Kröner, pp. 21–56. Amsterdam: Elsevier.

Head, J.W. & Crumpler, L.S. (1987). Evidence for divergent plate-boundary characteristics and crustal spreading on Venus. *Sci.*, **238**, 1380–5.

Head, J.W. & Solomon, S.C. (1981). Tectonic evolution of the terrestrial planets. *Sci.*, **213**, 62–76.

Herzberg, C.T. (1987). Magma density at high pressure. Part 1: the effect of composition on the elastic properties of silicate liquids. Part 2: a test of the olivine flotation hypothesis. In *Magmatic processes: physicochemical principles*, ed. B.O. Mysen, pp. 25–58. Geochem. Soc. Special Pub. No. 1.

Holloway, J.R. & Burnham, C.W. (1972). Melting relations of basalt with equilibrium water pressure less than total pressure. *Jour. Petrol.*, **13**, 1–29.

Ito, K. & Kennedy, G.C. (1967). Melting and phase relations in a natural peridotite to 40 kilobars. *Amer. Jour. Sci.*, **265**, 519–38.

Jackson, M.J. & Pollack, H.N. (1987). Mantle devolatilization and convection: implications for the thermal history of the earth. *Geophys. Res. Lett.*, **14**, 737–40.

Jacobsen, S.B. (1984). Isotopic constraints on the development of the early crust. *Trans. Amer. Geophys. Union*, **65**, 550.

Jacobsen, S.B. & Wasserburg, G.J. (1981). Transport models for crust and mantle evolution. *Tectonophys.*, **75**, 163–79.

Kaula, W.M. (1979). Thermal evolution of Earth and Moon growing by planetesimal impacts. *Jour. Geophys. Res.*, **84**, 999–1008.

Knoll, A.H. (1984). The Archean/Proterozoic transition: a sedimentary and paleobiological perspective. In *Patterns of change in Earth evolution*, ed. H.D. Holland & A.F. Trendall, pp. 221–42. Dahlem Konferenzen 1984. Berlin: Springer–Verlag.

Kröner, A. (1981). Precambrian plate tectonics. In *Precambrian plate tectonics*, ed. A. Kröner, pp. 57–90. Amsterdam: Elsevier.

Kushiro, I., Syono, Y. & Akimoto, S. (1968). Melting of a peridotite nodule at high pressures and high water pressures. *Jour. Geophys. Res.*, **73**, 6023–9.

Lambert, I.B. & Wyllie, P.J. (1972). Melting of gabbro (quartz eclogite) with excess water to 35 kilobars, with geological applications. *Jour. Geol.*, **80**, 693–708.

Lambert, R. St J. (1976). Archean thermal regimes, crustal and upper mantle temperatures, and a progressive evolutionary model for the earth. In *The early history of the Earth*, ed. B.F. Windley, pp. 363–73. New York: Wiley.

— (1980). The thermal history of the earth in the Archean. *Precambrian Res.*, **11**, 199–213.

— (1981). Earth tectonics and thermal history: review and a hot spot model for the Archean. In *Precambrian plate tectonics.* ed. A. Kröner, pp. 453–67. Amsterdam: Elsevier.

Lee, W.H.K. (1970). On the global variations of terrestrial heat flow. *Phys. Earth Planet. Inter.*, **2**, 332–41.

Longhi, J. (1978). Pyroxene stability and the composition of the lunar magma ocean. *Proc. Lunar Planet. Sci. Conf.*, **9**, 285–306.

Luth, W.C., Jahns, R.H. & Tuttle, O.F. (1964). The granite system at pressures of 4 to 10 kilobars. *Jour. Geophys. Res.*, **69**, 759–73.

MacGregor, I.D. (1968). Mafic and ultramafic inclusions as indicators of the depth of origin of basaltic magmas. *Jour. Geophys. Res.*, **73**, 3737–45.

Martin, H. (1986). Effect of steeper Archean geothermal gradient on geochemistry of subduction-zone magmas. *Geol.*, **14**, 753–6.

McGill, G.E. (1983). Geology and tectonics of Venus. In *Revolution in the Earth sciences*, ed. S.J. Boardman, pp. 68–79. Dubuque, Iowa: Kendall-Hunt.

McKenzie, D.P. & Richter, F.M. (1981). Parameterized thermal convection in a layered region and the thermal history of the Earth. *Jour. Geophys. Res.*, **86**, 11,667–80.

McKenzie, D.P. & Weiss, N. (1975). Speculations on the thermal and tectonic history of the earth. *Geophys. Jour. Roy. Astron. Soc.*, **42**, 131–74.

McLennan, S.M. & Taylor, S.R. (1982). Geochemical constraints on the growth of the continental crust. *Jour. Geol.*, **90**, 347–61.

Miyashiro, A. (1961). Evolution of metamorphic belts. *Jour. Petrol.*, **2**, 277–311.

— (1972). Metamorphism and related magmatism in plate tectonics. *Amer. Jour. Sci.*, **272**, 629–56.

Moorbath, C. (1975). Evolution of Precambrian crust from strontium isotopic evidence. *Nature*, **254**, 395–9.

— (1984). Patterns and geological significance of age determinations in continental blocks. In *Patterns of change in Earth evolution*, ed. H.D. Holland & A.F. Trendall, pp. 207–19. Dahlem Konferenzen 1984. Berlin: Springer–Verlag.

Moorbath, C., O'Nions, R.K., Pankhurst, J.R., Gale, N.H. & McGregor, V.R., (1972). Further Rb–Sr age determinations on the Early Precambrian rocks of the Godthaab district, West Greenland. *Nature, Phys. Sci*, **240**, 78–82.

Moore, R.O. & Gurney, J.J. (1985). Pyroxene solid solution in garnets included in diamond. *Nature*, **318**, 553–5.

Morse, S.A. (1986). Origin of earliest planetary crust: role of compositional convection. *Earth Planet. Sci. Lett.*, **81**, 118–26.

Myers, J.S. & Williams, I.R. (1985). Early Precambrian crustal evolution at Mount Narryer, Western Australia. *Precambrian Res.*, **27**, 153–63.

Mysen, B.O. & Boettcher, A.L. (1975). Melting in a hydrous mantle, II: geochemistry of crystals and liquids formed by anatexis of mantle peridotite with controlled activities of H_2O, CO_2, and O_2. *Jour. Petrol*, **16**, 549–93.

Nelson, B.K. & DePaolo, D.J. (1984). 1,700 Myr greenstone volcanic successions in southwestern North America and isotopic evolution of Proterozoic mantle. *Nature*, **311**, 143–6.

—— (1985). Rapid production of continental crust 1.7 to 1.9 b.y. ago: Nd isotopic evidence from the basement of the North American mid-continent. *Geol. Soc. Amer. Bull.*, **96**, 746–54.

Newton, R.C. (1978). Experimental and thermodynamic evidence for the operation of high pressures in Archaean metamorphism. In *Archean geochemistry*, ed. B.F. Windley & S.M. Naqvi, pp. 221–40. Amsterdam: Elsevier.

— (1983). Geobarometry of high-grade metamorphic rocks. *Amer. Jour. Sci.*, **283A**, 1–28.

Nisbet, E.G., Bickle, M.J. & Martin, A. (1977). Mafic and ultramafic lavas of Belingwe greenstone belt, Rhodesia. *Jour. Petrol.*, **18**, 521–66.

O'Hara, M.J. (1967). Mineral parageneses in ultrabasic rocks. In *Ultramafic and related rocks*, ed. P.J. Wyllie, pp. 393–403. New York: Wiley.

— (1968). The bearing of phase equilibria studies in synthetic and natural systems on the origin and evolution of basic and ultrabasic rocks. *Earth Sci. Rev.*, **4**, 69–133.

O'Nions, R.K., Evensen, N.M. & Hamilton, P.J. (1980). Differentiation and evolution of mantle. *Proc. Roy. Soc. London*, **297**, 479–93.

Papanastassiou, D.A. & Wasserburg, G. (1971). Lunar chronology and evolution from Rb–Sr studies of Apollo 11 and 12 samples. *Earth Planet. Sci. Lett.*, **11**, 37–62.

Patterson, C.C. (1956). Age of meteorites and the earth. *Geochim. Cosmochim. Acta*, **10**, 230–7.

Phillips, R.J., Kaula, W.M., McGill, G.E. & Malin, M.C. (1981). Tectonics and evolution of Venus. *Sci.*, **212**, 879–87.

Reymer, A. & Schubert, G. (1986). Rapid growth of some major segments of continental crust. *Geol.*, **14**, 299–302.

Richter, F.M. (1984). Time and space scales of mantle convection. In *Patterns of change in Earth evolution*, ed. H.D. Holland & A.F. Trendall, pp. 271–89. Dahlem Konferenzen 1984. Berlin: Springer–Verlag.

Ringwood, A.E. (1975). Composition and petrology of the Earth's crust, ocean and atmosphere. *Geochem. Int.*, **1**, 713–37.

— (1979). *Origin of the Earth and Moon*. New York: Springer–Verlag.

Ronov, A.B. (1964). Common tendencies in the chemical evolution of the Earth's crust, ocean, atmosphere. *Geochem. Int.*, **1**, 713–37.

Ronov, A.B. & Migdisov, A.A. (1971). Geochemical history of the crystalline basement and the sedimentary cover of the Russian and North American platforms. *Sedimentology*, **16**, 137–85.

Safronov, V.S. (1972). Evolution of the protoplanetary cloud and formation of the Earth and planets. N.A.S.A Tech. Trans., TTF-667, Washington, D.C.

Schubert, G. (1979). Subsolidus convection in the mantles of terrestrial planets. *Ann. Rev. Earth Planet. Sci.*, **7**, 289–342.

Schubert, G. & Reymer, A.P.S. (1985). Continental volume and freeboard through geologic time. *Nature*, **316**, 336–9.

Shaw, D.M. (1976). Development of the early continental crust. Part 2. Prearchean, Protoarchean and later eras. In *The early history of the Earth*, ed. B.F. Windley, pp. 33–54. New York: Wiley.

Shoemaker, E.M. (1984). Large body impacts through geologic time. In *Patterns of change in Earth evolution*, ed. H.D. Holland & A.F. Trendall, pp. 15–40. Dahlem Konferenzen 1984. Berlin: Springer–Verlag.

Smith, J.V. (1979). Mineralogy of the planets: a voyage in space and time. *Mineral. Mag.*, **43**, 1–89.

— (1981). The first 800 million years of earth's history. *Philos. Trans. Roy. Soc. London*, **301A**, 401–22.

Spohn, T. & Schubert, G. (1982). Modes of mantle convection and the removal of heat from the Earth's interior. *Jour. Geophys. Res.*, **87**, 4682–96.

Stevenson, D.J. (1983). The nature of the Earth prior to the oldest known rock record: the Hadean Earth. In *The Earth's earliest biosphere: its origins and evolution*, ed. J.W. Schopf, pp. 32–40. Princeton, N.J.: Princeton Univ. Press.

Stolper, E., Walker, D., Hagar, B. & Hays, J. (1981). Melt segregation from partially molten source regions: the importance of melt density and source region size. *Jour. Geophys. Res.*, **86**, 6261–71.

Takahashi, E. (1986). Melting of a dry peridotite KLB-1 up to 14 GPa: implications on the origin of peridotitic upper mantle. *Jour. Geophys. Res.*, **91**, 9367–82.

Turcotte, D.L. & Kellogg, L.H. (1986). Isotopic modeling of the evolution of the mantle and crust. *Rev. Geophys.*, **24**, 311–28.

Turcotte, D.L. & Schubert, G. (1982). *Geodynamics*. New York: Wiley.

Veizer, J. (1976). $^{87}Sr/^{86}Sr$ evolution of seawater during geologic history and its significance as an index of crustal evolution. In *The early history of the Earth*, ed. B.F. Windley, pp. 569–78. New York: Wiley.

Wanke, H., Dreibus, G. & Jagoutz, E. (1984). Mantle chemistry and accretion history of the Earth. In *Archean geochemistry*, ed. A. Kröner, G.N. Hanson & A.M. Goodwin, pp. 1–24. Berlin: Springer–Verlag.

Wetherill, G.W. (1972). The beginning of continental evolution. *Tectonophys.*, **13**, 31–45.

— (1976). The role of large bodies in the formation of the Earth and Moon. *Proc. Lunar Sci. Conf.*, **7**, 3245–57.

Wyllie, P.J. (1970). Ultramafic rocks and the upper mantle. Mineral. Soc. America Special Paper no. 3, pp. 3–32.

— (1979). Magmas and volatile components. *Amer. Mineral.*, **64**, 469–500.

— (1981). Plate tectonics and magma genesis. *Geol. Rundschau*, **70**, 128–53.

15

Temperatures in and around cooling magma bodies

MARK S. GHIORSO

Introduction

In recent years, much insight has been gained into the nature of magmatic crystallization and assimilation phenomena through the use of equilibrium thermodynamic
models of multicomponent silicate melts (Ghiorso et al., 1983, Ghiorso & Carmichael,
1985, Ghiorso & Kelemen, 1987). As a consequence of these studies, it has become
apparent that theoretical simulation of chemical reactions in magmas *requires* the
incorporation of numerical techniques that allow the boundaries of the system to be
open to oxygen transfer (Ghiorso, 1985, Ghioso & Carmichael, 1985, Carmichael &
Ghiorso, 1986). These techniques rely upon the definition of a thermodynamic
potential which is minimal, at thermodynamic equilibrium, in systems subject to
boundary conditions including fixed temperature, pressure, and bulk composition of
all components save oxygen, which is constrained by specification of fixed chemical
potential. The creation of a thermodynamic potential which satisfies these require-
ments was first described by Korzhinskii (1949, 1956, 1959) and later introduced to the
western scientific community by Thompson (1970). Ghiorso & Kelemen (1987) have
recently extended the Korzhinskii approach to allow the specification of arbitrary
thermodynamic potentials, minimal at equilibrium in thermodynamic systems subject
to very general boundary constraints. In particular, Ghiorso & Kelemen (1987) have
derived a potential function which allows for the modeling of chemical equilibrium in
magmatic systems as a function of pressure, the heat content of the system (specified as
the enthalpy), the system bulk composition and the chemical potential of oxygen. They
use this potential to investigate the compositional effects accompanying isenthalpic
assimilation of a variety of solids into basaltic magmas. In this paper it is demonstrated
that the function derived by Ghiorso & Kelemen (1987) can be used to investigate heat
flow from magma bodies of quite arbitrary geometry and composition. Analytical
solutions to the one-dimensional heat flow equation are provided for the case of a

rapidly convecting magma chamber and numerical simulations of the conductively cooling case are discussed. As demonstrated below, the main advantage of these calculations is the incorporation of a variable latent heat of fusion which is a function of temperature, and hence both space and time, in the cooling body of magma. Comparison with traditional methods of modeling the heat effects of magmatic crystallization (e.g. Jaeger, 1957), demonstrates that the effect of variable latent heat is significant on the temperature–time history of the magma and the surrounding country rock.

Magmatic phase relations as a function of heat content

Modeling the irreversible process of magmatic crystallization, due to the withdrawal of heat from a magma body, can be achieved by regarding the process as consisting of a series of steps in reaction progress, with each step characterized by thermodynamic equilibrium. Though this procedure neglects the importance of crystallization kinetics and melt diffusion, it does provide a useful description of the bulk chemical effects which accompany crystallization (Ghiorso & Carmichael, 1985). If the increments of heat withdrawn from the magma are sufficiently small, then the stepwise results can be readily linked to describe the overall continuous crystallization event. Consequently, for calculation purposes, it remains sufficient to describe the method of obtaining the magmatic phase relations and temperature as a function of a particular pressure, heat content and chemical potential of oxygen; that is, the calculation of equilibrium at any step in the 'irreversible' process.

The thermodynamic potential which is minimal in a magmatic system at equilibrium and subject to these constraints can be derived using the methods of Korzhinskii (1956). Let G denote the Gibbs free energy of the magmatic system under consideration, H the enthalpy, S the entropy and T the absolute temperature. From the definition of the Gibbs function,

$$\frac{G}{T} = \frac{H}{T} - S,$$

(1)

the total derivative of G/T can be obtained:

$$d\left(\frac{G}{T}\right) = H \, d\left(\frac{1}{T}\right) + \frac{1}{T} dH - dS.$$

(2)

Given that the derivative of H can be expressed in terms of the pressure, system volume, and the chemical potentials (μ_i) and mole numbers (n_i) of a set of system components, equation (2) may be written

$$d\left(\frac{G}{T}\right) = H \, d\left(\frac{1}{T}\right) + \frac{V}{T} dP + \frac{1}{T} \sum_i \mu_i \, dn_i.$$

(3)

Table 15.1. *Bulk compositions of starting liquids*

	Thingmuli olivine tholeiite	Boninite (U liquid)
SiO_2	48.55	56.38
TiO_2	1.71	0.28
Al_2O_3	15.34	10.65
FeO_2	11.41	9.67
MgO	8.78	14.38
CaO	11.59	5.94
Na_2O	2.31	1.48
K_2O	0.20	0.78
H_2O	0.10	0.00

Source: Ghiorso & Carmichael (1985).

From examination of equation (3), it is clear that the potential G/T is a function of the independent variables $1/T$, P and all the n_i. To obtain the desired potential from G/T, the dependent- variable H must be exchanged for the independent variable $1/T$ and in an equivalent manner μ_{O_2} must be exchanged for n_{O_2}. This exchange is effected via a Legendre transform of G/T to create the appropriate potential, $\Phi(H, P, n_i^*, \mu_{O_2})$:

$$\Phi(H, P, n_i^*, \mu_{O_2}) = \frac{G}{T} - \frac{1}{T}\left(\frac{\partial \frac{G}{T}}{\partial \frac{1}{T}}\right)_{P, n_i} - n_{O_2}\left(\frac{\partial \frac{G}{T}}{\partial \frac{n_{O_2}}{T}}\right)_{P, 1/T, n^*_i}$$

$$= \frac{G}{T} - \frac{1}{T}H - n_{O_2}\mu_{O_2}$$

$$= -S - n_{O_2}\mu_{O_2}$$

(4)

where n_i^* denotes a subset of the original system components excluding oxygen. Numerical methods for minimizing the potential defined by equation (4) are discussed in some detail in Ghiorso & Kelemen (1987). The thermodynamic data necessary for the application of this modeling technique to magmatic systems are provided by Ghiorso *et al.* (1983) and Ghiorso (1985, appendix). Similar techniques for the thermodynamic analysis of magmatic systems have been the subject of numerous investigations by Russian workers (e.g. Frenkel & Ariskin, 1984a, b, Ariskin *et al.*, 1986).

In order to demonstrate the application of this modeling technique to heat flow calculations, two examples, reflecting rather different bulk compositions and crystallization sequences, will be illustrated. Compositions are tabulated in Table 15.1 and crystallization characteristics as a function of heat content are displayed in Figs. 15.1 and 15.2. The results shown in these figures were computed using an algorithm based

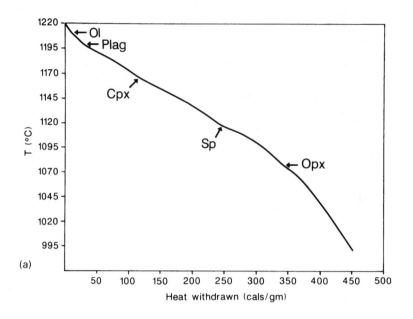

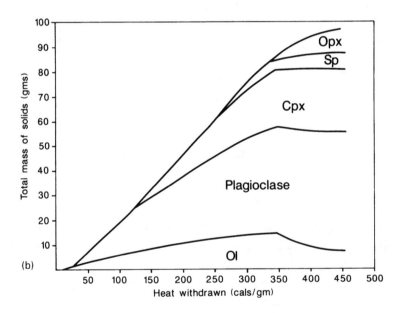

Fig. 15.1.

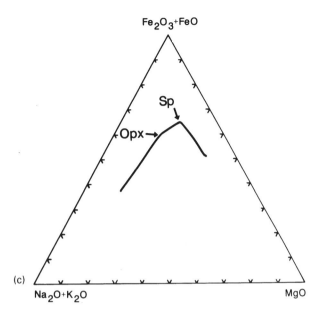

Fig. 15.1.   Results of thermodynamic modeling of the equilibrium crystallization of
(*cont.*)   the Olivine Tholeiite of Table 15.1. (a) Temperature as a function of heat
content (expressed in terms of cal withdrawn per gram of magma). The
liquidus temperature is 1220 °C. Arrows indicate first appearance of the
phases Olivine (Ol), Plagioclase (Plag), Clinopyroxene (Cpx), Spinel (Sp)
and Orthopyroxene (Opx). (b) Phase proportions as a function of the
heat content of the magma. Quantities are referenced to 100 grams of
magma. (c) Projected residual liquid compositions for the crystallization
range corresponding to (a) and (b).

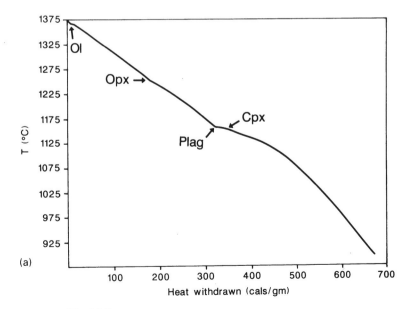

Fig. 15.2.

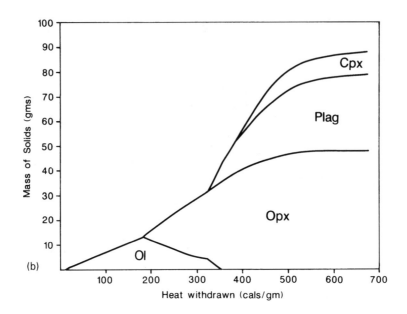

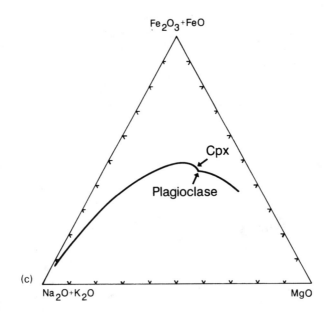

Fig. 15.2. Results of the thermodynamic modeling of the equilibrium crystallization
(*cont.*) of the Boninite of Table 15.1 (a) Temperature as a function of heat
content (expressed in terms of cal withdrawn per gram of magma). The
liquidus temperature is 1375 °C. Arrows indicate first appearance of the
phases Olivine (Ol), Orthopyroxene (Opx), Plagioclase (Plag) and
Clinopyroxene (Cpx). (b) Phase proportions as a function of the heat
content of the magma. Note the complete reaction between olivine and
orthopyroxene. Quantities are referenced to 100 grams of magma. (c)
Projected residual liquid compositions for the crystallization range
corresponding to (a) and (b).

upon the method described in the previous paragraph with the chemical potential of oxygen constrained to follow the quartz–fayalite–magnetite buffer throughout the crystallization history. Figs. 15.1(a) and 15.2(a) demonstrate rather dramatically that the heat output accompanying crystallization of a tholeiitic and boninitic liquid is not a uniform function of the temperature of the magma. This is due in part to the change in heat capacity accompanying the transformation from liquid to solid, but more importantly, arises from non-uniform liberation of latent heat over the crystallization interval. This is most apparent in Fig. 15.2(b) where a substantial heat effect associated with the liquid reaching the plagioclase–clinopyroxene–orthopyroxene cotectic is indicated. Close comparison of Figs. 15.1(a) and 15.1(b) and 15.2(a) and 15.2(b) reveal that a 'burst' of heat is liberated each time a new solid phase appears on the liquidus. This seems to be a general phenomena and has been reported previously (Ghiorso & Carmichael, 1985).

The results displayed in Figs. 15.1 and 15.2 comprise a base upon which a number of magmatic heat flow scenarios can be built. To illustrate this point two end-member calculations are considered below.

Convection mode

Heat flow from a rapidly convecting magma chamber, and the associated temperature distribution in the surrounding country rock, can be easily modeled if the assumption is made that convection is so rapid as to maintain a spatially uniform temperature profile in the magma body as a function of time (Bartlett, 1969). Under this rather restrictive, end-member condition, the heat flow problem can be reduced to modeling the conductive temperature evolution in an infinite half-plane of country rock, where the temperature at the magma/country rock boundary (the origin) is the spatially uniform temperature of the magma. The magma temperature, in turn, is a function only of the total amount of heat which has been released to the wall rock (i.e. which has traversed the boundary). For this problem, it is possible to obtain an analytical solution to the one-dimensional heat conduction equation if knowledge of the heat flow across the interface as a function of time (not temperature) is known. This solution is provided in the appendix (equation A-8) where it is also demonstrated that the heat flow per unit area per unit time can be approximated from the crystallization calculations of the previous section, if some approximate geometry is assumed for the magma body. In the calculations reported below it is assumed for convenience that the magma body is a cube, 10 km on edge, and that heat is released uniformly through all six faces of the cube into the country rock. Though this geometry represents a zeroth order description of a realistic magma body, it allows a simple comparison of end-member convective and conductive calculations to be made. Edge effects of the cube (i.e. Jaeger, 1961) will be ignored in the cooling simulations reported below. In addition, the magma is assumed to be intruded instantaneously at its liquidus temperature.

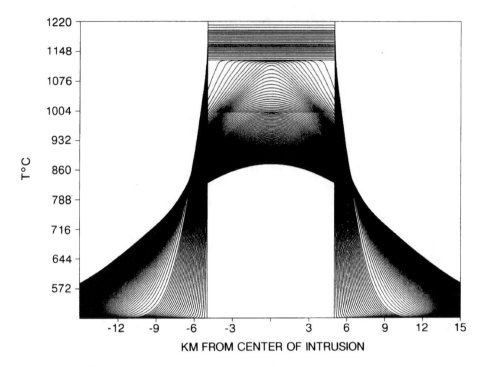

Fig. 15.3.   Temperatures in the magma and the country rock for a combined
convectively–conductively cooled magma body of Olivine Tholeiitic
composition (Table 15.1). Isochrons for the convectively cooled portion
of the plot are taken from Table 15.2. Conductively cooled isochrons are
calculated using a time interval of 10 000 years. The first 800 000 years of
cooling is displayed. The magma body is assumed to be a cube 10 km on
edge with equal heat flow through all faces. The initial temperature of the
magma is that of the liquidus (1220°C) and the temperature of the
country rock is 500 °C.

Realistically, a body of magma will remain convecting until solidified to approximately 50–55% crystals. After this point, cooling must be modeled (both within the
magma and the country rock) by numerical integration of the heat conduction
equation, utilizing methods which incorporate the effects of variable latent heat of
fusion. The formulation of the differential equation and a description of the method of
integration for this phase of the cooling history are discussed below in the section on the
conduction models.

The complete convective/conductive cooling history of a 10 km cube of tholeiitic
magma is displayed in Fig. 15.3. The initial temperature of the wall rock is taken to be
500 °C and the thermal conductivity, K, is approximated as that of average crustal rock
(0.0042; Carslaw & Jaeger, 1959, p. 497; cal/s-cm-°C) as is the thermal diffusivity, κ
(0.0118, Carslaw & Jaeger, 1959, p. 497; cm²/s). For the convective phase of these
calculations, isochrons report temperature profiles at successive increments of heat
flow corresponding to 1436 kcal/cm² of the magma cube surface. This increment is
equivalent to withdrawing 3 cal per gram of magma for each isochron. The calculated

Table 15.2. *Magma/country rock temperatures and heat fluxes for the Thingmuli convective cooling model*

Percent crystallized	Temperature (°C)	Time (years)	Flux $K \cdot 10^5$ (cal/s-cm²)
0.0	1220.00	0.00	∞
0.0	1211.10	67.38	33.22
0.8	1205.49	275.37	16.62
1.8	1199.56	620.71	10.29
4.0	1196.68	1125.66	8.164
6.1	1193.70	1763.82	6.245
8.3	1190.61	2563.68	5.244
10.4	1187.42	3512.69	4.412
12.5	1184.13	4623.45	3.827
14.6	1180.72	5896.09	3.356
16.7	1177.20	7336.95	2.984
18.8	1173.55	8950.16	2.676
20.8	1169.79	10741.42	2.419
22.8	1165.89	12716.25	2.201
25.4	1163.23	14868.21	2.043
28.0	1160.52	17190.53	1.879
30.6	1157.71	19703.45	1.749
33.1	1154.79	22404.38	1.626
35.7	1151.75	25303.00	1.518
38.2	1148.60	28404.15	1.421
40.7	1145.33	31714.71	1.332
43.3	1141.93	35241.92	1.251
45.8	1138.40	38993.62	1.178
48.3	1134.73	42978.27	1.109
50.7	1130.93	47204.96	1.047
53.2	1126.99	51683.23	0.9881

temperature–time history of the convecting magma is reported in Table 15.2 as well as the calculated flux of heat across the magma/country rock boundary. Similar calculations for the boninitic melt are displayed in Fig. 15.4 and tabulated in Table 15.3 (heat flow unit is 1413 k/cal/cm² or 3 cal/g). Isochrons for the conductive cooling interval are spaced at 10 000-year intervals, with the first 800 000 years of cooling displayed. For the conductive phase of these heat flow calculations, the thermal conductivity (K) of the magma and the country rock were assumed to be identical. Fig. 15.3 clearly demonstrates the irregular liberation of heat during the convective interval. This is easily seen through the spacing of the convective isochrons which correspond to equal increments of heat withdrawn from the chamber. The conductive phase of cooling allows the heat to be liberated more slowly when compared to the convective phase, and consequently provides for more of the country rock to be heated to a lower temperature.

In Fig. 15.4 a similar convective/conductive cooling history is displayed for the boninitic magma. The effect of the disproportionate amount of heat released at the plagioclase–clinopyroxene–orthopyroxene cotectic is dramatically indicated. Unlike the tholeiitic magma, the cooling interval of this boninitic liquid is protracted and the

Table 15.3. *Magma/country rock temperatures and heat fluxes for the boninite convective cooling model*

Percent crystallized	Temperature (°C)	Time (years)	Flux $K \cdot 10^5$ (cal/s-cm²)
0.0	1375.00	0.00	∞
0.0	1366.48	43.98	50.02
1.5	1363.06	178.49	25.32
2.1	1357.36	401.09	15.80
2.8	1351.62	727.25	12.33
3.5	1345.81	1144.00	9.438
4.1	1339.96	1667.46	7.832
4.8	1334.06	2293.49	6.568
5.4	1328.11	3029.62	5.661
6.1	1322.10	3877.76	4.941
6.7	1316.05	4842.76	4.371
7.4	1309.94	5928.42	3.902
8.1	1303.79	7139.37	3.512
8.7	1297.58	8480.13	3.182
9.4	1291.32	9955.80	2.899
10.0	1285.02	11571.36	2.654
10.7	1278.66	13332.37	2.439
11.3	1272.25	15244.72	2.250
12.0	1265.79	17314.38	2.082
12.6	1259.28	19547.88	1.932
13.3	1252.79	21951.22	1.799
14.6	1247.81	24511.72	1.708
16.0	1242.70	27223.26	1.593
17.2	1237.47	30116.51	1.504
18.5	1232.12	33187.09	1.414
19.7	1226.64	36449.39	1.333
21.0	1221.04	39910.08	1.257
22.2	1215.32	43578.86	1.186
23.3	1209.47	47465.44	1.120
24.5	1203.51	51580.15	1.058
25.6	1197.42	55933.84	1.000
26.7	1191.20	60539.30	0.9456
27.8	1184.87	65408.63	0.8946
28.9	1178.41	70556.96	0.8463
29.9	1171.84	75995.07	0.8010
30.9	1165.15	81742.42	0.7580
31.9	1158.33	87815.69	0.7173
35.1	1156.53	93994.56	0.7430
38.5	1155.09	100133.78	0.7093
41.9	1153.49	106489.27	0.6990
44.6	1150.63	113052.20	0.6633
47.3	1147.33	119926.44	0.6409
49.9	1144.01	127055.13	0.6164
52.8	1141.20	134423.09	0.6005

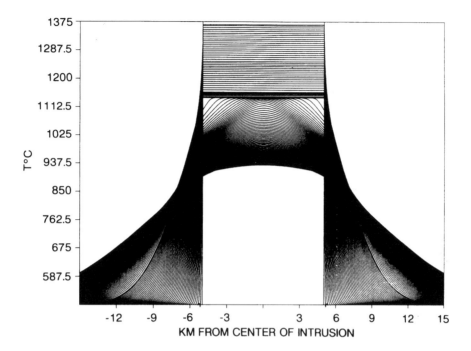

Fig. 15.4.   Temperatures in the magma and the country rock for a combined
convectively–conductively cooled magma body of Boninitic composition
(Table 15.1). Isochrons for the convectively cooled portion of the plot are
taken from Table 15.3. Conductively cooled isochrons are calculated
using a time interval of 10 000 years. The first 800 000 years of cooling is
displayed. The magma body is assumed to be a cube 10 km on edge with
equal heat flow through all faces. The initial temperature of the magma is
that of the liquidus (1375 °C) and the temperature of the country rock is
500 °C.

regularity of the heat output is interrupted substantially by just this one event. As with
the tholeiite, the rate of heat withdrawal slows during the conductive interval,
increasing the volumetric extent of country rock heating.

Several aspects of these calculations should be noted. Firstly, the magma is allowed
to convect from the indicated liquidus temperature until it is substantially rigid (50 to
55% crystallization). This combined convection/conduction scenario provides an end-
member case of extremely rapid heat withdrawal from the chamber, and the calculation
demonstrates maximal temperatures in the wall rock close to the magma body. The
second point is to remark upon the rapidity of heat loss from the magma body in
comparison to the conductive mode of heat transport illustrated below, and thirdly to
note that the maximal temperatures displayed in Figs. 15.3 and 15.4 do not extend a
great distance into the country rock. As is illustrated below, and has been documented
previously by Jaeger (1957, 1959, 1961), Lovering (1935, 1936, 1955) and Carslaw &
Jaeger (1959), one-stage conductive cooling models never yield near-contact temperatures higher than roughly half of the difference between the initial magma temperature

and that of the wall rock. Indeed, Jaeger (1959) was forced to call upon preheating as a means of generating elevated temperatures within meters of the intrusive contact. Thus sanidinite facies metamorphic events in the contact aureoles of intrusive gabbros give some indication of a convective regime during part of the magma cooling history. The calculations presented in Figs. 15.2 and 15.4 demonstrate that extremely elevated wall rock temperatures are not maintained for extensive periods and place constraints on the rates of metamorphic reactions in these high-temperature zones.

The cooling scenarios displayed in Figs. 15.3 and 15.4 are meant to be illustrations of end-member calculations involving rapid convection in magma bodies containing two distinctly different magma types. These scenarios are best viewed in reference to more traditional conductive models calculated for similar melts, magma dimensions and initial conditions. Comparison will then demonstrate those aspects of the cooling profile most dependent upon the cooling mechanism and defocus attention from the more arbitrary aspects of a particular magma geometry or country rock composition/ initial temperature. In order to provide this comparison, the calculation of conductive cooling profiles is taken up in the next section.

Conduction model

Methods of modeling the conductive temperature evolution of magma bodies and their surrounding wall rock have been the subject of numerous papers, many of which are of historic importance in representing some of the first substantive mathematical analysis applied to fundamental petrological problems (Lovering, 1935, 1936, 1955, Larsen, 1945, Jaeger, 1957, 1959, 1961). The great difficulty in modeling conductive heat transfer in magmas centers on the production of latent heat, which enters the differential equation of heat flow as a source term, and therefore must be known *a priori* as a function of both distance and time. Solutions to this problem generally involve the simplification that the latent heat is released from the magma at a definite crystallization temperature rather than over a crystallization interval. This allows the differential equation to be solved as a Stefan moving boundary problem. Both Lovering (1936) and Larsen (1945) realized that one of the key aspects of modeling heat transport in magma bodies is the fact that the latent heat is released over an extended temperature interval of crystallization. They proposed a modification of the Fourier differential equation to account for this latent heat production in the following manner.

The differential equation for conductive heat transport in the absence of a source term (latent heat production) may be written

$$\frac{\partial U}{\partial t} = \kappa \frac{\partial^2 U}{\partial x^2} \tag{5}$$

where κ, the thermal diffusivity, is defined to be

$$\kappa = \frac{K}{\rho\,c} \tag{6}$$

In equation (6), K is the thermal conductivity, ρ the density of the medium and c is the specific heat. Lovering (1936) first suggested a procedure for using equation (5) to model the evolution of latent heat. Essentially, the latent heat production is averaged over the entire crystallization interval and divided by the temperature difference between the liquidus and the solidus. This yields an effective heat capacity which accounts in a crude way for latent heat production as a function of temperature. Upon adding this effective heat capacity to the known heat capacity of the magma, a thermal diffusivity can be computed using equation (6) which accounts for the evolution of the latent heat. Thus, analytical solutions of equation (5) can be used to model latent heat production in magmatic liquids that are cooling conductively. The only difficulty that remains is that a distinction must now be made between the thermal properties of the magma and those of the country rock. Fortunately, Lovering (1936) provides a solution to the heat flow equation for this case. Let $2l$ be the width of the magma body and let κ_1 and K_1 denote the thermal properties of the magma and κ_2 and K_2 those of the country rock. If the difference between the initial magmatic temperature and that of the country rock is taken to be U_0, then temperatures in the magma relative to the initial wall rock temperature are given by

$$U(x,\,t) = \frac{U_0}{2}\left\{ -(1+p)\mathrm{erf}\left(\frac{x}{2\sqrt{\kappa_1}}\right) + (1+p)\sum_{n=1}^{n=\infty}(-p)^{n-1}\left[\mathrm{erf}\left(\frac{2ln+x}{2\sqrt{(\kappa_1\tau)}}\right) - \right.\right.$$

$$\left.\left. p\,\mathrm{erf}\left(\frac{2ln-x}{2\sqrt{(\kappa_1\tau)}}\right)\right]\right\} \tag{7}$$

where p is defined as

$$p = \frac{\sqrt{\kappa_1}\cdot K_2 - \sqrt{\kappa_2}\cdot K_1}{\sqrt{\kappa_1}\cdot K_2 + \sqrt{\kappa_2}\cdot K_1} \tag{8}$$

and where erf denotes the error function. Distance is measured from the right boundary of the magma body which is centered about x equal to $-l$. Lovering (1936) also provides an equation for temperatures in the country rock. These are given by

$$U(x,\,t) = \frac{U_0}{2}\left\{ -(1-p)\mathrm{erf}\left(\frac{x}{2\sqrt{(\kappa_2 t)}}\right) + (1+p^2)\sum_{n=1}^{n=\infty}(-p)^{n-1}\left[\mathrm{erf}\left(\frac{2bn+x}{2\sqrt{(\kappa_2 t)}}\right)\right]\right\} \tag{9}$$

where b is defined as

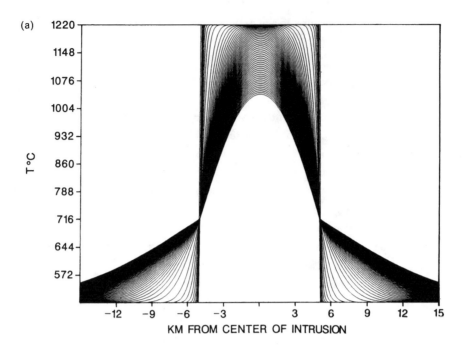

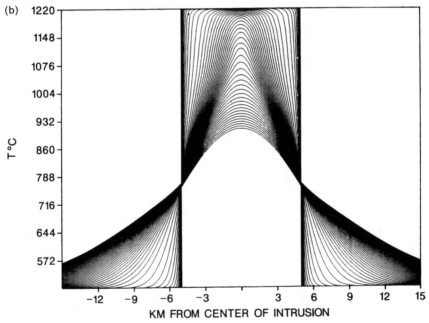

Fig. 15.5. Temperatures in the magma and the country rock for a conductively
cooled magma body of Olivine Tholeiitic composition (Table 15.1).
Isochrons are spaced at 1 year, 2 years, 4, 8, 16, up to 16 384 years and
thereafter at 20 000-year intervals. The first million years of cooling is
displayed. Magma dimensions and initial temperatures are the same as in
Fig. 15.3.(a) A temperature independent averaged latent heat of fusion is
used to model the heat flow. This is the Lovering (1936) approximation
with the average value of κ taken to be 0.002 24 cm^2/s. (b) A temperature
dependent (i.e. spatially and temporally variable) latent heat of fusion is
used to model the heat flow. Upper diagram is a and lower diagram is b.

400

$$b = \frac{l\sqrt{\kappa_2}}{\kappa_1} \tag{10}$$

Jaeger (1957, 1959, 1961) used the approach of Lovering in his thermal modeling with great success, treating the solidification process as that of a boundary layer moving toward the interior of the magma body.

Equations (7) through (10) have been evaluated to model the conductive cooling of the olivine tholeiitic magma of Table 15.1 and the results are displayed in Fig. 15.5(a). The thermal conductivity and diffusivity of the wall rock are identical to the convective/conductive model displayed in Fig. 15.3. The thermal conductivity of the magma is assumed to be identical to that of the country rock, but the average magma density and 'adjusted specific heat' were used to compute an effective thermal diffusivity consistent with the evolution of the latent heat. As with the convective/conductive model, the figure shows the first million years of cooling; the isochron spacing is provided in the caption. Before making a detailed comparison between the temperature profiles displayed in Figs. 15.3 and 15.5(a), it is useful to consider numerical solutions of the conductively cooling model for spatially and temporally variable heats of fusion. Lovering's approach of modifying the system heat capacity is still valid in this case except that the 'effective specific heat due to latent heat release' is no longer defined as an average over the entire crystallization interval but rather as the derivative of the latent heat production as a function of temperature. Consequently, through equation (6), the thermal diffusivity is now a function of temperature and equation (5) becomes a non-linear equation which can no longer be solved analytically. Numerical integration of equation (5) is straightforward, however, and solutions can be explored graphically. In Fig. 15.5(b) the conductive cooling profiles for the tholeiitic magma are shown, where account has been made of variable latent heat of fusion and melt + solid density (equation (6)) as a function of extent of crystallization.

Comparison of Figs. 15.3, 15.5(a) and 15.5(b) reveal some remarkable differences. Neither of the conductive cooling calculations raises the wall rock temperature above 750 °C, the least effect being demonstrated by the conductive cooling model with constant latent heat. The contact temperature between magma and country rock in the conductive cooling models is a complicated function of the ratio of the thermal properties of the two media (Lovering, 1936). In Fig. 15.5(a) the contact temperature is low because κ is small over the entire crystallization range, corresponding to an averaging of the latent heat effect. By contrast, in Fig. 15.5(b) κ varies as a function of T and is small at temperatures corresponding to cotectics or (through equation (6)) when the density of the system is high. The average value of κ used in the second calculation (Fig. 15.5(b)) will be approximately the same as that used in the first (Fig. 15.5(a)), however, the variability in κ means that the rate of heat loss will sometimes be faster than the average rate. The net result is to produce lower magma temperatures after similar cooling intervals in the variable latent heat model.

The contrast between the two conductive models seen in Fig.15.5 is not as

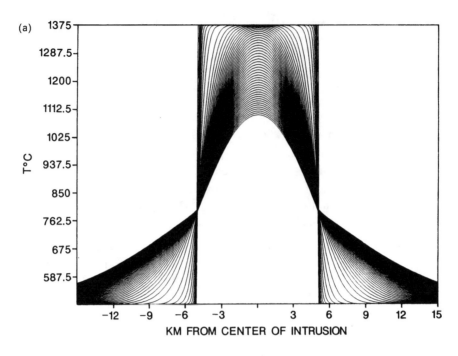

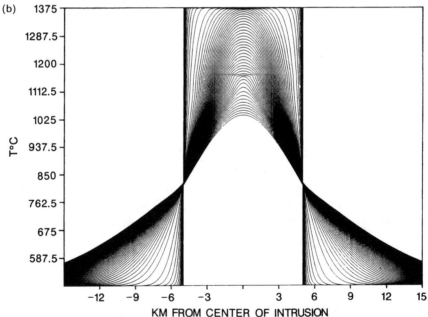

Fig. 15.6.   Temperatures in the magma and the country rock for a conductively
cooled magma body of Boninitic composition (Table 15.1). Isochrons are
spaced at 1 year, 2 years, 4, 8, 16, up to 16 384 years and thereafter at
20 000-year intervals. The first million years of cooling is displayed.
Magma dimensions and initial temperatures are the same as in Fig. 15.4.
(a) A temperature independent averaged latent heat of fusion is used to
model the heat flow. This is the Lovering (1936) approximation with the
average value of κ taken to be 0.003 18 cm^2/s. (b) A temperature
dependent (i.e. spatially and temporally variable) latent heat of fusion is
used to model the heat flow. Upper diagram is a and lower diagram is b.

402

pronounced for the boninitic melt (Fig. 15.6). Here the average value of κ used in calculating Fig. 15.6(a) generates a reasonably accurate approximation to the variable latent heat model of Fig. 15.6(b). Details of the latent heat effect associated with the plagioclase–clinopyroxene–orthopyroxene cotectic (seen as a 'ridge' in Fig. 15.6(a)) are lost, but the overall temperature profiles are within 50 °C in both figures. Clearly, the continuous release of latent heat over a protracted crystallization interval swamps the cotectic heat effect and the effective heat capacity rather than the density variations define the behavior of κ as a function of percent crystallization.

To generalize the results of Figs. 15.5 and 15.6, the constant latent heat approximation of Lovering (1936) is more applicable to the boninitic magma and is clearly a bad approximation to the conductive cooling history of a tholeiitic intrusion. This runs contrary to Jaeger's (1957, 1959) fundamental assumption that the non-uniform liberation of latent heat is largely a second-order effect in predicting the temperature evolution of magma bodies. The differences between the convective and the two conductive models should be independent of magma body geometry and details of initial wall rock composition and temperature. Though Figs. 15.5 and 15.6 clearly demonstrate magma temperature effects, variations in wall rock temperature are more difficult to discern. Figs. 15.7 and 15.8 have been prepared to scrutinize the result of the various modeling attempts regarding predictions of wall rock temperatures.

In Figs. 15.7 and 15.8 are plotted temperature as a function of time, for the convective/conductive and the two conductive models, at the center of the magma body (a), one kilometer from the magma–country rock boundary (b), and ten kilometers from the magma–country rock boundary (c). At 10 km from the intrusion the various models have similar effects on the temperature–time history of the rock. Model temperatures are in closest agreement for the tholeiitic case, but it would be difficult to distinguish convective or conductive heat transfer on the basis of metamorphic equilibria or varying rates of isograd reactions. Closer to the magma body (Figs. 15.7(b), 15.8(b)) differences in the modeling results are more pronounced. In particular, the convective regime heats the country rock up faster and allows more time to be spent at elevated temperatures during the conductive phase of the cooling model. For both lava types, the difference between the conductive and convective end-member models could result in rather different mineral parageneses in the contact aureole.

Conclusions

The calculations reported above represent a very preliminary attempt to model heat transport in magmatic systems with realistic multicomponent silicate melts and temperature dependent latent heats of fusion. These attempts are meant to be illustrative and to emphasize the need to include natural system compositional constraints on any fluid dynamical model of magma chamber evolution. From the perspective of chemical reactions in the magma body, the next step is to include

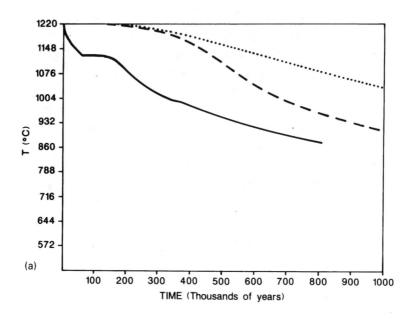

(a)

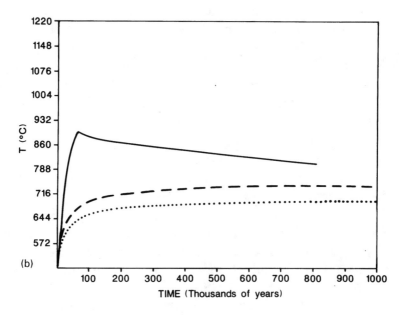

(b)

Fig. 15.7.

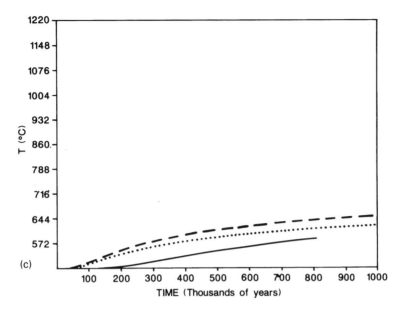

Fig. 15.7. Comparison of temperature–time curves for the three scenarios displayed
(*cont.*) in Figs. 15.3 and 15.5 in the case of cooling of the Olivine Tholeiitic
magma body (Table 15.1). The solid curve corresponds to the
convective–conductive model, the dashed curve to the conductive model
with variable latent heat of fusion and the dotted curve to the Lovering-
type conductive model with an averaged constant latent heat of fusion
over the crystallization interval. Magma dimensions and initial
temperatures are the same as in Fig. 15.3. (a) Profile at the center of the
intrusion. (b) Profile in the country rock one kilometer from the
intrusion. (c) Profile in the country rock ten kilometers from the
intrusion.

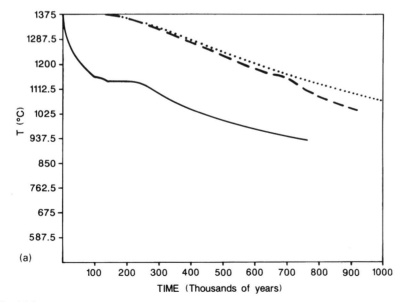

Fig. 15.8.

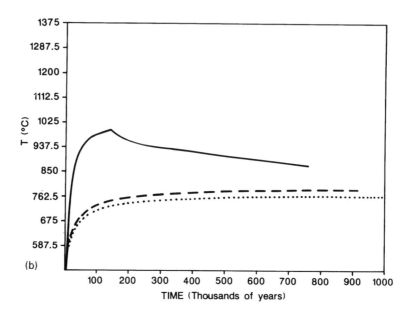

(b)

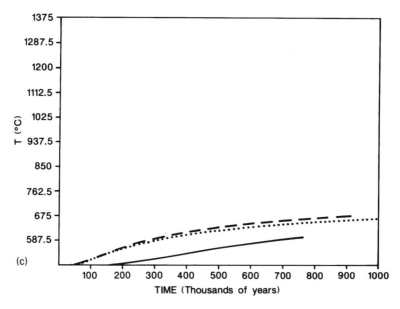

(c)

Fig. 15.8.   Comparison of temperature–time curves for the three scenarios displayed
(*cont.*)   in Figs. 15.4 and 15.6 in the case of cooling of the Boninitic magma body
              (Table 15.1). The solid curve corresponds to the convective–conductive
              model, the dashed curve to the conductive model with variable latent
              heat of fusion and the dotted curve to the Lovering-type conductive
              model with an averaged constant latent heat of fusion over the
              crystallization interval. Magma dimensions and initial temperatures are
              the same as in Fig. 15.4.(a) Profile at the center of the intrusion. (b)
              Profile in the country rock one kilometer from the intrusion. (c) Profile in
              the country rock ten kilometers from the intrusion.

information on the rate of crystal growth. This will allow the full coupling of the thermodynamic models with the transport of liquid and solids due to convection and will facilitate the investigation of related important phenomena such as wall rock assimilation and crystal fractionation. Though sufficient data are not currently available to model the fully coupled system, it may be possible to use the temperature–time profiles in the country rock (i.e. Figs. 15.3, 15.4, 15.7 and 15.8) to model certain aspects of the kinetics of metamorphic reactions in the contact aureole. To this end, more realistic calculations should incorporate advective heat transport due to fluid circulation in the wall rock. The cooling scenarios described above represent idealized end-members of the cooling histories of magma bodies found in nature. Though these examples may be incorrect in detail, the methods upon which the calculations are based can readily be adapted to numerically model the thermal evolution of a magma body of arbitrary geometry, intruded into country rocks of quite general stratigraphic complexity.

Acknowledgements   I am indebted to L. Perchuk for the pleasure of contributing to a volume in honor of D.S. Korzhinskii. Professor Korzhinskii's contributions to the understanding of chemical equilibria in the earth have yielded profound insight into our understanding of geochemical processes. Helpful reviews were provided by David Shirley and M.Y. Frenkel. Material support for this investigation was provided by the National Science Foundation (EAR-8451694) and an equipment grant from DIGITAL Equipment Corporation.

Appendix

We seek solutions, $U(x, t)$ of the heat conduction equation,

$$\frac{\partial U}{\partial t} = \kappa \frac{\partial^2 U}{\partial x^2}, \, x > 0, \, t > 0, \tag{A-1}$$

for temperature profiles in the country rock ($x > 0$), a medium which is characterized by constant thermal diffusivity (κ). The initial temperature is specified by

$$U(x, 0) = U_0, \, x > 0, \tag{A-2}$$

where U_0 is a constant, not necessarily equal to zero. Equation (A-1) is solved subject to the boundary conditions

$$\left. \begin{array}{l} U(0, t) = F(t), \\ \lim_{x \to \infty} U(x, t) = U_0, \end{array} \right\} \, t > 0 \tag{A-3}$$

where the magma temperature, $F(t)$, is specified as a function of t on the left boundary. $F(t)$ is not initially known.

Equation (A-1) is most easily solved using the method of Laplace transforms. Denoting the Laplace transform of $U(x, t)$ by $u(x, s)$, equation (A-1) and its associated boundary conditions (equation (A-3)) become

$$s\, u(x, s) - U(x, 0) = \kappa \frac{d^2 u}{dx^2}, \tag{A-4}$$

$$u(x, s) = f(s),$$
$$\lim_{x \to \infty} u(x, s) = \frac{U_0}{s} \tag{A-5}$$

where $f(s)$ is the Laplace transform of $F(t)$. A solution to equation (A-4) is given by

$$u(x, s) = c_1 \exp\left(\sqrt{\frac{s}{\kappa}} \cdot x\right) + c_2 \exp\left(-\sqrt{\frac{s}{\kappa}} \cdot x\right) + \frac{U_0}{s}, \tag{A-6}$$

where c_1 and c_2 are constants. The boundary conditions demand that c_1 be zero $(x \to \infty)$ and that c_2 be given by $f(s) - U_0/s$, thus equation (A-4) has the specific solution

$$u(x, s) = \left(f(s) - \frac{U_0}{s}\right) \exp\left(-\sqrt{\frac{s}{\kappa}} x\right) \frac{U_0}{s}. \tag{A-7}$$

The inverse transform of equation (A-7) can be determined from the standard tables (e.g. Churchill, 1972) and some minor manipulation. Our final solution is

$$U(x, t) = \frac{x}{2\sqrt{(\pi \kappa)}} \int_0^t \frac{F(t - \tau)}{\tau^{3/2}} \exp\left(-\frac{x^2}{4\kappa\tau}\right) d\tau - U_0 \operatorname{erfc}\left(\frac{x}{2\sqrt{(\kappa t)}}\right) + U_0, \tag{A-8}$$

where the erfc denotes the complementary error function. Equation (A-8) describes the temperature distribution in the wall rock provided that the temperature of the magma is uniform and is known as a function of time (i.e. magma temperature equals $F(t)$). $F(t)$ may be determined by recognizing that the total heat flow per unit area $(Q(t))$ across the magma/country rock boundary $(x = 0)$ is given by the time integral of the flux $(f(t))$:

$$Q(t) = \int_0^t \Phi(\tau) d\tau. \tag{A-9}$$

The flux across the boundary may be written

$$\Phi(t) = -K \frac{\partial U}{\partial x}\bigg|_{x=0}, \tag{A-10}$$

where K is the thermal conductivity of the country rock. Taking the Laplace transform of equations (A-9) and (A-10) and denoting the transformed functions by lower case equivalents, we obtain

$$q(s) = \frac{\phi(s)}{s} = \frac{K}{\sqrt{(\kappa s)}} f(s) - \frac{K}{\sqrt{(\kappa s^{3/2})}} U_0. \tag{A-11}$$

The inverse transform of equation (A-11) yields an expression for the total heat flow from the magma per unit area at time t:

$$Q(t) = \frac{K}{\sqrt{(\pi \kappa)}} \left[\int_0^t \frac{F(\tau)}{\sqrt{(t - \tau)}} d\tau - 2U_0\sqrt{t} \right]. \tag{A-12}$$

The function $F(t)$ can be determined from equation (A-12) by recognizing that thermodynamic modeling yields f as a function of heat content; the magma temperature is known explicitly as a function of the enthalpy content of the body. Adopting a magma volume and geometry and assuming spatially uniform heat flow yields Q as a function of F. A numerical method to extract the function $F(t)$ from equation (A-12) can now be proposed. Suppose we know the initial magma temperature, $F(0)$, and the temperature, $F(t_1)$, after some time interval $[0, t_1]$ corresponding to a small, but specified, discharge of heat. Assuming F is a linear function of time between 0 and t_1, equation (A-12) can be integrated numerically and solved for the value of t which makes the right-hand and left-hand sides equal. This is t_1. Now suppose an additional increment of heat is discharged which lowers the magma temperature to $F(t_2)$. Values of t_2 are guessed and the integral in equation (A-12) is evaluated by quadratic interpolation in t between the temperatures $F(0)$, $F(t_1)$ and $F(t_2)$. t_2 is modified until the right-hand side equals the known heat flow ($Q(t_2)$) corresponding to the current temperature ($F(t_2)$). Thus, t_2 is determined. The procedure continues, guessing a value of t_3, t_4, etc. corresponding to successive increments in total heat flow ($Q(t_3)$, $Q(t_4)$, etc.) for which the temperature after each increment ($F(t_3)$, $F(t_4)$, etc.) is known. The previously determined time dependence of F is interpolated and extrapolated using cubic splines at each step in order to evaluate the integral in equation (A-12). Thus, with each successive heat flow increment, knowledge of the time dependence of the temperature of the magma is obtained. The accuracy of the integral approximation is increased with each step as our knowledge of the form of the function is based upon more and more numerical values, and hence our ability to interpolate and extrapolate the function becomes more secure. Once a 'table' of $F(t)$ is obtained in this fashion for the entire magma cooling history, it can be used to evaluate equation (A-8) which allows country rock temperatures to be calculated.

References

Ariskin, A.A., Barmina, G.S. & Frenkel, M.Ya. (1986). Computer simulation of basalt magma crystallization at a fixed oxygen fugacity. *Geokhimiya*, **11**, 1614–27.

Bartlett, R.W. (1969). Magma convection, temperature distribution, and differentiation. *Am. J. Sci.*, **267**, 1067–82.

Carmichael, I.S.E. & Ghiorso, M.S. (1986). Oxidation-reduction relations in basic magma: a case for homogeneous equilibria. *Earth Planet. Sci. Letts.*, **78**, 200–10.

Carslaw, H.S. & Jaeger, J.C. (1959). *Conduction of heat in solids*, 2nd edn. Oxford: Oxford Univ. Press.

Churchill, R.V. (1972). *Operational mathematics*, 3rd edn. New York: McGraw-Hill Book Company.

Frenkel, M.Ya. & Ariskin, A.A. (1984a). A computer algorithm for equilibration in a crystallizing basalt magma. *Geokhimiya*, **5**, 679–90.

—— (1984b). Computer simulation of basalt-magma equilibrium and fractional crystallization. *Geokhimiya*, **10**, 1419–31.

Ghiorso, M.S. (1985). Chemical mass transfer in magmatic processes. I. Thermodynamic relations and numerical algorithms. *Contrib. Mineral. Petrol.*, **90**, 107–20.

Ghiorso, M.S. & Carmichael, I.S.E. (1985). Chemical mass transfer in magmatic processes. II. Applications in equilibrium crystallization, fractionation and assimilation. *Contrib. Mineral. Petrol.*, **90**, 121–41.

Ghiorso, M.S., Carmichael, I.S.E., Rivers, M.L. & Sack, R.O (1983). The Gibbs free energy of mixing of natural silicate liquids; an expanded regular solution approximation for the calculation of magmatic intensive variables. *Contrib. Mineral. Petrol.*, **84**, 107–45.

Ghiorso, M.S. & Kelemen, P.B. (1987). Evaluating reaction stoichiometry in magmatic systems evolving under generalized thermodynamic constraints: examples comparing isothermal and isenthalpic assimilation. In *Magmatic processes; physicochemical principles*, ed. B.O. Mysen, pp. 319–36. Geochemical Society, *Sp. Publ. No.* **1**.

Jaeger, J.C. (1957). The temperature in the neighborhood of a cooling intrusive sheet. *Am. J. Sci.*, **255**, 306–18.

— (1959). Temperatures outside a cooling intrusive sheet. *Am. J. Sci.*, **257**, 44–54.

— (1961). The cooling of irregularly shaped igneous bodies. *Am. J. Sci.*, **259**, 721–34.

Korzhinskii, D.S. (1949). The phase rule and systems with fully mobile components. *Dokl. Akad. Nauk. SSSR*, **64**, 361–4.

— (1956). Deduction of thermodynamic potentials for systems with perfectly mobile components. *Dokl. Akad. Nauk. SSSR*, **106**, 295–8.

— (1959). *Physicochemical basis of the analysis of the paragenesis of minerals*. New York: Consultants Bureau, Inc.

Larsen, E.S. (1945). Time required for the crystallization of the great batholith of Southern and Lower California. *Am. J. Sci.*, **243A**, 399–416.

Lovering, T.S. (1935). Theory of heat conduction applied to geological problems. *Bull. Geol. Soc. Am.*, **46**, 69–94.

— (1936). Heat conduction in dissimilar rocks and the use of thermal models. *Bull. Geol. Soc. Am.*, **47**, 87–100.

— (1955). Temperatures in and near intrusions. In *Econ. Geol. Fiftieth Anniversary Volume*, pp. 249–79.

Thompson, J.B., Jr (1970). Geochemical reaction and open systems. *Geochim. Cosmochim. Acta.*, **34**, 529–51.

16

Experimental studies of the system Mg_2SiO_4– SiO_2–H_2 at pressures 10^{-2}–10^{-10} bar and at temperatures to 1650 °C: application to condensation and vaporization processes in the primitive solar nebula

I. KUSHIRO and B.O. MYSEN

Introduction

Forsterite and enstatite (or forsteritic olivine and enstatitic pyroxene) are two major constituent minerals in chondrites. These minerals could have been formed in the primitive solar nebula by condensation of gas or crystallization of liquid or recrystallization associated with vaporization. The phase relations involving forsterite, enstatite, H_2-rich gas and liquid, particularly those near the 'triple point', therefore, are necessary for understanding the conditions of formation of these minerals.

The pressure–temperature conditions required for condensation of forsterite and enstatite from the primitive solar nebula gas have been investigated by means of thermodynamic calculations (e.g., Wood, 1963, Lord, 1965, Larimer, 1967, Blander & Katz, 1967, Grossman, 1972, Grossman & Larimer, 1974, Saxena & Eriksson, 1983) and by the suggested phase relations in the system forsterite–H_2 under solar nebula conditions (Yoder, 1976). The results of thermodynamic calculations (Grossman, 1972) indicate that with cooling of gas of solar composition at total pressure of 10^{-3} atm, forsterite condenses at 1444 K and begins to react with gas to form enstatite at 1349 K. The ratio of forsterite to enstatite decreases with further cooling. A reaction relation between Si-rich gas and forsterite to produce enstatite was also suggested for experimental observations (Sata *et al.*, 1978). The reaction relation of forsterite and gas to produce enstatite is important for understanding Mg/Si fractionation in the nebula.

The ratio of Mg and Si varies significantly among chondrites (e.g., Urey, 1961, Ahrens, 1964, 1965, Larimer & Anders, 1970, Wasson, 1974, Dodd, 1981) and between chondrules and matrix of unequilibrated ordinary chondrites (e.g., Huss *et al.*, 1981, Ikeda *et al.*, 1981, Grossman & Wasson, 1983, Nagahara, 1984). Estimated bulk chemical compositions of terrestrial planets also show significant variation in Mg/Si ratio (e.g., Lewis, 1973, Weidenschilling, 1976, Turekian & Clark, 1969, Wänke *et al.*, 1974, Morgan & Anders, 1979, 1980). Fractionation of Mg and Si is, therefore, an important problem relevant to the origin of chondrites and probably of planets. The fractionation of Mg and Si may have been achieved efficiently in the primitive solar nebula during the condensation and vaporization processes involving forsterite and enstatite. The condensation and vaporization processes may have occurred in large scale or only locally in the primitive solar nebula. In either case, the phase relations involving forsterite and enstatite and H_2-rich gas must be known to understand the chemical fractionation during these processes.

The thermodynamic calculations for the phase relations involving forsterite, enstatite and vapor depend on the thermochemical data of these minerals as well as those of the vapor species, which include some uncertainties. For example, initial calculations of condensation temperatures of forsterite and enstatite vary significantly and their relative positions are even reversed depending on which set of data is used (Larimer, 1967), although this problem was subsequently more clarified (Grossman, 1972). The experiments of Sata *et al.* (1978) were made only at 1465°, 1400° and 1345 °C and are not adequate to establish the phase relations. In the present studies the system $MgSiO_3$–H_2, which is a portion of the system Mg_2SiO_4–SiO_2–H_2, has been investigated experimentally in the pressure range from 10^{-2} to 10^{-10} bar and in the temperature range from 1300° to 1650 °C. The system Mg_2SiO_4–SiO_2 has been studied experimentally in detail at 1 atm (Bowen & Andersen, 1914) and the effect of H_2 as well as low pressure can be evaluated. Based on the experimental results, the fractional vaporization and condensation processes applicable to the primitive solar nebula are discussed.

Experimental methods

All experiments were conducted in a high-vacuum ($P \geq 10^{-12}$ bar), high-temperature (≤ 2000 °C) furnace described previously (Mysen *et al.*, 1985). The cylindrical chamber is 35 cm in length and 18 cm in inside diameter. All the junctions were sealed with copper rings. A furnace located coaxially inside the vacuum chamber is a W wound alumina tube 5.0 cm long and 0.8 cm in inside diameter (1.0 cm in outside diameter) equipped with a programmable temperature controller.

Mo capsules 1.6 mm in inside diameter and 3.5 mm in length with a 0.5 mm thick tightfitting lid were used as sample containers. W capsules were also used for several runs. The results are essentially the same as those obtained with Mo capsules. Two 0.5 mm orifices were drilled in the sides of the capsule walls to generate a Knudsen cell. A 0.25 mm diameter Mo hanging wire is inserted through these orifices to facilitate suspension of the capsule in the highest temperature part of the furnace. Thus, the effective area of each orifice is 0.147 mm² (0.295 mm² as total area). The vapor pressures of Mo and W are several orders of magnitude lower than those of silicates (Stull & Prophet, 1971). Starting materials were finely ground (~ 0.5 μm on average), synthetic enstatite crystallized from mixtures of MgO and SiO_2 at 15 kbar at 1450°–1500 °C for 6–24 hours under dry conditions. One batch of the sample was synthesized at 15 kbar at 1200 °C in the presence of about 2 wt.% water. All starting materials are well recrystallized orthoenstatite without any excess silica or forsterite. In each run a 5–7 mg sample was loaded in the capsule.

Pressure inside the vacuum chamber was measured with a Varian Ratiomatic 842 ionization gauge calibrated against nitrogen with correction factor of 1.83 times the nominal value for hydrogen. Hydrogen pressures were achieved by bleeding H_2 into the vacuum chamber through a valve placed in front of one of the ports into the main vacuum chamber. Consequently, during an experiment the measured pressure, P_{tot}, is equal to $P_{H_2} + P_{vac}$, where P_{vac} is the pressure in the vacuum chamber with the hydrogen valve closed. In the experiments reported here, P_{vac} was near 10^{-11} bar and $P_{tot} > 10^{-10}$ bar so that $P_{tot} \sim P_{H_2}$. Except at $P_{H_2} > 10^{-6}$ bar, where the total pressure fluctuated by as much as 2% during experiments lasting several hours (probably owing to convection inside the vacuum chamber), the P_{tot} is precise to at least 1%.

Pressure inside the capsule is not necessarily the same as that outside the capsule; the area of the orifice is smaller than that of the surface area of the powdered sample ($< 1/1000$), so that when vapor pressure over the sample is larger than the outside pressure, pressure builds up inside the capsule. Vapor pressure of the sample can be measured by the Knudsen method (Knudsen, 1909, Paule & Margrave, 1967) with the aid of measuring the rates of vaporization of the sample, the area of orifice and the Clausing factor for the orifice. The measured vapor pressure (P_m) is given by the following equation:

$$P_m = \frac{1}{AC} \cdot \frac{dw}{dt} \sqrt{\frac{2\pi RT}{M}} \qquad (1)$$

where dw/dt is the rate of vaporization (weight loss of sample as a function of time), A area of orifice, C Clausing factor, M average molecular weight of the gas molecules, R gas constant, and T absolute temperature. The rate of vaporization of enstatite was measured in the present experiments under several different temperature and pressure conditions. The Clausing factor was given by Iczkowskii et $al.$ (1963) for orifices with different shapes. The shape of orifice, however, is not exactly the same as those given by Iczkowskii et $al.$ and the measured vapor pressure is not equal to the equilibrium vapor pressure. Calibration of the present measurements is, therefore, required. Calibration has been made by measuring the vapor pressures of liquid silver and copper, the values for which have been determined over a wide temperature range (Jones et $al.$, 1927). Correction factors obtained by these calibrations are $4.0(\pm 1.0) \times 10^{-4}$. Using this correction factor, vapor pressure of silica was measured (Mysen & Kushiro, 1988). The results are compared with the previous determination by Hidalgo (1960), as shown in

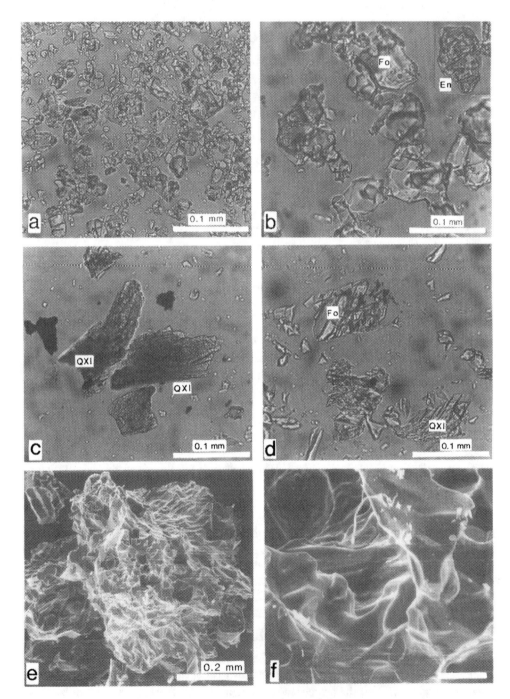

Fig. 16.1. (a) Photomicrograph of enstatite (most probably protoenstatite) formed at 1425 °C at 4.4×10^{-10} bar total pressure (outside pressure) (Run No. 2702). (b) Photomicrograph of forsterite and enstatite formed at 1525 °C at 4.4×10^{-8} bar (2651). Fo, forsterite; and En, enstatite (note cracks perpendicular to elongation). (c) Quench crystals of pyroxene formed at 1550 °C at 4.4×10^{-10} bar (2688). Note very fine vesicles in quench

Fig. 16.2. The average molecular weight of gas molecules is taken as 50, because species in the gas phase in the system Mg_2SiO_4–SiO_2–H_2 are essentially SiO_2, SiO and MgO under present experimental conditions (Mysen & Kushiro, 1988). The products of both vaporization and condensation have been observed under both optical and scanning microscopes and analyzed with the electron microprobe analyzer.

Results

Descriptions of run products

The experimental results are summarized in the Appendix. Phases encountered in the experiments are enstatite*, forsterite, quench crystals (mostly pyroxene) and glass. At temperatures below 1540 °C the experimental charges consist of granular forsterite and enstatite; the charges formed at temperatures between 1500° and 1540 °C contain large amounts of forsterite relative to enstatite, whereas those formed at lower temperature consist dominantly of enstatite with a very small amount of forsterite even for long runs. Evidently incongruent vaporization of enstatite to forsterite and silica-rich vapor takes place.

Enstatite is equant to short prismatic and often shows curved cracks perpendicular to its elongation (Fig. 16.1(a) and (b)). This feature is characteristic of protoenstatite inverted to clinoenstatite upon quenching (Boyd & Schairer, 1964). It shows oblique extinction and is most probably clinoenstatite, although polysynthetic twinning is not always seen and sharp terminal edges are not observed. The grain size increases with increasing temperature; it is about 20 μm on the average at about 1450 °C and increases to 1 mm at 1540 °C. Forsterite is rounded to subrounded and ranges in size from 50 μm to 1.5 mm in the temperature range between 1480° and 1540 °C (Fig. 16.1(b)). Forsterite in the forsterite-rich charges often includes vesicles and shows concave surfaces. These features are demonstrated in the SEM photographs of the charges (Figs. 16.1(e) and (f)). The granular enstatite and forsterite are considered to be subsolidus phases. No periclase was found by the SEM in any charges.

Above 1550 °C, all the charges except for one at 4.4×10^{-9}† bar consist of long-elongated or fibrous enstatite crystal, often aggregated radially. Fig. 16.2(c) shows broken fragments of such aggregates. These crystals usually make a thin coating of the

---

* Enstatite obtained in the experiments is most likely protoenstatite, as discussed below; however, because this has not been confirmed, the term enstatite is used throughout the paper.

† The pressure described here as well as those below is pressure in the vacuum chamber measured by the ion gauge and not pressure inside capsules, unless otherwise mentioned.

Fig. 16.1. (*cont.*)

crystals. (d) Forsterite with quench crystals of pyroxene formed at 1575 °C at 1.7×10^{-7} bar (2704). QX1, quench crystal. (e) SEM (scanning electron microscope) photograph of a run product consisting dominantly of forsterite formed at 1550 °C at 4.4×10^{-9} bar (2680). (f) The same as Fig. 16.1(e) with a larger magnification.

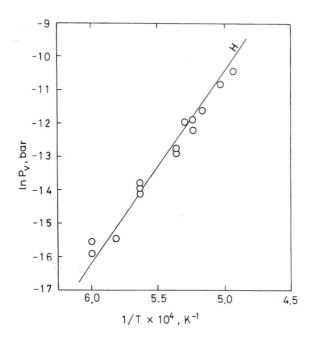

Fig. 16.2. Vapor pressure (P_v) of silica determined by Mysen & Kushiro (1988) plotted as a function of $1/T$. Line H is the vapor pressure of silica given by Hidalgo (1960).

inner surface of the capsules. In some runs they grew outside the capsules near the orifices or along the Mo wire used to suspend the capsules inside the furnace. These crystals are presumed to have been formed from liquid during quenching. The charges that contained these crystals are, therefore, interpreted to be in the liquid-present region. Quench crystals formed at low total pressures (e.g. 4.4×10^{-10} bar at 1550 °C and 1575 °C), however, contain many small vesicles (Fig. 16.1(c)). Such quench crystals containing vesicles indicate that liquid contained vesicles before quenching. The vesiculated glass or quench crystals are interpreted as evidence of vesiculation of melt and, therefore, the pressure–temperature conditions were within the vapor-present region (vapor or vapor + liquid region).

At pressures above 1.7×10^{-8} bar, granular forsterite coexists with quench crystals of pyroxene in a narrow temperature range 1550°–1575 °C (Fig. 16.1(d)). This region is considered to be a region of incongruent melting of enstatite.

Preliminary experiments were also made on a natural enstatite $(En_{90}Fs_{10})$ at 4.4×10^{-9} and 4.4×10^{-7} bar and at 1475 °C and 1350 °C. The results indicate that the enstatite vaporizes incongruently to almost pure forsterite and vapor (Appendix).

Rate of vaporization

The rate of vaporization is essential for determining the vapor pressure of samples and the pressure inside capsules. It may also be used to distinguish the regions of solid, liquid and vapor (Mysen *et al.*, 1985). In the present experiments, the rate of

Table 16.1. *Vapor pressure of enstatite + fosterite assemblage*

Pressure (bar)	Temperature (°C)	Vapor pressure[a] (bar)
4.4×10^{-10}	1475	6.6×10^{-7}
4.4×10^{-10}	1450	1.1×10^{-7}
4.4×10^{-10}	1350	1.9×10^{-9}
4.4×10^{-8}	1525	1.4×10^{-6}
4.4×10^{-7}	1525	7.3×10^{-7}
2.0×10^{-6}	1525	1.1×10^{-6}
2.0×10^{-6}	1490	6.1×10^{-7}
2.4×10^{-9}	1475[b]	3.4×10^{-7}

Notes:

[a] Maximum error for vapor pressure introduced by uncertainties of molecular weight of gas species, correction factor, and weight loss measurements are estimated as $\pm 50\%$.

[b] This run was made without hydrogen.

vaporization has been measured under several different conditions. The weight loss of enstatite was first measured with both Mo and W capsules which may generate different oxygen fugacites. The results are, however, not significantly different from each other. Subsequent measurements were, therefore, made only with Mo capsules. The table in the Appendix shows the weight loss of samples in wt. %. At all pressures, the rate of weight loss increases with increasing temperature. At 4.4×10^{-9} bar, for example, the weight loss increases from about 25% at 1475 °C to about 68% at 1550 °C during 24 hours. It should be mentioned that in these experiments, the ratio of forsterite to enstatite increases with time during the vaporization process, especially at high temperatures. Both forsterite and enstatite were present in most of the charges, however, so that the rate of vaporization observed is not that of enstatite, but that of a mixture of forsterite and enstatite in most of the runs.

The vapor pressure of the enstatite + forsterite assemblage has been obtained by the Knudsen method from the rates of vaporization of this assemblage, as described in the method section. The results are shown in Table 16.1. One measurement was made without H_2 at 2.4×10^{-9} bar P_{vac} at 1475 °C (0.72 mg wt. loss during 1320 min.). The vapor pressure obtained is 3.4×10^{-7} bar (Table 16.1). The value agrees well with those obtained from the runs with H_2 (Fig. 16.3), indicating that H_2 does not affect significantly the vapor pressures measured.

The heat of vaporization, ΔH, of the enstatite + forsterite assemblage can be obtained from the slope of a curve for vapor pressure in the $\ln P - 1/T$ plot. The value obtained is 573 ± 73 kJ/mol.

Phase relations

On the basis of the microscopic observations of the run products and vapor pressure over the enstatite + forsterite assemblage, the phase relations in this system can be

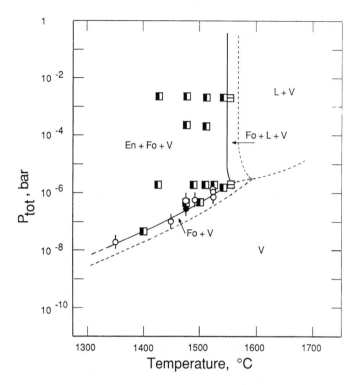

Fig. 16.3. Equilibrium phase diagram of the system Mg_2SiO_4–SiO_2–H_2. Vaporus curve for the forsterite + enstatite assemblage has been determined by the vapor pressure of this assemblage shown by open circles (with H_2) and a solid circle (without H_2). The runs made near and in the higher pressure side of the vaporus are shown. The dashed curve between L + V and Fo + L + V assemblages have been deduced on the basis of the runs made at 4.4×10^{-7} bar (especially runs 2674, 2668, 2669, 2703), which are close to the vaporus curves. Abbreviations: En, enstatite (most probably protoenstatite); Fo, forsterite; L, liquid; V, vapor.

deduced. The phase assemblages obtained in the experiments, however, cannot be used immediately to discuss the equilibrium phase relations. The experiments are made under continuously evacuating conditions. Although the hydrogen gas is continuously supplied into the chamber to keep the total pressure constant, the bulk composition of the system changes with time, particularly in the region where enstatite vaporizes incongruently. In the region of incongruent vaporization, the ratio of forsterite to enstatite increases with time, and in some long experiments at high temperatures, enstatite disappears completely. This feature is due to the fractional vaporization, by which the bulk composition became forsterite-rich during vaporization as discussed below. This region is interpreted as the Fo + En + V region because of the fact that enstatite crystals in this region are mostly large, euhedral crystals evidently grown from the fine-grained starting material and rarely rimmed by forsterite, as shown in Fig. 16.1(b). At lower temperatures (< 1450 °C), the amount of forsterite is usually very small (less than a few %) even after extensive vaporization of enstatite ($>30\%$). Apparently incongruent vaporization is less extensive at lower temperatures than at higher temperatures. The lower limit of the Fo + En + V region has not been determined in the present study. At 2.4×10^{-3} bar, however, the amount of forsterite is much larger than that in the experimental charges from pressures lower than 10^{-4} bar at the same temperatures. The rate of vaporization in this region is also larger than that at lower pressures at the same temperature. The reason for the higher rate of vaporization and enrichment of silica in vapor is not clear. In this region, partial pressure of hydrogen outside capsules is much (at least one order of magnitude) greater than the vapor pressures of silica, enstatite and forsterite, so that partial pressure of hydrogen in the capsule is much higher than that at lower total pressures. Such a high hydrogen pressure may facilitate fractional vaporization of Si from enstatite.

The upper stability limit of the solid region (solidus) can be determined accurately (i.e., 1550 °C). The melting takes place almost instantly (as observed even in a 1 minute run), so that changes in the bulk composition are negligible. This conclusion was confirmed by the microprobe analyses of the bulk quench crystals. Incongruent melting of enstatite to produce forsterite and liquid extends to the 'triple point' where the solidus intersects the vaporus.

The vaporus is difficult to determine from observations of run products because of metastable melting and persistence of crystals in the vapor region. The vaporus of a material at a given pressure–temperature condition is defined, however, as a point where vapor pressure of the material is equal to total pressure. The vaporus curve in the pressure–temperature plane can be drawn, therefore, by determining vapor pressure of condensed materials at different temperatures. The vapor pressure over enstatite + forsterite has been measured in the temperature range between 1350° and 1525 °C (Table 16.1 and Fig. 16.3). The charges for which the vapor pressure was measured contain forsterite in addition to enstatite. The measured vapor pressure is, therefore, that for the enstatite + forsterite assemblage. Vapor pressure of the enstatite + forsterite assemblage does not depend on the proportion of these two phases, as required from the phase rule. Therefore, even if the bulk composition changes under continuously evacuating conditions, the equilibrium relations can be obtained from the equilibrium vapor pressures. The vaporus curve has been drawn from these vapor pressure values. However, because of the relatively large uncertainties of measurements ($\pm 50\%$), the vaporus curve is not well defined as shown in Fig. 16.3.

The 'triple point' where the vaporus intersects the solidus exists at about 2.0 (± 1.0) $\times 10^{-6}$ bar at about 1550 °C. At this point, enstatite, forsterite, liquid and vapor coexist. Strictly speaking, the present system is a three-component system, so that this 'point' is not an invariant point but is a singular point. The Fo + L + V region terminates at another point that lies at a slightly higher temperature. The vaporus curve for liquids originates from this point; however, its slope has not been determined.

Experimental data obtained at pressures higher than the equilibrium vapor pressure of the sample are shown in Fig. 16.3, because the measured total pressure of these runs

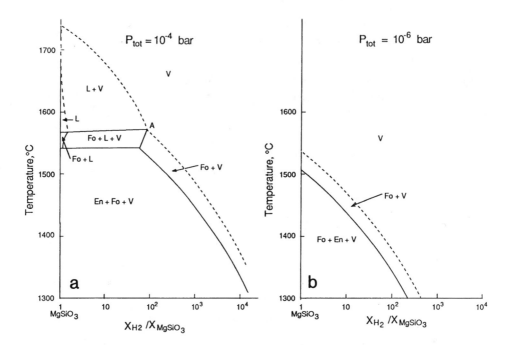

Fig. 16.4. (a) and (b). Possible equilibrium phase diagrams of the system $MgSiO_3$–H_2 at 10^{-4} and 10^{-6} bar total pressure, respectively. Abbreviations as in Fig. 16.3. Solid lines are based on the phase diagram of Fig. 16.3 and calculations made with the heat of vaporization of forsterite + enstatite assemblage. Dashed lines have been deduced.

must be equal to or very close to pressure inside the capsules in this pressure range. On the other hand, in runs made at pressure lower than the equilibrium vapor pressure, the inside pressure should be larger than the measured total pressure. Those runs are not plotted in the diagram.

The phase relations of the system $MgSiO_3$–H_2 at 10^{-4} and 10^{-6} bar are shown in Figs. 16.4(a, b), which have been constructed from Fig. 16.3 and heat of vaporization for enstatite + forsterite assemblage and the Van't Hoff equation. It is assumed that the system is ideal and the gas species does not change in the pressure and temperature ranges of calculation. If the composition of vapor is more $MgSiO_3$-rich than A in Fig. 16.4(a), liquid condenses directly from vapor as the temperature is lowered, whereas if the composition of vapor is more H_2-rich than A, forsterite condenses directly from vapor and is followed by enstatite. This composition effect must be taken into account when the present phase relations for the $MgSiO_3$–H_2 system are applied to the condensation process in the solar nebula.

Condensation of vapor

In the present experiments, condensation of vapor took place on the wire used to suspend the sample capsules in the furnace (Fig. 16.5). The pressure of condensation is

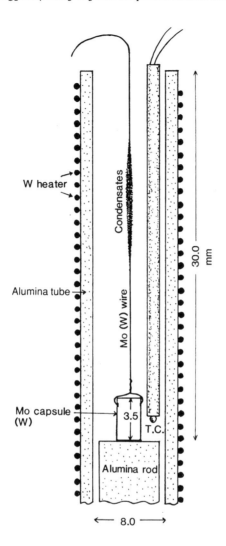

Fig. 16.5. A schematic figure showing the furnace assembly and the condensation products formed along the wire hanging a sample container.

not defined because of gas flow from the sample. It would be somewhat higher than the imposed hydrogen pressure. Condensation products are observed in most of the relatively long runs. In some of the runs they are visible even with the naked eye. The condensed products identified by the microprobe analysis are silica, enstatite and molybdenum. Tsuchiyama et al. (1987) observed the condensed materials in one of the runs (#2662) with the analytical transmission electron microscope and found that silica is quartz and enstatite is low clinoenstatite. No forsterite is detected. The amount of condensed materials depends on temperature as well as on duration of the runs; it is generally larger at higher temperatures and longer runs. The runs made above the solidus or its metastable extension were very short and no significant condensation

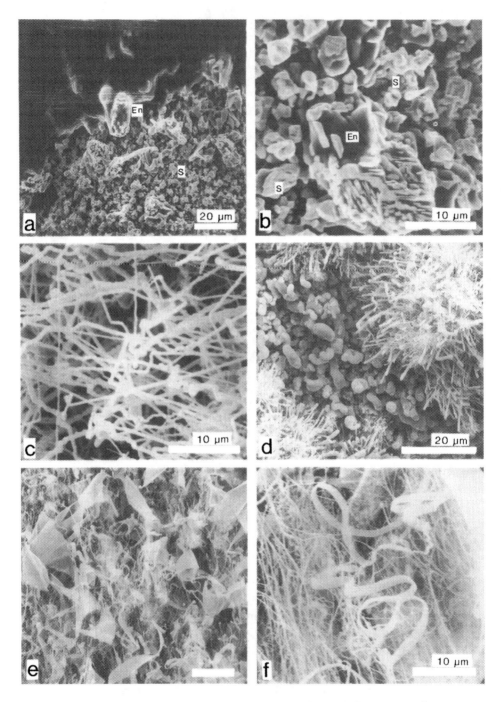

Fig. 16.6. SEM photographs of the condensation products. (a) Equant enstatite (En) and equant-elongated silica mineral (quartz) (S) condensed near the highest temperature part (~1100 °C) of the condensed zone in the run made at 4.4×10^{-10} bar and 1525 °C for 5720 min. (2662). (b) Enstatite and silica mineral in the same part as Fig. 16.6(a). (c) Fibrous silica

products were observed. The condensation products in three runs made at high temperatures have been observed in detail with the scanning microscope.

Fig. 16.6(a-f) are SEM photographs of the condensation products formed in 4-day experiments at 1525° and 1450 °C at 4.4×10^{-10} bar and a one-day experiment at 1510 °C at 2.4×10^{-4} bar. The condensation products start to grow at about 1.2 cm above the top of the capsule and the zone of condensation extends to about 1.0 cm. Fig. 16.6(a) shows the condensation products formed near the lower end (highest temperature part) of the condensed zone. The temperature of the lower end estimated from the temperature gradient along the wire is about 1100 °C. The condensates consist of loose aggregates of a small amount of enstatite and a larger amount of quartz (Fig. 16.6(b)). Both enstatite and quartz are granular and nearly equant crystals 10–50 μm in size. They protrude into free space as nearly euhedral crystals and cannot be recrystallization products of precursor materials. Near the middle of the zone (about 1000 °C), the products consist of fibrous quartz as shown in Fig. 16.6(c). Tsuchiyama *et al.* (1987) observed molybdenum in the core of fibrous quartz. Toward the lower temperature end of the condensed zone (about 600 °C), the amount of molybdenum increases.

The morphology of quartz varies with the temperature of condensation. In the condensation products formed in the run made at 1450 °C and 4.4×10^{-10} bar, granular silica is observed in the higher temperature zone and fibrous and platy quartz becomes dominant with decreasing temperature. Fig. 16.6(d) shows the aggregates of fibrous quartz growing on the layer of granular quartz.

At high P_{H_2} (2.4×10^{-4} bar), condensates of silica are observed as low as 500 °C. They show platy and ribbon-like features. Fig. 16.6(e) shows thin platy silica mineral observed in the lower temperature part of the condensed zones and Fig. 16.6(f) shows a spectacular ribbon-like silica. It appears that the surface area per unit mass of silica increases with decreasing temperature of condensation.

The vapor phase that condensed silica mineral and enstatite would have been enriched in silica relative to $MgSiO_3$. In the Fo + En + V region, where condensation is extensive, vapor must be more silica-rich than $MgSiO_3$ and consists of $MgSiO_3$ and SiO_2 components in addition to H_2. From such a vapor enstatite and silica are expected to condense with lowering temperature. The amount of the condensed enstatite estimated by the SEM analysis is, however, much smaller than that expected from the mass balance calculation. Apparently, only a part of the Mg in vapor condenses on the wire, whereas most of the Si in the vapor may condense on it.

Fig. 16.6. (*cont.*)

mineral condensed in the middle of the condensed zone in the same run as Figs. 16.6(a) and (b). (d) Granular and fibrous silica mineral condensed in the middle of the condensed zone in the run made at 4.4×10^{-10} bar at 1450°C for 5760 min. (2692). (e) Platy silica mineral condensed at a low temperature part ($\sim$600 °C) in the run made at 2.4×10^{-4} bar at 1510 °C for 1440 min. (2705). (f) Ribbon-like silica mineral condensed near but slightly to the lower temperature side of Fig. 16.6(e).

Discussion

Conditions for stability of liquid in the primitive solar nebula

Stability of liquid in the primitive solar nebula is relevant to the origin of chondrules and some inclusions in chondrites. This problem has been discussed by many investigators (e.g., Wood, 1963, Whipple, 1966, Blander & Katz, 1967, Blander & Abdel-Gawad, 1969, Wasson, 1972, Nagahara, 1981, Dodd, 1981). The 'triple point', which gives the lowest pressure limit of stability of liquid, may be inferred for the system Mg_2SiO_4–SiO_2–H_2 based on the present experimental results. Using the heat of vaporization of the enstatite + forsterite assemblage, an approximate pressure–temperature condition of the vaporus curve for this assemblage can be calculated when the H_2/ (forsterite + enstatite) ratio in the vapor is close to that in the primitive solar nebula (i.e., $\sim 10^4$ atm). The pressure–temperature condition of the 'triple point' which is an intersection of the solidus (Fig. 16.3) and the vaporus curve calculated for the enstatite + forsterite assemblage, is obtained at about 2×10^{-2} bar and 1550 °C. This calculation would be valid only if the vaporization mechanism under the experimental conditions is not significantly different from that under the conditions at the place in the primitive solar nebula where these minerals were formed. Based on this assumption, it is suggested that liquid could be stable only at pressures higher than 2×10^{-2} bar. In the primitive solar nebula, the total pressure is suggested to have been less than 10^{-3} atm (Grossman & Clark, 1973, Arrhenius & Alfven, 1971, Cameron & Pine, 1973, Whipple, 1966), so that no stable liquid would have been formed directly from gas. However, this conclusion is based on the results of experiments in the simple 4-component system as well as the assumption and should be examined in more complex systems closer to the solar composition.

Liquid, however, could be formed metastably in the primitive solar nebula. Blander & Katz (1967) suggested on the basis of nucleation theory that metastable liquid would have been formed directly from gas well below the solidus and the vaporus of minerals. More likely, metastable liquid would have been formed by instant, metastable melting of solids well below their vaporus. In the present experiments, liquid was formed instantly in the vapor region at temperatures above the metastable extension of the solidus, although the liquid formed was vaporized rapidly under such conditions. It has been suggested that most chondrules are products of remelting of precursor solid materials by impact or lightning (Wood, 1967, Whipple, 1966, Wasson, 1972, Nagahara, 1981). Such melting is possible at pressures well below the lower stability limit of liquid; however, the time for melting must have been very short (within a few minutes) for melts to survive, unless pressure was locally high enough to decrease the rate of vaporization of melts. Tsuchiyama et al. (1981) suggested from their experiments on the vaporization of Na from chondrule-like melt spheres that the time for melting to produce barred olivine and some porphyritic olivine chondrules must have

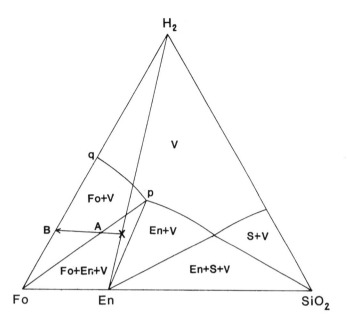

Fig. 16.7.   An isobaric, isothermal section of the system Mg$_2$SiO$_4$–SiO$_2$–H$_2$ showing possible compositional changes of the charge and coexisting vapor by fractional vaporization.

been less than 1 minute at temperatures higher than 1550 °C. Such a short-term melting process could produce metastable melts over a wide pressure range in the primitive solar nebula.

Fractional vaporization

'Fractional vaporization' in this paper means that the bulk composition of solid phase(s) and the composition of vapor change during vaporization. It is essentially a dynamic process. In the vaporization of enstatite, incongruent vaporization of enstatite results in forsterite and Si-rich vapor shown as the Fo + En + V region in Fig. 16.3. Continuous fractional vaporization in this region changes the bulk composition of the solid toward silica-poor (or forsterite-rich) compositions. At constant pressure–temperature conditions, the composition of vapor coexisting with forsterite and enstatite is fixed. If extraction of vapor continues, all enstatite in the starting material would convert to forsterite. This statement is supported by the present experiments; in the experiments made in the incongruent vaporization region, the ratio of forsterite to enstatite increases with time and in a long run the resulting charge consists entirely of forsterite. This observation is illustrated in an isothermal, isobaric section in Fig. 16.7. The bulk composition shifts from the original composition X toward A by fractional vaporization, assuming that the ratio of H$_2$ to solid is nearly constant. Until the bulk composition reaches A, the composition of the vapor is fixed at P. When the bulk composition reaches A, all enstatite is consumed. As the bulk composition changes

I. Kushiro and B.O. Mysen

from A to B the composition of vapor changes from p to q. When the bulk composition reaches B or any composition between forsterite and q, the composition of vapor is fixed at q unless the temperature changes. Strictly speaking, the ratio of H_2 to the sample would increase slightly during vaporization because of continuous removal of vapor from the sample and continuous supply of H_2 gas (to keep total pressure constant). The position of A and B, therefore, is not fixed. The composition of vapor, however, is fixed at p regardless of the ratio of forsterite to enstatite and that of forsterite + enstatite to H_2, and it changes from p to q regardless of the ratio of forsterite to H_2.

Sata et al. (1978) mentioned incongruent vaporization of forsterite to produce MgO; however, no convincing evidence of MgO has been found in the present experimental conditions. Mysen & Kushiro (1988) showed that forsterite vaporizes congruently at least in the temperature range from 1500° to 1700 °C. Vapor formed by vaporization of forsterite would, therefore, have a Mg:Si ratio equal or very close to 2:1. In this case, almost no fractional vaporization takes place; the compositions of the solid and vapor phases remain nearly constant during vaporization.

Incongruent vaporization occurs not only in enstatite, but also in akermanite and anorthite (Mysen & Kushiro, unpublished data). It is expected that some other silicate minerals may also vaporize incongruently. Incongruent vaporization of solid solutions also occurs. Natural enstatite ($En_{90}Fs_{10}$) also vaporizes incongruently to produce iron- and silica-rich vapor and nearly pure forsterite as shown before. Thus, it might be expected that fractional vaporization of Fe–Mg silicates would result in more magnesian residues. The fractional vaporization of meteorites or primordial materials of the planets could have occurred in the solar nebula. Hashimoto et al. (1979) and Hashimoto (1983) made vaporization experiments on the matrix materials of a C2 carbonaceous chondrite and synthetic mixtures, respectively, and showed the changes in chemical composition of the residual materials and gas with time. The latter experiments showed that fractional vaporization of melts of FeO, MgO, CaO, Al_2O_3 and SiO_2 mixtures also produces vapor enriched in silica and iron relative to the residues. They called the changes of composition and mineral assemblage by vaporization 'evaporation metamorphism'.

Fractional vaporization would also occur in melts, because the compositions of melts and vapor are usually different in a polycomponent system (Hashimoto, 1983). Because the vapor is most probably enriched in silica relative to the coexisting liquid as discussed above, the composition of liquid would become more forsterite-rich by continuous vaporization. With further vaporization, forsterite crystallizes and finally liquid disappears.

Fractional condensation

Fractional condensation is defined in this paper as a condensation process by which the compositions of the condensation products and vapor deviate from the initial composition by processes such as removal of condensed phases from the system or failure of reaction between condensed phases and gas. Fractional condensation defined here is, therefore, essentially a disequilibrium process. In the case of equilibrium condensation of gas consisting of $MgSiO_3$ and H_2 components, the final products are enstatite and H_2 gas regardless of the appearance of forsterite or liquid during cooling. On the other hand, by fractional condensation, the final products have various different phase assemblages. Larimer & Anders (1970) suggested 'incomplete condensation' including removal of early condensates as a cause of fractionation of elements among different chondrite classes.

Fractional condensation would be more enhanced if there is a reaction relation between condensed phase and vapor. The reaction relation between forsterite- and silica-enriched gas to produce enstatite has been suggested by Wood (1963) and Grossman (1972) on the basis of thermodynamic calculations of condensation of a gas of solar composition. The present experiments, as well as those by Sata *et al.* (1978), support such a reaction. To the extent this reaction holds, and if forsterite is removed from the system or forsterite fails to react with gas, the composition of the gas and the bulk composition of the residue become more Si-rich than the initial gas composition. In the case of condensation of gas consisting of $MgSiO_3$ and H_2 components, if forsterite is removed from the system in the Fo + En + V region, the composition of gas as well as the bulk composition of the residue becomes more Si-rich than Mg:Si = 1:1 and the final crystalline product would be a mixture of enstatite and silica.

Fractionation of Mg/Si in chondrites

The Mg:Si ratio varies significantly among chondrites as noted by a number of investigators as mentioned in the Introduction; carbonaceous chondrites have an average ratio of 1.04, ordinary chondrites 0.93, and enstatite chondrites 0.78. The primitive solar nebula would have had a composition close to the solar abundances. The fact that the Mg:Si ratios of most chondrites are smaller than that of the solar abundance of carbonaceous chondrites indicates that some fractionation must have been taking place in the nebula before the formation of chondrites. There is a possibility that small portions of chondrites, especially carbonaceous chondrites, may contain extra-solar components as suggested from the anomalies of oxygen and Mg isotopes (Clayton *et al.*, 1973, 1977, Lee *et al.*, 1976) and isotopes of other minor elements (i.e., Cr, Ne and Xe); however, the major part of the components in chondrites would have existed in the solar nebula. Fractionation of Mg/Si could have occurred by fractional condensation involving forsterite and enstatite, which are two major Mg- and Si-bearing silicate minerals in chondrites and terrestrial planets. As discussed in the previous section, if incomplete reaction between forsterite and gas occurs, the forsterite:enstatite ratio in the condensed products becomes higher. Such incomplete reaction could occur, when forsterite was removed to different places in the nebula such as gravitational settling toward the median plane of the nebula (Larimer & Anders,

I. Kushiro and B.O. Mysen

1970, Wood, 1967), or aggregated to form planetesimal bodies and failed to react with the surrounding gas. By incomplete reaction, the composition of the residual gas would have changed to more enstatite- or silica-rich compositions, having resulted in lower Mg:Si of the products. By extreme fractionation, the Mg:Si of the residual gas would have become less than 1 (Mg:Si in this case is that after subtracting Mg and Si in the early condensates such as melilite and clinopyroxene). If there is no such a reaction relation involving gas, it is difficult, if not impossible, to produce materials (either gas or condensation products) with Mg:Si less than 1 from the gas of solar composition unless crystal fractionation involving forsterite and liquid was an important process. Larimer & Anders (1970) demonstrated that removal of forsterite, spinel and melilite results in a decrease of Mg:Si from type C1 to ordinary chondrites and further to enstatite chondrites. Such a derivation is possible only if there are reaction relationships among them.

Matrices of unequilibrated ordinary chondrites

Bulk chemical compositions of fine-grained, dark matrices of unequilibrated ordinary chondrites have been shown to be different from those of chondrules and bulk chondrites: the fine-grained, dark matrices are usually enriched in Fe and Si relative to coexisting chondrules and bulk chondrites (Huss et al., 1981, Grossman & Wasson, 1983, Nagahara, 1984, Matsunami, 1984, Wlotzka, 1983). Particularly, the least metamorphosed chondrite, Semarkona (type 3.0) clearly shows such a relation even after subtraction of iron metal and troilite from the matrix composition (Nagahara, 1984). The bulk composition of chondrules, therefore, should be more magnesian and forsteritic than the bulk chondrite, although the difference is small because the amount of the matrix is usually small compared to those of chondrules. Some fractionation must have taken place to produce such compositional differences. Minerals in the matrices of type 3 ordinary and carbonaceous chondrites are highly heterogeneous and show disequilibrium relations among them (Huss et al., 1981, Nagahara, 1984, Rambaldi et al., 1981, Peck, 1983). For example, very magnesian- and iron-rich olivines and pyroxenes coexist together, either in contact with each other or occurring within a narrow area of the matrices. Nagahara & Kushiro (1984) suggested that iron-rich olivine (Fa > 65) could have formed by reactions involving free silica or Si-rich gas. In any case, the presence of a Si-rich phase as well as iron is essential to form such iron-rich olivine. As discussed above, a fractionation process to produce free silica or Si-rich gas is possible by fractional condensation of nebula gas involving a reaction relation between forsterite and gas to produce enstatite. Furthermore, preliminary experiments on a natural enstatite ($En_{90}Fs_{10}$) strongly suggest that Si- and Fe-rich residual gas would have been formed by fractional condensation in the solar nebula. On the other hand, magnesian olivine-rich chondrules, which are most abundant in the ordinary chondrites, might have been formed from relatively early condensates enriched in forsterite and enstatite.

428

Conclusions

The phase relations in the system Mg_2SiO_4–SiO_2–H_2 at pressures between 10^{-9} and 10^{-2} bar and at temperatures between 1350° and 1650 °C indicate that the 'triple point', which is the lowest pressure–temperature condition for stability of liquid, exists at about 2×10^{-6} bar and 1550 °C. Fractional vaporization of solids consisting of enstatite with or without forsterite would have produced residues enriched in forsterite. Fractional condensation of the nebular gas would have produced forsterite-rich early condensates and a residual gas enriched in Si and Fe. The observed variations in Mg:Si among chondrites can be explained by this fractional condensation process; carbonaceous chondrites may have formed from relatively unfractionated gas, ordinary chondrites from moderately fractionated gas, and enstatite chondrites from highly fractionated gas. The place of formation of these chondrites in the primitive solar nebula would have been different; however, this important problem is beyond the scope of this paper. The fine-grained matrices of unequilibrated ordinary chondrites, which are enriched in Fe and Si relative to the coexisting chondrules and the bulk chondrites, may have been formed from the residual gas enriched in Si and Fe. The direct condensation of forsterite from the nebular gas and formation of enstatite by reaction of forsterite and gas could have occurred at pressures as high as 10^{-2} bar. Liquid could be formed metastably, however, at pressures well below the stability region of liquid at temperatures above the metastable extension of the solidus. Some of these conclusions have already been suggested by the previous investigators; however, the present experimental study gives strong support for them and provides additional constraints. Also, the present experiments are useful for an understanding of the principal phase relations relevant to vaporization and condensation that may have occurred in the primitive solar nebula.

Acknowledgements   The authors are grateful to Drs A. Hashimoto, H. Nagahara, C.W. Prewiit, and H.S. Yoder for critical reading of the manuscript, and Dr J.A. Wood for comments.

Appendix. *Results of Runs on $MgSiO_3$*

Run no.	Pressure (bar)	Temperature (°C)	Time (min)	Phases	Weight loss (wt%)
2727	4.4×10^{-10}	1525	30	En + Fo(s)	2.3
2681	4.4×10^{-10}	1575	15	QXl + Gl	94.3
2682	4.4×10^{-10}	1575	5	QXl + Gl(b)	88.4
2683	4.4×10^{-10}	1575	1	QXl + Gl(b)	67.3
2684	4.4×10^{-10}	1550	480	—	100
2688	4.4×10^{-10}	1550	5	QXl + Gl(b)	84.0

Appendix. (*cont.*)

Run no.	Pressure (bar)	Temperature (°C)	Time (min)	Phases	Weight loss (wt%)
2641	4.4×10^{-10}	1525	400	En + Fo	36.6
2640	4.4×10^{-10}	1525	100	En + Fo	17.7
2643	4.4×10^{-10}	1525	2880	En(s) + Fo	70.6
2662	4.4×10^{-10}	1525	5720	—	100
2667	4.4×10^{-10}	1525	4320	—	100
2655	4.4×10^{-10}	1475	1800	En + Fo	41.5
2764	4.4×10^{-10}	1475	3800	En + Fo	62.4
2766	4.4×10^{-10}	1475	900	En + Fo	17.8
2690	4.4×10^{-10}	1450	2330	En + Fo(s)	8.3
2692	4.4×10^{-10}	1450	5760	En + Fo(s)	20.0
2702	4.4×10^{-10}	1425	4320	En + Fo(s)	23.4
2728	4.4×10^{-10}	1350	17423	En + Fo(s)	9.4
2768	4.4×10^{-10}	1350	10080	En + Fo(s)	3.1
2718	1.7×10^{-9}	1550	30	QXl + Gl	
2723	1.7×10^{-9}	1550	5	QXl + Gl	
2721	4.4×10^{-9}	1650	2	QXl(b)	
2685	4.4×10^{-9}	1575	15	QXl	86.1
2687	4.4×10^{-9}	1575	5	QXl(b)	
2714	4.4×10^{-9}	1560	120	—	100
2717	4.4×10^{-9}	1560	30	QXl + Gl	
2680	4.4×10^{-9}	1550	1440	En + Fo	68.3
2719	4.4×10^{-9}	1475	1440	En + Fo	25.3
2663	4.4×10^{-9}	1450	4324	En + Fo(s)	17.9
2726	4.4×10^{-9}	1350	26085	En + Fo(s)	10.8
2722	2.0×10^{-8}	1650	2	QXl(b)	
2736	2.4×10^{-8}	1600	5	QXl + Gl(b)	
2676	4.4×10^{-8}	1575	30	—	100
2686	4.4×10^{-8}	1575	15	QXl + Gl	63.4
2712	4.4×10^{-8}	1565	30	—	100
2713	4.4×10^{-8}	1565	10	QXl + Gl	
2708	4.4×10^{-8}	1560	10	En + Fo(s) + QXl	4.6
2709	4.4×10^{-8}	1560	120	—	100
2710	4.4×10^{-8}	1560	30	En + Fo(s) + QXl	
2651	4.4×10^{-8}	1525	1440	En(s) + Fo	65.0
2659	4.4×10^{-8}	1525	480	En + Fo	18.5
2670	4.4×10^{-8}	1475	4320	En + Fo	47.2
2725	4.4×10^{-8}	1400	4320	En + Fo(s)	14.0
2647	4.9×10^{-8}	1525	4320	—	100
2724	1.2×10^{-7}	1650	2	QXl(b)	
2704	1.7×10^{-7}	1575	5	QXl + Fo	73.6
2700	1.7×10^{-7}	1550	30	En + Fo	5.9
2674	4.4×10^{-7}	1590	30	QXl	
2666	4.4×10^{-7}	1575	360	—	100
2668	4.4×10^{-7}	1575	120	QXl + Fo(s)	90.0
2669	4.4×10^{-7}	1575	30	QXl + Fo(s)	21.1
2703	4.4×10^{-7}	1562	5	En + Fo(s) + QXl	
2706	4.4×10^{-7}	1562	15	QXl	
2707	4.4×10^{-7}	1562	10	QXl + En + Fo(s)	
2730	4.4×10^{-7}	1550	15	QXl + En	13.2
2731	4.4×10^{-7}	1550	60	Fo + QXl	
2657	4.4×10^{-7}	1525	1440	En + Fo(s)	28.5
2664	4.4×10^{-7}	1525	4320	Fo + En(s)	81.1
2675	4.4×10^{-7}	1500	4320	En + Fo	40.0
2677	4.4×10^{-7}	1475	4320	En + Fo(s)	20.4
2720	4.9×10^{-7}	1650	5	QXl + Gl	

Appendix. (*cont.*)

Run no.	Pressure (bar)	Temperature (°C)	Time (min)	Phases	Weight loss (wt%)
2699	4.9×10^{-7}	1550	5	En + QXl	4.1
2733	1.5×10^{-6}	1540	1020	En + Fo	83.2
2693	2.0×10^{-6}	1550	1440	—	100
2694	2.0×10^{-6}	1550	120	—	100
2695	2.0×10^{-6}	1550	30	En + QXl	
2698	2.0×10^{-6}	1550	5	En + Fo(s) + QXl + Gl	
2696	2.0×10^{-6}	1525	30	En + Fo(s)	3.7
2763	2.0×10^{-6}	1525	820	En + Fo	34.4
2691	2.0×10^{-6}	1510	2880	En + Fo	60.0
2701	2.0×10^{-6}	1490	5250	En + Fo(s)	79.1
2765	2.0×10^{-6}	1490	3840	Fo + En(s)	55.9
2767	2.0×10^{-6}	1490	1440	En + Fo	25.7
2716	2.0×10^{-6}	1425	4320	En + Fo(s)	48.7
2705	2.4×10^{-4}	1510	1440	Fo	97.9
2715	2.4×10^{-4}	1475	1440	En + Fo(s)	32.4
2756	2.4×10^{-3}	1550	15	Fo + QXl	
2755	2.4×10^{-3}	1540	30	En + Fo	7.1
2711	2.4×10^{-3}	1510	1380	—	100
2729	2.4×10^{-3}	1510	300	En + Fo	36.4
2734	2.4×10^{-3}	1475	1380	En + Fo	70.8
2745	2.4×10^{-3}	1425	1000	En + Fo(s)	49.5
2737	2.4×10^{-3}	1400	2880	—	100
2750	2.4×10^{-3}	1375	4400	—	100
Natural orthopyroxene ($En_{90}Fs_{10}$)					
2735	4.4×10^{-9}	1475	2880	Ol	80.0
2732	4.4×10^{-9}	1350	4320	Opx($En_{98.2}Fs_{1.8}$) + Ol($Fo_{99.8}$)	26.0
2738	4.4×10^{-7}	1475	2880	—	100

Notes:

Abbreviations: En, enstatite (protoenstatite); Fo, forsterite; Gl, glass; Ol, olivine; Opx, orthopyroxene; QXl, quench crystals; b, bubbles; s, small amount.

The weight loss for the runs made above the solidus (1550 °C) includes significant uncertainties because in many of these runs melt partially flowed out from capsule.

References

Ahrens, L.H. (1964). Si–Mg fractionation in chondrites. *Geochim. Cosmochim. Acta.*, **28**, 411–23.

— (1965). Observations on the Fe–Si–Mg relationship in chondrites. *Geochim. Cosmochim. Acta*, **29**, 801–6.

Arrhenius, G. & Alfven, H. (1971). Fractionation and condensation in space. *Earth Planet. Sci. Letters*, **10**, 253–67.

Blander, M. & Abdel-Gawad, M. (1969). The origin of meteorites and the constrained equilibrium condensation theory. *Geochim. Cosmochim. Acta*, **33**, 701–16.

Blander, M. & Katz, J.L. (1967). Condensation of primordial dust. *Geochim. Cosmochim. Acta*, **31**, 1025–34.

Bowen, N.L. & Andersen, O. (1914). The binary system MgO–SiO₂. *Am. J. Sci.*, **37**, 487–500.

Boyd, F.R. & Schairer, J.F. (1964). The system MgSiO₃–CaMgSi₂O₆. *J. Petrol.*, **5**, 275–309.

Cameron, A.G.W. & Pine, M.R. (1973). Numerical models of the primitive solar nebula. *Icarus*, **18**, 377–406.

431

Clayton, R.N., Grossman, L. & Maeda, T.K. (1973). A component of primitive nuclear composition in carbonaceous meteorites. *Sci.*, **182**, 485–9.

Clayton, R.N., Onuma, N., Grossman, L. & Maeda, T.K. (1977). Distribution of the presolar component in Allende and other carbonaceous chondrites. *Earth Planet. Sci. Letters*, **34**, 209–24.

Dodd, R.T. (1981). *Meteorites: a petrologic-chemical synthesis.* Cambridge Univ. Press.

Grossman, L. (1972). Condensation in primitive solar nebula. *Geochim. Cosmochim. Acta*, **36**, 597–619.

Grossman, L. & Clark, S.P. (1973). High-temperature condensates in chondrites and the environment in which they formed. *Geochim. Cosmochim. Acta*, **37**, 635–49.

Grossman, L. & Larimer, J.W. (1974). Early chemical history of the solar system. *Rev. Geophys. Space Phys.*, **12**, 71–101.

Grossman, J.N. & Wasson, J.T. (1983). Refractory precursor components of Semarkona chondrite and the fractionation of refractory elements among chondrites. *Geochim. Cosmochim. Acta*, **47**, 759–71.

Hashimoto, A. (1983). Evaporation metamorphism in the early solar nebula – evaporation experiments on the melt FeO–MgO–SiO$_2$–CaO–Al$_2$O$_3$ and chemical fractionations of primitive materials. *Geochem. J.*, **17**, 111–45.

Hashimoto, A., Kumazawa, M. & Onuma, N. (1979). Evaporation metamorphism of primitive dust material in early solar nebula. *Earth Planet. Sci. Letters*, **43**, 13–21.

Hidalgo, H. (1960). Ablation of glassy material around blunt bodies of revolution. *Am. Rocket Soc. J.*, **30**, 806–14.

Huss, G.R., Keil, K. & Taylor, G.J. (1981). The matrices of unequilibrated ordinary chondrites; implications for the origin and history of chondrites. *Geochim. Cosmochim. Acta*, **45**, 33–51.

Iczkowskii, R.P., Margrave, J.L. & Robinson, S.M. (1963). Effusion of gases through conical orifices. *J. Phys. Chem.*, **67**, 229–33.

Ikeda, Y., Kimura, M., Mori, H. & Takeda, H. (1981). Chemical compositions of matrices of unequilibrated ordinary chondrites. *Mem. Nat. Inst. Polar Res.* Spec. Issue No. 20, 124–44.

Jones, H., Langmuire, I. & Mackay, G.M.J. (1927). The rates of evaporation and the vapor pressures of tungsten, molybdenum, platinum, nickel, iron, copper and silver. *Phys. Rev.*, **30**, 201–14.

Knudsen, M. (1909). Die Molekularströmung der Gase durch Offnungen und die Effusion. *Ann. Phys.*, **28**, 999–1016.

Larimer, J.W. (1967). Chemical fractionations in meteorites – I. Condensation of the elements. *Geochim. Cosmochim. Acta*, **31**, 1215–38.

Larimer, J.W. & Anders. E. (1970). Chemical fractionation in meteorites – III. Major element fractionations in chondrites. *Geochim. Cosmochim. Acta*, **34**, 367–87.

Lee, T., Papanastassiou, D.A. & Wasserburg, G.T. (1976). Demonstration of [26]Mg excess in Allende and evidence for [26]Al. *Geophys. Res. Letters*, **3**, 109–12.

Lewis, J.S. (1973). Chemistry of the planets. *Ann. Rev. Phys. Chem.*, **24**, 339–51.

Lord, H.C., III (1965). Molecular equilibria and condensation in solar nebula and cool stellar atmospheres. *Icarus*, **4**. 279–88.

Matsunami, S. (1984). The chemical compositions and textures of matrices and chondrule rims of eight unequilibrated ordinary chondrites: a preliminary report. *Mem. Nat. Inst. Polar Res.* Spec. Issue No. 35, 126–48.

Morgan, J.W. & Anders, E. (1979). Chemical composition of Mars. *Geochim. Cosmochim. Acta*, **43**, 1601–10.

—— (1980). Chemical composition of Earth, Venus, and Mercury. *Proc. Nat. Acad. Sci.*, **77**, 6973–7.

Mysen, B.O. & Kushiro, I. (1988). Condensation, evaporation, melting, and crystallization in the primitive solar nebula: experimental data in the system MgO–SiO_2–H_2 to 1.0×10^{-9} bar and 1870 °C with variable oxygen fugacity. *Am. Mineral.*, **73**, 1–19.

Mysen, B.O., Virgo, D. & Kushiro, I. (1985). Experimental studies of condensation processes of silicate materials at low pressures and high temperatures, I. Phase equilibria in the system $CaMgSi_2O_6$–H_2 in the temperature range 1200°–1500 °C and the pressure range (P_{H_2}) 10^{-6} to 10^{-9} bar. *Earth Planet. Sci. Letters*, **75**, 139–46.

Nagahara, H. (1981). Evidence for secondary origin of chondrules. *Nature*, **292**, 135–6.

— (1984). Matrices of type 3 ordinary chondrites – primitive nebular records. *Geochim. Cosmochim. Acta*, **48**, 2581–95.

Nagahara, H. & Kushiro, I. (1984). Reaction in the matrix of chondrites: preliminary experiments [abstract]. *Lunar Planet. Sci.*, **15**, 583–4.

Paule, R.C. & Margrave, J.L. (1967). Free evaporation and effusion techniques. In *The characterization of high-temperature vapor*, ed. J.L. Margrave, pp. 130–50. Wiley.

Peck. J.A. (1983). Mineral chemistry and fabric of CV3 meteorite matrix [abstract]. *Meteoritics*, **18**, 373–4.

Rambaldi, E.R., Fredriksson, B.J. & Fredriksson, K. (1981). Primitive ultrafine matrix in ordinary chondrites. *Earth Planet. Sci. Letters*, **56**, 107–26.

Sata, T., Sasamata, T., Lee, H.L. & Maeda, E. (1978). Vaporization processes from magnesia minerals. *Rev. Int. Hautes Temp. Réfract. Fr.*, **15**, 237–48.

Saxena, S.K. & Eriksson, G. (1983). High temperature phase equilibria in a solar-composition gas. *Geochim. Cosmochim. Acta*, **47**, 1865–74.

Stull, D.R. & Prophet, H. (1971). *JANAF thermochemical tables*. Nat. Bureau of Standards.

Tsuchiyama, A., Nagahara, H. & Kushiro, I. (1981). Vaporization of sodium from silicate melt spheres and its application to the formation of chondrules. *Geochim. Cosmochim. Acta*, **45**, 1357–1367.

Tsuchiyama, A., Kushiro, I. & Morimoto, N. (1987). An electron microscopic study of gas condensates in the system Mg–Si–O–H [abstract]. In *12th Symposium on Antarctic Meteorites, Tokyo*, pp. 70–2.

Turekian, K.K. & Clark, S.P. (1969). Inhomogeneous accumulation of the earth from the primitive solar nebula. *Earth Planet. Sci. Letters*, **6**, 346–8.

Urey, H.C. (1961). Criticism of Dr. B. Mason's paper on 'The origin of meteorites'. *J. Geophys. Res.*, **66**, 1988–91.

Wänke, H., Baddenhausen, H., Palme, H. & Spettel, B. (1974). On the chemistry of the Allende inclusions and their origin as high temperature condensates. *Earth Planet. Sci. Letters*, **23**, 1–7.

Wasson, J.T. (1972). Formation of ordinary chondrites. *Rev. Geophys. Space Phys.*, **10**, 711–59.

— (1974). *Meteorites*. New York, Heidelberg and Berlin: Springer–Verlag.

Weidenschilling, S.J. (1976). Accretion of the terrestrial planets. II. *Icarus*, **27**, 161–70.

Whipple, F.L. (1966). Chondrules: suggestions concerning their origin. *Sci.*, **153**, 54–6.

Wlotzka, J.A. (1983). Composition of chondrules, fragments and matrix in the unequilibrated ordinary chondrite Tieschitz and Sharps. In *Chondrules and their origin*, ed. E.A. King, pp. 296–318. Houston: LPI.

Wood, J.A. (1963). The origin of chondrules and chondrites. *Icarus*, **2**, 152–80.

— (1967). Olivine and pyroxene composition in Type II carbonaceous chondrites. *Geochim. Cosmochim. Acta*, **31**, 2095–108.

Yoder, H.S. (1976). Generation of basaltic magma. Washington, D.C.: National Acad. Sci.

17

Volatiles in magmatic liquids

BJORN O. MYSEN

Introduction

Volatile contents and distribution of volatile species in magmatic systems can be inferred by direct analysis of volcanic gas, analysis of fluid inclusions and gas contents of glass inclusions in phenocryst and xenocrysts. Information can also be obtained by indirect methods based on observed phase relations and phase chemistry and by theoretical analysis of activity – composition relations in appropriate systems. The principal volatiles in magmatic systems can be described with the system C–H–O–S–F. The volatiles in volcanic gases generally are quite oxidized and CO_2, H_2O and SO_2 are the main gas species (e.g., Anderson, 1975, Gerlach & Nordlie, 1975, Casadewall *et al.*, 1987). Gas compositions from volcanoes along convergent plate boundaries generally are water-rich with carbon dioxide as the second most important volatile component, (Muenow *et al.*, 1977, Helgeson *et al.*, 1978, Rutherford *et al.*, 1984), whereas the gases in mid-ocean ridge basalts and basalts from oceanic islands contain principally SO_2 and CO_2 (e.g., Mathez & Delaney, 1981, Greenland, 1987) although others (see, for example, Gerlach, 1980) have suggested that H_2O is more important than previously recognized.

Information on volatile compositions at depth in the earth is less direct and relies on analysis of fluid inclusions in phenocrysts (Roedder, 1965, Murck *et al.*, 1978), gas content of glass inclusions in phenocrysts (Delaney *et al.*, 1977, 1978) and activity–composition relations derived from volatile-containing mineral parageneses in igneous rocks (e.g., amphibole, mica, sulfide and carbonate minerals). Recent determinations of intrinsic oxygen fugacities of mantle-derived minerals have led to suggestions that at least portions of the upper mantle might be more reduced than the crust of the earth (e.g., Arculus *et al.*, 1984; see also Rhyabchikov *et al.*, 1981) although the evidence for sufficiently low f_{O_2}-values to stabilize reduced carbon-bearing gases (near or slightly above the f_{O_2} of the iron-wustite oxygen buffer) is conflicting (Eggler, 1983, Haggerty,

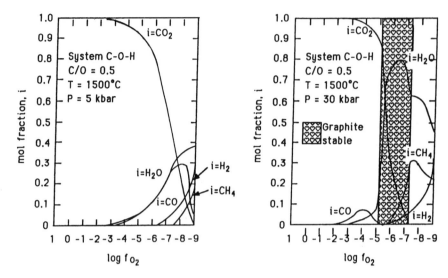

Fig. 17.1. Calculated abundance of gas species in the system C–O–H at 5 and 30 kbar at 1500 °C with C/O = 0.5 (calculated with the modified Redlich–Kwong equation of state).

1978, 1986, Haggerty & Tompkins, 1983, Eggler & Baker, 1982). Holloway & Jakobsson (1986) suggested that even if a mantle-source of volatiles in the system C–H–O was sufficiently reducing to stabilize significant amounts of CH_4 and H_2, fractionation of the gas species in this system between a partial melt and residual fluid results in the most oxidizing gas species entering the melt. A thermodynamic analysis based on composition of gases in such melt would, therefore, provide erroneously high f_{O_2} values.

In any event, gas species in the system C–O–H are significantly dependent on oxygen fugacity at magmatic temperatures at f_{O_2} levels at or below that of the nickel – nickel oxide oxygen buffer (Fig. 17.1). Moreover, there is a profound pressure effect on the proportions of gas species at and below the f_{O_2} of the quartz–magnetite–fayalite buffer. Thus, even if the mantle is not as reduced as suggested by the results from some of the intrinsic f_{O_2} studies, reduced carbon and hydrogen species may occur in significant abundance at the pressures and temperatures believed to exist during partial melting in the upper mantle.

Fluorine in magmatic rocks has generally been considered important only in late-stage, felsic igneous rocks (e.g., pegmatite) where concentrations as high as between 1 and 2 wt% have been recorded (e.g., Pichavant et al., 1987). Mantle-derived igneous rocks generally contain only several hundred ppm F, although some potassium-rich rocks, such as lamproites, exhibit significantly greater amounts (up to about 0.5 wt%, Jaques et al., 1984). Whereas in the former case, fluorine may not have affected upper mantle melting processes significantly, an important petrogenetic role of fluorine has been suggested for the formation of K-rich lamproitic melts in the upper mantle (Jaques et al., 1984).

It is evident, therefore, that the major volatile species in igneous rocks are H_2O, CO_2, CH_4, SO_2, H_2S, H_2 and F, and to a lesser extent CO. In order to characterize the influence of these volatiles on physical and chemical properties of magmatic systems, a detailed characterization of their interaction with silicate liquids is necessary. This information will be addressed here.

Water

Bowen (1928) recognized that solution of H_2O in magmatic liquids might significantly alter fractionation paths of the magmas and numerous solubility mechanisms for water in silicate melts have been proposed since that time. The simplest model is simply interaction between H_2O and bridging oxygens in melts to form OH groups

$$SiO_2(melt) + 2H_2O(gas) \Leftrightarrow Si(OH)_4. \tag{1}$$

An essential consequence of this simple concept is that solution of water in silicate melts results in its depolymerization. This depolymerization, in turn, has been held responsible, for example, for greatly enhanced fluidity and depression of liquidus surfaces of highly polymerized crystalline phases as well as several-hundred-degree depression of liquidus temperatures. Wasserburg (1957) pointed out, however, that a simple mechanism such as depicted in equation (1) is insufficient to explain the solubility behavior of water in more complex aluminosilicate melts. This latter observation and subsequent solubility studies (e.g., Burnham & Jahns, 1962, Hamilton et al., 1964, Burnham & Davis, 1971) culminated in a model for water solubility mechanisms in silicate melts by Burnham (1975) who suggested that in a model aluminosilicate melt such as that of $NaAlSi_3O_8$ composition, H_2O interacts by exchanging Na^+ in the silicate with an H^+ in dissolved water to form one OH group together with the formation of an additional OH group. Thus, an expression for solution of up to 1 mol of H_2O in $NaAlSi_3O_8$ composition melt may be written (Burnham, 1975)

$$NaAlSi_3O_8(melt) + H_2O(gas) \Leftrightarrow [NaOH]^0(melt) + AlSi_3O_7(OH)(melt). \tag{2}$$

The principal supporting evidence for this model was the P–V–T data for molten $NaAlSi_3O_8$–H_2O (Fig. 17.2) where Burnham & Davis (1974) found that in the H_2O concentration range up to about 5 wt% H_2O, the fugacity of water in the melt, $f^m_{H_2O}$, is about proportional to the square of its mol fraction, $X^m_{H_2O}$

$$f^m_{H_2O} = k(X^m_{H_2O})^2. \tag{3}$$

An analogous relationship holds for water solubility in basaltic melts (Hamilton et al., 1964). This water solubility model implies, therefore, that within the water concentration range where a solution mechanism such as (2) is valid, the dissolved water is completely dissociated. Further support for complete dissociation of H_2O into OH groups was found (Burnham, 1975) in the value of partial molar volume of water in

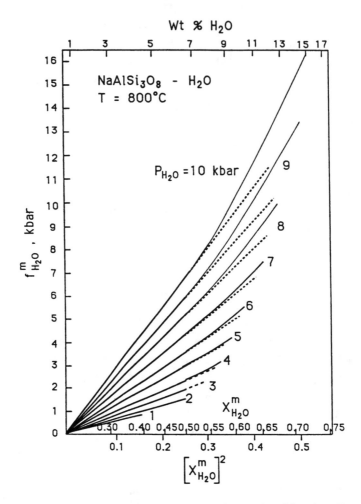

Fig. 17.2. Relationships between water fugacity, $f_{H_2O}^m$, and mol fraction, $X_{H_2O}^m$, in NaAlSi$_3$O$_8$ composition melt (data from Burnham & Davis, 1971). Dashed lines represent extensions of straight line relationships in the $X_{H_2O}^m$-range between 0 and 0.5.

solution in albite composition melt as compared with the molar volume of pure H_2O (11.2 cm³/mol at 1 bar and 20 °C versus 18.2 cm³/mol for pure water).

At water contents greater than 50 mol%, the curves in Fig. 17.2 no longer are straight (dashed lines indicate straight line extension) and different solubility mechanisms are required. Burnham & Davis (1974) suggested further OH-formation through interaction between the AlSi$_3$O$_7$(OH)-group in equation (2). Eggler & Burnham (1984), on the other hand, proposed that water in this high water concentration range was significantly 'less dissociated', a feature also observed by them for hydrous CaMgSi$_2$O$_6$ composition melt in all water concentration ranges. It should be noted, however, that in an earlier study of water solubility and solubility mechanisms on the join CaMg-Si$_2$O$_6$–H$_2$O, Eggler & Rosenhauer (1978) suggested that in the $X_{H_2O}^{Di\ melt}$ range between 0

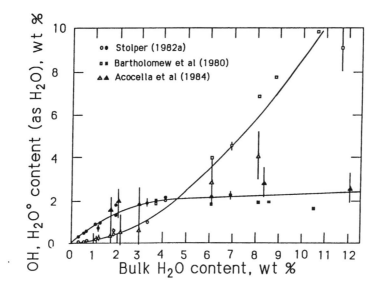

Fig. 17.3. Weight percent H_2O as OH and molecular water (H_2O^0) as a function of bulk water content of various silicate melts and glasses (data from Stolper, 1982a, Bartholomew *et al.*, 1980, Acocella *et al.*, 1984).

and about 0.3, that data were interpreted as consistent with the formation of OH groups only. The partial molar volume of water in diopside melt in their model is near 17 cm³/mol, a value near that suggested by Hodges (1974) for dissolved water in Mg_2SiO_4 composition melt. Such large partial molar volumes lend credence to the suggestion that H_2O in these melts might exhibit a structural resemblance to molecular H_2O (Hodges, 1974).

In contrast to the models discussed above, Stolper (1982a) proposed that an equilibrium relation of the form

$$H_2O_{molecular}(melt) + O^0(melt) \Leftrightarrow 2OH(melt), \tag{4}$$

where O^0 denotes bridging oxygen, operates throughout all water concentration ranges in all silicate melts and glasses. In subsequent models (e.g., Stolper *et al.*, 1983, Silver & Stolper, 1985), it was suggested that O^0 (bridging oxygen) could be replaced with all oxygen in the melt not bonded to hydrogen. This model implies that at least some dissolved water in silicate melts exists in molecular form in *all* concentration ranges, a conclusion supported by observations from infrared spectroscopic determinations of proportions of water dissolved in molecular form and as OH in various silicate glasses (Fig. 17.3). The abundance ratio, $OH/(H_2O)^0$, is, however, a function of total water content (Fig. 17.3; see also Stolper, 1982a, b, Stolper *et al*, 1983, Silver & Stolper, 1985, Epel'baum, 1985). This solubility model suggests, therefore, that water is dissolved by different forms of interaction with the melt. Competing solubility mechanisms of water as a function of water content may also be inferred from the enthalpy of mixing data in melts on the join $NaAlSi_3O_8–H_2O$ by Navrotsky (1987) (see also Fig. 17.4). Her data

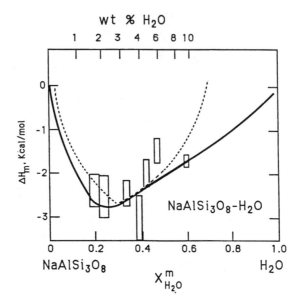

Fig. 17.4. Enthalpy of mixing, ΔH_m, on the join $NaAlSi_3O_8$–H_2O (data from
Navrotsky, 1987).

indicate that whereas at low water content ($X^m_{H_2O} < 0.3$), the ΔH_m (enthalpy of mixing)
decreases with increasing water content, this trend reverses at higher water concentrations. This observation is consistent with the suggestion (Stolper, 1982a, b, Epel'b-
aum, 1985, Stolper *et al*, 1983, Silver & Stolper, 1985) that water dissolves both as OH
groups and as molecular H_2O and that the proportion of molecular H_2O increases as
the water content increases.

There are, however, inconsistencies between the two sets of observations. The
equilibrium constant for equation (4),

$$K = \frac{(a_{OH})^2}{a_{H_2O}(\text{molecular}) \cdot a_O^0},$$

(5)

was reported by Stolper (1982a) to be between 0.17 and 0.2. The ΔH for reaction (4) is
near $-140 \, kJ/mol$ (Navrotsky, 1987). The ΔG^0 of reaction (4) must be positive if $K < 1$.
Consequently, the entropy of reaction (4) must be of the order of -120 to -140 J/
mol·K. Navrotsky (1987) suggested this to be an unreasonably large negative number.
Although Epel'baum (1985) did not report K-values for equation (5), his data are
consistent with greater values as the concentration ratio, OH/H_2O, from his data is
consistently higher than that of Stolper (1982a, b). Thus, from the data of Epel'baum
(1985), one would infer that the entropy of reaction (4) would be less negative than that
inferred by Navrotsky (1987) from her thermochemical data and the equilibrium
constant suggested by Stolper (1982a).

The solubility models discussed above do not address in detail *how* the OH groups
interact with the silicate melt structure. In general, it has been assumed, either explicitly

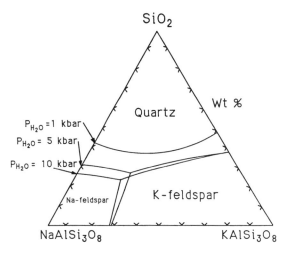

Fig. 17.5.   Simplified liquidus diagram for the system $NaAlSi_3O_8$–$KAlSi_3O_8$–H_2O as a function of water pressure (data from Tuttle & Bowen, 1958, Luth *et al.*, 1964).

or implicitly, that all OH groups are energetically equivalent and that one may assume ideal mixing (Stolper, 1982a) or Henrian behavior (Burnham, 1975, Burnham & Davis, 1974). These assumptions may be tested experimentally. Manning *et al.* (1980) observed that in the system SiO_2–$KAlSi_3O_8$–$NaAlSi_3O_8$–H_2O, the quartz–alkali feldspar liquidus boundary shifts away from the SiO_2 apex with increasing water pressure (Tuttle & Bowen, 1958, Luth *et al.*, 1964; see also Fig. 17.5). Manning *et al.* (1980) suggested that this shift occurs because of increasing alumina activity in the melt resulting from transformation of some tetrahedrally-coordinated Al^{3+} to octahedral coordination with increasing water content of the melts. Such a possible coordination transformation in Al^{3+} was also suggested by Epel'baum (1973) and Epel'baum *et al.* (1975). This explanation is consistent with a suggestion by Oxtoby & Hamilton (1978), who concluded from solubility experiments in the system Na_2O–Al_2O_3–SiO_2–H_2O that dissolved water is associated with Na and Si, but because their water solubility data indicated no dependence on alumina content, Al^{3+} was not a part of the solubility mechanism. This concept may be illustrated with a formulation such as (Mysen *et al.*, 1980)

$$14NaAlSi_3O_8(melt) + 13H_2O(vapor) = 9Si_4O_7(OH)_2(melt) +$$
$$8NaOH^0(melt) + 8Al^{3+}(melt) + 6NaAlSiO_6^{4-}(melt). \qquad (6)$$

In this expression, the superscript, ^{0}, is used to indicate association of Na^+ with OH^- in the melt. The Al^{3+} is aluminium in a network-modifying position. The partial transformation of Al^{3+} from tetrahedral coordination in anhydrous $NaAlSi_3O_8$ composition melt (Taylor & Brown, 1979, Mysen *et al*, 1980) to a network-modifying role in hydrous melts is a direct result of some of the Na^+ required for electrical charge-balance of tetrahedrally-coordinated Al^{3+}, is associated with OH^- in the hydrous

$SiO_2 + 5\ wt\%\ H_2O$

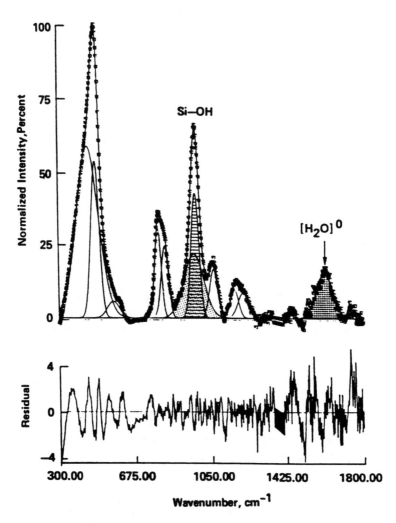

Fig. 17.6. Unpolarized Raman spectrum of quenched SiO_2–H_2O (5 wt% H_2O) (data from Mysen & Virgo, 1986a).

melt. An equivalent proportion of Al^{3+} will, therefore, be expelled from tetrahedral coordination.

In order to reconcile the various water solubility models and to produce a solubility mechanism that is *consistent* with all available information on physical and chemical properties of water-bearing silicate systems, it is necessary first to address the simplest possible system, SiO_2–H_2O. Raman and infrared spectroscopic data on quenched melts in this system are available (Stolen & Walrafen, 1976, McMillan & Remmele, 1986, Mysen & Virgo, 1986a), an example of which is shown in Fig. 17.6. In this system, the dissolved H_2O results in vibrational bands associated with OH-complexes in the

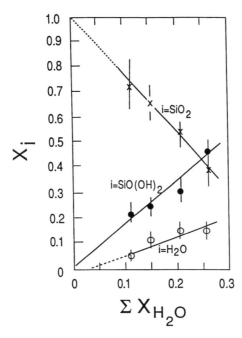

Fig. 17.7. Distribution of species in melts on the join SiO_2–H_2O as a function of mol fraction of water ($\sum X_{H_2O}$) (data from Mysen & Virgo, 1986a).

melts (hachured in Fig. 17.6). The lowest frequency band, or bands, near $970\ cm^{-1}$ is unambiguously assigned to Si–OH stretching and the band near $1600\ cm^{-1}$ to H–OH bending. The latter spectral feature is only possible if molecular water $[(H_2O)^0]$ is present in the melt (e.g., McMillan et al., 1983).

The mol fraction of molecular water in the silica melts ($X_{H_2O}^0$) increases systematically with increasing total water content (Fig. 17.7) even though the proportion of molecular water relative to total amount of water dissolved in the melts remains near 0.6 at least in the relatively high water concentration range of 4–10 wt% H_2O (Mysen & Virgo, 1986a). For a simple hydroxylation equilibrium such as

$$SiO_2(melt) + H_2O(melt) \Leftrightarrow SiO(OH)_2(melt), \tag{7}$$

the standard free energy of the reaction is insensitive to total water content within analytical uncertainty ($-71 \pm 20\ kJ/mol$ at the experimental temperature of $1550\ ^\circ C$), a value slightly less negative than that suggested by Stolper (1982a) for hydrous silicate melts and glasses in general. With the equilibrium constant in equation (5) between 0.17 and 0.2, the ΔG^0 at $1550\ ^\circ C$ is between -102 and $-112\ kJ/mol$.

From the solubility data of Oxtoby & Hamilton (1978) one might expect that (i) at least some of the H_2O would interact with the metal cations to form metal hydroxide complexes thus effectively polymerizing the metal, (ii) the water might interact with nonbridging oxygens to form Si–OH bonds, (iii) water might dissolve in molecular form, or (iv) combinations of the three mechanisms might operate. The three Raman spectroscopic studies that focussed on this problem (Mysen *et al.*, 1980, McMillan & Remmele, 1986, Mysen & Virgo, 1986a) accord in the sense that similar Raman spectra were obtained. There is molecular water as evidenced by H–OH bending vibrations (distinct Raman bands near 1600 cm^{-1}; see also Fig. 17.6). Furthermore, in sodium silicate melts, the spectra indicate either a decreased abundance of depolymerized structural units (principally $Si_2O_5^{2-}$ with a Si–O$^-$ stretch band near 1100 cm^{-1}), or the frequency of Si–O$^-$ stretching bands shifted as a function of increased water content. This postulated shift could, presumably, be caused by formation of OH groups in replacement for nonbridging oxygen. Such a replacement of nonbridging oxygen with OH groups would, however, also give rise to Si–OH bonds that will be reflected in Si–OH stretch bands in the Raman spectra near 970 cm^{-1}. In the systems Na_2O–SiO_2–H_2O, both McMillan & Remmele (1986) and Mysen & Virgo (1986a) observed a weak band near 940 cm^{-1} that could possibly result from the presence of Si–OH bonds although the 940 cm^{-1} frequency is slightly too low compared with the anticipated value near 970 cm^{-1}. There is, however, also an Si–O$^-$ stretch band near 950 cm^{-1} in *anhydrous* sodium silicate melts (e.g., Brawer & White, 1975) so this assignment is ambiguous. Thus, reactions either of the form

$$Si_2O_5^{2-} + 5H_2O \Leftrightarrow 2Si_4O_7(OH)_2 + 4SiO_2 + 8OH^-, \tag{8}$$

or of the form

$$Si_2O_5^{2-} + H_2O \Leftrightarrow 2SiO_2 + 2OH^-, \tag{9}$$

may be used to describe the solubility mechanism of water in metal oxide – silica melts. In these reactions the $Si_2O_5^{2-}$ unit schematically represents a depolymerized unit (1 nonbridging oxygen per silicon), and OH$^-$ represents hydroxyl groups associated with metal cations in the melt. In either case, solution of water results in polymerization of the melt.

For hydrous metal oxide – silica melt systems involving stoichiometric proportions of metal cations and water to form neutral hydroxyl complexes (e.g., NaOH, Ca(OH)$_2$ and Al(OH)$_3$) Mysen & Virgo (1986a) concluded that a reaction analogous to equation (9), and not equation (8), describes the solution mechanism. This conclusion was based on the observation that the frequency of Raman bands near 950 cm^{-1} was insensitive to substitution of ^{2}H for ^{1}H, a substitution that should result in a 20 cm^{-1} frequency reduction of an Si–OH stretch band (Freund, 1982). The ΔG^0 of reactions involving NaOH, Ca(OH)$_2$ and Al(OH)$_3$ are compared in Table 17.1 indicating that among the cations investigated, the NaOH complex is the most stable. The proportions of molecular to hydroxylated water in the melts vary as a function of melt composition as do the standard free energies of the hydroxylation reactions.

The least likely complex to form is with Al^{3+} in accord with the suggestion by Oxtoby & Hamilton (1978), Manning *et al.* (1980) and Mysen *et al.* (1980) that water dissolving in metal aluminosilicate melts tends to expel Al^{3+} from tetrahedral coordination. Whether, or the extent to which, this occurs can be explored through

Table 17.1. *Thermodynamic data, silica–metal hydroxyl systems*

System	H$_2$O (wt%)	$X_{\Sigma H_2O}$	$X_{SiO(OH)_2}$	X_{MOH}^a	$X_{H_2O}^{H2O}$	ΔG^0
SiO$_2$–H$_2$O	4.0	0.12	0.21	0.00	0.06	−17.0
SiO$_2$–H$_2$O	5.0	0.15	0.24	0.00	0.10	−13.0
SiO$_2$–H$_2$O	7.5	0.21	0.31	0.00	0.14	−14.0
SiO$_2$–H$_2$O	10.0	0.27	0.44	0.00	0.16	−24.0
SiO$_2$–Al(OH)$_3$	7.5	0.22	0.25	0.21	0.08	−1.8
SiO$_2$–Ca(OH)$_2$	7.5	0.19	0.00	0.29	0.12	−5.0
SiO$_2$–Na(OH)	2.5	0.08	0.00	0.06	0.06	−9.5

Notes:
[a] M = Na, Ca, Al
ΔG^0 calculated at 1550 °C.
Source: Mysen & Virgo (1986a).

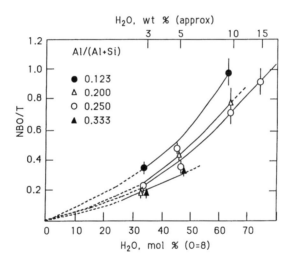

Fig. 17.8. Degree of polymerization, NBO/T, of melts on the join NaAlO$_2$–SiO$_2$–
H$_2$O as a function of water content (mol % calculated on the basis of 8
oxygen) (data from Mysen & Virgo, 1986b).

spectroscopic studies of water solubility mechanism in melts on the join NaAlO$_2$–SiO$_2$.
For such a study from a practical point of view, the most accessible information
attainable by spectroscopic methods is variation in degree of polymerization (nonbridging oxygen per tetrahedrally-coordinated cations, NBO/T) of a hydrous aluminosilicate melt as a function of water content (Fig. 17.8). In hydrous melts on the join
NaAlO$_2$–SiO$_2$, there is little evidence for (Si, Al)–OH bonding at least in the Al/
(Al + Si) range relevant to magmatic processes. A Raman band near 900 cm^{-1} that
could be ascribed to such bonding (McMillan *et al.*, 1983) does not exhibit the
anticipated frequency shift as a result of ^{2}H substitution for ^{1}H (Mysen & Virgo,
1986b) and this band probably results from nonbridging oxygen in the melts. Thus,
most likely water in aluminosilicate melts enters the melts in the form of OH-complexes

bonded perhaps to both Al^{3+} (non-tetrahedral) and Na^+ as well as H_2O but not as Si–OH and not to tetrahedrally coordinated Al^{3+}. The proportion of molecular water can be inferred from information from the systems $NaOH–SiO_2$ and $Al(OH)_3–SiO_2$.

In order to elucidate further details of the solution mechanism of water, one may consider two extreme mechanisms of OH-formation in aluminosilicate melts. The albite stoichiometry will be used as an example, but analogous expressions may be written for other compositions. One extreme case is that all OH is associated with Na^+ in the melt:

$$2NaAlSi_3O_8 + H_2O^0 \Leftrightarrow 2NaOH^0 + 3SiO_3^{2-} + 3SiO_2 + 2Al^{3+}. \qquad (10)$$

And the other extreme is association of OH with Al^{3+}:

$$2NaAlSi_3O_8 + 3H_2O^0 \Leftrightarrow 2Al(OH)_3^0 + SiO_3^{2-} + 5SiO_2 + 2Na^+. \qquad (11)$$

In view of the fact that the spectral results from hydrous aluminosilicate melts indicate that Si–OH complexing is unimportant, this mechanism is not considered further. In equations (10) and (11), the hydroxyl complexes are written as illustrations of the OH-association in the melts. The depolymerized silicate units are shown as SiO_3^{2-} because from the Raman spectra (Mysen & Virgo, 1986b), there was only evidence for nonbridging oxygen in SiO_3^{2-} units. The notation, H_2O^0 indicates molecular water.

In the solution mechanism illustrated with equation (10), an amount of Al^{3+} equivalent to the proportion of Na^+ associated with OH groups will be expelled from the network and becomes a network-modifier. In equation (11), an amount of Na^+ equivalent to the proportion of Al^{3+} associated with OH will become a network-modifier. The rate of depolymerization of the aluminosilicate network associated with such a mechanism is shown as the lines marked Na and Al in Fig. 17.9. The length of these lines is limited by the amount of Na^+ or Al^{3+} in the compound of interest. The slopes of the lines express the rate of depolymerization (expressed as $\partial(NBO/T)/\partial H_2O$), and is a function of the aluminosilicate stoichiometry. The more aluminous the melt, the greater is this rate. Under all conditions, the rate of depolymerization is greater for sodium hydroxyl complexing than for aluminium hydroxyl complexing.

It is evident from Fig. 17.9 that the observed rate of depolymerization falls somewhere between the extreme lines suggesting, therefore, that both Na^+ and Al^{3+} are associated with OH. This Al^{3+} is not, however, a part of the aluminosilicate network (Mysen et al., 1980, Mysen & Virgo, 1986b). For a given aluminosilicate melt, the NBO/T versus mol% dissolved H_2O trajectories become steeper as the water content is increased suggesting, therefore, that the Na/Al of the hydroxyl complexes increases with increasing water content. It is also evident from the data in Fig. 17.9, that the Na/Al of the hydroxyl complexes at given total water content increases with decreasing $Al/(Al+Si)$ of the melt. Thus, it would appear that in addition to some molecular H_2O, the principal form of water solubility is the formation of OH groups with alkali metals and aluminium. The latter portion of the mechanism governs depolymerization of the silicate melts.

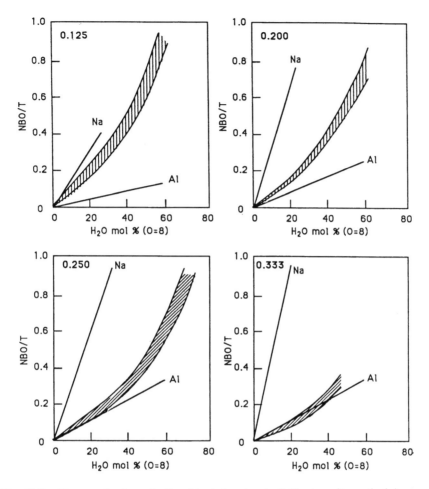

Fig. 17.9. Degree of polymerization (shaded regions) of silicate melts on the join
NaAlO$_2$–SiO$_2$–H$_2$O as a function of bulk water content (mol %
calculated on the basis of 8 oxygen) at Al/(Al + Si) indicated and
compared with theoretical NBO/T evolution under the assumption of all
hydroxylated water associated with Na (Na(OH)^0) or Al (Al(OH)$_3^0$) (data
from Mysen & Virgo, 1986b).

Hydrogen

Although relatively little is known about interaction between hydrogen and natural
magmatic liquids, results of calculations of vapor compositions in the system C–O–H
(Fig. 17.1) indicate that, under upper mantle pressure, temperature and oxygen
fugacity conditions, as much as 10% of the fluid may be H$_2$. The solubility of hydrogen
in silicate melts has only been determined in the presence of H$_2$O–H$_2$ mixtures (Luth &
Boettcher, 1986) where the proportion of hydrogen in the vapor phase was controlled
with different hydrogen buffers (from magnetite–hematite to iron–wustite).

B.O. Mysen

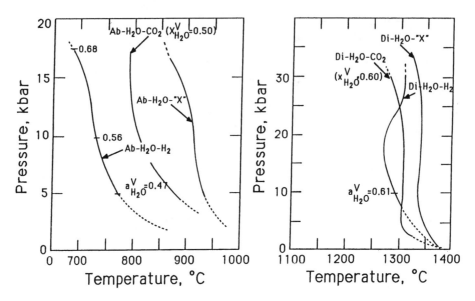

Fig. 17.10. Pressure–temperature trajectories of NaAlSi$_3$O$_8$ (Ab)–H$_2$O–H$_2$ and CaMgSi$_2$O$_6$ (Di)–H$_2$O–H$_2$ solidi (data from Luth & Boettcher, 1986) compared with selected solidi with (CO$_2$ + H$_2$O)-bearing vapor and (CO$_2$ + X)-bearing vapor, where X denotes theoretical dilution with mol fraction of water equivalent to those of the hydrogen-bearing vapors. Carbon-dioxide-bearing solidi from Bohlen *et al.* (1982) and Eggler & Rosenhauer (1978).

In the study of Luth & Boettcher (1986) the solidus temperatures of diopside–H$_2$O–H$_2$ and albite–H$_2$O–H$_2$ were compared (Fig. 17.10) with solidus temperatures calculated with mol fractions of an inert diluent (denoted 'X' in Fig. 17.10) in proportion similar to $X_{H_2}[=H_2/(H_2 + H_2O)]$ and with the solidus temperatures of diopside–H$_2$O–CO$_2$ and albite–H$_2$O–CO$_2$ with comparable mol fractions of water in coexisting CO$_2$ + H$_2$O vapor (Luth & Boettcher, 1986). In the system NaAlSi$_3$O$_8$–H$_2$O–H$_2$ the temperature depression of the solidus is considerably greater than would be expected if hydrogen simply acted as an inert component to lower the water activity in the system. Although there are several assumptions and uncertainties in these calculations (in particular in regard to mixing properties of H$_2$O–H$_2$ vapors; see Luth & Boettcher (1986) for discussion), it is evident from the data that the presence of hydrogen in the vapor affects the solidus temperature more than would be expected if hydrogen simply acted as a diluent. In the diopside system, the solidus temperature depression is comparable to that for CaMgSi$_2$O$_6$–H$_2$O–CO$_2$ with similar $X^v_{H_2O}$. This observation might be consistent with a suggestion of comparable solubilities of hydrogen and carbon dioxide in hydrous diopside melt. From the available carbon dioxide solubility data (see also below), this similarity corresponds to a hydrogen solubility of about 20–30 mol% (Eggler & Rosenhauer, 1978, Mysen *et al.*, 1976). In the case of NaAlSi$_3$O$_8$–H$_2$O–H$_2$, the temperature depression of the solidus by

($H_2O + H_2$) vapor is significantly greater than for an equivalent amount of CO_2 or 'inert component' of equivalent mol fraction. Thus, one might suggest that in fully polymerized aluminosilicate melts, the hydrogen solubility is significantly greater than that of carbon dioxide. The CO_2-solubility in $NaAlSi_3O_8$-composition melt, even in the presence of H_2O is, however, less than 50% of that of diopside melt at similar pressures and temperatures (Mysen *et al.*, 1976). Thus, it is possible that whereas the carbon dioxide solubility is sensitive to melt composition (structure; see below), the solubility of hydrogen is much less so.

The solidus phase relations for hydrogen-bearing vapor (Fig. 17.10) indicate that some interaction between the silicate melts and hydrogen might take place. In the only attempt to determine the structure of hydrogen-bearing silicate melts (Luth *et al.*, 1987), the Raman spectra of quenched hydrogen-bearing $NaAlSi_3O_8$ composition melt showed evidence of molecular H_2. Furthermore, the spectra were interpreted to suggest that some interaction between the hydrogen molecule and the silicate network does take place. No conclusive evidence was found, however, for reconstructive changes in either the silicate melt or the hydrogen species (to form, for example, OH groups).

Sulfur

Sulfur dioxide is among the three most common volcanic gases (e.g., Gerlach & Nordlie, 1975, Casadewall *et al.*, 1987) and the solubility of sulfur in natural magma under near surface pressure conditions appears to be near 1000 ppm (e.g., Moore & Fabbi, 1971, Mathez, 1976). Sulfur may, however, occur in several oxidation states under magmatic conditions (SO_3, SO_2, S^{2-}). Thus, if the solubility of sulfur in magmatic liquids depends on its oxidation state, sulfur solubility depends on both oxygen and sulfur fugacity. Although ubiquitous in extrusive basaltic magmas, the intensive and extensive variables that control solubility of sulfur (in its various possible oxidation states) are, however, surprisingly poorly known.

Reduced sulfur

Reduced sulfur, principally in the S^{2-} form, plays an important role in magmatic processes at depth as evidenced, for example, by massive Ni–Fe sulfide deposits associated with mafic and ultramafic igneous rocks (e.g., Naldrett, 1973, Shima & Naldrett, 1975). Experimental data on silicate melt – sulfide melt transition metal partition coefficients (e.g., Rajamani & Naldrett, 1975, Mysen & Kushiro, 1976) have demonstrated enrichments by factors of several hundred into a sulfide melt relative to coexisting silicate melt. This latter observation is also important in regard to magma genesis in the upper mantle. If, during partial melting of a sulfide-bearing peridotite ((Fe, Ni, Cu) monosulfide phases are common in mantle-derived peridotite nodules; White, 1966; Vakrushev & Sobolev, 1971; Bishop *et al.*, 1975), sulfide remains in the

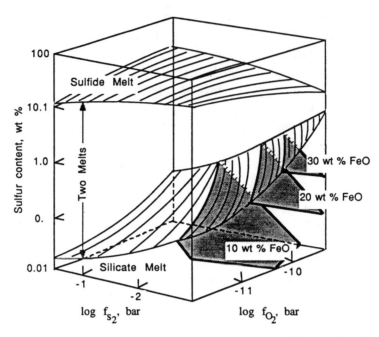

Fig. 17.11. Sulfur solubility and sulfide–silicate melt immiscibility as a function of
sulfur and oxygen fugacity and wt% FeO in the system (data from
Haughton *et al.*, 1974).

residue, then the bulk partition coefficient, $D_{\text{transition metal}}^{\text{partial melt–crystalline residue}}$, can be greatly
lowered compared with values for sulfide-free peridotite sources of partial melts (for
Ni, for example, D-values near 30–50 rather than 5–10 might be expected; Mysen &
Popp, 1980). Moreover, separation of tenths of weight percent of immiscible sulfide
during ascent of magma will similarly affect Ni and Co contents of the silicate melts
even though other major element components may not provide any evidence of sulfide
fractionation. It is evident, therefore, that sulfide solubilities and solubility mechanisms require determination in order to characterize interaction between sulfide and
silicate systems.

The principal sulfide solubility mechanisms in silicate melts may be illustrated with
an equilibrium of the form (Richardson & Fincham, 1956)

$$O^{2-} + \tfrac{1}{2}S_2 \Leftrightarrow S^{2-} + \tfrac{1}{2}O_2. \tag{12}$$

The sulfur solubility is, therefore, a function of sulfur fugacity, f_{S_2}, oxygen fugacity, f_{O_2},
and the activity of oxide anion in the melt. For example, for ferrous iron, the
equilibrium

$$2\text{FeO} + \text{S}_2 \Leftrightarrow 2\text{FeS} + \text{O}_2 \tag{13}$$

has been studied extensively, mostly in mafic composition melts (e.g., Shima &
Naldrett, 1975, Buchanan & Nolan, 1979). Relationships between sulfur fugacity and
sulfide solubility are illustrated in Fig. 17.11. Increasing oxygen fugacity drives

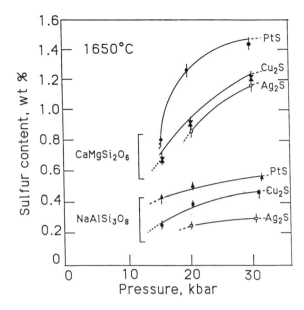

Fig. 17.12. Sulfur solubilities in $CaMgSi_2O_6$ and $NaAlSi_3O_8$ composition melts as a function of pressure at the f_{O_2} of the $C–CO_2–CO–O_2$ oxygen buffers and sulfide buffers as indicated (data from Mysen & Popp, 1980).

equilibrium (12) to the left (decreasing solubility), whereas increasing sulfur fugacity drives this equilibrium to the right.

Similar relationships hold at pressure corresponding to those of the upper mantle (Fig. 17.12) where data have been obtained with the f_{O_2} held at that buffered by the graphite–$CO–CO_2–O_2$ oxygen buffer. These data also indicate a positive pressure effect on the solubility. The exact relationship with pressure is, however, not clear because the pressure effect on the f_{S_2} of the sulfide buffers used in these experiments is unknown.

The relationships between melt composition and sulfur solubility have been extensively studied (e.g., Richardson & Fincham, 1956, Rosenquist, 1951, 1954, Haughton *et al.*, 1974, Abraham *et al.*, 1960, Abraham & Richardson, 1960). From these data it appears that sulfide principally is dissolved in silicate melts in association with various network-modifying cations (e.g., Fe^{2+}, Ni^{2+}, Ca^{2+}, Na^+). In order to quantify the bulk compositional dependence on sulfide solubility, it has been found convenient to define a parameter, sulfide capacity (C_s), that relates sulfur solubility to bulk composition at constant $\sqrt{\left(\dfrac{f_{O_2}}{f_{S_2}}\right)}$:

$$C_s = Wt_i \sqrt{\frac{f_{O_2}}{f_{S_2}}}, \qquad (14)$$

where Wt_i is the weight percent of a given metal oxide. Strong correlations exist between the sulfide capacity and several important major element, network-modifying cations (Fig. 17.13). Although these relationships do not explicitly demonstrate sulfide

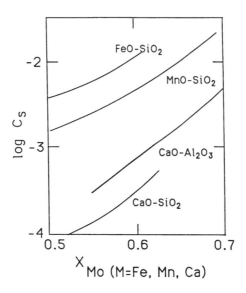

Fig. 17.13. Relationship between sulfide capacity (C_s) and mol fraction of metal
oxide (MO) for binary joins indicated (data from Abraham &
Richardson, 1960).

association with the metal cations, the observations of the sulfur solubility behavior are
consistent with such a principle.

Oxidized sulfur

Near-surface magmatic processes involve interaction between silicate melt and oxidized sulfur. It is possible, therefore, that redox equilibria involving sulfur are common
during ascent and extrusion of magmatic rocks. In a closed system, oxidation of sulfur
involves the silicate melt. In principle, this process results in melt depolymerization as
illustrated by a formalized equilibrium involving di- and tectosilicate structural units in
the melt:

$$S^{2-} + 2SiO_2 \Leftrightarrow SO_2 + Si_2O_5^{2-}. \tag{15}$$

This relationship also illustrates that fractional crystallization processes that affect
melt polymerization (e.g., olivine fractionation results in decreasing NBO/T), will also
affect the oxidation state of sulfur at constant f_{S_2} and f_{O_2}. As a magmatic system
evolves, the SO_2/S^{2-} will increase.

Most likely, the solubility of oxidized sulfur in magmatic liquids differs from that of
reduced sulfur. Unfortunately, experimental studies at 1 bar on the solubility and
solubility mechanisms of oxidized sulfur species (SO_2 and SO_3) in magmatic liquids are
essentially non-existent. The only solubility data available are for the system NaAl-
Si_3O_8–SO_2 (Mysen, 1977) in the pressure range from 10 to 30 kbar (Fig. 17.14). These
data, when extrapolated to low pressure, indicate a maximum solubility near 0.3 wt%
(Fig. 17.14). As for other volatile components, increasing gas pressure ($P_{gas} = P_{total}$)
results in increasing volatile content.

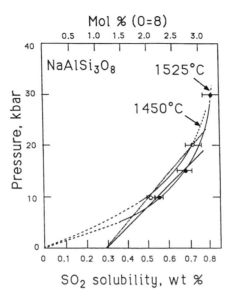

Fig. 17.14. SO_2 solubility in $NaAlSi_3O_8$ composition melt as a function of temperature and pressure (data from Mysen, 1977).

Sulfur dioxide probably enters silicate melts through the formation of SO_3^{2-} and SO_4^{2-} complexing. For a depolymerized sodium silicate melt, one may write, for example

$$Na_2Si_2O_5(melt) + SO_2(gas) \Leftrightarrow Na_2SO_3(melt) + 2SiO_2(melt), \qquad (16)$$

where the chemical entities have no physical meaning except to show that such a solubility mechanism results in polymerization of the silicate melt. Analogous reactions may be written to form sulfate complexes. In this case, the rate of polymerization, expressed as $\partial(NBO/T)/\partial X_S$, is greater for sulfate than for sulfite complexes.

In contrast, solution of SO_2 (or SO_3) in fully polymerized melts (e.g., $NaAlSi_3O_8$-composition melt) may result in depolymerization of the melt itself. This difference is because in such melt compositions the only cations available to stabilize SO_3^{2-} or SO_4^{2-} complexes are also required for electrical charge-balance of Al^{3+} in tetrahedral coordination. Thus, one may write a formalized expression

$$NaAlSi_3O_8(melt) + SO_2(gas) \Leftrightarrow Al^{3+}(melt) + Na^+(melt)$$
$$+ SO_3^{2-}(melt) + SiO_2(melt) + Si_2O_5^{2-}(melt), \qquad (17)$$

where the $Si_2O_5^{2-}$ units illustrate how nonbridging oxygens may be formed in the melt (the $Si_2O_5^{2-}$ unit has $NBO/Si = 1$). Other depolymerized units may be substituted for the disilicate unit, but in the absence of experimental data, characterization of this aspect of sulfur dioxide solubility in aluminosilicate melts awaits further study. The Na^+ or Al^{3+} cations, or both, may be associated with the sulfite anion. From experimental studies at 1 bar in the system $Na_2O–SiO_2–FeS–FeO$ (Shamazaki & Clark, 1973), most likely the sodium is associated with the sulfur complex, and aluminium is a network-modifier.

Carbon

Carbon speciation in the system C–H–O varies significantly in the oxygen fugacity range of magmatic processes in the upper mantle and the crust of the earth (Fig. 17.1). The main species are methane (CH_4), carbon monoxide (CO) and carbon dioxide (CO_2). Evidence from volcanic gases (e.g., Gerlach & Nordlie, 1975), fluid inclusions in phenocrysts (and xenocrysts?) in alkali basalt (e.g., Roedder, 1965) and the existence of carbonatitic magmas and carbonate inclusions in mantle-derived minerals (pyrope) (McGetchin & Besancon, 1973) suggest that CO_2 does exist in the entire pressure range at least under certain circumstances.

Evidence for methane is more ambiguous. Bulk chemical analysis of gases released from crushed diamonds generally yields H/O > 2 (Melton & Giardini, 1974), an observation consistent with the presence of H_2 or CH_4, or both, in the source regions of these diamonds. Some of the intrinsic oxygen fugacity data on mantle-derived minerals (e.g., Arculus $et\ al.$, 1984) indicate sufficiently low f_{O_2} so that if a C–H–O gas is present, the principal carbon-bearing gas species is CH_4.

Direct evidence for CO in mantle-derived rocks is lacking. Inferences in regard to its presence and importance in magmatic processes are based exclusively on calculated speciation in the C–H–O system under appropriate f_{O_2} conditions (Holloway, 1977, Eggler $et\ al.$, 1979b). Carbon monoxide is an unimportant gas species in the system C–H–O at these oxygen fugacity conditions and plays a significant role only in the oxygen fugacity range between that of the quartz–fayalite–magnetite and magnetite–wustite oxygen buffers (see Fig. 17.1). If, as suggested by, for example, Eggler (1983), Eggler & Baker (1982) and Haggerty (1978, 1986), this is the typical oxygen fugacity range of the source regions of partial melts in the upper mantle, carbon monoxide can be an important gas species.

The natural observations most probably indicate that a range of oxygen fugacities governed magmatic processes. The oxygen fugacity range probably is from above that of the nickel–nickel oxide buffer for felsic, water-bearing rocks in the crust (e.g., Carmichael, 1967, Fudali, 1965) to values near the iron–wustite oxygen buffer at least in portions of the upper mantle. Thus, there is a need to address the interaction of all three carbon species with silicate melts.

Carbon dioxide

Pearce (1964) recognized that the equilibrium

$$CO_2(gas) + O_2^-(melt) \Leftrightarrow CO_3^{2-}(melt) \tag{18}$$

could be used as a measure of oxygen ion activity in silicate melts. This observation was not, however, applied to characterize igneous processes in CO_2-bearing systems until Hill & Boettcher (1970) observed that carbonate minerals formed upon quenching of

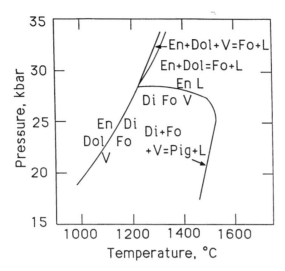

Fig. 17.15. Simplified phase relations in the system $CaO–MgO–SiO_2–CO_2$ (data
from Eggler, 1975).

hydrous, carbon-dioxide-bearing basalts from high pressure. That observation indicated that some of the CO_2 in the system may have existed as CO_3^{2-} in the melt.

An important effect of silicate–carbonate equilibria on melt compositions at upper
mantle is now well established. At low pressures (< 25 kbar) liquidus phase relations
among two pyroxenes and olivine are only marginally affected by CO_2 (Fig. 17.15). At
pressures above 25–30 kbar, liquidus temperature are depressed by several hundred
degrees centigrade (Fig. 17.15) and the composition of the melt is changed from
basaltic to carbonatitic (Eggler, 1975, 1976, Wyllie & Huang, 1975, Wendlandt &
Mysen, 1980) as the univariant equilibrium curve (Fig. 17.15) is crossed:

$$2Forsterite + Diopside + 2CO_2 \Leftrightarrow 4Enstatite + Dolomite. \qquad (19)$$

Exsolution of carbon dioxide from ascending carbonate-rich magma as the pressure
corresponding to this reaction is reached has been proposed as a driving mechanism for
explosive kimberlitic magmatism (Wyllie, 1980, 1987).

Important transition and alkaline earth metals have been shown to have significant
solubility in carbon dioxide rich fluids. Equilibrium relationships between CO_2-rich
fluid and magmatic liquids in the upper mantle become important, therefore, in
understanding metasomatic enrichment processes (e.g., Wendlandt & Harrison, 1979,
Mysen, 1983). If, for example, melting in the presence of CO_2 occurs at depth exceeding
about 30 kbar, a free CO_2 gas phase is unlikely, and these geochemically important
trace elements will dissolve in the melt. On the other hand, melting at shallower depth,
where CO_2 gas may be present, may result in significant retention of the same elements
in a coexisting vapor phase because of the apparent low carbon dioxide solubility in
silicate melts under such pressure conditions.

The solubility of CO_2, in contrast to hydrogen and water, is significantly dependent

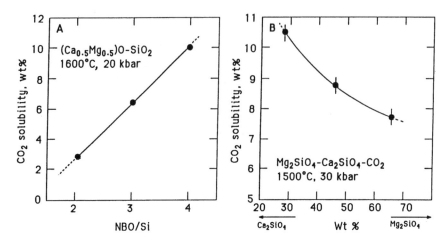

Fig. 17.16. Carbon dioxide solubility in the system CaO–MgO–SiO$_2$–CO$_2$ as a function of NBO/Si and Ca/Mg (data from Holloway *et al.*, 1976).

on bulk composition on the melts. There are two structural aspects of this bulk compositional dependence that require attention. The degree of polymerization (NBO/Si or NBO/T) is positively correlated with the activity of nonbridging oxygen. Thus, one would expect that the CO$_2$ solubility, driven by CO$_3^{2-}$ formation in the melt, would increase with increasing a_{O_2-} as is clearly evidenced for melts along the join CaMgSi$_2$O$_6$–(Ca$_{0.5}$Mg$_{0.5}$)O (Fig. 17.16). The data in that figure (Holloway *et al.*, 1976) also show, however, that in melts on the join CaSiO$_3$–MgSiO$_3$, where the degree of polymerization is not affected by Ca/Mg, the CO$_2$ solubility is also a distinct function of Ca/Mg with increasing solubility as the Ca/Mg increases. Furthermore, there is a significant CO$_2$ solubility in fully polymerized aluminosilicate melts (e.g., melts on the join NaAlO$_2$–SiO$_2$ and CaAlSi$_2$O$_8$ melt), where the solubility at constant pressure and temperature is correlated with Al/(Al+Si) of the melt (Mysen, 1977). Even in the system SiO$_2$–H$_2$O–CO$_2$, the CO$_2$ solubility is significant (Boettcher, 1984).

   The speciation of C–O complexes in silicate melts can be identified with Raman or infrared spectroscopy as C–O stretching in CO$_3^{2-}$ complexes appear as two characteristic bands between 1400 and 1550 cm^{-1} (e.g., Sharma, 1979). With an undistorted planar CO$_3^{2-}$ group, antisymmetric C–O stretching results in a single band near 1420 cm^{-1}. A split into two bands results from distortion of the CO$_3^{2-}$ group as observed in crystalline carbonates where the size of this split might be related to the degree of distortion of the CO$_3^{2-}$ group (White, 1974). Thus, as suggested first by Sharma (1979), carbonate groups in silicate glasses are significantly more distorted than in crystalline materials. Fine & Stolper (1985) observed that whereas in calcium-bearing systems the split is approximately 100 cm^{-1}, in sodium-bearing, calcium-free systems, the split can be as wide as 235 cm^{-1}. In the bulk compositional range studied by Fine & Stolper (1985) (Al/(Al+Si) = 0.200–0.333) with CO$_2$-undersaturated glasses, this split was not affected by bulk composition. Interestingly, in soda-melilite composition melt

($CaNaAlSi_2O_7$), the split of this band is similar to that observed in spectra of CO_2-saturated $CaMgSi_2O_6$ composition quenched melts. Thus, one might suggest that the CO_3^{2-} groups in this mixed (Na, Ca) melt are associated with Ca^{2+} in preference over Na^+.

These spectroscopic data indicate, therefore, that CO_3^{2-} groups formed by solution of CO_2 in silicate melts are associated with metal cations such as Ca^{2+} and Na^+. Most probably, the more stable of the two complexes is $(CaCO_3)^0$. Although spectroscopic data do not exist for melts on joins such as $CaSiO_3$–$MgSiO_3$–CO_2, one might speculate that because the CO_2 solubility decreases with decreasing Ca/Mg, this decrease reflects relative stabilities of Mg- and Ca-complexed carbonate at least in the pressure range investigated. In view of the enhanced stability of crystalline $MgCO_3$ and $CaMg(CO_3)_2$ relative to $CaCO_3$ with further pressure increase (e.g., Eggler *et al.*, 1979a) it may be speculated that the relationship between Ca/Mg of the melt and the CO_2 solubility (Fig. 17.16) might be reversed with increasing pressure.

Formation of metal carbonate complexes in depolymerized silicate melts results in melt polymerization as network-modifying cations become associated with the carbonate complex. Qualitatively, this solution mechanism is consistent with the experimental observation that partial melts in equilibrium with a mantle mineral assemblage become increasingly depleted in silica as the activity of CO_2 increases (Mysen & Boettcher, 1975). Metal carbonate formation by CO_2 solution in fully polymerized melts results, however, in depolymerization. If, as indicated by the spectroscopic data (e.g., Fine & Stolper, 1985), Na–carbonate complexes are formed in sodium aluminosilicate melts, the sodium required for carbonate complexing no longer can serve as a charge-balancing cation of tetrahedrally-coordinated Al^{3+}, and the solubility mechanism, using $NaAlSi_3O_8$ composition as an example, may be written as (Mysen & Virgo, 1980)

$$10NaAlSi_3O_8 + 2CO_2 \Leftrightarrow 6NaAlSiO_5^{2-} +$$
$$24SiO_2 + 4Al^{3+} + 2(Na_2CO_3)^0. \qquad (20)$$

In this equation, Al^{3+} represents aluminium not in tetrahedral coordination, and the other entities may be viewed as different types of structural units in the melt. Analogous expressions can be written for calcium aluminosilicate melts. For $CaAl_2Si_2O_8$ composition, for example, the CO_2 solubility is comparable to that in $NaAlSiO_4$ composition melt (Mysen, 1976, Mysen & Virgo, 1980), and the solubility mechanism is

$$10CaAl_2Si_2O_8 + 4CO_2 \Leftrightarrow 6CaAl_2Si_2O_{10}^{4-} +$$
$$8SiO_2 + 8Al^{3+} + 4(CaCO_3)^0. \qquad (21)$$

Under the assumption that all CO_2 is dissolved in these melts in the form of CO_3^{2-}, the proportion of aluminum not in tetrahedral coordination is a simple function of CO_2 content (Fig. 17.17). It has been suggested, however, that at least in certain melts on the join $NaAlO_2$–SiO_2–CO_2, some of the carbon dioxide may exist in molecular form (Mysen, 1976, Fine & Stolper, 1985). If so, the data shown in Fig. 17.17 represent

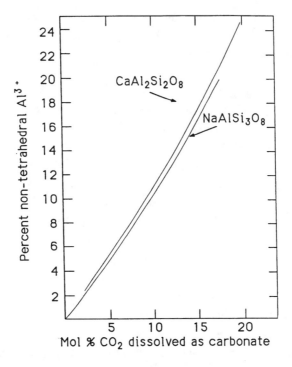

Fig. 17.17. Percent Al^{3+} not in tetrahedral coordination in melts on the joins $NaAlSi_3O_8$–CO_2 and $CaAlSi_2O_8$–CO_2 as a function of CO_2 dissolved as carbonate (data from Mysen & Virgo, 1980).

the proportion of Al^{3+} not in tetrahedral coordination as a function of amount of CO_2 dissolved as CO_3^{2-}.

Fine & Stolper (1985) suggested, on the basis of infrared spectroscopic information, that for a given composition on the $NaAlO_2$–SiO_2 join, the CO_3^{2-}/CO_2 is independent of total carbon dioxide content, but that the proportion of molecular CO_2 relative to total amount of dissolved carbon dioxide decreases substantially with increasing $Al/(Al+Si)$. They suggested that the equilibrium constant for an equilibrium of the form

$$CO_2^{vapor} \Leftrightarrow CO_2^{melt} \tag{22}$$

is only weakly dependent on composition, and that variations in total carbon dioxide solubility in silicate melts reflected the compositional dependence of the CO_3^{2-} solubility in the melt. In this sense, the solubility mechanism differs significantly from that suggested by Stolper (1982a) for water (see equation (4)), where the proportion of the molecular component was reported to be sensitive to the total amount of water present, and where the equilibrium constant for the solubility reaction was concluded not to depend on the bulk composition of the system.

Carbon monoxide

As indicated above, there is no direct observational information from rocks pointing to the presence of significant amounts of CO in igneous processes. Thus, its abundance in

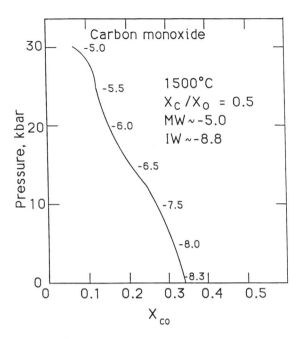

Fig. 17.18. Maximum abundance of CO (mol fraction, X_{CO}) in C–H–O gas as a function of pressure at 1500 °C with C/O = 0.5. Numbers indicate -log f_{O_2} at the maximum at given pressure.

natural C–O–H gas systems must be inferred from pressure–temperature–oxygen fugacity relations as illustrated in Fig. 17.1. At given temperature, pressure and C/O, the oxygen fugacity corresponding to the maximum abundance of CO can be estimated (Fig. 17.18). This oxygen fugacity decreases from near that of the iron–wustite oxygen buffer at 1 bar pressure to near that of the magnetite–wustite buffer at 30 kbar with the particular constraints chosen for the example in that figure. It is also evident that the relative abundance of CO in the gas phase at the f_{O_2} corresponding to the maximum carbon monoxide content increases with decreasing pressure.

The oxygen fugacity range of significant CO is well within that of igneous processes. Thus, there is a need to consider the effect of CO on phase relations and the solubility mechanisms of CO in silicate melts. The only experimental study available is that of Eggler *et al.* (1979b) who determined carbon solubility in several silicate melts in equilibrium with a mixed CO–CO$_2$ gas phase. At 1700 °C the solubilities in the 20–30 kbar range are compared with the solubilities of pure, or nearly pure CO$_2$. The X_{CO} (mol fraction: CO/(CO + CO$_2$)) is a linear function of pressure in this pressure range (0.28 at 30 kbar and 0.44 at 20 kbar as calculated with the modified Redlich–Kwong equation of state).

The solubility of carbon in equilibrium with (CO + CO$_2$) vapor is less than in equilibrium with CO$_2$ (Fig. 17.19). Because the (CO + CO$_2$) solubility is less tempera- ture dependent than the CO$_2$ solubility (Eggler *et al.*, 1979b), this difference decreases with decreasing temperature. Moreover, the relative solubilities in albite and diopside

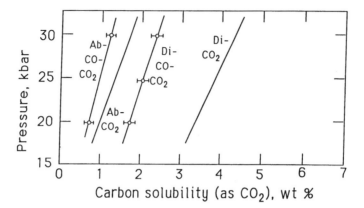

Fig. 17.19. Carbon solubility in $NaAlSi_3O_8$ and $CaMgSi_2O_6$ composition melts with CO_2 (data from Mysen, 1976, Mysen *et al.*, 1976) and CO_2–CO vapor (data from Eggler *et al.*, 1979b).

composition melts remain the same, although the effect of partial substitution of CO for CO_2 in the vapor becomes greater as the melt is increasingly depolymerized.

Infrared spectra of the quenched melts from these experiments showed that the only carbon-bearing species was CO_3^{2-}. This observation is consistent with a simple solubility mechanism of the form (Eggler *et al.*, 1979b)

$$CO(gas) + 2O^-(nonbridging\ oxygen) \Leftrightarrow CO_3^{2-}. \tag{23}$$

Interaction of nonbridging oxygen (O^-, as contrasted with free oxygen, O^{2-}, as shown in equation (18)) and carbon dioxide requires the formation of bridging oxygens in the process. Eggler *et al.* (1979b) concluded, therefore, that carbon monoxide less efficiently polymerizes silicate melts than does CO_2, a conclusion consistent with experimental liquidus phase relations.

Methane

To a considerable extent evidence for CH_4 being an important gas component in igneous processes is indirect and stems from calculated speciation of gases in the system C–H–O under appropriate pressure, temperature, gas composition and oxygen fugacity conditions. Whereas the maximum CO abundance increases with decreasing pressure at constant temperature (Fig. 17.18), the abundance of methane (Fig. 17.20) exhibits the opposite pressure dependence. The higher the pressure, the larger the maximum abundance of methane. Furthermore, the oxygen fugacities necessary to produce a significant amount of CH_4 in the C–H–O system are somewhat lower than for CO. The f_{O_2} at the maximum CH_4 abundance at 30 kbar is, for example, only an order of magnitude above that defined by the iron–wustite buffer, whereas for CO the maximum is encountered at f_{O_2} about three orders of magnitude higher. In fact, as illustrated in Fig. 17.1, coexistence of significant amounts of CO and CH_4 is unlikely

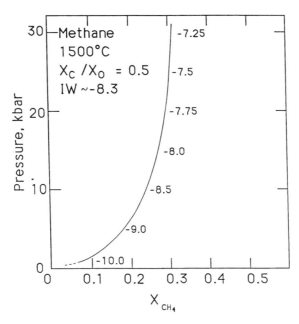

Fig. 17.20. Maximum abundance of CH_4 (mol fraction, X_{CH_4}) in C–H–O gas as a function of pressure at 1500 °C with C/O = 0.5. Numbers indicate -log f_{O_2} at the maximum at given pressure.

under the pressure, temperature and oxygen fugacity conditions corresponding to those of partial melting of the upper mantle.

Methane solubility mechanisms have been inferred from its influence on liquidus phase equilibria (Eggler & Baker, 1982, Taylor & Green, 1987). Eggler & Baker (1982) determined liquidus phase relations in the system $CaO–MgO–Al_2O_3–SiO_2$ at the oxygen fugacity defined by the $Si–SiO_2$ buffer

$$\log f_{O_2} = 9.06 - \frac{47171}{T} + 0.055\frac{P-1}{T},$$
(24)

where T is temperature (K) and P is pressure (GPa), so that the only two gas species in the system C–H–O are CH_4 and H_2, and methane constitutes more than 80% of the vapor phase. Most notable in these experiments was the expansion of the forsterite, and to a lesser extent, pyrope, liquidus phase volume (Fig. 17.21). In the volatile absent $CaO–MgO–Al_2O_3–SiO_2$ system, at pressure of about 26 kbar, olivine is no longer a liquidus phase (Kushiro & Yoder, 1974). In the presence of CO_2 or $CO + CO_2$, olivine disappears from the liquidus near 20 kbar (Eggler *et al.*, 1979b). In the presence of nearly pure CH_4, olivine remains a liquidus phase to pressures of at least 40 kbar (Fig. 17.21). Analogous observations were made in an experimental determination of the location of the forsterite–enstatite liquidus boundary in the system $SiO_2–Mg_2SiO_4–KAlSiO_4$ (Taylor & Green, 1987; see also Fig. 17.21). According to Taylor & Green (1987), the vapor in their experiments was essentially pure CH_4. Their liquidus phase

461

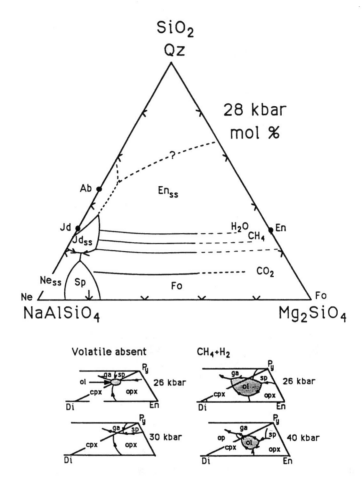

Fig. 17.21. Selected portions of liquidus phase equilibria in the systems NaAlSiO₄–
Mg₂SiO₄–SiO₂ (data from Taylor & Green, 1987) and CaO–MgO–
Al₂O₃–SiO₂ (data from Eggler & Baker, 1982, Kushiro & Yoder, 1974)
with volatiles as indicated.

equilibrium data illustrate the same effect as those of Eggler & Baker (1982) for the system CaO–MgO–Al₂O₃–SiO₂, in that the forsterite liquidus volume expands as methane is added to the dry system, in contrast to oxidized carbon species that have the opposite effect (Fig. 17.21).

In an attempt to measure the CH₄ solubility directly, Holloway & Jakobsson (1986) studied gas solubilities in the system C–H–O at the oxygen fugacity defined by the iron–wustite oxygen buffer. Those authors concluded that among the major relevant gas species in the system C–H–O, CH₄ was the least soluble.

From Raman spectra of quenched basanite and tholeiite melts, Holloway & Jakobsson (1986) report (p. D506) that evidence of CH₄ was detected although no spectra or details of the spectroscopic investigation were provided. Their observation contrasts with that of Taylor & Green (1987), who from infrared spectra of vapor-saturated soda-melilite (NaCaAlSi₂O₇) and jadeite (NaAlSi₂O₆) composition melts, concluded that there was no evidence of C–H bonds. Although the bulk compositions of the systems differ, if evidence for CH_4 was obtained in quenched basanite and tholeiite melt (Holloway & Jakobsson, 1986), one might also expect to see this in soda-melilite composition melt. The reason for this possible discrepancy cannot be ascertained with the limited amount of structural information currently available.

The liquidus phase equilibrium data by both Eggler & Baker (1982) and Taylor & Green (1987) are consistent with a suggestion that solution of CH_4 results in depolymerization of the silicate melts. Depolymerization would explain the expansion of the liquidus volume of depolymerized phases such as forsterite compared with that of enstatite, for example. Eggler & Baker (1982) suggested that in light of the analogous effect of H_2O and CH_4 on liquidus phase relations, solution of CH_4 might result in breakage of bridging oxygen bonds to form Si–OH and Si–CH_3 bonds together with dissolved C in the melt. Taylor & Green (1987) from their infrared spectra suggested a different mechanism involving reduction of Si^{4+} to, perhaps, Si^{2+} together with OH groups and elemental carbon dissolved in the melts. Both mechanisms require the formation of hydroxyl groups and elemental carbon. The Eggler & Baker (1982) model does not, however, call for reduction of silicon and suggests the formation of nonbridging oxygen instead. Their model suggested, however, the presence of methyl groups in the melt for which no spectra evidence has been found (Taylor & Green, 1987). Although the infrared spectra of Taylor & Green (1987) are consistent with their interpretation, reduction of Si^{4+} to Si^{2+} is not the only interpretation. Thus, it would appear at present that details of the solubility mechanism of CH_4 in silicate melts are not well characterized. It does seem clear, however, that solution involves formation of OH groups and results in depolymerization of the melt.

Fluorine

Wyllie & Tuttle (1961) first observed that addition of fluorine to silicate systems resulted in solidus temperature depressions comparable to or greater than that of water. This and other more recent studies of liquidus phase equilibria as well as physical properties of fluorine-bearing aluminosilicate melts have led to suggestions that the solubility mechanism of fluorine may resemble that of water (e.g., Bailey, 1977, Collins et al., 1982, Dingwell et al., 1985). There is, however, one significant difference in the behavior of fluorine and water silicate melts. Whereas water solubilities decrease rapidly with decreasing pressures, fluorine solubilities of several wt% can be achieved even at 1 bar. This is an important difference because whereas water-rich felsic magmas will expel H_2O upon ascent and, thus, experience changing fluidity and liquidus phase relations, fluorine-bearing magma (e.g., rhyolite) may retain its fluorine content and, thus, the comparatively high fluidity even upon extrusion. Moreover, with mixed fluorine + water systems, fluorine is preferentially partitioned into the melt phase. Thus, exsolution of fluid from a magma with a mixture of fluorine and water in solution

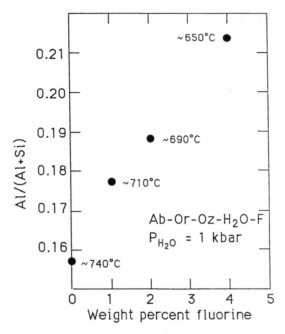

Fig. 17.22. Al/(Al + Si) of the 'granite' minimum in the system $NaAlSi_3O_8$–$KAlSi_3O_8$–SiO_2–H_2O–F at 1 kbar water pressure. Temperatures represent approximate temperatures of the minimum (data from Tuttle & Bowen, 1958, Manning, 1981).

does not result in the same dramatic changes in fluidity and solidus temperatures (with attendant explosive nature of an eruption) as would be the case for a water-bearing, fluorine-free system.

As has been the case for water, most studies of the solubility behavior of fluorine have been in highly polymerized aluminosilicate melt compositions. For example, Manning *et al.* (1980), on the basis of studies of liquidus phase relations in the system $NaAlSi_3O_8$–$KAlSi_3O_8$–SiO_2 in the presence of excess water (Manning, 1981), found that addition of only a few percent fluorine to the $NaAlSi_3O_8$–$KAlSi_3O_8$–SiO_2–H_2O system profoundly affects the temperatures and composition of the minimum in the system (Fig. 17.22). Those investigators drew particular attention to the observation that upon addition of fluorine, the composition of the minimum became increasingly aluminous (Fig. 17.22). Manning *et al.* (1980) concluded that this behavior could be explained with a solubility mechanism of fluorine involving formation of alumino-fluoride complexes in the melt:

$$AlSi_3O_8^- + 6F^- \Leftrightarrow AlF_6^{3-} + 2O^{2-} + 3SiO_2. \qquad (25)$$

In principle, this schematic expression shows that aluminum from an aluminosilicate network (with Al^{3+} in tetrahedral coordination) interacts with fluorine to form aluminofluoride complexes thus resulting in enhanced silica activity. Presumably, the interaction of aluminum with fluorine results in an equivalent amount of charge-

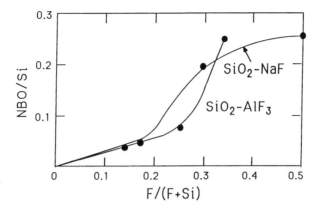

Fig. 17.23. Degree of polymerization of systems indicated as a function of fluorine contents (expressed as F/(F + Si)) (data from Mysen & Virgo, 1985a).

balancing cations to become a network-modifier. Consequently, fully polymerized aluminosilicate melts would become depolymerized as fluorine is dissolved.

The solubility mechanism suggested by Manning *et al.* (1980), summarized in equation (25), differs from other recent suggestions which involve interaction between fluorine and silicate melts to form either Si(IV)–F or Si(VI)–F bonds, or both, in the melts (e.g., Bailey, 1977, Collins *et al.*, 1982, Dingwell *et al.*, 1985). Formation of fluorinated silicate tetrahedra leads to depolymerization of a silicate network, and would be consistent with observed changes in fluidity. This mechanism can also explain depressions of solidus temperatures. It is, however, difficult to reconcile this model with the observed liquidus phase relations in the haplogranite system (Manning, 1981).

One might suggest, however, that the observations reported by Manning (1981) reflect the presence of *both* water and fluorine in the melts. Whether or not this is the case can be ascertained with the aid of spectroscopic studies of fluorine-bearing, fully polymerized (when volatile-free) aluminosilicate melts. From a study of quenched melts on the joins SiO_2–AlF_3 and SiO_2–NaF, Mysen & Virgo (1985a) observed that melts along both joins contained nonbridging oxygens (Fig. 17.23) and the Raman spectra showed evidence of Si(IV)–F bonds (with a distinct Raman band near 935 cm^{-1} assigned to Si–F stretching) although some portion of the Na^+ and Al^{3+} was also associated with fluorine in the melt. Fluorine in six-fold coordination with silicon was conclusively ruled out. This solubility behavior is qualitatively similar to that of OH in melts on the joins SiO_2–Na(OH) and SiO_2–Al(OH)$_3$ discussed above (see summary in Table 17.1).

As also pointed out by Foley *et al.* (1986), these systems are, however, compositionally quite far removed from natural aluminosilicate melts and the results obtained might not be applicable to natural magmatic systems. From a study of fluorine solubility in $NaAlO_2$–SiO_2 melts in the Al/(Al + Si) range comparable to that used for water solubility mechanisms (Fig 17.8), there was no spectroscopic evidence for either Si(IV)–F or Si (VI)–F. The spectra did, however, show evidence of melt depolymeriza-

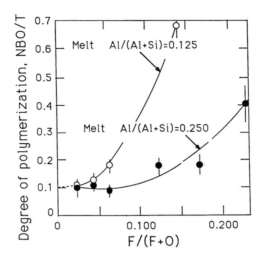

Fig. 17.24.   Degree of polymerization of two melts with Al/(Al + Si) as indicated on the join NaAlO$_2$–SiO$_2$–F as a function of fluorine content (data from Mysen & Virgo, 1985b).

tion (Fig 17.24), where the rate of depolymerization, $\partial(NBO/T)/\partial(F/F+O)$, was greater the less aluminous the melt:

$$\frac{Al}{Al+Si}=0.125: \partial\left(\frac{NBO}{T}\right)\bigg/\partial\left(\frac{F}{F+O}\right)=-0.90+68.0\frac{F}{F+O}, \qquad (26)$$

$$\frac{Al}{Al+Si}=0.250: \partial\left(\frac{NBO}{T}\right)\bigg/\partial\left(\frac{F}{F+O}\right)=-1.18+20.2\frac{F}{F+O}. \qquad (27)$$

In this system, fluorine is dissolved by forming either aluminum fluoride or sodium fluoride complexes or mixtures of the two. The rate of depolymerization of the melts can be calculated for the two extreme cases and can then be compared with observations (Fig. 17.25). It is evident from the results that most likely fluorine forms complexes with a mixture of both Na$^+$ and Al^{3+} where the proportion of Na and Al in the fluoride complexes is a function of both total fluorine content and Al/(Al + Si) of the melt (Mysen & Virgo, 1985a):

$$X_{F-}^{Na^+}=0.69-9.8\frac{F}{F+O}+23.8\left(\frac{F}{F+O}\right)^2+$$

$$0.19\left(\frac{F}{F+O}\right)\left(\frac{Al}{Al+Si}\right)+0.45\left(\frac{F}{F+O}\frac{Al}{Al+Si}\right)^2. \qquad (28)$$

From the results summarized in Figs. 17.9 and 17.25, there is a qualitative similarity between OH- and F-complexing in fully polymerized aluminosilicate melts. In both cases, with Al/(Al + Si) comparable with that of natural magmatic liquids, neither fluorine nor water interacts with Si^{4+} to form Si–OH or Si–F bonds. In both cases, Na$^+$ and Al^{3+} do interact with OH and F to form hydroxyl and fluoride complexes. In

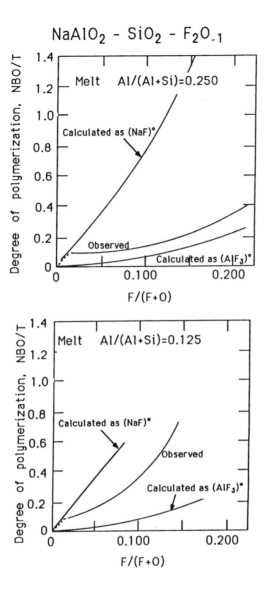

Fig. 17.25. Degree of polymerization of two melts (with $Al/(Al+Si)$ as indicated) on the join $NaAlO_2$–SiO_2–F as a function of fluorine content compared with anticipated NBO/T resulting from fluorine association with Na or Al (data from Mysen & Virgo, 1985b).

addition, some of the water is dissolved in molecular form and has no effect on the degree of polymerization of the melt. The proportion of Na and Al in these complexes is a function of the amount of the volatile component present and the $Al/(Al+Si)$ of the melt. Furthermore, the Na/Al in the hydroxylated and fluorinated complexes at the same $Al/(Al+Si)$ and $X/(X+O)$ ($X=F$, OH) differs. As a result, the rate of depolymerization, $\partial(NBO/T)\partial(X/X+O)$ ($X=OH$, F), is significantly greater for $X=F$ than for $X=OH$ (Fig. 17.26). Thus, for the same amount of fluorine or water added to a fully polymerized aluminosilicate melt, fluorine solution results in more pronounced depolymerization of the melt.

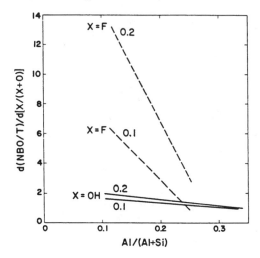

Fig. 17.26. Rate of depolymerization as a function of $Al/(Al+Si)$ of melts on the join $NaAlO_2$–SiO_2 with either water or fluorine added, expressed as $X/(X+O)$ with mol fraction as indicated (data from Mysen & Virgo, 1985b, 1986b).

There may be a significant difference between solution of both fluorine and water in fully polymerized and in depolymerized melts. Whereas in fully polymerized melts, nonbridging oxygens are formed either by replacement of bridging oxygen with OH or F, in a depolymerized melt, OH or F may replace nonbridging oxygens and, thus, effectively polymerize the melt (see equations (8), (9) above for H_2O). An indication of such solubility behavior for fluorine was provided by Foley et al. (1986) in their determination of liquidus phase relations in the system $KAlSiO_4$–Mg_2SiO_4–SiO_2 (Fig. 17.27). In that study, it was shown that the forsterite–enstatite liquidus boundary shifted to less silica-rich composition with the solution of fluorine. In fact, fluorine had similar effects on the liquidus phase equilibria as did CO_2. From Fourier Transform Infrared spectroscopy of selected liquid compositions, Foley et al. (1986) concluded that the principal fluorine complexes involved potassium, magnesium and aluminum with no indication of Si–F bonding

$$SiO_4^{4-} + Mg^{2+} + 2HF \Leftrightarrow SiO_3^{2-} + (MgF_2)^0 + H_2O, \tag{29}$$

$$SiO_4^{4-} + 2K^+ + 2HF \Leftrightarrow SiO_3^{2-} + 2(KF)^0 + H_2O \tag{30}$$

and

$$3SiO_4^{4-} + 2Al^{3+} + 6HF \Leftrightarrow 3SiO_3^{2-} + 2(AlF_3)^0 + 3H_2O. \tag{31}$$

In equations (29)–(31), structural units with 4 and 2 NBO/Si are used for illustrative purposes. One could, however, write analogous reactions for structural units with different degrees of polymerization. The notations $(KF)^0$ etc. are used to denote association with a specific cation. In principle, the solubility mechanism is analogous to

468

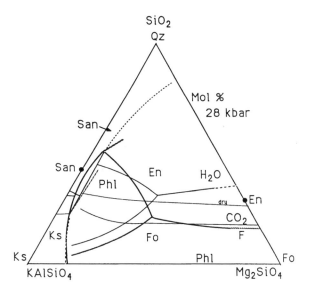

Fig. 17.27. Simplified liquidus diagram at 28 kbar of the system KAlSiO$_4$–Mg$_2$SiO$_4$–SiO$_2$ dry, with CO$_2$, with H$_2$O and with F added as indicated (data from Foley *et al.*, 1986).

that suggested from Raman spectroscopic studies of quenched melts on the join CaO–SiO$_2$–CaF$_2$ (Tsunawaki *et al.*, 1981). Equations (29)–(31) illustrate how HF interacts with a depolymerized silicate melt to form more polymerized structures. The melts become polymerized because portions of all of the metal cations acting as network-modifiers in the fluorine-free system interact with fluorine to form fluoride complexes. Thus, in the silicate network, the metal–silicon ratio is lowered and the melts become polymerized.

Those reactions are consistent with the liquidus phase relations in Fig. 17.27 and also with the infrared spectroscopic data that exhibit the possible presence of the metal fluoride complexes. The reaction mechanisms as written also require the presence of H$_2$O in the melts. Unfortunately, Foley *et al.* (1986) did not publish IR spectra in the frequency ranges relevant to confirm this latter requirement.

Summary

1. Water dissolves in silicate melts by forming bonds between OH groups and metal cations and as molecular H$_2$O. In compositions relevant to magmatic processes, Si–OH bonding is unimportant. In highly polymerized melts, this solubility mechanism leads to melt depolymerization. In depolymerized melts, on the other hand, the metal cations associated with OH have been removed from their network-modifying role, and hydrous, depolymerized melts are more polymerized than their anhydrous equivalents.

2. Fluorine solubility mechanisms qualitatively resemble water solubility mechanisms although molecular fluorine does not exist, whereas molecular H_2O does. Moreover, the relative stabilities of the individual fluoride complexes differ from the analog OH complexes. In polymerized melts, fluorine is a more effective depolymerizing agent than H_2O. Conversely, in depolymerized melts, fluorine more effectively polymerizes the melts than does water.

3. Hydrogen dissolves in silicate liquids in molecular form. There is, however, some interaction between the silicate network and hydrogen leading to significant solidus temperature depressions.

4. Sulfur in magmatic liquids occurs both as sulfide and as oxidized sulfur (SO_2 or SO_3). As sulfide, the S^{2-} anion exhibits pronounced preference for ferrous iron and to a lesser extent with other metal cations. The sulfur solubility is positively correlated with f_{S_2} and negatively correlated with f_{O_2}. Oxidized sulfur is dissolved as sulfite (SO_3^{2-}) or sulfate (SO_4^{2-}). In either case, solution of oxidized sulfur leads to melt polymerization. The solubility of oxidized sulfur is less than as reduced S^{2-}.

5. The carbon species, CO_2, CO and CH_4, all show pronounced solubility. The former two dissolve principally in the form of carbonate complexes thus leading to polymerization of depolymerized melts and depolymerization of highly polymerized melts. The exact form of methane solubility is not known, but involves most likely the stabilization of elemental carbon and OH-complexes in the melts. This mechanism is consistent with inferred depolymerization of silicate melts as CH_4 is dissolved.

References

Abraham, K.P. & Richardson, F.D. (1960). Sulphide capacities in silicate melts. Part II. *J. Iron Steel Inst.*, **196**, 313–17.

Abraham, K.P., Davis, M.W. & Richardson, F.D. (1960). Sulphide capacities of silicate melts. *J. Iron Steel Inst.*, **196**, 309–12.

Acocella, J., Tomozawa, M. & Watson, E.B. (1984). The nature of dissolved water in silicate glasses and its effect on various properties. *J. Non-Cryst. Solids*, **65**, 355–72.

Anderson, A.T. (1975). Some basaltic and andesitic gasses. *Rev. Geophys.*, **13**, 37–57.

Arculus, R.J., Dawson, J.B., Mitchell, R.H., Gust, D.A. & Holmes, R.D. (1984). Oxidation state of the upper mantle recorded by megacryst ilmenite in kimberlite and type A and B spinel lherzolite. *Contr. Mineral. Petrol.*, **85**, 85–94.

Bailey, J.C. (1977). Fluorine in granitic rocks and melts. *Chem. Geol.*, **19**, 1–42.

Bartholomew, R.F., Butler, B.L., Hoover, H.L. & Wu, C.-K. (1980). Infrared spectra of water-containing glasses. *J. Amer. Ceram. Soc.*, **63**, 481–5.

Bishop, F.C., Smith, J.V. & Dawson, J.B. (1975). Pentlandite–magnetite intergrowth in de Beers spinel lherzolite: review of sulfides in nodules. *Phys. Chem. Earth*, **9**, 323–37.

Boettcher, A.L. (1984). The system SiO_2–H_2O–CO_2: melting solubility mechanisms of carbon and liquid structure at high pressure. *Amer. Mineral.*, **69**, 823–34.

Bohlen, S.R., Boettcher, A.L. & Wall, V.J. (1982). The system albite–H_2O–CO_2: a model for melting and activities of water at high pressures. *Amer. Mineral.*, **67**, 451–62.

Bowen, N.L. (1928). *The evolution of the igneous rocks.* Princeton, NJ: Princeton Univ. Press.

Brawer, S.A. & White, W.B. (1975). Raman spectroscopic investigation of the structure of silicate glasses. *J. Chem. Phys.*, **63**, 2421–32.

Buchanan, D.L. & Nolan, J. (1979). Solubility of sulfur and sulfide immiscibility in synthetic tholeiite melts and their relevance to Bushveld complex rocks. *Can. Mineral.*, **17**, 483–94.

Burnham, C. W. (1975). Thermodynamics of melting in experimental silicate-volatile systems. *Geochim. Cosmochim. Acta*, **39**, 1077–84.

Burnham, C.W. & Jahns, R.H. (1962). A method for determining the solubility of water in silicate melts. *Amer. J. Sci.*, **260**, 721–45.

Burnham, C.W. & Davis, N.F. (1971). The role of H_2O in silicate melts: I. *P–V–T* relations in the system $NaAlSi_3O_8–H_2O$ to 10 kilobars and 1000 °C. *Amer. J. Sci.*, **270**, 54–79.

—— (1974). The role of H_2O in silicate melts: II. Thermodynamic and phase relations in the system $NaAlSi_3O_8–H_2O$ to 10 kilobars, 700 °C to 1100 °C. *Amer. J. Sci.*, **274**, 902–40.

Carmichael, I.S.E. (1967). The iron–titanium oxides of salic volcanic rocks and their associated ferromagnesian silicates. *Contr. Mineral. Petrol.*, **14**, 36–64.

Casadewall, T.J., Stokes, J.B., Greenland, L.P., Malinconico, L.L., Casadewall, J.R. & Furukawa, B.T. (1987). SO_2 and CO_2 emission rates at Kialuea Volcano, 1979–1984. *U.S. Geol. Survey Prof. Pap. 1350*, 771–80.

Collins, W.J., Beams, S.D., White, A.J.R. & Chappell, B.W. (1982). Nature and origin of A-type granites with particular reference to Southeastern Australia. *Contr. Mineral. Petrol.*, **80**, 189–200.

Delaney, J.R., Muenow, D., Ganguly, J. & Royce, D. (1977). Anhydrous glass-vapor inclusions from phenocrysts in oceanic tholeiite pillow basalt. *EOS*, **58**, 530.

Delaney, J.R., Muenow, D.W. & Graham, D.G. (1978). Abundance and distribution of water, carbon and sulfur in the glassy rims of submarine pillow basalts. *Geochim. Cosmochim. Acta*, **42**, 581–94.

Dingwell, D.B., Scarfe, C.M. & Cronin, D.J. (1985). The effect of fluorine on viscosities in the system $Na_2O–Al_2O_3–SiO_2$: implications for phonolites, trachytes and rhyolites. *Amer. Mineral.*, **70**, 80–7.

Eggler, D.H. (1975). CO_2 as a volatile component in the mantle: the system $Mg_2SiO_4–SiO_2–H_2O–CO_2$. *Phys. Chem. Earth*, **9**, 869–81.

— (1976). Does CO_2 cause partial melting in the low-velocity layer of the mantle? *Geology*, **2**, 69–72.

— (1983). Upper mantle oxidation state: evidence from olivine or pyroxene–ilmenite assemblages. *Geophys. Res. Lett.*, **10**, 365–8.

Eggler, D.H. & Rosenhauer, M. (1978). Carbon dioxide in silicate melts. II. Solubilities of CO_2 and H_2O in $CaMgSi_2O_6$ (diopside) liquids and vapors at pressure to 40 kb. *Amer. J. Sci.*, **278**, 64–94.

Eggler, D.H. & Baker, D.R. (1982). Reduced volatiles in the system C–H–O: implications to mantle melting, fluid formation and diamond genesis. *Adv. Earth Planet. Sci.*, **12**, 237–50.

Eggler, D.H. & Burnham, C.W. (1984). Solution of H_2O in diopside melts: a thermodynamic model. *Contr. Mineral. Petrol.*, **85**, 58–67.

Eggler, D.H., Kushiro, I. & Holloway, J.R. (1979a). Free energies of decarbonation reactions at mantle pressures. I. Stability of the assemblage forsterite–enstatite–magnesite in the system $MgO–SiO_2–CO_2–H_2O$ to 60 kbar. *Amer. Mineral.*, **64**, 288–94.

Eggler, D.H., Mysen, B.O., Hoering, T.C. & Holloway, J.R. (1979b). The solubility of carbon monoxide in silicate melts at high pressures and its effect on silicate phase relations. *Earth Planet. Sci. Lett.*, **43**, 321–30.

Epel'baum, M.B. (1973). The alteration of the basicity of silicate melts under the effect of water-dissolution under high pressures. *Izv. A. N. SSSR*, **8**, 33–45.

— (1985). The structure and properties of hydrous granitic melts. *Geologica Carpathica*, **36**, 491–8.

Epel'baum, M.B., Salova, T.P. & Varshal, B.G. (1975). Influence of water on the coordination of Al^{3+} in water-containing aluminosilicate glasses. *Izv. A.N. SSSR*, **8**, 1496–500.

Fine, G. & Stolper, E. (1985). The speciation of carbon dioxide in sodium aluminosilicate glasses. *Contr. Mineral. Petrol.*, **91**, 105–212.

Foley, S.F., Taylor, W.R. & Green, D.H. (1986). The effect of fluorine on phase relationships in the system $KAlSiO_4$–Mg_2SiO_4–SiO_2 at 28 kbar and the solution mechanism of fluorine in silicate melts. *Contr. Mineral. Petrol.*, **93**, 46–55.

Freund, D. (1982). Solubility mechanisms of H_2O in silicate melts at high pressures and temperatures: a Raman spectroscopic study: discussion. *Amer. Mineral.*, **67**, 153–4.

Fudali, F. (1965). Oxygen fugacity of basaltic and andesitic magmas. *Geochim. Cosmochim. Acta*, **29**, 1063–75.

Gerlach, T.M. & Nordlie, B.E. (1975). The C–O–H–S gaseous system, Part I. Composition limits, trends and basaltic cases. *Amer. J. Sci.*, **275**, 353–76.

Gerlach, T.M. & Nordlie, B.E. (1975). the C–O–H–S gaseous system, Part I. Composition limits, trends and basaltic cases. *Amer. J. Sci*, **275**, 353–76.

Greenland, L.P. (1987). Composition of gases from the 1984 eruption of Mauna Loa Volcano. *U.S. Geol. Survey Prof. Pap. 1350*, 781–90.

Haggerty, S.E. (1978). The redox state of planetary basalts. *Geophys. Res. Lett.*, **5**, 443–6.

— (1986). Diamond genesis in a multiply-constrained model. *Nature*, **320**, 34–8.

Haggerty, S.E. & Tompkins, L.A. (1983). Redox state of the earth's upper mantle from kimberlitic ilmenites. *Nature*, **303**, 295–300.

Hamilton, D.L., Burnham, C.W. & Osborn, E.F. (1964). The solubility of water and the effect of oxygen fugacity and water content on crystallization of mafic magmas. *J. Petrol.*, **5**, 21–39.

Haughton, D.R., Roeder, P.L. & Skinner, B.J. (1974). Solubility of sulfur in mafic magmas. *Econ. Geol.*, **69**, 451–67.

Helgeson, H.C., Delaney, J.M. & Nesbitt, H.W. (1978). Calculation of thermodynamic consequences of dehydration in subducting oceanic crust to 10 kb and 1000 °C. *Amer. J. Sci.*, **278**.

Hill, R.E.T. & Boettcher, A.L. (1970). Water in the earth's mantle: melting curves of basalt–water and basalt–water–carbon dioxide. *Science*, **167**, 980–2.

Hodges, F.N. (1974). The solubility of H_2O in silicate melts. In *Carnegie Institution, Washington, Year Book 73*, pp. 251–5.

Holloway, J.R. (1977). Fugacity and activity of molecular species in supercritical fluids. In *Thermodynamics in geology*, ed. D.G. Fraser, pp. 161–82. Dordrecht, Netherlands: Reidel.

Holloway, J.R. & Jakobsson, S. (1986). Volatile solubilities in magmas: transport of volatiles from mantles to planet surfaces. *J. Geophys. Res.*, **91**, D505–D508.

Holloway, J.R., Mysen, B.O. & Eggler, D.H. (1976). The solubility of CO_2 in liquids on the join CaO–MgO–SiO_2–CO_2. In *Carnegie Institution, Washington, Year Book 75*, pp. 626–31.

Jaques, A.L., Lewis, J.D., Smith, C.B., Gregory, G.P., Ferguson, J., Chappell, B.W. & McCulloch, M.T. (1984). The diamond bearing ultrapotassic (lamproitic) rock of the West Kimberley Region, Western Australia. In *Kimberlites I: kimberlites and related rocks*, ed. J. Kornprobst, pp. 225–54.

Kushiro, I. & Yoder, H.S. (1974). Formation of eclogite from garnet ilherzolite: Liquidus relations in a portion of the system $MgSiO_3$–$CaSiO_3$–Al_2O_3 at high pressure. In *Carnegie Institution, Washington, Year Book 73*, pp. 266–9.

Luth, R.W. & Boettcher, A.L. (1986). Hydrogen and the melting of silicates. *Amer. Mineral.*, **71**, 264–76.

Luth, R.W., Mysen, B.O. & Virgo, D. (1987). A Raman spectroscopic study of the solubility behavior of H_2 in the system Na_2O–Al_2O_3–SiO_2–H_2. *Amer. Mineral.*, in press.

Luth, W.C., Jahns, R.H. & Tuttle, O.F. (1964). The granite system at pressures of 4 to 10 kilobars. *J. Geophys. Res.*, **69**, 759–73.

Manning, D.A.C. (1981). The effect of fluorine on liquidus phase relationships in the system Qz–Ab–Or with excess water at 1 kb. *Contr. Mineral. Petrol.*, **76**, 206–11.

Manning, D.A.C., Hamilton, C.M.B., Henderson, C.M.B & Dempsey, M.J. (1980). The probable occurrence of interstitial Al in hydrous F-bearing and F-free aluminosilicate melts. *Contr. Mineral. Petrol.*, **75**, 257–62.

Mathez, E.A. (1976). Sulfur solubility and magmatic sulfides in submarine basalt glass. *J. Geophys. Res.*, **81**, 4269–77.

Mathez, E.A. & Delaney, J.R. (1981). The nature and distribution of carbon in submarine basalts and peridotite nodules. *Earth Planet. Sci. Lett.*, **56**, 217–32.

McGetchin, T.R. & Besancon, J.R. (1973). Carbonate inclusions in mantle-derived pyropes. *Earth Planet. Sci. Lett.*, **18**, 408–10.

McMillan, P., Jakobsson, S., Holloway, J.R. & Silver, L. (1983). A note on the Raman spectra of water-bearing albite glasses. *Geochim. Cosmochim. Acta*, **47**, 1937–45.

McMillan, P.F. & Remmele, R.L. (1986). Hydroxyl sites in SiO_2 glass: a note on infrared and Raman spectra. *Amer. Mineral.*, **71**, 772–8.

Melton, C.E. & Giardini, A.A. (1974). The composition and significance of gas release from natural diamonds from Africa and Brazil. *Amer. Mineral.*, **59**, 775–82.

Moore, J.G. & Fabbi, B.P. (1971). An estimate of the juvenile sulfur content of basalt. *Contrib. Mineral. Petrol.*, **33**, 118–27.

Muenow, D., Delaney, J.R., Meijer, A. & Liu, N. (1977). Water-rich glass–vapor inclusions in phenocrysts from tholeiitic pillow basalt rims in the Marianas back arc basin. *EOS*, **58**, 530.

Murck, B.W., Burruss, R.C. & Hollister, L.S. (1978). Phase equilibria in fluid inclusions in ultramafic rocks. *Amer. Mineral.*, **63**, 40–6.

Mysen, B.O. (1976). The role of volatiles in silicate melts: solubility of carbon dioxide and water in feldspar, pyroxene, feldspathoid melts to 30 kb and 1625 °C. *Amer. J. Sci.*, **276**, 969–96.

— (1977). Solubility of volatiles in silicate melts under the pressure and temperature conditions of partial melting in the upper mantle. In *Magma genesis*, Proc. AGU Chapman Conference on Partial melting in the Earth's Upper Mantle, H.J.B. Dick, pp. 1–14. State of Oregon Dept. Geology and Mineral Resources.

— (1983). Rare earth element partitioning between ($H_2O + CO_2$) vapor and upper mantle minerals: experimental data bearing on the conditions of formation of alkali basalt and kimberlite. *N. Jb. Mineral. Abh.*, **146**, 41–65.

Mysen, B.O. & Boettcher, A.L. (1975). Melting of a hydrous mantle. II. Geochemistry of crystals and liquids formed by anatexis of mantle peridotite at high pressures and high temperatures as a function of controlled activities of water, hydrogen and carbon dioxide. *J. Petrol.*, **16**, 549–90.

Mysen, B.O., Eggler, D.H., Seitz, M.G. & Holloway, J.R. (1976). Carbon dioxide in silicate melts and crystals. I. Solubility measurements. *Amer. J. Sci.*, **276**, 455–79.

Mysen, B.O. & Kushiro, I. (1976). Partitioning of iron, nickel and magnesium between metal oxide, and silicates in Allende meteorite as a function of f_{O_2}. In *Carnegie Institution, Washington, Year Book 75*, pp. 678–84.

Mysen, B.O. & Popp, R.K. (1980). Solubility of sulfur in $CaMgSi_2O_6$ and $NaAlSi_3O_8$ melts at high pressure and temperature with controlled f_{O_2} and f_{S_2}. *Amer. J. Sci.*, **280**, 78–92.

Mysen, B.O. & Virgo, D. (1980). Solubility mechanisms of carbon dioxide in silicate melts: a Raman spectroscopic study. *Amer. Mineral.*, **65**, 885–99.

—— (1985a). Interaction between fluorine and silica in quenched melts on the joins $SiO_2–AlF_3$ and $SiO_2–NaF$ determined by Raman spectroscopy. *Phys. Chem. Mineral.*, **12**, 77–85.

—— (1985b). Structure and properties of fluorine-bearing aluminosilicate melts: the system $Na_2O-Al_2O_3-SiO_2-F$ at 1 atm. *Contr. Mineral. Petrol.*, **91**, 205–22.

—— (1986a). Volatiles in silicate melts at high pressure and temperature. 1. Interaction between OH groups and Si^{4+}, Al^{3+}, Ca^{2+}, Na^+ and H^+. *Chem. Geol.*, **57**, 303–31.

—— (1986b). Volatiles in silicate melts at high pressure and temperature. 2. Water inmelts along the join $NaAlO_2-SiO_2$ and a comparison of solubility mechanisms of water and fluorine. *Chem. Geol.*, **57**, 333–58.

Mysen, B.O., Virgo, D., Harrison, W.J. & Scarfe, C.M. (1980). Solubility mechanisms of H_2O in silicate melts at high pressures and temperatures: a Raman spectroscopic study. *Amer. Mineral.*, **65**, 900–14.

Naldrett, A.J. (1973). Nickel-sulfide deposits – their classification and genesis with special emphasis on volcanic association. *Can. Mining Metall. Bull.*, **66**, 45–63.

Navrotsky, A. (1987). Calorimetric studies of melts, crystals and glasses, especially in hydrous systems. In *Magmatic processes: physicochemical principles*, ed. B.O. Mysen, pp. 411–22. Geochem. Soc. Spec. Publ. No. 1, University Park.

Oxtoby, S. & Hamilton, D.L. (1978). The discrete association of water with Na_2O and SiO_2 in Na Al silicate melts. *Contrib. Mineral. Petrol.*, **66**, 185–8.

Pearce, M.L. (1964). Solubility of carbon dioxide and variation of oxygen ion activity in soda-silicate melts. *J. Amer. Ceram. Soc.*, **47**, 342–7.

Pichavant, M., Herrera, J.V., Boulmier, S., Briqueu, L., Joron, J.-L., Juteau, L.M., Michard, A., Sheppard, S.M.F., Treuil, M. & Vernet, M. (1987). The Macusani glasses, SE Peru: evidence of chemical fractionation of peraluminous magmas. In *Magmatic processes: physicochemical principles*, ed. B.O. Mysen, pp. 359–74. Geochem. Soc. Spec. Publ. No. 1, University Park.

Rajamani, V. & Naldrett, A.J. (1975). Partitioning of Fe, Co, Ni and Cu between sulfide liquid and basaltic melts and the composition of Ni–Cu sulfide deposits. *Econ. Geol.*, **73**, 82–94.

Rhyabchikov, I.D., Green, D.H., Wall, V.J. & Brey, G.P. (1981). The oxidation state of carbon in the reduced-velocity zone. *Geochem. Intern.*, **18**, 148–58.

Richardson, F.D. & Fincham, C.J.B. (1956). Sulfur and silicate in aluminate slags. *J. Iron Steel Inst.*, **178**, 4–15.

Roedder, E. (1965). Liquid CO_2 inclusions in olivine-bearing nodules and phenocrysts from basalts. *Amer., Mineral.*, **50**, 1746–82.

Rosenquist, T. (1951). A thermodynamic study of the reaction $CaS + H_2O = CaO + H_2S$ and the desulfurization of liquid metals with lime. *Amer. Inst. Mining Metall. Petroleum Engineers Trans.*, **191**, 535–40.

— (1954). A thermodynamic study of the iron, cobalt and nickel sulfides. *J. Iron Steel Inst.*, **176**, 37–57.

Rutherford, M.J., Sigurdsson, H., Carey, S. & Davis, A. (1984). The May 18, 1980, eruption of Mount St Helens, 1, Melt composition and experimental phase equilibria. *J. Geophys. Res.*, **90**, 2929–48.

Shamazaki, H. & Clark, L.A. (1973). Liquidus relations in the system $FeS-FeO-SiO_2-Na_2O$ and geological implications. *Econ. Geol.*, **68**, 79–96.

Sharma, S.K. (1979). Structure and solubility of carbon dioxide in silicate glasses of diopside and sodium melilite compositions at high pressures from Raman spectroscopic data. In *Carnegie Institution, Washington Year Book* **78**.

Shima, H. & Naldrett, A.J. (1975). Solubility of sulfur in ultramafic melt and the relevance of the system Fe–S–O. *Econ., Geol.*, **70**, 960–7.

Silver, L. & Stolper, E. (1985). A thermodynamic model for hydrous silicate melts. *J. Geol.*, **93**, 161–78.

Stolen, R.H. & Walrafen, G.E. (1976). Water and its relation to broken bond defects in fused silica. *J. Chem. Phys.,* **64**, 2623–31.

Stolper, E. (1982a). Water in silicate glasses: an infrared spectroscopic study. *Contr. Mineral. Petrol.,* **81**, 1–17.

— (1982b). The speciation of water in silicate melts. *Geochim. Cosmochim. Acta,* **46**, 2609–20.

Stolper, E., Silver, L.A. & Aines, R.D. (1983). The effect of quenching rate and temperature on the speciation of water in silicate glasses. *EOS,* **64**, 339.

Taylor, M. & Brown, G.E. (1979). Structure of mineral glasses. II. The SiO_2–$NaAlSiO_4$ join. *Geochim. Cosmochim. Acta,* **43**, 1467–75.

Taylor, W.R. & Green, D.H. (1987). The petrogenetic role of methane: effect on liquidus phase relations and the solubility mechanisms of reduced C–H volatiles. In *Magmatic processes: physicochemical principles,* ed. B.O. Mysen, pp. 121–38. Geochem. Soc. Spec. Publ. No. 1, University Park.

Tsunawaki, Y., Iwamoto, N., Hattori, T. & Mitsuishi, A. (1981). Analysis of CaO–SiO_2 and CaO–SiO_2–CaF_2 glasses by Raman-spectroscopy. *J. Non-Cryst. Solids,* **44**, 369–78.

Tuttle, O.F. & Bowen, N.L. (1958). *Origin of granite in light of experimental studies in the system* $NaAlSi_3O_8$–$KAlSi_3O_8$–SiO_2–H_2O. Geol. Soc. Amer. Mem. 74.

Vakrushev, V.A. & Sobolev, N.V. (1971). Sulfide intergrowths (sulfidization) in deep-seated xenoliths from the kimberlite pipes of Yakutia (in Russian). *Geologiiya i Geophizika,* **11**, 3–11.

Wasserburg, G.J. (1957). The effects of H_2O in silicate systems. *J. Geol.,* **65**, 15–23.

Wendlandt, R.F. & Harrison, W.J. (1979). Rare earth partitioning between immiscible carbonate and silicate liquids and CO_2 vapor: results and implications for the formation of light rare earth-enriched rocks. *Contrib. Mineral. Petrol.,* **69**, 409–19.

Wendlandt, R.F. & Mysen, B.O. (1980). Melting phase relations of natural peridotite + CO_2 as a function of degree of partial melting at 15 and 30 kbar. *Amer. Mineral.,* **65**, 37–44.

White, R.W. (1966). Ultramafic inclusions in basaltic rocks from Hawaii. *Contr. Mineral. Petrol.,* **12**, 245–314.

White, W.B. (1974). The Carbonate Minerals. In *Infrared spectra of minerals,* ed. V.C. Farmer, ch. 12. London: Monograph Min. Soc. London.

Wyllie, P.J. (1980). The origin of kimberlites. *J. Geophys. Res.,* **85**, 6902–10.

— (1987). Transfer of subcratonic carbon into kimberlites and rare earth carbonatites. In *Magmatic processes: physicochemical principles,* ed. B.O. Mysen, pp. 107–20. Geochem. Soc. Spec. Publ. No. 1, University Park.

Wyllie, P.J. & Huang, W.-L. (1975). Influence of mantle CO_2 in the generation of carbonatites and kimberlites. *Nature,* **257**, 297–9.

Wyllie, P.J. & Tuttle, O.F. (1961). Experimental investigation of silicates containing two volatile components. II. The effects of NH_3 and HF in addition to water on the melting temperatures of granite and albite. *Amer. J. Sci.,* **259**, 128–43.

18

Magmatic consequences of volatile fluxes from the mantle

PETER J. WYLLIE

Introduction

The 1959 translation of Korzhinskii's book *Physicochemical Basis of the Analysis of the Paragenesis of Minerals* introduced me to the concept of inert and perfectly mobile components in open systems. At that time, O.F. Tuttle and I were studying phase relationships in the systems $CaO-CO_2-H_2O$ and $MgO-CO_2-H_2O$ (Wyllie & Tuttle, 1960, Walter *et al.*, 1962, Wyllie, 1962), with the volatile components contained securely inside closed gold capsules and therefore thermodynamically inert. The results were to be applied to carbonatites, igneous rocks through which there is no doubt that volatile components have flowed influentially. In Korzhinskii's book I discovered how to represent the volatile components CO_2 and H_2O in chemical potential diagrams, applicable to both closed and open systems. The method was also applied to many other systems, including granitic rocks with mineralogy controlled by the chemical potentials of sodium and potassium. Korzhinskii's work has provided the basis for quantitative treatment of metasomatism.

Metasomatic processes, originally studied in connection with crustal rocks, are now believed to be important in the mantle, as well. There is evidence that peridotite nodules brought to the surface in kimberlites or alkali basalts were metasomatized within the mantle before being transported by their igneous hosts (e.g. Boettcher & O'Neill, 1980, Dawson, 1980, pp. 183–5; Harte, 1983), and mantle metasomatism is commonly assumed to explain the observation that many basaltic magmas have trace element and isotope geochemistry that is difficult to explain in terms of partial melting of upper mantle rocks with compositions considered to be normal (Walker, 1983). Extension of phase equilibrium studies involving volatile components to mantle pressures (e.g. $CaO-MgO-SiO_2-CO_2-H_2O$; Wyllie & Huang, 1976, Eggler, 1978, Ellis & Wyllie, 1980) have provided applications to the petrogenesis of kimberlites, probably the most volatile-charged magmas rising from the mantle.

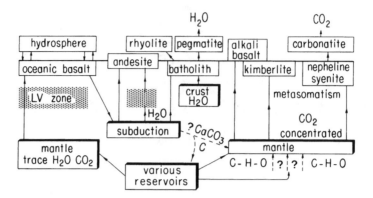

Fig. 18.1.   Schematic diagram illustrating magmatic sources and products at
divergent plate boundaries, convergent plate boundaries, and continental
plates. Note especially H_2O and CO_2 (Wyllie, 1987a).

When Dawson (1980) reviewed hypotheses for the origin of kimberlites he concluded
that experimental evidence supported their generation by the partial melting of
phlogopite–carbonate–garnet lherzolite. By this time I had concluded (Wyllie, 1980)
that the existence of such rocks in the kimberlite source regions was unlikely, unless
temperatures were extraordinarily low, and I was impressed by the evidence that the
mantle oxygen fugacity may be too low for the formation or survival of such rocks in
the base of the lithosphere. This led me to formulate a model involving the migration
from the deep mantle of reduced vapors with major components C–H–O, which
generated kimberlite melts in the asthenosphere, with the melts subsequently being
erupted from the deep lithosphere. This contribution outlines the possible effects of
intermittent fluxes of reduced volatile components from the mantle, into and through
the lithosphere. Similar themes have been developed by Perchuk (1976), Bailey (1985)
and Green *et al.* (1987), although Bailey (1985) assigned a major role to oxidized CO_2.

Volatile components in mantle systems

The sources of and migrations of volatile components through the mantle are
important problems still poorly understood. There is abundant evidence from fluid
inclusions in the minerals of mantle xenoliths for the passage of H_2O and CO_2 through
the lithosphere, indicating oxidized conditions, and for the occasional concentration of
other components such as S (e.g., Andersen *et al.*, 1984, Kovalenko *et al.*, 1987). These
components may have been derived from subducted oceanic lithosphere, or from some
deep reservoir retaining primordial volatile components, or from both.

Mantle–magma relationships

The variety of magmatic sources and volcanic products in different tectonic environments is outlined in Fig. 18.1. The mantle comprises several chemical reservoirs that

have apparently remained physically distinct from each other for more than a billion years. At divergent plate boundaries and in oceanic plates, basalts containing small concentrations of H_2O and CO_2 are certainly derived from the mantle. The submarine basalts react with sea water, with the formation of calcite and hydrated minerals. The oceanic crust when subducted may carry with it pelagic sediments, including slabs of limestone (during the relatively brief period of geological history that pelagic lime-stones has been formed). During subduction it appears that most of the water in oceanic crust is released to participate in convergent boundary processes, but some may be carried deeper as intergranular films or in minerals not yet adequately studied. Some of the carbonate may escape dissociation and magmatic processes and be carried deeper into the mantle for long-term storage of carbon. Andesites and basalts are erupted at convergent plate boundaries with the magmatic processes probably being initiated in the subducted slab, and at ocean–continent plate boundaries these are joined by rhyolites escaping from batholiths. In these environments the volatile flux is probably dominated by components rising directly from the subducted slab. Away from convergent plate boundaries, volatile components from deeper mantle become more influential. Kimberlites rise from volatile-enriched mantle near the astheno-sphere–lithosphere boundary beneath cratons. Nephelinitic magmas are normally associated with rift zones in continental platforms, and carbonatites are formed by their differentiation.

Mantle fluids

The word 'fluid' has been widely used in the context of mantle metasomatism, although it is evident that some chemical changes are caused by magmas, and others by aqueous solutions. Fluid is a descriptive term applicable in geological context to liquid, vapor, or supercritical solution, or to an undefined fluid phase. Fluid also has a more restricted meaning in phase equilibria; it is the phase that is indeterminate between liquid and vapor, where critical phenomena occur.

I consider it to be important in petrological discussions to retain the distinction between liquid (melt), vapor (or dense gas, solution) and the descriptive term 'fluid' that can refer to either phase. Liquid and vapor are two distinct phases with different compositions and different properties that can coexist with each other. At high pressures and temperatures, the compositions and properties of these two fluid phases approach each other, but all experimental data so far available indicate that for normal rock compositions with H_2O and CO_2 in the upper mantle, liquid and vapor retain their separate identities. Chemical changes caused by the intrusion of magma are quantitatively different from those caused by the passage of a dense vapor phase. A magma may react with its wall rocks effecting metasomatic exchanges, but eventually the magma solidifies and a significant mass of new material is thus introduced. The passage of vapors or solutions, however, causes reactions with wall rocks that may entail either leaching or precipitation. It takes much more vapor or solution to cause significant metasomatic changes. In crustal processes, rocks that have experienced

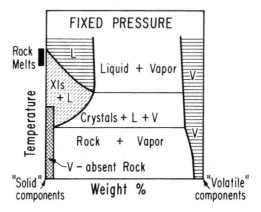

Fig. 18.2. Schematic isobaric phase diagram illustrating the relationships between volatile components and solid mantle components at high pressures. Note in particular the two separate fields for fluids: L = liquid (melt or magma); Xls = crystals; and V = vapor (dense gas or solution). 'Rock melts' shows the temperature interval between solidus and liquidus for the rock without volatile components (Wyllie, 1987c).

chemical change through reactions with magmas are not normally described as metasomatic rocks but as hybrids, or injection gneisses. It is not inappropriate to extend the term 'metasomatism' to hidden mantle rocks that have experienced chemical change through the passage or intrusion of magmas, but it is appropriate to keep in mind the fact that these two kinds of metasomatism are quantitatively different in style and scale.

Fig. 18.2 is a generalized isobaric phase diagram for peridotite or eclogite under upper mantle conditions, in the presence of volatile components H_2O, CO_2 and others, or their reduced equivalents CO, CH_4 and H_2. The horizontal axis projects all components except the volatiles at the left end, and all volatile components at the right end. The vapor field at the right hand side contains a significant proportion of dissolved volatile components, the solubility increasing with increasing temperature. At subsolidus temperatures, through a limited range of volatile compositions, there is no vapor phase, because the volatile components react with the solid components to generate minerals such as phlogopite, amphibole, DHMS (dense hydrated magnesian silicates), or carbonate. The narrow field of vapor-absent, volatile-bearing rock ends at the volatile content corresponding to the maximum formation of such minerals for the bulk composition. In the presence of reduced volatile components, this interval may be narrow. There is a wide subsolidus interval where the rock coexists with vapor.

Note the three different solidus temperatures. There is a high-temperature melting interval for the solid, volatile-absent rock components. The volatile-bearing vapor-absent rock begins to melt at a lower temperature where the minerals react to release volatile components directly into a liquid phase producing the vapor-absent assemblage crystals + liquid, which occupies a large area on the diagram. The solidus for the rock in the presence of vapor is the horizontal line in Fig. 18.2, the lowest melting temperature.

The fields for liquid and vapor remain distinct from each other, and the two phases can coexist with each other. Liquids can exist at temperatures only above the solidus (which may be one of three temperatures depending upon the amount and compositions of volatile components), whereas a concentrated vapor can exist through a wide range of temperature below the solidus, as well. A liquid may crystallize and exsolve a separate vapor phase. Vapor passing through a rock may react with it, changing the compositions of minerals by reaction, or forming new minerals, possibly including hydrous or carbonated minerals. The vapor may also cross a solidus boundary, cause partial melting, and dissolve in the newly formed liquid. The various reactions involving the two fluid phases have different geochemical consequences, which is why it is important to distinguish liquid and vapor from 'fluid' whenever the distinction can be made.

Much effort has been devoted to determination of the compositions of liquids coexisting with mantle rocks at various pressures and with various volatile components, but much less is known about the solubilities of the solid components in vapor phases of various compositions under mantle conditions. Wyllie & Sekine (1982) reviewed briefly data applicable to conditions above subducted oceanic crust, and Schneider & Eggler (1984, 1986) reviewed existing data and presented new data on the compositions of solutes coexisting with peridotite at high pressures. The solubility of components is sensitive to the mineralogy of the host rocks as well as to depth and temperature. Metasomatic vapors rising from greater to lower depths commonly follow paths of decreasing temperature, but in subduction zones they may rise from lower temperature subducted oceanic slab into higher temperature mantle wedge. Metasomatic vapors either scavenge peridotite for additional components, or change its composition by exchange or precipitation of components. Schneider & Eggler (1986) demonstrated that addition of even small amounts of CO_2 to a hydrous solution causes a significant decrease in solubilities of peridotite components. We need much more experimental data to supplement geochemical algebra.

Oxygen fugacity of mantle vapors

The oxygen fugacity in the mantle and its variation with depth and tectonic environment remain controversial. Woermann & Rosenhauer (1985), Green *et al.* (1987), Taylor & Green (1987) and Wyllie (1987a) reviewed data and recent opinions about the redox state of the upper mantle. The molecular constituents of vapors in the system C–H–O and their reactions with peridotite at low oxygen fugacities have been discussed by Deines (1980), Ryabchikov *et al.* (1981), Eggler & Baker (1982), Woermann & Rosenhauer (1985), Holloway & Jakobsson (1986), Taylor & Green (1987) and Green *et al.* (1987). Carbon may be dissolved in silicate minerals in the form of carbon atoms (Freund *et al.*, 1980). Tingle & Green (1987) have reported measurable solubility of carbon in olivine at high pressures, with diffusivities sufficient to cause exsolution in

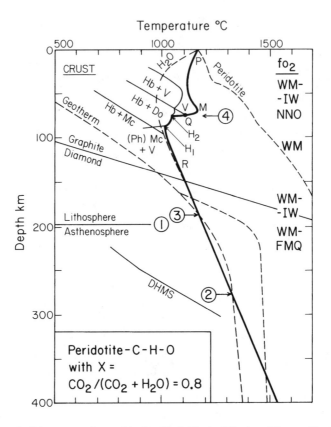

Fig. 18.3. Solidus curves for peridotite–H_2O (dashed line) and for peridotite–CO_2–H_2O with defined ratio of CO_2/H_2O (PMQR) extrapolated to high pressures, and compared with the cratonic geotherm, with and without inflection (Boyd & Gurney, 1986). The positions of selected phase boundaries relevant to mantle processes and the origin of kimberlites and nephelinites are given, and discussed in the text. Level (1) represents the lithosphere–asthenosphere boundary layer, points (2) and (3) are depth levels limiting the interval within which magma can be generated (in the presence of fluids) for the particular geotherm, and level (4) corresponds to the change in slope of the solidus curve near point Q. The oxygen fugacities at various levels are indicated by the buffers listed (Haggerty, 1987). The peridotite solidus is from Takahashi (1986). For DHMS (dense hydrated magnesian silicates) see Ringwood (1975, p. 295). Abbreviations: Hb = amphibole, Ph = phlogopite, Do = dolomite, Mc = magnesite, V = vapor (Wyllie, 1987a).

bubble precipitates upon decrease of pressure. Views that carbonate and CO_2 can exist deep in the lithosphere (e.g., Eggler & Baker, 1982, Schneider & Eggler, 1986) compete with views that the redox state of the deeper mantle may be such that the components C–H–O exist as carbon and H_2O with CH_4, rather than as CO_2 and H_2O (e.g., Green *et al.*, 1987). Haggerty (1986) considered that fluids as deep as the asthenosphere would be relatively oxidized, between FMQ and WM buffers, and that the lithosphere is more reduced, between the WM and IW buffers. He also concluded (Haggerty, 1987) that a

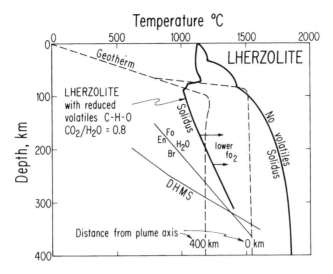

Fig. 18.4.   Solidus curves for lherzolite with and without volatile components, and
deep dehydration reactions, compared with geotherms located at center
of plume axis and 400 km away. See Figs. 18.3 and 18.5.

layer of lithosphere between 60 and 100 km has been metasomatized and oxidized up to
the NNO buffer, where carbonate would be stable (Fig. 18.3).

The mantle

The mantle is heterogeneous and composed of various kinds of peridotite and eclogite,
with distributions much debated. These complications are neglected in the following
discussions, and the application of phase boundaries for lherzolite to the processes
described is a simplification. The solidus curve for volatile-free mantle lherzolite is
shown in Figs. 18.3 and 18.4. A flux of volatiles either causes metasomatism or initiates
melting where the vapor crosses the solidus for lherzolite–vapor. In order to locate
regions of metasomatism by vapors and sites of melting it is necessary to know the
thermal structure and the solidus curve for the material in any region.

Phase relationships for lherzolite–C–H–O

The phase relationships for the system lherzolite–CO_2–H_2O, involving the minerals
amphibole, phlogopite, dolomite and magnesite, provide the framework for upper
mantle petrology, and for evaluation of the phase relations with reduced oxygen
fugacity. There remains uncertainty about the position of the solidus, which also varies
as a function of composition.

Two experimental studies on amphibole–dolomite–peridotite indicate that the
amphibole stability volume overlaps the solidus with dolomite. There is a difference of
several kilobars between the results of Brey *et al.* (1983) and those from the detailed

investigation of Olafsson & Eggler (1983). I have interpreted these two sets of experimental results in terms of the topology of overlapping phase volumes (Wyllie, 1978, 1987a, b). Despite differences in detail, the existence of the solidus ledge near Q, where the solidus changes slope and becomes subhorizontal (Eggler, 1987), appears to be established (level 4 in Fig. 18.3). An estimated (but constrained) and extrapolated solidus curve for amphibole–dolomite–peridotite with a high ratio of CO_2–H_2O is compared with those for peridotite dry and H_2O in Figs. 18.3 and 18.4. This is modified from a previous version (Wyllie, 1980) incorporating new data (Olafsson & Eggler, 1983, Wyllie, 1987a).

Important levels in the mantle are those where volatile components would be released by activation of a dissociation reaction. One such reaction is represented in Figs. 18.3 and 18.4 by the line DHMS giving the boundary for the reaction of olivine to form dense hydrous magnesian silicates in the presence of H_2O. The estimated position of the reaction for the conversion of forsterite to brucite and enstatite in MgO–SiO_2–H_2O (Fig. 18.4) is close to the DHMS curve at 300 km (Ellis & Wyllie, 1979).

The redox state of the deeper mantle appears to be such that the components C–H–O exist as carbon and H_2O with CH_4, rather than as CO_2 and H_2O (Deines, 1980), Ryabchikov et al., 1981). Fig. 18.3 summarizes Haggerty's (1986, 1987) picture of the possible variations of oxygen fugacity in terms of standard buffers down to 250 km.

With low oxygen fugacites at high pressures, C–H–O exists as H_2O with CH_4 or graphite/diamond. Under these conditions, it appears that the peridotite–C–H–O solidus will be raised in temperature compared with the oxidized solidus as indicated by the arrows in Fig. 18.4. According to some investigators, the solidus may remain close to that for peridotite–H_2O, with carbonate ions being generated in the melt when CH_4 or graphite/diamond dissolves (Eggler & Baker, 1982, Ryabchikov et al., 1981, Woermann & Rosenhauer, 1985). According to D.H. Green (personal communication, 1987), however, recent experiments with pyrolite–C–H–O at various H_2O activities with reduced oxygen fugacities do move the solidus to higher temperatures (Fig. 18.4). In the following discussion the extrapolated solidus for peridotite–H_2O, which is believed to remain very close to that for peridotite–CO_2–H_2O at pressures greater than about 35 kbar, is adopted as the solidus at depth for peridotite–C–H–O for a system with volatile components in the ratio $CO_2/(CO_2 + H_2O) = 0.8$, if oxidized. With variation in this ratio the solidus temperature changes at pressures shallower than point Q at about 75 km, but remains unchanged at higher pressures. Under reducing conditions, however, all solidus boundaries in the petrological models can be moved to higher temperature locations once the relationships among oxygen fugacity, water activity, solidus curves and actual redox conditions in the mantle have been established.

The lithosphere and thermal structure

The thickness of the lithosphere is an important variable because of the change in rheology (from ductile to brittle) occurring through the asthenosphere–lithosphere

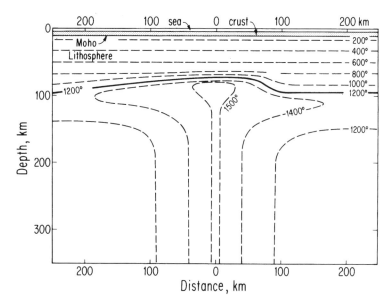

Fig. 18.5.   Schematic isotherms for mantle plume based on plumes of Courtney &
White (1986), the requirement for melting at M in Fig. 18.4, and with
asymmetry caused by motion of lithosphere plate above the plume. The
change in rheology associated with the asthenosphere–lithosphere
boundary layer is arbitrarily represented by the heavy line for the 1200 °C
isotherm.

boundary layer. This layer is depicted as a line near 200 km depth beneath a continental
craton in Figs. 18.3 and 18.6, and near 90 km beneath an ocean in Fig. 18.5. The
asthenosphere–lithosphere boundary layer is commonly assumed to be near the 1200
°C isotherm, and this temperature is adopted in the following diagrams.

From the geothermometry and geobarometry of mantle nodules we have a geotherm
for cratons consistent with that calculated from heat loss (Boyd & Gurney, 1986). The
geotherm for many kimberlites is inflected to higher temperatures at a depth of about
175 km, somewhat deeper than the graphite to diamond transition. Both normal and
inflected geotherms are shown in Fig. 18.3. As reported by Boyd & Gurney (1986) the
inflection has been interpreted in terms of uprise of mantle diapirs associated with the
generation of kimberlites or in terms of local magma chambers. Nickel & Green (1985)
refined empirical garnet–orthopyroxene geobarometry, and presented a distinctive
pattern for South African xenoliths where the high-temperature xenoliths give near-
isobaric estimates, corresponding to a depth of 150–160 km, at 900–1400 °C.

If a thermal plume rises below the lithosphere, the asthenosphere–lithosphere
boundary (represented by the 1200 °C isotherm) rises to shallower levels, causing
thinning of the lithosphere, as indicated by the 1200 °C point on the inflected geotherm
of Fig. 18.3. The effect of a plume on the isotherms is depicted in Fig. 18.5, which
represents an oceanic plate moving from right to left. The asymmetry of the plume and
of lithosphere thinning is caused by movement of the plate.

One interpretation of the geochemistry of Hawaiian lavas requires that they be derived from a source rock containing garnet (Feigenson, 1986), which requires that melting occurred at the cusp M in Fig. 18.4, or deeper. Therefore, the geotherm located in the center of the plume in Fig. 18.5 must reach M in Fig. 18.4, exceeding 1500 °C. The lithosphere is heated above the plume, and associated with the rise in isotherms is thinning of the lithosphere.

The isotherms in Fig. 18.5 provide geotherms for any given distance from the plume center. Fig. 18.4 compares two geotherms with the solidus curves for lherzolite: the geotherm for the center of the plume, and the geotherm for a position 400 km from the plume axis (where the asymmetry is greatly reduced).

Critical depth levels

In addition to the depths of dissociation reactions, there are four levels in the upper mantle, identified in Fig. 18.3, where critical changes occur in the physical processes that control the chemistry and mode of migration of the volatile-rich magmas emplaced in cratonic environments. The first critical level, (1), is the depth of the asthenosphere–lithosphere boundary layer, through which the mantle flow regime changes from convective (ductile) to static (brittle). The two depths where the solidus is intersected by the local geotherm, (2) and (3), limit the depth interval within which magmas can be generated. The solidus ledge at the fourth level, (4), is the narrow depth interval near Q where the solidus changes slope and becomes sub-horizontal, with low dP/dT. Levels (2), (3) and (4) are different for lherzolites and harzburgites (Wyllie et al., 1983); compared with lherzolite the solidus for harzburgite is higher, and therefore the levels (2) and (3) are deeper, and level (4) is also deeper. The depths of levels (1), (2) and (3) vary from place to place and from time to time, as a function of the geotherm and local history. This can be seen by locating the corresponding levels and intersection points for the suboceanic mantle in Fig. 18.4.

Fig. 18.3 compares the subcratonic geotherm with and without inflection with the solidus peridotite–C–H–O. There is no melting if no volatile components are present. If the geotherm intersects the solidus, and if vapors are introduced by some process between the depth levels (2) and (3), then partial melting occurs. The depth interval within which melting can occur is strongly dependent on the geotherm. Furthermore, if the solidus intersects the geotherm, then only liquid can exist between levels (2) and (3), because all volatile components would dissolve in the liquid. This limits the depth intervals within which vapor phases can cause metasomatism by vapors as discussed and illustrated in detail by Wyllie (1980).

The estimated solidus and the uninflected geotherm only just intersect. If conditions were such that they did not intersect, then vapors could migrate through the mantle section without generating melts (Bailey, 1985). This condition might be achieved by elevation of the solidus in a reduced mantle (see Fig. 18.4). There is good geological evidence that some kimberlite magmas rise from near the base of the lithosphere shown in Fig. 18.6, indicating that in this region the geotherm does exceed the solidus.

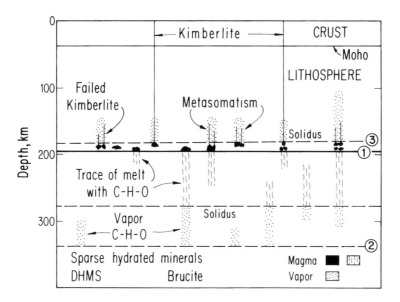

Fig. 18.6. Craton section with phase boundaries and normal geotherm transferred from Fig. 18.3, showing depth limits for the migration of deep volatile components, the generation of melts in the asthenosphere and their entrapment in the lower lithosphere. Escape of volatile components as magmas approach the solidus leads to metasomatism, crack propagation, and intrusion of kimberlites, only some of which reach the surface. Various stages are illustrated, representing events distributed through a long period of time.

Consider the craton in Fig. 18.3 with normal, uninflected geotherm. No magma is generated unless volatile components are present or are introduced into the depth level between 270 km and 185 km (levels 2 and 3). The generation of new melt is dependent on the uprise of volatile components from depths greater than (2). Note that the reactions of olivine with water to form **DHMS** or brucite probably intersect with the geotherm somewhere between 300–400 km, (Figs. 18.3 and 18.4) suggesting that perhaps no H_2O, but only reduced gases, CO, CH_4, and H_2 can exist at deeper levels. The depths (2) and (3) are variable, depending on conditions.

Kimberlites

Dawson (1980) reported that experimental data supported an origin for kimberlites by partial melting of carbonated lithosphere, but I concluded at the same time (Wyllie, 1980) that this process was less likely than one where migration of reduced volatile components C–H–O from the deep mantle generated kimberlite melts in the asthenosphere, with the melts being subsequently erupted from the deep lithosphere. I suggested, alternatively, that the reduced vapors reached a shallower mantle horizon

487

with higher oxygen fugacity where they were partly oxidized into a condition where the solidus for the system migrated to a lower temperature, causing partial melting. Green *et al.* (1987), using data of Taylor & Green (1987), have since elaborated the latter process in elegant detail with a major role assigned to 'redox melting' where reduced fluids enter the lithosphere. Smith *et al.* (1985) now have evidence that isotopically defined Group I (basaltic) and Group II (micaceous) kimberlites of South Africa (Smith, 1983) have distinctive major and trace element signatures. They concluded that Group I kimberlites are derived from asthenospheric sources (similar to ocean island basalts), and Group II kimberlites originate from sources within ancient lithosphere characterized by time-average incompatible element enrichment.

Static petrological structures for the mantle can be constructed if the rock compositions, the phase relationships and the thermal structure are known. The phase boundaries in Figs. 18.3 and 18.4 are mapped in terms of pressure and temperature. Each point on the standard geotherm in Fig. 18.3 is specified in terms of pressure and temperature. Therefore, the phase boundaries from Figs. 18.3 and 18.4 can be plotted as a function of depth as in Fig. 18.6. Similarly, the phase boundaries can be mapped as lines of variable depth as in Fig. 18.9 for the more complex thermal structure of Fig. 18.5.

Fig. 18.6 represents a stable craton with uninflected geotherm. Phase boundaries at the specified depths are drawn at intersections of the uninflected geotherm with the phase boundaries in Fig. 18.3. For simplicity, the two dehydration boundaries of Figs. 18.3 and 18.4 were merged into a single dehydration boundary. Intermittent local disturbances could release vapors from the deep dissociation front (Woermann & Rosenhauer, 1985). The volatile components would rise through mantle to the solidus at level (2), where they would dissolve in melt. The trace of interstitial melt so developed would rise through the asthenosphere to the asthenosphere–lithosphere boundary at level (1). The rate of percolation of magmas entering the depleted lithosphere would be reduced. Small magma chambers could form and remain sealed within the more rigid lithosphere, maintained at temperatures above the solidus for lherzolite–C–H–O. The magmas have little tendency to crystallize or to evolve vapors until they approach level (3), 10–15 km above the asthenosphere–lithosphere boundary. Contact of the lherzolite-derived magma with harzburgite, however, should cause reaction and the precipitation of minerals through magma contamination. This slow process may cause the more or less isothermal growth of large minerals resembling the discrete nodules in some kimberlites.

Those magmas managing to insinuate their way near to level (3), the solidus, will evolve H_2O-rich vapors. CO_2-rich vapors cannot exist in this part of the mantle. The vapors may promote crack propagation, permitting rapid uprise of the kimberlite magma (Perchuk, 1973, p. 296). Many intrusions from this level will solidify through thermal death before rising far (Spera, 1984), but others will enter the crust as kimberlite intrusions (Artyushkov & Sobolev, 1984).

The depleted, refractory base of the lithosphere, 150–200 km deep, has probably

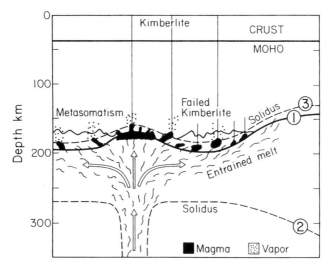

Fig. 18.7. Upper mantle section with thermal plume and with phase boundaries transferred from Fig. 18.3. The mantle plume (with volatile components) melts, with concentration of magma above the plume, and entrainment of melts in flowing lherzolite of the asthenosphere. Magma in the boundary layer of the lithosphere approaches the solidus, level (3), with consequences as illustrated in Fig. 18.6 (Wyllie, 1987d).

been invaded intermittently by small bodies and dikes of kimberlite, through billions of years. Most of these aborted and gave off vapors. The combination of failed kimberlite intrusions and their vapors have contributed to the generation of a heterogeneous layer in this depth interval, characterized by many local volumes metasomatized in diverse ways. The aqueous vapors may percolate upward through many kilometers to shallower levels, perhaps generating the sequence of metasomatized garnet peridotites described in detail by Erlank *et al.* (1987). Schneider & Eggler (1986) suggested on the basis of their experimental data that aqueous vapors and solutions rising through the lithosphere would leach the mantle with minor precipitation until they reached a level of about 70 km where a 'region of precipitation' is associated with the formation of amphibole and consequent change in vapor composition toward CO_2.

  If a thermal plume rises below the cratonic lithosphere, the petrological structure of Fig. 18.6 becomes modified in the way illustrated in Fig. 18.7. The depth of intersection of the geotherm with a dissociation reaction (e.g., DHMS, Fig. 18.3) is displaced to greater depths, releasing vapor (Woermann & Rosenhauer, 1985). Interstitial melt is generated at level (2) where the lherzolite is transported across the solidus curve. With continued convection, the geotherm rises to a higher, inflected position as depicted in Fig. 18.3, and the depth (2) at which melting begins is increased considerably, as shown in Figs. 18.7 and 18.8. The intersection points of geotherm with solidus and dissociation boundary could overlap, causing the hydrated solid to melt directly without the opportunity for a vapor phase to exist through a discrete depth interval (see Fig. 18.9).

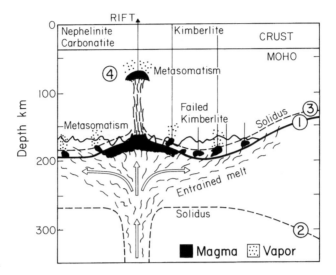

Fig. 18.8. An upper mantle section corresponding to conditions in Fig. 18.7 followed by lithospheric thinning beneath a rift valley. Thinning of the lithosphere above the plume permits upward migration of the magma to level (4), where magma chambers develop, and vapor is evolved with metasomatic effects. This level is the source of parent magmas for nephelinitic volcanism, and the associated carbonatites (Wyllie, 1987a).

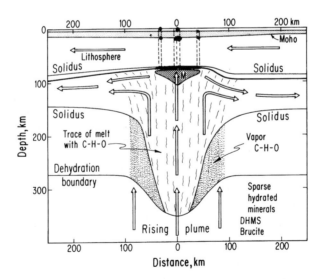

Fig. 18.9. The isotherms in Fig. 18.5 and the phase boundaries in Fig. 18.4 define the positions of solidus curves and dehydration boundaries as shown in this mantle cross-section. Material following flow lines crosses the phase boundaries, causing the distribution of vapor and melt in rising plume as shown. Picritic magma enters the lithosphere from the region of major plume melting at **M**, with magma chambers forming at various levels. Fig. 18.10 shows more details in the region of the asthenosphere–lithosphere boundary layer (heavy line).

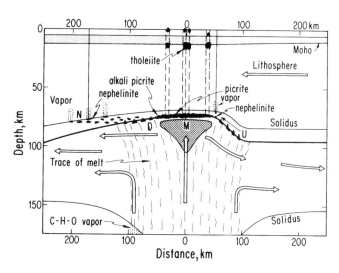

Fig. 18.10. Petrological structure of lithosphere and upper plume corresponding to
Fig. 18.9, with expanded vertical scale. Picritic magmas enter
lithosphere from melting region M. Volatile-charged magmas enter the
lithosphere on either side of M, with subsequent percolation retarded.
As they approach the solidus, vapor is evolved (e.g., at N) enhancing
the prospect of explosive eruption of nephelinite or basanite directly
through the lithosphere. Paths followed by these magmas to the solidus
differ according to their position on the upstream (U) or downstream
(D) side of the plume.

As the plume diverges below the lithosphere, the melt becomes concentrated in layers
or chambers in the boundary layer above the plume, as shown in Fig. 18.7. Ribe &
Smooke (1987) concluded from their analysis of a dynamic model for melt extraction
from a mantle plume that layers of segregated melt are likely to form at the base of the
lithosphere above a plume. Lateral divergence of the asthenosphere transports some of
the entrained plume melt, and this penetrates the lithosphere, forming small dikes or
magma chambers. Kimberlites may rise either from the early magma accumulating
above the plume or from the lateral magma chambers in the lithosphere base. If the
plate is moving across the plume, modifications are required as suggested by Figs. 18.9
and 18.10.

Nephelinites and carbonatites of rifted cratons

There is evidence that beneath major rifts the lithosphere is thinned. The continued
heat flux from a rising plume, and the concentration of hotter magma at the
asthenosphere–lithosphere boundary (see Fig. 18.7), will promote further thinning of
the lithosphere. According to Gliko et al. (1985), it takes only several million years for
lithosphere thickness to be halved when additional heat flow of appropriate magnitude

is applied to the base of the lithosphere. Fig. 18.8 depicts the petrological structure in a craton with rifting crust above a thinning lithosphere.

The magma near the boundary layer at levels (1)–(3) in Fig. 18.7 will rise with the layer as the lithosphere thins, either percolating through the newly deformable rock matrix, or rising as a series of diapirs with the amount of liquid increasing in amount as the diapirs extend further above the solidus for peridotite–C–H–O. This magma approaches the solidus ledge at level (4) in the depth range 90–70 km (Fig. 18.3). Magma chambers may be formed as the magma solidifies (Fig. 18.8), and vapors will be evolved causing metasomatism in the overlying mantle and causing intermittent crack propagation, which permits the escape of magmas through the lithosphere. The metasomatic vapors will be aqueous if released at depth greater than Q, and CO_2-rich if released at depths shallower than Q (Fig. 18.3; Wyllie, 1980, Fig. 8), which is consistent with Haggerty's (1987) account of metasomes at these depths.

A variety of alkalic magmas may be generated at level (4), depending sensitively upon conditions. There are large changes in the compositions of melts and vapors in this region for small changes in pressure and temperature (Wyllie, 1978, Wendlandt & Eggler, 1980, Wendlandt, 1984). Magmas rising from this level may include the parents of olivine nephelinites, melilite-bearing lavas, and other igneous associations differentiating at shallow depths to carbonatites. The igneous sequence associated with the thinning of the lithosphere between the conditions in Figs. 18.7 and 18.8 was discussed by Wendlandt & Morgan (1982).

Hawaii, example of ocean island nephelinites (and basanites)

The thermal structure for a plume rising beneath Hawaii is represented in Fig. 18.5, and the relationship of geotherms to the phase boundaries for lherzolite–C–H–O is shown in Fig. 18.4. Mapping the phase boundaries from Fig. 18.4 in terms of the pressure–temperature points of Fig. 18.5 yields the petrological structure in Fig. 18.9, with an expanded view of the asthenosphere–lithosphere relations in Fig. 18.10. These phase boundaries are static, fixed in position as long as the isotherms do not change. Material migrates through the framework, following flow lines represented by the arrows in Figs. 18.9 and 18.10. The positions of the phase boundaries in this structure become modified by energy changes associated with melting and the extraction of melts, but those plotted provide a guide as to what happens where within the cross-section.

If there are volatile components in the deep mantle, H_2O is released from the hydrates as the plume material crosses the dehydration boundary (Woermann & Rosenhauer, 1985), joining the carbon species (Deines, 1980). The mantle plume with volatiles rises until it reaches the solidus boundary, where the volatiles become dissolved in a trace of melt. In the central part of the plume, for the thermal structure in Fig. 18.5, the hydrated peridotite melts directly without an interval for a separate vapor phase. Significant melting of the plume occurs in the shaded area M, which is situated above the volatile-free solidus (see M in Fig. 18.4). In the outer envelope of the plume, the entrained volatile-charged melt bypasses the region M, entering the lithosphere near the 1200 °C isotherm. At this level, the change in rheology would significantly retard the upward movement of the small percentage of interstitial melt.

Fig. 18.10 shows in more detail what may happen in the lower lithosphere. The compositions of the magmas reaching the lithosphere are labelled according to the experimentally determined compositions of melts under the specified conditions (Jaques & Green, 1980). The picritic magmas enter the lithosphere, and remain there in magma chambers (Ribe & Smooke, 1987) until suitable tectonic conditions permit their uprise through cracks to shallower magma chambers, where they differentiate to the tholeiites and alkali basalts erupted at the surface. The concentration of magmas in chambers at the base of the lithosphere may cause partial melting of the overlying lithosphere (Chen & Frey 1985).

The fate of the volatile-charged magma differs according to its position within the plume. The melt in the central region is incorporated into the much larger volume of picritic melt. The melt on either side of the region M reaches the lithosphere, and then follows different routes according to its location on the upsteam (U) or downstream (D) side of M. Depending upon the relative rates of mantle flow and melt percolation in the region U (Fig. 18.10), magma on the upstream side could percolate slowly through the lithosphere with decreasing temperature to the solidus, or percolate upwards and downstream from U along the asthenosphere–lithosphere boundary layer, and become incorporated into the hotter alkali picrite magma above the upstream margin of the melting region M. Melt entering the lithosphere on the downstream side, D, would be carried away from the plume along lithosphere flow lines as shown in Fig. 18.10, remaining above its solidus as temperature decreased along DN. The distance downstream that it would reach the solidus depends on the relative rates of lithosphere movement, and of migration of small bodies of magma through the lithosphere. When magma approaches the solidus N, vapor is released, and this may enhance crack propagation and permit explosive eruption of nephelinites or basanites directly from the asthenosphere–lithosphere boundary region to the surface.

Magmatic sequence and sources

If the Hawaiian Islands were generated over a thermal plume, the sequence of magmatic events should correspond to those illustrated in Fig. 18.10. A specific surface area receives magmas in succession following the processes depicted in Fig. 18.10 from right to left. The first event could be explosive eruption of nephelinites (or basanites), followed by, or almost contemporaneously with, the eruption of alkali basalts. Then comes major eruption of tholeiites, with no time break. The width of M, and the time interval for the eruption of tholeiites, is extensive compared with the initial alkalic magma phase. The third phase is the eruption of alkali basalts generated at the downstream edge of the melting region M; if some of the alkali picrite becomes trapped in the lithosphere and is transported in the direction downstream before escaping to shallower levels, this stage could be somewhat extended in time, and intermittent. The final stage occurs much later, when the small pockets of nephelinite magma are erupted from N. Thereafter, magmatic activity ceases until the lithosphere crosses some other thermal disturbance in the mantle capable of raising the geotherm to a level where it crosses the solidus. This sequence of events appears to satisfy the broad geological and petrological history of the Hawaiian Islands (Decker *et al.*, 1987). There have been many proposals to explain the geochemistry of the different lavas (Wright & Helz, 1987).

There are three distinct mantle reservoirs involved in the melting process (Wasserburg & de Paolo, 1979). The central plume material is assumed to rise from an undepleted lower mantle layer, the outer envelope of the plume may be derived from the depleted upper mantle layer, and the lithosphere is residual after removal of MORB constituents. The framework outlined in Figs. 18.9 and 18.10 indicates that different deep source materials are involved in the formation of the tholeiites and of the nephelinites, with a prospect that more than one deep source may provide material for the alkalic magmas near the margins of the melting zone. The lithosphere reservoir may contaminate nephelinites and alkali basalts on the upstream side of the plume, U, but this is less likely on the downstream side, where the thickening lithosphere is composed of plume material. The tholeiites from the deep mantle reservoir may be contaminated with the lithosphere to the extent that the hot picrite magmas can cause local melting in the lithosphere (Chen & Frey, 1985).

The migration and distribution of incompatible trace elements is controlled largely by the behavior of the volatile components, and their solution in and release from melt. The high concentrations of incompatible elements in the traces of melt in the rising plume will be swamped by the major melting in the center of region M, becoming more evident toward the margins of the region, where alkalic picrites are generated by a smaller degree of partial melting. The melts on both upsteam and downstream sides of the plume will carry the high concentrations of incompatible elements, along with the geochemical signatures of depleted sources. The release of vapors into the lower lithosphere depicted in Fig. 18.10 will transfer concentrations of incompatible elements into metasomatized lithosphere, which may contribute to local geochemical anomalies where the deep lithosphere is subsequently involved in melting.

Volatile flux at convergent plate boundaries

There are many variables in petrology and geophysics at convergent plate boundaries, but the emergent plutonic and volcanic rocks have common, distinctive chemical characteristics and organized structural geology (Gill, 1981). These characteristics indicate crustal differentiation modified by dissolved H_2O (Grove & Kinzler, 1986), but the evidence for multiple source materials is clear. The sources include subducted oceanic crust, overlying mantle, and continental crust if present.

Vapors released from deep dissociation fronts in subducted oceanic crusts migrate through large volumes of upper mantle. The geometry of the arrangement of metasomatic volumes, and volumes where the vapors dissolve in magmas, is strongly dependent on the geometry of subduction, the thermal history of the plate being subducted, and the thermal structure of the subduction zone.

There have been so many different calculated thermal structures published for subduction zones that we may conclude either that the thermal structures are not yet known in detail, or that there is considerable variation. I have explored the locations of metasomatism and magma generation in subduction for the limits represented by four thermal structures corresponding to combinations of relatively warm and cool mantle, or subducted lithosphere (Wyllie, 1983). The experimental phase boundaries for dehydration and melting of the rocks in the different sources have been reviewed by Gill (1981), Wyllie (1979, 1982), and Wyllie & Sekine (1982). By mapping the phase boundaries across the isotherms for each thermal structure (compare Figs. 18.4, 18.5 and 18.9) we obtain the four arrangements of dehydration and solidus boundaries shown in Fig. 18.11. Partial melting occurs if H_2O passes into the regions on the high-temperature sides of the dashed solidus boundaries for rock–H_2O. There is a fifth arrangement advocated by Marsh (1979) and Maaløe & Petersen (1981) with extreme induced convection cells in the mantle wedge raising the temperature of the subducted oceanic crust at 100 km depth to 1250–1450 °C, sufficient to generate primary basaltic andesite directly from the subducted crust.

Aqueous solutions rising from the dehydration front DG may migrate back toward the trench along the dislocation above the slab, but direct upward migration of solutions into the overlying mantle wedge is commonly assumed. These solutions may precipitate metasomatic amphibole in the large shaded areas of mantle and continental crust, and generate magma in the continental crust (C). For a cool mantle – warm crust (Fig. 18.11(A)), most solutions enter the mantle along DE, but some generate magma from subducted crust at M. For a cool mantle – cool crust (Fig. 18.11(B)), the metasomatic solutions do not enter the melting regions of either subducted crust or mantle peridotite. For a warm mantle – cool crust (Fig. 18.11(C)), induced convection brings the region for potential melting of the mantle above the dehydration fronts, and metasomatic fluids cause partial melting of peridotite. Note how dehydration of metasomatic mantle amphibole could localize enhanced melting at N, perhaps controlling the position of the volcanic front.

Fig. 18.11(D) represents an intermediate thermal structure with relatively warm mantle and relatively warm subducted crust. Magmas are generated simultaneously in all three source materials at M, N, C, and H' (Anderson et al., 1980). Nicholls & Ringwood (1973) proposed that hydrous siliceous magma from M would react with overlying mantle, producing hybrid olivine pyroxenite. Wyllie & Sekine (1982) concluded from their experiments (Sekine & Wyllie, 1982, 1983) that a series of discrete or overlapping bodies of hybrid phlogopite–pyroxenite would be transported downwards with the slab to R, where vapor-absent melting defines another possible site of

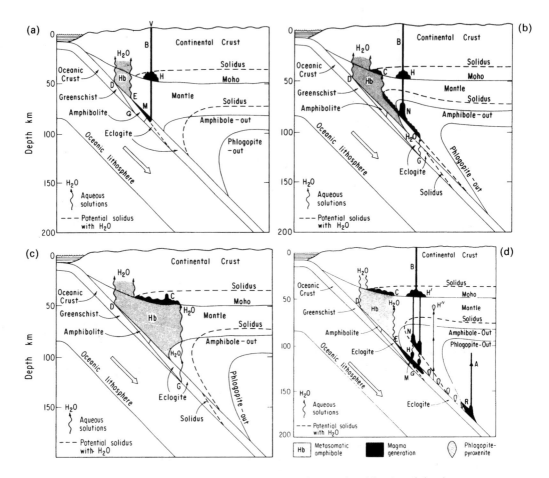

Fig. 18.11.   Locations of dehydration, metasomatism, and melting in subduction
zones from results of experimental petrology, according to four thermal
structures obtained by combinations of warm or cool (by endothermic
dehydration) subducted ocean crust, with cool or warm (by induced
convection) mantle. (A) Cool mantle–warm crust. (B) Cool mantle–cool
crust. (C) Warm mantle–cool crust. (D) Warm mantle–warm crust
(Wyllie & Sekine, 1982, Wyllie, 1983).

magma generation. Solutions released during hybridization at H cause metasomatism,
followed by partial melting in the shallower mantle at N. Another possibility is that the
hydrous magma from M could enter the region N for hydrous melting of peridotite.

Note in Fig. 18.11(D) the complicated sequence of events that could be experienced
by water stored initially in serpentine in the oceanic crust. Fig. 18.2 shows the types of
phase fields traversed by the water. It starts in the 'vapor-absent rock', then alternates
between fields for 'rock + V' and 'xls + L' (with very short intervals of 'crystals + L + V'
between). Following dehydration at about 100 km depth (Fig. 18.11(D)) the aqueous
vapor would rise through eclogite, cross the solidus and dissolve in interstitial melt near
M. The magma would solidify by hybridization, releasing H_2O at N, forming vapor

that would rise through peridotite until it crossed the mantle solidus near N, generating a second batch of magma. This would rise into the base of the continental crust at H', perhaps escaping in eruption B but more likely becoming involved in additional reactions between H' and the surface.

Metasomatism and magmatism

Magmatism involves the formation and migration of melts. Metasomatism, as discussed in a previous section, was normally considered to result from the interaction of solution or vapors with crustal rocks, but the term has now been extended to refer to chemical changes of uncertain cause in unseen source rocks deep in the mantle. Examples of the source rocks for magmas are the mantle nodules brought to the surface in kimberlite and alkali basalt eruptions, but for the most part the appeal to metasomatism in mantle rocks is an appeal to rocks unseen. Wilshire (1984) described in mantle nodules a hierarchy of successive metasomatic events involving melts and fluids or vapors.

It is easy to appeal to metasomatism by fluids. It is important to attempt to distinguish between metasomatism by vapors and by liquids. We can expect large differences in the nature of the chemical changes caused by the passage of a vapor phase compared with the intrusion of a liquid, or magmatic phase. For the vapor phase or solution, the fluid phase passes through the rock interacting with it. Material may be precipitated or dissolved. Passage of a large volume of vapor is therefore required in order to make significant chemical changes in the host rocks. The passage of a liquid, melt or magnetic phase, however, adds new material to the host rock. During the dynamic act of passage the magma may react with the wall rocks, forming hybrid products, and some of the melt may infiltrate the host rock, causing significant chemical changes. The final stage of such metasomatism is solidification of the magma with the resultant introduction of new material into the rock. The distinction between these two processes of passage and addition of magma on the one hand, and the passage of vapors or solution on the other, is a quantitative one involving mass of material added to the rock. Schneider & Eggler (1984, 1986), following their experimental studies on the solubility of peridotite components in H_2O and $H_2O–CO_2$ fluids at high pressures, calculated that high fluid/rock ratios would be required to effect significant changes in the chemistry of rocks with such fluids passing through them. They concluded that the fluids were not effective metasomatic agents and that most mantle metasomatism was accomplished by magmas. It is also established, however, that metasomatism of individual mantle samples has been accomplished both by the intrusions of magmas, and by the infiltration of solutions emanating from the magmas during solidification of the melt (Wilshire, 1984). In time we should be able to erect criteria to make the distinction even when the source rock is not in hand.

Two depth intervals of lithosphere metasomatism are indicated in Figs. 18.6–18.8. In regions of rifted craton, lithospheric thinning was accompanied by the release of vapors at depth near 75 km at level (4). Those released at somewhat greater depths were enriched in H_2O, and those at shallow depths were enriched in CO_2. Metasomatism of this kind is localized in rifted regions. Magmas passing from the deeper level (3) to level (4) may also have left metasomatic traces beneath former rift zones. A deeper level of metasomatism near the lithosphere base has developed through leakage of magma or vapors from the asthenosphere. The vapors released at level (3) may percolate upwards forming metasomatic rocks reaching the shallower level (4) (Erlank *et al.*, 1987). The level (4) coincides with the level where the results of Schneider & Eggler (1984, 1986) suggest that hydrous solutions rising from greater depth precipitate a large fraction of their solute as they become richer in CO_2.

Kimberlites, nephelinites, and other rift valley magmas, are melts with histories involving more than one mantle source material. They start as melts formed in fertile asthenosphere lherzolite, enriched by migration of volatile components from deeper levels. They may spend time residing in contact with the depleted keel of the lithosphere, comprising harzburgite, lherzolite, and eclogite. Many kimberlites are emplaced from this level (Figs. 18.6 and 18.7). Nephelinites and related magmas are developed by progressive evolution of an original kimberlite-like magma as it rises through the lithosphere, increasing in melt fraction until it reaches level (4), near 75 km. The magmas may here be enriched by solution of metasomatites formed during a previous occurrence of rifting and magmatic processes (Haggerty, 1987). Eruption occurs through cracks (Fig. 18.8).

There are parallels between the sequence of deep metasomatism and magmatism occurring beneath cratons and beneath oceanic lithosphere. Volatile-charged melts from the asthenosphere may enrich the lower oceanic lithosphere, with events similar to those depicted in Fig. 18.6 for the cratons, but there are significant differences because the oceanic lithosphere is younger (less time to accumulate metasomatic events) and thinner (melts and vapors of different composition are involved). The distribution of isotopes and other trace elements in the magmas associated with a suboceanic thermal plume depends not only on the mantle reservoirs represented by the source rocks, but also on the nature of the physical involvement of the volatile-charge melt with the plume and lithosphere, as indicated in Figs. 18.9 and 18.10.

Not all magmas are affected by the solidus ledge at level (4) in Figs. 18.3 and 18.4. For those magmas that have already escaped from equilibrium with a peridotite host, the solidus ledge has no significance. For those diapirs or melts of higher temperature, rising from greater depths, adiabatic paths would miss the solidus at M in Fig. 18.3, and these might reach the solidus for volatile-free peridotite, as shown at M in Fig. 18.4.

The magmas emplaced and erupted at convergent plate boundaries may carry the geochemical imprint of various source rocks, which may have been transferred by both metasomatism and magmatism, as illustrated by the possibilities outlined in Fig. 18.11. Although the major element chemistry of the calc-alkaline series is probably dominated by relatively shallow differentiation (Gill, 1981), there is still much debate about the source of the magmas and the roles of vapors and melts in the transfer of isotopes and other trace elements from one reservoir to another (McCallum, 1987).

The petrological cross-sections shown in Figs. 18.6–18.11, derived by transferring phase boundaries onto assumed thermal structures, provide physical frameworks for consideration of processes with specific boundaries located for magmatic and metasomatic events. Their positions will change in response to the dynamics of mantle flow and the chemical differentiation caused by separation of melts and vapors from rocks. The fluid dynamics of rock–melt–vapor systems is of paramount importance in deciding what will actually happen to the phases in the phase fields depicted in these petrological cross-sections (Marsh, 1987). The physical frameworks may be adjusted to satisfy the geochemical constraints as the geochemical models are themselves adjusted (Carlson, 1987). The frameworks appear to be robust enough to accommodate modifications, and it is precisely through such modifications arising from the interplay of geochemistry, fluid dynamics, thermodynamics and experimental petrology that understanding of the petrological consequences of volatile fluxes from the mantle may be advanced.

Acknowledgements   This research was supported by the Earth Sciences section of the U.S. National Science Foundation, Grants EAR84-16583 and EAR85-06857. California Institute of Technology Contribution Number 4525.

References

Andersen, T., O'Reilly, S.Y. & Griffin, W.L. (1984). The trapped fluid phase in upper-mantle xenoliths from Victoria, Australia: implications for mantle metasomatism. *Contr. Miner. Petrol.*, **88**, 72–85.

Anderson, R.N., de Long, S.E. & Schwartz, W.M. (1980). Dehydration, asthenospheric convection and seismicity in subduction zones. *J. Geol.*, **88**, 445–51.

Artyushkov, I.V. & Sobolev, S.V. (1984). Physics of the kimberlite magmatism. In *Kimberlites I: kimberlites and related rocks*, ed. J. Kornprobst, pp. 301–22. Amsterdam: Elsevier.

Bailey, D.K. (1985). Fluids, melts, flowage and styles of eruption in alkaline ultramafic magmatism. *Trans. Geol. Soc. S. Africa*, **88**, 449–57.

Boettcher, A.L. & O'Neill, J.R. (1980). Stable isotope, chemical and petrographic studies of high pressure amphiboles and mica: evidence for metasomatism in the mantle source regions of alkali basalts and kimberlites. *Amer. J. Sci.*, **280A**, 594–621.

Boyd, F.R. & Gurney, J.J. (1986). Diamonds on the African lithosphere. *Sci.*, **232**, 472–7.

Brey, G., Brice, W.R., Ellis, D.J., Green, D.H., Harris, K.L. & Ryabchikov, I.D. (1983). Pyroxene–carbonate relations in the upper mantle. *Earth Planet. Sci. Lett.*, **62**, 63–74.

Carlson, R.W. (1987). Geochemical evolution of the crust and mantle. *Rev. Geophys.*, **25**, 1011–20.

Chen, C.-Y. & Frey, F.A. (1985). Trace element and isotopic geochemistry of lavas from Haleakala Volcano, East Maui, Hawaii: implications for the origin of Hawaiian basalts. *J. Geophys. Res.*, **90**, 8743–68.

Courtney, R.C. & White, R.S. (1986). Anomalous heat flow and geoid across the Cape Verde Rise: evidence for dynamic support from a thermal plume in the mantle. *Geophys. J. R. Astr. Soc.*, **87**, 815–67.

Dawson, J.B. (1980). Kimberlites and their xenoliths. Berlin: Springer–Verlag.

Decker, R.W., Wright, T.L. & Stauffer, P.H. (1987). *Volcanism in Hawaii*, 2 vols. Prof. paper **1350**, U.S. Geol. Surv., Washington, D.C.

Deines, P. (1980). The carbon isotopic composition of diamonds: relationship to diamond shape, color, occurrence and vapor compositions. *Geochim. Cosmochim. Acta*, **44**, 943–61.

Eggler, D.H. (1978). The effect of CO_2 upon partial melting of peridotite in the system Na_2O–CaO–Al_2O_3–MgO–SiO_2–CO_2 to 35 kb, with an analysis of melting in a peridotite–H_2O–CO_2 system. *Amer. J. Sci.*, **278**, 305–43.

— (1987). Discussion of recent papers on carbonated peridotite, bearing on mantle metasomatism and magmatism: an alternative. *Earth Planet. Sci. Lett.*, **82**, 398–400.

Eggler, D.H. & Baker, D.R. (1982). Reduced volatiles in the system C–O–H: implications to mantle melting, fluid formation, and diamond genesis. In *High pressure research in geophysics*, Advances in Earth Sciences, **12**, pp. 237–50. Dordrecht: Reidel.

Ellis, D. & Wyllie, P.J. (1979). Hydration and melting reactions in the system MgO–SiO_2–H_2O at pressures up to 100 kilobars. *Amer. Miner.*, **64**, 41–8.

—— (1980). Phase relations and their petrological implications in the system MgO–SiO_2–CO_2–H_2O at pressures up to 100 kbar. *Amer. Miner.*, **65**, 540–56.

Erlank, A.J., Waters, F.G., Hawkesworth, C.J., Haggerty, S.E., Allsop, H.L., Rickard, R.S. & Menzies, M. (1987). Evidence for mantle metasomatism in peridotite nodules from the Kimberley pipes, South Africa. In *Mantle metasomatism*, ed. M. Menzies & C.J. Hawkesworth. London: Academic Press, in press.

Feigenson, M.D. (1986). Constraints on the origin of Hawaiian lavas. *J. Geophys. Res.*, **91**, 9383–93.

Freund, F., Kathrein, H., Wengler, H., Knobel, R. & Heinen, H.J. (1980). Carbon in solid solution in forsterite – a key to the untractable nature of reduced carbon in terrestrial and cosmogenic rocks. *Geochim. Cosmochim. Acta*, **44**, 1319–33.

Gill, J. (1981). *Orogenic andesites and plate tectonics*. New York: Springer–Verlag.

Gliko, A.L., Grachev, A.F. & Magnitsky, V.S. (1985). Thermal model for lithospheric thinning and associated uplift in the neotectonic phase of intraplate orogenic activity and continental rifts. *J. Geodynam. Res.*, **3**, 137–8.

Green, D.H., Falloon, T.J. & Taylor, W.R. (1987). Mantle-derived magmas – roles of variable source peridotite and variable C–H–O fluid compositions. In *Magmatic processes: physicochemical principles*, ed. B.O. Mysen, pp. 139–54. The Geochemical Society Spec. Publ. No. 1.

Grove, T.L. & Kinzler, R.J. (1986). Petrogenesis of andesites. *Ann. Rev. Earth Planet. Sci.*, **14**, 417–54.

Haggerty, S.E. (1986). Diamond genesis in a multiply constrained model. *Nature*, **320**, 34–8.

— (1987). Source regions for oxides, sulfides and metals in the upper mantle; clues to the stability of diamonds, and the genesis of kimberlites, lamproites and carbonatites. Proc. 4th Int. Kimb. Conf., Perth, Australia, 1986, in press.

Harte, B. (1983). Mantle peridotites and processes – the kimberlite sample. In *Continental basalts and mantle xenoliths*, ed. C.J. Hawkesworth & M.J. Norry, pp. 46–91. Shiva Publishing.

Holloway, J.R. & Jakobsson, S. (1986). Volatile solubilities in magmas: Transport of volatiles from mantles to planet surfaces. *J. Geophys. Res.*, **91**, D505–D508.

Jaques, A.L. & Green, D.H. (1980). Anhydrous melting of peridotite at 0–15 kb pressure and the genesis of tholeiitic basalts. *Contr. Miner. Petrol.*, **73**, 287–310.

Korzhinskii, D.S. (1959). *Physicochemical basis of the analysis of the paragenesis of minerals*. New York: Consultants Bureau.

Kovalenko, V.I., Solovova, I.D., Ryabchikov, I.D., Ionov, D.A. & Bogatikov, O.A. (1987). Fluidized CO_2–sulphide–silicate media as agents of mantle metasomatism and megacryst formation: evidence from a large druse in a spinel–lherzolite xenolith. *Phys. Earth Plan. Int.*, **45**, 280–93.

Maaløe, S. & Petersen, T.S. (1981). Petrogenesis of oceanic andesites. *J. Geophys. Res.*, **86**, 10273–86.

Marsh, B.D. (1979). Island-arc volcanism. *Amer. J. Sci.*, **67**, 161–72.

— (1987). Magmatic processes. *Rev. Geophys.*, **25**, 1043–53.

McCallum, I.S. (1987). Petrology of the igneous rocks. *Rev. Geophys.*, **25**, 1021–42.

Nicholls, I.A. & Ringwood, A.E. (1973). Effect of water on olivine stability in tholeiites and production of silica-saturated magmas in the island arc environment. *J. Geol.*, **81**, 285–300.

Nickel, K.G. & Green, D.H. (1985). Empirical geothermobarometry for garnet peridotites and implications for the nature of the lithosphere, kimberlites and diamonds. *Earth Planet. Sci. Lett.* **73**, 158–70.

Olafsson, M. & Eggler, D.H. (1983). Phase relations of amphibole–carbonate, and phlogopite–carbonate peridotite: petrologic constraints on the asthenosphere. *Earth Planet. Sci. Lett.*, **64**, 305–15.

Perchuk, L.L. (1973). Thermodynamic regime of deep-seated petrogenesis. Moscow: Nauka Press.

— (1976). Gas–mineral equilibria and a possible geochemical model of the Earth's interior. *Phys. Earth Planet. Sci.*, **13**, 232–9.

Ribe, N.M. & Smooke, M.D. (1987). A stagnation point flow model for melt extraction from a mantle plume. *J. Geophys. Res.*, **B7**, 6437–43.

Ringwood, A.E. (1975). Composition and petrology of the Earth's mantle. New York: McGraw-Hill.

Ryabchikov, I.D., Green, D.H., Wall, W.J. & Brey, G.P. (1981). The oxidation state of carbon in the reduced-velocity zone. *Geochem. Int.*, 148–58.

Schneider, M.E. & Eggler, D.H. (1984). Compositions of fluids in equilibrium with peridotite: implications for alkaline magmatism–metasomatism. In *Kimberlites I: kimberlites and related rocks*, ed. J. Kornprobst, pp. 383–94. Amsterdam: Elsevier.

—— (1986). Fluids in equilibrium with peridotite minerals: implications for mantle metasomatism. *Geochim. Cosmochim. Acta*, **50**, 711–24.

Sekine, T. & Wyllie, P.J. (1982). The system granite–peridotite–H_2O at 30 kbar, with applications to hybridization in subduction zone magmatism. *Contr. Miner. Petrol.*, **81**, 190–202.

—— (1983). Experimental simulation of mantle hybridization in subduction zones. *J. Geol.*, **91**, 511–28.

Smith, D.B. (1983). Pb, Sr and Nd isotopic evidence for sources of south African kimberlites. *Nature*, **304**, 51–4.

Smith, C.B., Gurney, J.J., Skinner, E.M.W., Clement, C.R. & Ebrahim, N. (1985). Geochemical character of Southern African kimberlites: a new approach based on isotopic constraints. *Trans. Geol. Soc. S. Africa*, **88**, 267–80.

Spera, F.J. (1984). Carbon dioxide in petrogenesis III: role of volatiles in the ascent of alkaline magma with special reference to xenolith-bearing mafic lavas. *Contr. Miner. Petrol.*, **88**, 217–32.

Takahashi, E. (1986). Melting of dry peridotite KLB-1 up to 14 GPa: implications on the origin of peridotitic upper mantle. *J. Geophys. Res.*, **91**, 9367–82.

Taylor, W.R. & Green, D.H. (1987). The petrogenetic role of methane: effect on liquidus phase relations and the solubility mechanism of reduced C–H volatiles. In *Magmatic processes: physicochemical principles*, ed. B.O. Mysen, pp. 131–38. The Geochemical Society, Spec. Publ. No. 1.

Tingle, T.N. & Green, H.W. (1987). Carbon solubility in olivine: implications for upper mantle evolution. *Geology*, **15**, 324–6.

Walker, D. (1983). New developments in magmatic processes. *Rev. Geophys. Space Phys.*, **21**, 1372–84.

Walter, L.S., Wyllie, P.J. & Tuttle, O.F. (1962). The system MgO–CO₂–H₂O at high pressures and temperatures. *J. Petrol.*, **3**, 49–64.

Wasserburg, G.J. & de Paolo, D.J. (1979). Models of earth structure inferred from neodymium and strontium isotopic abundances. *Proc. Nat. Acad. Sci.*, **76**, 3574–98.

Wendlandt, R.F. (1984). An experimental and theoretical analysis of partial melting in the system KAlSiO₄–CaO–MgO–SiO₂–CO₂ and applications to the genesis of potassic magmas, carbonatites and kimberlites. In *Kimberlites I: kimberlites and related rocks*, ed. J. Kornprobst, pp. 359–69. Amsterdam: Elsevier.

Wendlandt, R.F. & Eggler, D.H. (1980). The origins of potassic magmas. 2. Stability of phlogopite in natural spinel lherzolite and in the system KAlSiO₄–MgO–SiO₂–H₂O–CO₂ at high pressures and high temperatures. *Amer. J. Sci.*, **280**, 421–58.

Wendlandt, R.F. & Morgan, P. (1982). Lithospheric thinning associated with rifting in East Africa. *Nature*, **298**, 734–6.

Wilshire, H.G. (1984). Mantle metasomatism: the REE story. *Geology*, **12**, 395–8.

Woermann, E. & Rosenhauer, M. (1985). Fluid phases and the redox state of the Earth's mantle: extrapolations based on experimental, phase-theoretical and petrological data. *Fortschr. Miner.*, **63**, 263–349.

Wright, T.L. & Helz, R.T. (1987). Recent advances in Hawaiian petrology and geochemistry. In *Volcanism in Hawaii*, vol. 1, ed. R.W. Decker, T.L. Wright & P.H. Stauffer, pp. 625–40. Prof. Paper **1350**, U.S. Geol. Surv., Washington, D.C.

Wyllie, P.J. (1962). The petrogenetic model, an extension of Bowen's petrogenetic grid. *Geol. Mag.*, **94**, 558–69.

— (1978). Mantle fluid compositions buffered in peridotite–CO₂–H₂O by carbonates, amphibole, and phlogopite. *J. Geol.*, **86**, 687–813.

— (1979). Magmas and volatile components. *Amer. Miner.*, **64**, 469–500.

— (1980). The origin of kimberlites. *J. Geophys. Res.*, **85**, 6902–10.

— (1982). Subduction products according to experimental prediction. *Geol. Soc. Amer. Bull.*, **93**, 468–76.

— (1983). Experimental and thermal constraints on the deep-seated parentage of some granitoid magmas in subduction zones. In *Migmatites, melting, and metamorphism*, ed. M.P. Atherton & D.C. Gribble, pp. 37–51. Nantwich, Cheshire: Shiva Publishing.

— (1987a). Transfer of subcratonic carbon into kimberlites and rare earth carbonatites. In *Magmatic processes: physicochemical principles*, ed. B.O. Mysen, pp. 107–19. The Geochemical Society, Spec. Publ. No. 1.

— (1987b). Discussion of recent papers on carbonated peridotite, bearing on mantle metasomatism and magmatism. Also, Response. *Earth Planet. Sci. Lett.*, **82**, 391–7, 401–2.

— (1987c). Metasomatism and fluid generation in mantle xenoliths: experimental. In *Mantle xenoliths*, ed. P.H. Nixon, pp. 609–21. New York: Wiley.

— (1987d). The genesis of kimberlites and some low SiO₂, high-alkali magmas. Australian Geological Society, in press.

Wyllie, P.J. & Huang, W.L. (1976). Carbonation and melting reactions in the system CaO–MgO–SiO₂–CO₂ at mantle pressures with geophysical and petrological applications. *Contr. Miner. Petrol.*, **54**, 79–107.

Wyllie, P.J., Huang, W.L., Otto, J. & Byrnes, A.P. (1983). Carbonation of peridotites and decarbonation of siliceous dolomites represented in the system CaO–MgO–SiO₂–CO₂ to 30 kbar. *Tectonophys.*, **100**, 359–88.

Wyllie, P.J. & Sekine, T. (1982). The formation of mantle phlogopite in subduction zone hybridization. *Contr. Miner. Petrol.*, **79**, 375–80.

Wyllie, P.J. & Tuttle, O.F. (1960). The system CaO–CO_2–H_2O and the origin of carbonatites. *J. Petrol.*, **1**, 1–46.

Printed in the United States
By Bookmasters